Lecture Notes in Physics

The Lecture Notes in Physics

The series Lecture Notes in Physics (LNP), founded in 1969, reports new developments in physics research and teaching – quickly and informally, but with a high quality and the explicit aim to summarize and communicate current knowledge in an accessible way. Books published in this series are conceived as bridging material between advanced graduate textbooks and the forefront of research and to serve three purposes:

- to be a compact and modern up-to-date source of reference on a well-defined topic

- to serve as an accessible introduction to the field to postgraduate students and nonspecialist researchers from related areas

- to be a source of advanced teaching material for specialized seminars, courses and schools

Both monographs and multi-author volumes will be considered for publication. Edited volumes should, however, consist of a very limited number of contributions only. Proceedings will not be considered for LNP.

Volumes published in LNP are disseminated both in print and in electronic formats, the electronic archive being available at springerlink.com. The series content is indexed, abstracted and referenced by many abstracting and information services, bibliographic networks, subscription agencies, library networks, and consortia.

Proposals should be send to a member of the Editorial Board, or directly to the managing editor at Springer:

Christian Caron
Springer Heidelberg
Physics Editorial Department I
Tiergartenstrasse 17
69121 Heidelberg / Germany
christian.caron@springer.com

L. Papantonopoulos (Ed.)

The Invisible Universe:
Dark Matter
and Dark Energy

 Springer

Editor

Lefteris Papantonopoulos
National Technical University of Athens
Department of Physics
Zografou Campus
157 80 Athens, Greece
lpapa@central.ntua.gr

L. Papantonopoulos, *The Invisible Universe: Dark Matter and Dark Energy*, Lect. Notes
Phys. 720 (Springer, Berlin Heidelberg 2007), DOI 10.1007/978-3-540-71013-4

Library of Congress Control Number: 2007923172

ISSN 0075-8450
ISBN 978-3-540-71012-7 Springer Berlin Heidelberg New York

Springer is a part of Springer Science+Business Media
springer.com
© Springer-Verlag Berlin Heidelberg 2007

Typesetting: by the authors and Integra using a Springer LATEX macro package
Cover design: eStudio Calamar S.L., F. Steinen-Broo, Pau/Girona, Spain

Printed on acid-free paper   SPIN: 12023448   5 4 3 2 1 0

Preface

This book is an edited version of the review talks given in the Third Aegean School on the Invisible Universe: Dark Matter and Dark Energy, held in Karfas on Chios Island, Greece, from 26th of September to 1st of October 2005. The aim of this book is not to present another proceedings volume, but rather an advanced multiauthored textbook which meets the needs of both the postgraduate students and the young researchers, in the fields of Modern Cosmology and Astrophysics.

The issue of dark matter and dark energy is one of the central interest in Astroparticle Physics, Astrophysics, Astronomy, and Modern Cosmology. Much of observational data indicate that there is a missing matter and missing energy in the Universe. Evidence of the existence of this unknown form of matter and energy can be obtained from different sources. In Astrophysics, the dynamics of galaxy formation and galaxy clusters can give information on the amount of missing matter. In Astroparticle physics, particle candidates were proposed from string theory and supersymmetry to identify the unknown matter. In Cosmology, the recent data from Cosmic Microwave Background (CMB) and Supernovae Observations strongly indicate that there is a large amount of an unknown form of energy in the energy balance of the Universe. The purpose of this book is to present these issues and discuss in detail the physics involved.

The first part of the book presents the problem of missing matter of the Universe as seen by Astroparticle Physics and Astrophysics. G. Lazaride's chapter reviews the main proposals of particle physics for the composition of the dark matter in the universe. The lightest neutralino is the most popular candidate constituent of dark matter. Axinos and gravitinos can also contribute to dark matter. A model is presented which possesses a wide range of parameters consistent with the data on dark matter abundance as well as other phenomenological constraints. In view that many particle theories will be tested in the next round experiments in large accelerators, such as the Large Hadron Collider (LHC), a more phenomenological approach to dark matter in elementary particle physics is adopted in the next chapter by A. Lahanas. The interest in these experiments is that may provide candidates for dark matter of supersymmetric origin.

A more difficult task is undertaken in the next chapter: the direct detection of supersymmetric dark matter. J. Vergados after reviewing supersymmetric models with their parameters constrained from the recent data at low energies and cosmological observations, is suggesting experiments of direct detection of dark matter mainly through a neutralino-nucleus interaction.

The challenge of dark matter is addressed in the context of Astrophysics by J. Silk's chapter. It describes the confrontation of structure formation with observation and it focuses on the detection of the most elusive component, non–baryonic dark matter. It explains how galaxy formation theory is driven by phenomenology and by numerical simulations of dark matter clustering under gravity. Once the complications of star formation are incorporated, the theory becomes very complex. Semi-analytical perspectives of the theory are presented that may shed some insight into the nature of galaxy formation.

The second part of the book deals with the energy balance of the Universe. In the first chapter by P. Tozzi, the basic procedures are presented to constrain the cosmological parameters which they describe the energy content of the Universe. Data from clusters of galaxies and their X-ray properties are used as cosmological tools to deduce information on these parameters. The difficulties in analysing galaxy redshift surveys data like the 2dF Galaxy Redshift Survey (2dFGRS) and the Sloan Digital Sky survey (SDSS) are explained in W. Percival's chapter. A very interesting example is provided of joint analysis of the latest CMB and large-scale structure data, leading to a set of cosmological parameter constraints.

The chapter by R. Crittenden discusses the evidence for dark energy coming from a wide variety of data. After reviewing the physics of the CMB, it discusses the different methods that are used in determining the dark energy's density, evolution, and clustering properties and the crucial role the microwave background plays in all of these methods. L. Perivolaropoulos's chapter deals with another interesting manifestation of the presence of dark energy in the Universe: the late time acceleration. It presents of the recent observational data obtained from type Ia supernova surveys that support the accelerating expansion of the universe. The methods for the analysis of the data are reviewed and the theoretical implications obtained from their analysis are discussed.

The last chapter of the second part of the book by M. Sami is a presentation of current theoretical models for dark energy. These models rely on scalar field dynamics and this chapter focusses mainly on the underlying basic features rather than on concrete scalar field models. The cosmological dynamics of standard scalar fields, phantoms, and tachyon fields is developed in detail. Scaling solutions are discussed emphasizing their importance in modelling dark energy. The developed concepts are implemented in an example of quintessential inflation.

The third part of the book discusses the issue of dark matter and dark energy beyond the standard theory of General Relativity. Higher dimensional string and brane theories are employed and also theories that modify the usual Newtonian dynamics. An introduction to high dimensional theories is given in I. Antoniadis' chapter. The basic idea is that the apparent weakness of gravity can be accounted by the existence of large internal dimensions, in the submillimeter region, and transverse to a braneworld where our universe must be confined. The main properties of this scenario are reviewed and its implications for observations at both particle colliders and in non-accelerator gravity experiments are discussed.

These ideas are applied to Cosmology in R. Maartens' chapter. As explained in L. Perivolaropoulos' contribution an accelerating Universe requires the presence of a dark energy field with effectively negative pressure. An alternative to dark energy is that gravity itself may behave differently from general relativity on the largest scales, in such a way as to produce acceleration. In this chapter an example of modified gravity is presented which is provided by braneworld models that self-accelerate at late times. The challenges of dark matter and dark energy in the context of string theory are discussed in N. Mavromatos's article. In this chapter the resolution of these issues in string theory is briefly reviewed and a suggestion for the resolution of the dark energy issue is discussed.

The most successful alternative to dark matter in bound gravitational systems is the modified Newtonian dynamics, or MOND, which is discussed in R. Sanders' chapter. There, the various attempts to formulate MOND as a modification of General Relativity are presented and the covariant theories that have been proposed as a basis for this idea are explained. Finally, local modifications of general relativity by making the Lagrangian an arbitrary function of the Ricci scalar are presented in R. Woodard's contribution. The interest of such theories is that they can reproduce the current phase of cosmic acceleration without dark energy.

The Third Aegean School and consequently this book became possible with the kind support of many people and organizations. The School was organized by the Physics Department of the National Technical University of Athens, and supported by the Physics Department of King's College, University of London, the Institute of Cosmology and Gravitation, University of Portsmouth, the Physics and Astronomy Department, University of Tennessee. We also received financial support from the following sources and this is gratefully acknowledged: Ministry of National Education and Religious Affairs, Prefecture of Chios, Municipality of Chios.

We thank Giannis Gialas for his valuable assistance and help in organizing the School in Chios and the University of the Aegean for providing technical support. We thank also the other members of the Organizing Committee of the School, Alex Kehagias, George Koutsoumbas, George Siopsis, and Nikolas Tracas for their help in organizing the School. The administrative support of the Third Aegean School was again taken up with great care by Mrs. Evelyn Pappa. We acknowledge the help of Vasilis Zamarias, who designed and maintained the webside of the School and assisted us in resolving technical issues in the process of editing this book.

Last, but not least, we are grateful to the staff of Springer-Verlag, responsible for the *Lecture Notes in Physics*, whose abilities and help contributed greatly to the appearance of this book.

Athens, October 2006   Lefteris Papantonopoulos

Contents

4 Galaxy Formation and Dark Matter

Part II Dark Energy: The Energy Balance of the Universe within the Standard Cosmological Model

5 Cosmological Parameters from Galaxy Clusters: An Introduction

Part I

Dark Matter: The Missing Matter
of the Universe as Seen
by Astroparticle Physics and Astrophysics

1

Particle Physics Approach to Dark Matter

George Lazarides

Physics Division, School of Technology, Aristotle University of Thessaloniki,
Thessaloniki 54124, Greece
lazaride@eng.auth.gr

Abstract. We review the main proposals of particle physics for the composition of
the cold dark matter in the universe. Strong axion contribution to cold dark matter
is not favored if the Peccei-Quinn field emerges with non-zero value at the end of
inflation and the inflationary scale is superheavy since, under these circumstances,
it leads to unacceptably large isocurvature perturbations. The lightest neutralino
is the most popular candidate constituent of cold dark matter. Its relic abundance
in the constrained minimal supersymmetric standard model can be reduced to acceptable values by pole annihilation of neutralinos or neutralino-stau coannihilation.
Axinos can also contribute to cold dark matter provided that the reheat temperature
is adequately low. Gravitinos can constitute the cold dark matter only in limited
regions of the parameter space. We present a supersymmetric grand unified model
leading to violation of Yukawa unification and, thus, allowing an acceptable b-quark
mass within the constrained minimal supersymmetric standard model with $\mu > 0$.
The model possesses a wide range of parameters consistent with the data on the
cold dark matter abundance as well as other phenomenological constraints. Also, it
leads to a new version of shifted hybrid inflation.

1.1 Introduction

The recent measurements of the Wilkinson microwave anisotropy probe
(WMAP) satellite [1] on the cosmic microwave background radiation (CMBR)
have shown that the matter abundance in the universe is $\Omega_m h^2 = 0.135^{+0.008}_{-0.009}$,
where $\Omega_i = \rho_i/\rho_c$ with ρ_i being the energy density of the i-th species and
ρ_c the critical energy density of the universe and h is the present value of
the Hubble parameter in units of 100 km sec^{-1} Mpc^{-1}. The baryon abundance is also found by these measurements to be $\Omega_b h^2 = 0.0224 \pm 0.0009$.
Consequently, the cold dark matter (CDM) abundance in the universe is
$\Omega_{\text{CDM}} h^2 = 0.1126^{+0.00805}_{-0.00904}$. The 95% confidence level (c.l.) range of this quantity is then, roughly, $0.095 \lesssim \Omega_{\text{CDM}} h^2 \lesssim 0.13$. Taking $h \simeq 0.72$, which is its
best-fit value from the Hubble space telescope (HST) [2], and assuming that
the total energy density of the universe is very close to its critical energy

G. Lazarides: *Particle Physics Approach to Dark Matter*, Lect. Notes Phys. **720**, 3–34 (2007)
DOI 10.1007/978-3-540-71013-4_1

density (i.e. $\Omega_{\text{tot}} \simeq 1$), as implied by WMAP, we conclude that about 22% of the energy density of the present universe consists of CDM.

The question then is, what the nature, origin, and composition of this important component of our universe is. Particle physics provides us with a number of candidate particles out of which CDM can be made. These particles appear naturally in various particle physics frameworks for reasons completely independent from CDM considerations and are, certainly, not invented for the sole purpose of explaining the presence of CDM in the universe.

The basic properties that such a candidate particle must satisfy are the following: (i) it must be stable or very long-lived, which can be achieved by an appropriate symmetry, (ii) it should be electrically and color neutral, as implied by astrophysical constraints on exotic relics (like anomalous nuclei), but can be interacting weakly, and (iii) it has to be non-relativistic, which is usually guaranteed by assuming that it is adequately massive, although even very light particles such as axions can be non-relativistic for different reasons. So, what we need as constituent of CDM is a weakly interacting massive particle. There are several possibilities, but we will concentrate here on the major particle physics candidates which are the axion, the lightest neutralino, the axino, and the gravitino (for other candidates, see e.g. [3]). Note that the last three particles exist only in supersymmetric (SUSY) theories.

In Sect. 1.2, we examine the possibility that the axions are constituents of CDM. Section 1.3 is devoted to outlining the salient features of the minimal supersymmetric standard model (MSSM), which will be used as a basic frame for discussing SUSY CDM. In Sect. 1.4, we summarize the calculation of the relic abundance of the lightest neutralino, which is normally the lightest supersymmetric particle (LSP), and investigate the circumstances under which it can account for the CDM in the universe. In Sects. 1.5 and 1.6, we discuss, respectively, axinos and gravitinos as constituents of CDM. In Sect. 1.7, we present a SUSY grand unified theory (GUT) model which solves the bottom-quark mass problem by naturally and modestly violating the exact unification of the third generation Yukawa couplings. We study the parameter space of the model which is allowed by neutralino dark matter considerations as well as some other phenomenological constraints. Finally, in Sect. 1.8, we summarize our conclusions.

1.2 Axions

The most natural solution to the strong CP problem (i.e. the apparent absence of CP violation in strong interactions implied by the experimental bound on the electric dipole moment of the neutron) is the one provided by a Peccei-Quinn (PQ) symmetry [4]. This is a global U(1) symmetry, U(1)$_{\text{PQ}}$, which carries QCD anomalies and is spontaneously broken at a scale f_a, the so-called axion decay constant or simply PQ scale. Astrophysical [5] and cosmological constraints imply that 10^9 GeV $\lesssim f_a \lesssim 10^{12}$ GeV. The upper bound originates

[6, 7] from the requirement that the relic energy density of axions does not overclose the universe. It should be noted, however, that this upper bound can be considerably relaxed if the axions are diluted [7, 8, 9] by entropy generation after their production at the QCD phase transition (for more recent applications of this dilution mechanism, see e.g. [10]).

The axion is a pseudo Nambu-Goldstone boson corresponding to the phase of the complex PQ field, which breaks $U(1)_{PQ}$ by its vacuum expectation value (VEV). After the end of inflation [11], this phase appears homogenized over the universe (supposing that the PQ field is non-zero) with a value θ, which is known as the initial misalignment angle. Naturalness suggests that θ is of order unity. This angle remains frozen until the QCD phase transition, where the QCD instantons come into play. They break explicitly the PQ symmetry to a discrete subgroup [12] since this symmetry carries QCD anomalies. So, a sinusoidal potential for the phase of the PQ field is generated and this phase starts oscillating coherently about a minimum of the potential. The resulting state resembles pressureless matter consisting of static axions with mass $m_a \sim \Lambda^2_{QCD}/f_a$, where $\Lambda_{QCD} \sim 200$ MeV is the QCD scale. For $f_a \sim 10^{12}$ GeV, the mass of the axion $m_a \sim 10^{-5}$ eV. Note that axions, although very light, are good candidates for being constituents of the CDM in the universe since they are produced at rest. Also, they are very weakly interacting since their interactions are suppressed by the axion decay constant f_a.

The relic abundance of axions can be calculated by using the formulae of [13], where we take the QCD scale $\Lambda_{QCD} = 200$ MeV and ignore the uncertainties for simplicity. We find

$$\Omega_a h^2 \approx \theta^2 \left(\frac{f_a}{10^{12} \text{ GeV}} \right)^{1.175} \tag{1.1}$$

(note that a primordial magnetic helicity, may [14] influence this abundance). So, for natural values of $\theta \sim 0.1$ and $f_a \sim 10^{12}$ GeV, axions can contribute significantly to CDM, which can even consist solely of axions.

The main disadvantage of axionic dark matter is that it leads to isocurvature perturbations if the PQ field emerges with non-zero (homogeneous) value at the end of inflation. Indeed, during inflation, the angle θ acquires a superhorizon spectrum of perturbations as all the almost massless degrees of freedom. At the QCD phase transition, these perturbations turn into isocurvature perturbations in the axion energy density, which means that the partial curvature perturbation in axions is different than the one in photons. The recent results of WMAP [1] put stringent bounds [15, 16, 17] on the possible isocurvature perturbation. So, a large axion contribution to CDM is disfavored in models where the inflationary scale is superheavy (i.e. of the order of the SUSY GUT scale) and the PQ field is non-zero at the end of inflation.

We now wish to turn to the discussion of the main SUSY candidates for dark matter: the lightest neutralino $\tilde{\chi}$, the axino $\tilde{a}$ and the gravitino $\tilde{G}$. We will consider them mainly within the simplest SUSY framework, which is the MSSM. It is, thus, important to first outline the basics of MSSM.

1.3 Salient Features of MSSM

We consider the MSSM embedded in some general SUSY GUT model. We further assume that the GUT gauge group breaking down to the standard model (SM) gauge group G_{SM} occurs in one step at a scale $M_{\mathrm{GUT}} \sim 10^{16}$ GeV, where the gauge coupling constants of strong, weak, and electromagnetic interactions unify. Ignoring the Yukawa couplings of the first and second generation, the effective superpotential below M_{GUT} is

$$W = \epsilon_{ij}(-h_t H_2^i q_3^j t^c + h_b H_1^i q_3^j b^c + h_\tau H_1^i l_3^j \tau^c - \mu H_1^i H_2^j) \, , \qquad (1.2)$$

where $q_3 = (t, b)$ and $l_3 = (\nu_\tau, \tau)$ are the quark and lepton $SU(2)_L$ doublet left handed superfields of the third generation and t^c, b^c, and τ^c the corresponding $SU(2)_L$ singlets. Also, H_1, H_2 are the electroweak Higgs superfields and ϵ the 2×2 antisymmetric matrix with $\epsilon_{12} = +1$. The gravity-mediated soft SUSY-breaking terms in the scalar potential are given by

$$V_{\mathrm{soft}} = \sum_{a,b} m_{ab}^2 \phi_a^* \phi_b +$$

$$\left[\epsilon_{ij}(-A_t h_t H_2^i \tilde{q}_3^j \tilde{t}^c + A_b h_b H_1^i \tilde{q}_3^j \tilde{b}^c + A_\tau h_\tau H_1^i \tilde{l}_3^j \tilde{\tau}^c - B\mu H_1^i H_2^j) + \mathrm{h.c.}\right] \, , \quad (1.3)$$

where the sum is taken over all the complex scalar fields ϕ_a and tildes denote superpartners. The soft gaugino mass terms in the Lagrangian are

$$\mathcal{L}_{\mathrm{gaugino}} = \frac{1}{2}\left(M_1 \tilde{B}\tilde{B} + M_2 \sum_{r=1}^{3} \tilde{W}_r \tilde{W}_r + M_3 \sum_{a=1}^{8} \tilde{g}_a \tilde{g}_a + \mathrm{h.c.}\right) \, , \qquad (1.4)$$

where $\tilde{B}$, $\tilde{W}_r$ and $\tilde{g}_a$ are the bino, winos and gluinos respectively.

The Lagrangian of MSSM is invariant under a discrete Z_2 matter parity symmetry under which all "matter" (i.e. quark and lepton) superfields change sign. Combining this symmetry with the Z_2 fermion number symmetry under which all fermions change sign, we obtain the discrete Z_2 R-parity symmetry under which all SM particles are even, while all sparticles are odd. By virtue of R-parity conservation, the LSP is stable and, thus, can contribute to the CDM in the universe. It is important to note that matter parity is vital for MSSM to avoid baryon- and lepton-number-violating renormalizable couplings in the superpotential, which would lead to highly undesirable phenomena such as very fast proton decay. So, the possibility of having the LSP as CDM candidate is not put in by hand, but arises naturally from the very structure of MSSM.

The SUSY-breaking parameters m_{ab}, A_t, A_b, A_τ, B, and M_i ($i = 1, 2, 3$) are all of the order of the soft SUSY-breaking scale $M_{\mathrm{SUSY}} \sim 1$ TeV, but are otherwise unrelated in the general case. However, if we assume that soft SUSY breaking is mediated by minimal supergravity (mSUGRA), i.e. supergravity with minimal Kähler potential, we obtain soft terms which are universal "asymptotically" (i.e. at M_{GUT}). In particular, we obtain a common scalar mass m_0, a common trilinear scalar coupling A_0, and a common gaugino mass $M_{1/2}$. The MSSM supplemented by universal boundary conditions is known as constrained MSSM (CMSSM) [18]. It is true that mSUGRA implies two more asymptotic relations: $B_0 = A_0 - m_0$ and $m_0 = m_{3/2}$, where $B_0 = B(M_{GUT})$ and $m_{3/2}$ is the (asymptotic) gravitino mass. These extra conditions are usually not included in the CMSSM. Imposing them, we get the so-called very CMSSM [19], which is a very restrictive version of MSSM and will not be considered in these lectures.

The CMSSM can be further restricted by imposing asymptotic Yukawa unification (YU) [20], i.e. the equality of all three Yukawa coupling constants of the third family at M_{GUT}:

$$h_t(M_{GUT}) = h_b(M_{GUT}) = h_\tau(M_{GUT}) \equiv h_0 . \tag{1.5}$$

Exact YU, which makes the CMSSM considerably more predictive, can be obtained in GUTs based on a gauge group such as SO(10) or E_6 under which all the particles of one family belong to a single representation with the additional requirement that the masses of the third family fermions arise primarily from their unique Yukawa coupling to a single superfield representation which predominantly contains the electroweak Higgs superfields. It should be noted that exact YU in the CMSSM leads to unacceptable values for the bottom-quark mass m_b and, thus, must be corrected in order to become consistent with observations. We will ignore this problem for the moment, but we will return to it in Sect. 1.7.

Now, we assume that our effective theory below M_{GUT} is the CMSSM with YU. This theory depends on the following parameters ($\mu_0 = \mu(M_{GUT})$):

$$m_0, \ M_{1/2}, \ A_0, \ \mu_0, \ B_0, \ \alpha_{GUT}, \ M_{GUT}, \ h_0, \ \tan\beta ,$$

where $\alpha_{GUT} \equiv g_{GUT}^2/4\pi$ with g_{GUT} being the GUT gauge coupling constant and $\tan\beta \equiv \langle H_2 \rangle / \langle H_1 \rangle$ is the ratio of the two electroweak VEVs. The parameters α_{GUT} and M_{GUT} are evaluated consistently with the experimental values of the electromagnetic and strong fine-structure constants α_{em} and α_s, and the sine-squared of the Weinberg angle $\sin^2\theta_W$ at M_Z. To this end, we integrate [21] numerically the renormalization group equations (RGEs) for the MSSM at two loops in the gauge and Yukawa coupling constants from M_{GUT} down to a common but variable [22] SUSY threshold $M_{SUSY} \equiv \sqrt{m_{\tilde{t}_1} m_{\tilde{t}_2}}$ ($\tilde{t}_{1,2}$ are the stop-quark mass eigenstates). From M_{SUSY} to M_Z, the RGEs of the non-SUSY SM are used. The set of RGEs needed for our computation can be found in many references (see e.g. [23]). We take $\alpha_s(M_Z) = 0.12 \pm 0.001$ which, as it turns out, leads to gauge coupling unification at M_{GUT} with an accuracy better than 0.1%. So, we can assume exact unification once the appropriate SUSY particle thresholds are taken into account.

The unified third generation Yukawa coupling constant h_0 at M_{GUT} and the value of $\tan\beta$ at M_{SUSY} are estimated using the experimental inputs for the top-quark mass $m_t(m_t) = 166$ GeV and the τ-lepton mass

$m_\tau(M_Z) = 1.746$ GeV. Our integration procedure of the RGEs relies [21] on iterative runs of these equations from M_{GUT} to low energies and back for every set of values of the input parameters until agreement with the experimental data is achieved. The SUSY corrections to m_τ are taken from [24] and incorporated at M_{SUSY}.

Assuming radiative electroweak symmetry breaking, we can express the values of the parameters μ (up to its sign) and B (or, equivalently, the mass m_A of the CP-odd neutral Higgs boson A) at M_{SUSY} in terms of the other input parameters by minimizing the tree-level renormalization group (RG) improved potential [25] at M_{SUSY}. The resulting conditions are

$$\mu^2 = \frac{m_{H_1}^2 - m_{H_2}^2 \tan^2 \beta}{\tan^2 \beta - 1} - \frac{1}{2} M_Z^2 , \quad \sin 2\beta = \frac{2B\mu}{m_{H_1}^2 + m_{H_2}^2 + 2\mu^2} \equiv \frac{2B\mu}{m_A^2} ,$$
$$(1.6)$$

where m_{H_1}, m_{H_2} are the soft SUSY-breaking scalar Higgs masses. We can improve the accuracy of these conditions by including the full one-loop radiative corrections to the potential from [24] at M_{SUSY}. We find that the corrections to μ and m_A from the full one-loop effective potential are minimized [22, 26] by our choice of M_{SUSY}. So, a much better accuracy is achieved by using this variable SUSY threshold rather than a fixed one. Furthermore, we include in our calculation the two-loop radiative corrections to the masses m_h and m_H of the CP-even neutral Higgs bosons h and H. These corrections are particularly important for the mass of the lightest CP-even neutral Higgs boson h. Finally, the SUSY corrections to m_b are also included at M_{SUSY} using the relevant formulae of [24]. As already mentioned, the predicted value of the bottom-quark mass is not compatible with experiment. However, we will ignore this problem for the moment. The sign of μ is taken to be positive, since the $\mu < 0$ case is excluded because it leads [27, 28] to a neutralino relic abundance which is well above unity, thereby overclosing the universe, for all m_A's permitted by $b \to s\gamma$. We are left with m_0, $M_{1/2}$ and A_0 as free input parameters.

The LSP is the lightest neutralino $\tilde{\chi}$. The mass matrix for the four neutralinos is

$$\begin{pmatrix} M_1 & 0 & -M_Z s_W \cos\beta & M_Z s_W \sin\beta \\ 0 & M_2 & M_Z c_W \cos\beta & -M_Z c_W \sin\beta \\ -M_Z s_W \cos\beta & M_Z c_W \cos\beta & 0 & -\mu \\ M_Z s_W \sin\beta & -M_Z c_W \sin\beta & -\mu & 0 \end{pmatrix} \quad (1.7)$$

in the $(-i\tilde{B}, -i\tilde{W}_3, \tilde{H}_1, \tilde{H}_2)$ basis. Here, $s_W = \sin\theta_W$, $c_W = \cos\theta_W$, and M_1, M_2 are the mass parameters of $\tilde{B}$, $\tilde{W}_3$ in (1.4). In CMSSM, the lightest neutralino turns out to be an almost pure bino $\tilde{B}$.

The LSPs are stable due to the presence of the unbroken R-parity, but can annihilate in pairs since this symmetry is a discrete Z_2 symmetry. This reduces their relic abundance in the universe. If there exist sparticles with masses close to the mass of the LSP, their coannihilation [29] with the LSP leads to a further reduction of the LSP relic abundance. It should be noted that the number density of these sparticles is not Boltzmann suppressed relative to the LSP number density. They eventually decay yielding an equal number of LSPs and, thus, contributing to the relic abundance of the LSPs. Of particular importance is the next-to-LSP (NLSP), which, in CMSSM, turns out to be the lightest stau mass eigenstate $\tilde{\tau}_2$. Its mass is obtained by diagonalizing the stau mass-squared matrix

$$\begin{pmatrix} m_\tau^2 + m_{\tilde{\tau}_L}^2 + M_Z^2(-\frac{1}{2} + s_W^2)\cos 2\beta & m_\tau(A_\tau - \mu\tan\beta) \\ m_\tau(A_\tau - \mu\tan\beta) & m_\tau^2 + m_{\tilde{\tau}_R}^2 - M_Z^2 s_W^2 \cos 2\beta \end{pmatrix} \quad (1.8)$$

in the gauge basis $(\tilde{\tau}_L, \tilde{\tau}_R)$. Here, $m_{\tilde{\tau}_{L[R]}}$ is the soft SUSY-breaking mass of the left [right] handed stau $\tilde{\tau}_{L[R]}$ and m_τ the tau-lepton mass. The stau mass eigenstates are

$$\begin{pmatrix} \tilde{\tau}_1 \\ \tilde{\tau}_2 \end{pmatrix} = \begin{pmatrix} \cos\theta_{\tilde{\tau}} & \sin\theta_{\tilde{\tau}} \\ -\sin\theta_{\tilde{\tau}} & \cos\theta_{\tilde{\tau}} \end{pmatrix} \begin{pmatrix} \tilde{\tau}_L \\ \tilde{\tau}_R \end{pmatrix}, \quad (1.9)$$

where $\theta_{\tilde{\tau}}$ is the $\tilde{\tau}_L - \tilde{\tau}_R$ mixing angle.

The large values of b and τ Yukawa coupling constants implied by YU cause soft SUSY-breaking masses of the third generation squarks and sleptons to run (at low energies) to lower physical values than the corresponding masses of the first and second generation. Furthermore, the large values of $\tan\beta$ implied by YU lead to large off-diagonal mixings in the sbottom and stau mass-squared matrices. These effects reduce further the physical mass of the lightest stau, which is the NLSP. Another effect of the large values of the b and τ Yukawa coupling constants is the reduction of the mass m_A of the CP-odd neutral Higgs boson A and, consequently, the other Higgs boson masses to smaller values.

1.4 Neutralino Relic Abundance

We now turn to the calculation of the cosmological relic abundance of the lightest neutralino $\tilde{\chi}$ (almost pure $\tilde{B}$) in the CMSSM with YU according to the standard cosmological scenario (for non-standard scenaria, see e.g. [30]). In general, all sparticles contribute to $\Omega_{\tilde{\chi}}h^2$, since they eventually turn into LSPs, and all the (co)annihilation processes must be considered. The most important contributions, however, come from the LSP and the NLSP. So, in the case of the CMSSM, we should concentrate on $\tilde{\chi}$ (LSP) and $\tilde{\tau}_2$ (NLSP) and consider the coannihilation of $\tilde{\chi}$ with $\tilde{\tau}_2$ and $\tilde{\tau}_2^*$. The important role of the coannihilation of the LSP with sparticles carrying masses close to its mass in the calculation of the LSP relic abundance has been pointed out by many authors (see e.g. [21, 29, 31, 32, 33]). Here, we will use the method of [29], which was also used in [21]. Note that our analysis can be readily applied to any MSSM scheme where the LSP and NLSP are the bino and stau respectively. In particular, it applies to the CMSSM without YU, where we have $\tan \beta$ as an extra free input parameter.

The relevant quantity, in our case, is the total number density

$$n = n_{\tilde{\chi}} + n_{\tilde{\tau}_2} + n_{\tilde{\tau}_2^*} , \qquad (1.10)$$

since the $\tilde{\tau}_2$'s and $\tilde{\tau}_2^*$'s decay into $\tilde{\chi}$'s after freeze-out. At cosmic temperatures relevant for freeze-out, the scattering rates of these (non-relativistic) sparticles off particles in the thermal bath are much faster than their annihilation rates since the (relativistic) particles in the bath are considerably more abundant. Consequently, the number densities n_i ($i = \tilde{\chi}, \tilde{\tau}_2, \tilde{\tau}_2^*$) are proportional to their equilibrium values n_i^{eq} to a good approximation, i.e. $n_i/n \approx n_i^{\text{eq}}/n^{\text{eq}} \equiv r_i$. The Boltzmann equation (see e.g. [34]) is then written as

$$\frac{dn}{dt} = -3Hn - \langle \sigma_{\text{eff}} v \rangle (n^2 - (n^{\text{eq}})^2) , \qquad (1.11)$$

where H is the Hubble parameter, v is the "relative velocity" of the annihilating particles, $\langle \cdots \rangle$ denotes thermal averaging and σ_{eff} is the effective cross section defined by

$$\sigma_{\text{eff}} = \sum_{i,j} \sigma_{ij} r_i r_j \qquad (1.12)$$

with σ_{ij} being the total cross section for particle i to annihilate with particle j averaged over initial spin states. In our case, σ_{eff} takes the following form

$$\sigma_{\text{eff}} = \sigma_{\tilde{\chi}\tilde{\chi}} r_{\tilde{\chi}} r_{\tilde{\chi}} + 4\sigma_{\tilde{\chi}\tilde{\tau}_2} r_{\tilde{\chi}} r_{\tilde{\tau}_2} + 2(\sigma_{\tilde{\tau}_2\tilde{\tau}_2} + \sigma_{\tilde{\tau}_2\tilde{\tau}_2^*}) r_{\tilde{\tau}_2} r_{\tilde{\tau}_2} . \qquad (1.13)$$

For r_i, we use the non-relativistic approximation

$$r_i(x) = \frac{g_i(1 + \Delta_i)^{\frac{3}{2}} e^{-\Delta_i x}}{g_{\text{eff}}} , \qquad (1.14)$$

$$g_{\text{eff}}(x) = \sum_i g_i(1 + \Delta_i)^{\frac{3}{2}} e^{-\Delta_i x} , \quad \Delta_i = \frac{m_i - m_{\tilde{\chi}}}{m_{\tilde{\chi}}} . \qquad (1.15)$$

Here $g_i = 2, 1, 1$ ($i = \tilde{\chi}, \tilde{\tau}_2, \tilde{\tau}_2^*$) is the number of degrees of freedom of the i-th particle with mass m_i and $x = m_{\tilde{\chi}}/T$ with T being the photon temperature.

Using Boltzmann equation (which is depicted in (1.11)), we can calculate the relic abundance of the LSP at the present cosmic time. It has been found [29, 34] to be given by

$$\Omega_{\tilde{\chi}} h^2 \approx \frac{1.07 \times 10^9 \text{ GeV}^{-1}}{g_*^{\frac{1}{2}} M_{\text{P}} x_F^{-1} \hat{\sigma}_{\text{eff}}} \qquad (1.16)$$

with

$$\hat{\sigma}_{\text{eff}} \equiv x_F \int_{x_F}^{\infty} \langle \sigma_{\text{eff}} v \rangle x^{-2} dx .$$ (1.17)

Here $M_{\text{P}} \simeq 1.22 \times 10^{19}$ GeV is the Planck scale, $g_* \simeq 81$ is the effective number of massless degrees of freedom at freeze-out [34] and $x_F = m_{\tilde{\chi}}/T_F$ with T_F being the freeze-out photon temperature calculated by solving iteratively the equation [34, 35]

$$x_F = \ln \frac{0.038 \, g_{\text{eff}}(x_F) \, M_{\text{P}} \, (c+2) \, c \, m_{\tilde{\chi}} \, \langle \sigma_{\text{eff}} v \rangle(x_F)}{g_*^{\frac{1}{2}} \, x_F^{\frac{1}{2}}} .$$ (1.18)

The constant c is chosen to be equal to $1/2$ [35]. The freeze-out temperatures which we obtain here are of the order of $m_{\tilde{\chi}}/25$ and, thus, our non-relativistic approximation (see (1.14)) is *a posteriori* justified.

Away from s-channel poles and final-state thresholds, the quantities $\sigma_{ij} v$ are well approximated by applying the non-relativistic Taylor expansion up to second order in the relative velocity v:

$$\sigma_{ij} v = a_{ij} + b_{ij} v^2 .$$ (1.19)

Actually, this corresponds [31] to an expansion in s and p waves. The thermally averaged cross sections are then easily calculated

$$\langle \sigma_{ij} v \rangle(x) = \frac{x^{\frac{3}{2}}}{2\sqrt{\pi}} \int_0^{\infty} dv \, v^2 (\sigma_{ij} v) e^{-\frac{xv^2}{4}} = a_{ij} + 6 \frac{b_{ij}}{x} .$$ (1.20)

Here, we approximated the masses of the incoming particles by the neutralino mass, i.e. $m_i = m_j = m_{\tilde{\chi}}$. The reduced mass of the incoming particles is then equal to $m_{\tilde{\chi}}/2$. We also thermally averaged over the relative velocity rather than the separate velocities of the incoming particles, which would be more accurate. Using (1.12), (1.13), (1.17), and (1.20), one obtains

$$\hat{\sigma}_{\text{eff}} = \sum_{(ij)} (\alpha_{(ij)} a_{ij} + \beta_{(ij)} b_{ij}) \equiv \sum_{(ij)} \hat{\sigma}_{(ij)} ,$$ (1.21)

where we sum over $(ij) = (\tilde{\chi}\tilde{\chi})$, $(\tilde{\chi}\tilde{\tau}_2)$, and $(\tilde{\tau}_2 \tilde{\tau}_2^{(*)})$ with $a_{\tilde{\tau}_2 \tilde{\tau}_2^{(*)}} = a_{\tilde{\tau}_2 \tilde{\tau}_2} + a_{\tilde{\tau}_2 \tilde{\tau}_2^*}$, $b_{\tilde{\tau}_2 \tilde{\tau}_2^{(*)}} = b_{\tilde{\tau}_2 \tilde{\tau}_2} + b_{\tilde{\tau}_2 \tilde{\tau}_2^*}$, and $\alpha_{(ij)}$, $\beta_{(ij)}$ given by

$$\alpha_{(ij)} = c_{(ij)} x_F \int_{x_F}^{\infty} \frac{dx}{x^2} r_i(x) r_j(x) , \quad \beta_{(ij)} = 6 c_{(ij)} x_F \int_{x_F}^{\infty} \frac{dx}{x^3} r_i(x) r_j(x) .$$ (1.22)

Here $c_{(ij)} = 1, 4, 2$ for $(ij) = (\tilde{\chi}\tilde{\chi})$, $(\tilde{\chi}\tilde{\tau}_2)$, and $(\tilde{\tau}_2 \tilde{\tau}_2^{(*)})$ respectively.

It should be emphasized that, near s-channel poles or final-state thresholds, the Taylor expansion in (1.19) fails [29, 36] badly and, thus, the thermal average in (1.20) has to be calculated accurately by numerical methods. Also, for better accuracy, we should use fully relativistic formulae instead of the non-relativistic expressions in (1.13), (1.14), and (1.20). Finally, in (1.20), we must take the thermal average over the two initial particle velocities v_i and v_j separately and not just over their relative velocity v. The masses of the incoming particles should also be taken different $m_i \neq m_j$. After all these improvements, (1.20) takes [16] the form

$$\langle \sigma_{ij} v \rangle = \frac{1}{2m_i^2 m_j^2 T K_2\left(\frac{m_i}{T}\right) K_2\left(\frac{m_j}{T}\right)} \int_{(m_i+m_j)^2}^{\infty} ds\, K_1\left(\frac{\sqrt{s}}{T}\right) p_{ij}^2(s) \sqrt{s}\, \sigma_{ij}(s)\,,$$

(1.23)

where K_n are Bessel functions, s the usual Mandelstam variable,

$$p_{ij}^2(s) = \frac{s}{4} - \frac{m_i^2 + m_j^2}{2} + \frac{(m_i^2 - m_j^2)^2}{4s}\,,$$

(1.24)

and

$$\sigma_{ij}(s) = \frac{1}{4\sqrt{s}p_{ij}(s)} \int \frac{d^3p'}{(2\pi)^3 E'} \frac{d^3p''}{(2\pi)^3 E''} (2\pi)^4 \delta^4(p_i + p_j - p' - p'') |\mathcal{T}_{ij}|^2 \quad (1.25)$$

with p', p'', E', E'' being the 3-momenta and energies of the outgoing particles and $|\mathcal{T}_{ij}|^2$ the squared transition matrix element summed over final-state spins and averaged over initial-state spins. Summation over all final states is implied.

The relevant final states and Feynman diagrams for $\tilde{\chi} - \tilde{\tau}_2$ (co)annihilation are listed in Table 1.1. The exchanged particles are indicated for each pair

Table 1.1. Feynman diagrams

Initial State	Final States	Diagrams
$\tilde{\chi}\tilde{\chi}$	$f\bar{f}$	$s(h,H,A,Z)$, $t(\tilde{f})$, $u(\tilde{f})$
	hh, hH, HH, HA, AA, ZA, ZZ	$s(h,H)$, $t(\tilde{\chi})$, $u(\tilde{\chi})$
	hA, hZ, HZ	$s(A,Z)$, $t(\tilde{\chi})$, $u(\tilde{\chi})$
	H^+H^-, W^+W^-	$s(h,H,Z)$, $t(\tilde{\chi}^\pm)$, $u(\tilde{\chi}^\pm)$
	$W^\pm H^\mp$	$s(h,H,A)$, $t(\tilde{\chi}^\pm)$, $u(\tilde{\chi}^\pm)$
$\tilde{\chi}\tilde{\tau}_2$	τh, τH, τZ	$s(\tau)$, $t(\tilde{\tau}_{1,2})$
	τA	$s(\tau)$, $t(\tilde{\tau}_1)$
	$\tau\gamma$	$s(\tau)$, $t(\tilde{\tau}_2)$
$\tilde{\tau}_2\tilde{\tau}_2$	$\tau\tau$	$t(\tilde{\chi})$, $u(\tilde{\chi})$
$\tilde{\tau}_2\tilde{\tau}_2^*$	hh, hH, HH, ZZ	$s(h,H)$, $t(\tilde{\tau}_{1,2})$, $u(\tilde{\tau}_{1,2})$, c
	AA	$s(h,H)$, $t(\tilde{\tau}_1)$, $u(\tilde{\tau}_1)$, c
	H^+H^-, W^+W^-	$s(h,H,\gamma,Z)$, $t(\tilde{\nu}_\tau)$, c
	$\gamma\gamma$, γZ	$t(\tilde{\tau}_2)$, $u(\tilde{\tau}_2)$, c
	$t\bar{t}$, $b\bar{b}$	$s(h,H,\gamma,Z)$
	$\tau\bar{\tau}$	$s(h,H,\gamma,Z)$, $t(\tilde{\chi})$
	$u\bar{u}$, $d\bar{d}$, $e\bar{e}$	$s(\gamma,Z)$

of initial and final states. The symbols $s(x, y, ...)$, $t(x, y, ...)$, and $u(x, y, ...)$ denote tree-level graphs in which the particles $x, y, ...$ are exchanged in the s-, t-, and u-channel respectively. The symbol c stands for "contact" diagrams with all four external legs meeting at a vertex. The charged Higgs bosons are denoted as $H^\pm$, while f stands for all the matter fermions (quarks and leptons) and e, u, and d represent the first and second generation charged leptons, up-, and down-type quarks respectively. The bars denote the anti-fermions, $\tilde{\chi}^\pm$ are the charginos, and $\tilde{\nu}_\tau$ is the superpartner of the τ-neutrino. We have included all possible $\tilde{\chi} - \tilde{\chi}$ annihilation processes (see e.g. [38]), but only the most important $\tilde{\chi} - \tilde{\tau}_2$, $\tilde{\tau}_2 - \tilde{\tau}_2$, and $\tilde{\tau}_2 - \tilde{\tau}_2^*$ coannihilation processes from [21, 39] (for a complete list see e.g. [40]), which are though adequate for giving accurate results for all values of $\tan\beta$, including the large ones. Some of the diagrams listed here have not been considered in previous works [32, 33] with small $\tan\beta$.

The $\tilde{\chi} - \tilde{\chi}$ annihilation via an A- or H-pole exchange in the s-channel can be [41] very important especially in the CMSSM with large $\tan\beta$. As $\tan\beta$ increases, the Higgs boson masses m_A and m_H decrease due to the fact that h_b increases and, thus, its influence on the RG running of these masses is enhanced. As a consequence, m_A and m_H approach $2m_{\tilde{\chi}}$ and the neutralino pair annihilation via an A- or H-pole exchange in the s-channel is resonantly enhanced. The contribution from the H pole is p-wave suppressed as one can show [31] using CP invariance (recall that the p wave is suppressed by $x_F \sim 25$). Therefore, the dominant contribution originates from the A pole with the dominant decay mode being the one to $b\bar{b}$ since, for large $\tan\beta$, the $Ab\bar{b}$ coupling is enhanced. We find [42] that there exists a region in the parameter space of the CMSSM corresponding to large values of $\tan\beta$ where the $\tilde{\chi} - \tilde{\chi}$ annihilation via an A pole reduces drastically the relic neutralino abundance and, thus, makes it possible to satisfy the WMAP constraint on CDM (note that, generically, $\Omega_{\tilde{\chi}}h^2$ comes out too large).

As we already mentioned, near the A pole, the partial wave (or Taylor) expansion in (1.19) and (1.20) fails [29, 36] badly. So, the thermal averaging must by performed exactly using numerical methods and employing the formulae in (1.23), (1.24), and (1.25). In order to achieve good accuracy, it is also important to include the one-loop QCD corrections [43] to the decay width of the A particle entering its propagator as well as to the quark masses.

Another phenomenon which helps reducing drastically $\Omega_{\tilde{\chi}}h^2$ and, thus, satisfying the CDM constraint is strong $\tilde{\chi} - \tilde{\tau}_2$ coannihilation [21, 32, 33] which operates when $m_{\tilde{\tau}_2}$ gets close to $m_{\tilde{\chi}}$. This yields [32, 33] a relatively narrow allowed region in the $m_0 - M_{1/2}$ plane (for fixed A_0 and $\tan\beta$), which stretches just above the excluded region where the LSP is the $\tilde{\tau}_2$.

There exists [42] also a "bulk" region at $m_0 \sim M_{1/2} \sim$ few $\times 100$ GeV which is allowed by CDM considerations. The (co)annihilation is enhanced in this region due to the low values of the various sparticle masses. However, this region is, generally, excluded by other phenomenological constraints (see Sect. 1.7.4). So, the A-pole annihilation of neutralinos and the $\tilde{\chi} - \tilde{\tau}_2$ coannihilation are the two basic available mechanisms for obtaining acceptable values for the neutralino relic abundance in the CMSSM.

There are publicly available codes such as the `micrOMEGAs` [44] or the `DarkSUSY` [20] for the calculation of $\Omega_{\tilde{\chi}} h^2$ in MSSM which, among other improvements, include all the relevant (co)annihilation channels between all the sparticles (neutralinos, charginos, squarks, sleptons, gluinos), use exact tree-level cross sections, calculate accurately and relativistically the thermal averages, treat poles and final-state thresholds properly, integrate the Boltzmann equation numerically, and include the one-loop QCD corrections to the decay widths of the Higgs particles and the fermion masses. These codes apply to any composition of the neutralino and also include other phenomenological constraints such as the accelerator bounds on certain (s)particle masses and the bounds on the anomalous magnetic moment of the muon and the branching ration of the process $b \to s\gamma$ (see Sect. 1.7.4).

1.5 Axinos

Another SUSY particle that could account for the CDM in the universe is [46] (see also [47]) the axino $\tilde{a}$. This particle, which is the superpartner of the axion field, is a neutral Majorana chiral fermion with negative R-parity. Its mass $m_{\tilde{a}}$ is [48] strongly model-dependent and can be anywhere in the range $1 \text{ eV} - M_{\text{SUSY}}$. In the limit of unbroken SUSY, the axino mass is obviously equal to the axion mass, which is tiny. Soft SUSY breaking, however, generates suppressed corrections to $m_{\tilde{a}}$ via non-renormalizable operators of dimension five or higher. So, the corrected mass is at most of order $M_{\text{SUSY}}^2/f_a \sim 1 \text{ keV}$ (note that no dimension-four soft mass term is allowed for the axino since this particle is a chiral fermion). In specific SUSY models, there also exist one-loop contributions to $m_{\tilde{a}}$, which are typically $\lesssim M_{\text{SUSY}}$. When the axion is a linear combination of the phases of more than one superfields, we can even have tree-level contributions to the axino mass which can easily be as large as M_{SUSY}. In conclusion, $m_{\tilde{a}}$ is basically a free parameter ranging between 1 eV and M_{SUSY}. This means that the axino can easily be the LSP in SUSY models.

The axino couplings are suppressed by f_a with the most important of them being the dimension-five axino $(\tilde{a})$–gaugino $(\tilde{\lambda})$–gauge boson (A) Lagrangian coupling:

$$\mathcal{L}_{\tilde{a}\tilde{\lambda}A} = i\frac{3\alpha_Y C_{aYY}}{8\pi f_a}\bar{\tilde{a}}\gamma_5[\gamma^\mu,\gamma^\nu]\tilde{B}B_{\mu\nu} + i\frac{3\alpha_s}{8\pi f_a}\bar{\tilde{a}}\gamma_5[\gamma^\mu,\gamma^\nu]\tilde{g}^b F_{\mu\nu}^b , \quad (1.26)$$

where B and $\tilde{B}$ are, respectively, the gauge boson and gaugino corresponding to $U(1)_Y$, F^b and $\tilde{g}^b$ the gluon and gluino fields, α_Y and α_s the $U(1)_Y$ and strong fine-structure constants, and C_{aYY} a model-dependent coefficient of order unity.

Inflation dilutes utterly any pre-existing axinos, which, after reheating, are *not* in thermal equilibrium with the thermal bath because of their very weak couplings (suppressed by f_a). They can, however, be thermally produced from the bath by 2-body scattering processes or the decay of (s)particles. The so-produced axinos are initially relativistic, but out of thermal equilibrium. This thermal production (TP) of axinos is [46] predominantly due to 2-body scattering processes of strongly interacting particles (because of the relative strength of strong interactions) involving the $\tilde{a}\tilde{g}F$ coupling in (1.26). Such processes are

$$g + g \rightarrow \tilde{a} + \tilde{g} \,, \quad g + \tilde{g} \rightarrow \tilde{a} + g \,, \quad g + \tilde{q} \rightarrow \tilde{a} + q \,, \quad g + q \rightarrow \tilde{a} + \tilde{q} \,,$$
$$\tilde{q} + q \rightarrow \tilde{a} + g \,, \quad \tilde{g} + \tilde{g} \rightarrow \tilde{a} + \tilde{g} \,, \quad \tilde{g} + q \rightarrow \tilde{a} + q \,, \quad \tilde{g} + \tilde{q} \rightarrow \tilde{a} + \tilde{q} \,,$$
$$q + \bar{q} \rightarrow \tilde{a} + \tilde{g} \,, \quad \tilde{q} + \tilde{q} \rightarrow \tilde{a} + \tilde{g} \,, \tag{1.27}$$

where gluons and quarks are denoted by g and q respectively. There exists [46] also TP of axinos from the decay of thermal gluinos ($\tilde{g} \rightarrow \tilde{a} + g$) or thermal neutralinos ($\tilde{\chi} \rightarrow \tilde{a} + \gamma$ [or Z]). The latter proceeds through the dimension-five Lagrangian coupling $\tilde{a}\tilde{B}B$ in (1.26) provided that the neutralino possesses an appreciable bino component. These two decay processes are important only for reheat temperatures T_r of the order of the gluino mass $m_{\tilde{g}}$ or the neutralino mass $m_{\tilde{\chi}}$ respectively.

There is also non-thermal production (NTP) of axinos resulting from the decays of sparticles which are out of thermal equilibrium. Indeed, due to the suppressed couplings of the axino, the sparticles first decay to the lightest ordinary sparticle (LOSP), i.e. the lightest sparticle with non-trivial SM quantum numbers, which is the NLSP in this case. The LOSPs then freeze out of thermal equilibrium and eventually decay into axinos.

If the LOSP happens to be the lightest neutralino, the relevant decay process is [46] $\tilde{\chi} \rightarrow \tilde{a} + \gamma$ [or Z] through the coupling $\tilde{a}\tilde{B}B$ in (1.26) provided that $\tilde{\chi}$ has a $\tilde{B}$ component. If, alternatively, the LOSP is the lightest stau mass eigenstate, the decay process for the NTP of axinos is [49] $\tilde{\tau}_2 \rightarrow \tau + \tilde{a}$ via the one-loop Feynman diagrams in Fig. 1.1, which contain the effective vertex $\tilde{\chi}\tilde{a}\gamma$ [or $\tilde{\chi}\tilde{a}Z$] from the coupling $\tilde{a}\tilde{B}B$ in (1.26). In the decay of $\tilde{\chi}$, γ's and $q\bar{q}$ pairs are produced. The latter originate from virtual γ and Z, or real Z exchange and lead to hadronic showers. In the $\tilde{\tau}_2$ case, the resulting τ decays immediately into light mesons yielding again hadronic showers. The electromagnetic and hadronic showers emerging from the LOSP decay in both cases, if they are generated after big bang nucleosynthesis (BBN), can cause destruction and/or overproduction of some of the light elements, thereby jeopardizing the successful predictions of BBN. This implies some constraints on the parameters of the model which, in the present case where the axino is the LSP, come basically from the hadronic showers alone due to the relatively short LOSP lifetime. In the case of a neutralino LOSP, we obtain [46] the bound $m_{\tilde{a}} \gtrsim 360$ MeV for low values of the neutralino mass $m_{\tilde{\chi}}$ ($\lesssim 60$ GeV), but no bound on the axino mass is obtained for higher values of $m_{\tilde{\chi}}$ ($\gtrsim 150$ GeV).

We must further impose the following constraints: (a) the predicted axino abundance $\Omega_{\tilde{a}}h^2$ should lie in the 95% c.l. range for the CDM abundance

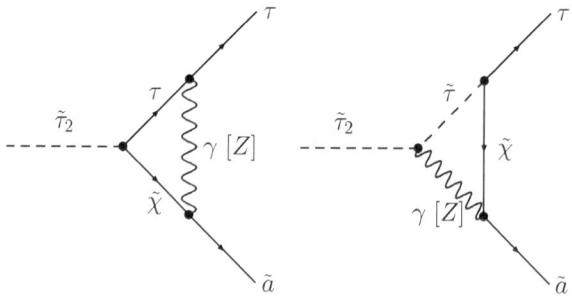

Fig. 1.1. The one-loop diagrams for the decay $\tilde{\tau}_2 \to \tau + \tilde{a}$

in the universe derived by the WMAP satellite [1], (b) both the TP and NTP axinos must become non-relativistic before matter domination so as to contribute to CDM, and (c) the NTP axinos should not contribute too much relativistic energy density during BBN since this can destroy its successful predictions. For both $\tilde{\chi}$ or $\tilde{\tau}_2$ LOSP, the requirements (b) and (c) imply that $m_{\tilde{a}} \gtrsim 100$ keV or, equivalently, $T_r \lesssim 5 \times 10^6$ GeV. For large values of the reheat temperature ($T_r \gtrsim 10^4$ GeV), TP of axinos is more efficient than NTP and the cosmologically favored region in parameter space where the requirement (a) holds is quite narrow. For smaller T_r's, NTP dominates yielding a much wider favored region with $m_{\tilde{a}} \gtrsim 10$ MeV. The upper bound on $m_{\tilde{a}}$ increases as T_r decreases towards $m_{\tilde{\chi}}$. For $m_{\tilde{q}} \ll m_{\tilde{g}}$, TP of axinos via the process $\tilde{q} \to q + \tilde{a}$ becomes [50] very efficient leading to a reduction of the upper limit on T_r. As a result, the cosmologically favored region from NTP is reduced in this case. The Feynman diagrams for the process $\tilde{q} \to q + \tilde{a}$ are depicted in Fig. 1.2. The restrictions on the $m_{\tilde{a}} - T_r$ plane from axino CDM considerations are presented in Fig. 1.3.

We find [49] that, for the CMSSM, with appropriate choices of $m_{\tilde{a}}$ and T_r, almost any pair of values for m_0 and $M_{1/2}$ can be allowed. This holds for both $\tilde{\chi}$ or $\tilde{\tau}_2$ as LOSP. However, the required T_r's for achieving the WMAP bound on CDM turn out to be quite low ($\lesssim$ few $\times$ 100 GeV).

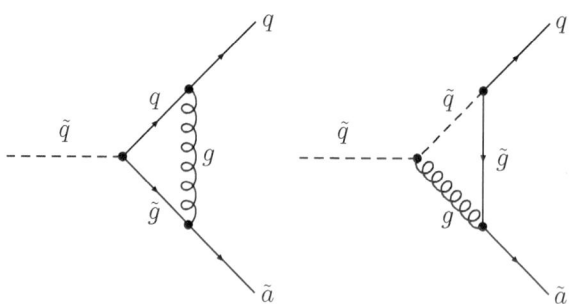

Fig. 1.2. The one-loop diagrams for the decay $\tilde{q} \to q + \tilde{a}$

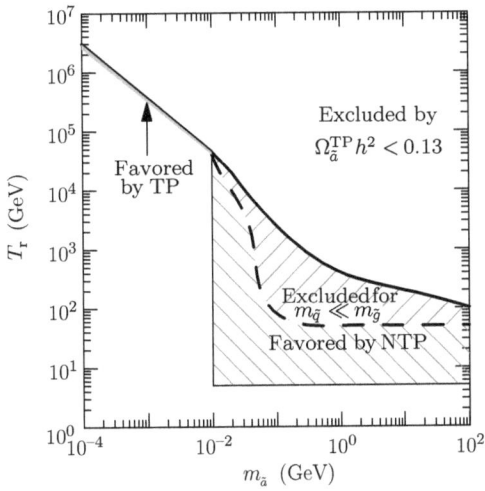

Fig. 1.3. The restrictions on the $m_{\tilde{a}} - T_r$ plane from axino CDM considerations for $\tilde{\chi} = \tilde{B}$, $m_{\tilde{\chi}} = 100$ GeV, $m_{\tilde{g}} = m_{\tilde{q}} = 1$ TeV, and $f_a = 10^{11}$ GeV. The solid almost diagonal line corresponds to $\Omega_{\tilde{a}}^{TP}h^2 \approx 0.13$, where $\Omega_{\tilde{a}}^{TP}h^2$ is the TP axino abundance. So, the area above this line is cosmologically excluded. The narrow shaded area just below the thin part of this line for $m_{\tilde{a}} \lesssim 10$ MeV is cosmologically favored by TP. The hatched areas are favored by NTP. For $m_{\tilde{q}} \ll m_{\tilde{g}}$, the solid line is replaced by the dashed one, whose position is strongly dependent on the actual values of $m_{\tilde{q}}$, $m_{\tilde{g}}$ and is only indicative here. The area favored by NTP is then limited only to the "back-hatched" region which lies below the dashed line

1.6 Gravitinos

It has been proposed [51, 52] that CDM could also consist of gravitinos. The gravitino $\tilde{G}$ is the superpartner of the graviton and has negative R-parity. It can be the LSP in many cases and, thus, contribute to CDM. In the very CMSSM, its mass $m_{\tilde{G}}$ is fixed by the asymptotic condition $m_{3/2} = m_0$. In the general CMSSM, however, it is a free parameter ranging between 100 GeV and 1 TeV. It can, thus, very easily be the LSP in this case.

The couplings of the gravitino are suppressed by the Planck scale. The most important of them are given by the dimension-five Lagrangian terms

$$\mathcal{L} = -\frac{1}{\sqrt{2}m_P} \mathcal{D}_\nu \phi^{i*} \bar{\tilde{\psi}}_\mu \gamma^\nu \gamma^\mu \psi^i - \frac{1}{\sqrt{2}m_P} \mathcal{D}_\nu \phi^i \bar{\psi}^i \gamma^\mu \gamma^\nu \tilde{\psi}_\mu$$
$$-\frac{i}{8m_P} \bar{\tilde{\psi}}_\mu [\gamma^\nu, \gamma^\rho] \gamma^\mu \tilde{\lambda}^a F_{\nu\rho}^a , \qquad (1.28)$$

where $\tilde{\psi}_\mu$ denotes the gravitino field, ϕ^i are the complex scalar fields, ψ^i are the corresponding chiral fermion fields, $\tilde{\lambda}^a$ are the gaugino fields, $m_P \simeq 2.44 \times 10^{18}$ GeV is the reduced Planck scale, and $\mathcal{D}_\nu$ denotes the covariant derivative. From these Lagrangian terms, we obtain scalar–fermion–gravitino vertices ($\phi f \tilde{G}$) such as $q \tilde{q} \tilde{G}$, $l \tilde{l} \tilde{G}$, and $H \tilde{H} \tilde{G}$, as well as gaugino–gauge boson–gravitino vertices ($\tilde{\lambda} F \tilde{G}$) such as $g \tilde{g} \tilde{G}$ and $B \tilde{B} \tilde{G}$ (in this section, l and H represent any lepton and Higgs boson respectively).

The gravitinos are thermally produced after reheating by $2 \to 2$ scattering processes involving the above vertices. Such processes are [51, 52]

$$
\begin{array}{llll}
g + g \to \tilde{G} + \tilde{g}, & g + \tilde{g} \to \tilde{G} + g, & g + \tilde{q} \to \tilde{G} + q, & g + q \to \tilde{G} + \tilde{q}, \\
q + \tilde{q} \to \tilde{G} + g, & \tilde{g} + \tilde{g} \to \tilde{G} + \tilde{g}, & \tilde{g} + q \to \tilde{G} + q, & \tilde{g} + \tilde{q} \to \tilde{G} + \tilde{q}, \\
& q + \bar{q} \to \tilde{G} + \tilde{g}, & \tilde{q} + \tilde{q} \to \tilde{G} + \tilde{g}.
\end{array}
\tag{1.29}
$$

There is [52, 53] also NTP of gravitinos via the decay of the NLSP. For neutralino NLSP, the relevant decay processes are $\tilde{\chi} \to \tilde{G} + \gamma$ [or Z] from the $\tilde{\lambda} F \tilde{G}$ coupling and $\tilde{\chi} \to \tilde{G} + H$ from the $H \tilde{H} \tilde{G}$ coupling. In the case of $\tilde{\tau}_2$ NLSP, the relevant decay process is $\tilde{\tau}_2 \to \tau + \tilde{G}$ from the vertex $l \tilde{l} \tilde{G}$. There is an important difference between the NTP of gravitinos and axinos. In the former case, the NLSP has a large lifetime (up to about 10^8 sec). Consequently, it gives rise mostly to electromagnetic, but also to hadronic showers well after BBN. The electromagnetic showers cause destruction of some light elements (D, ^{4}He, ^{7}Li) and/or overproduction of ^{3}He and ^{6}Li, thereby disturbing BBN. The hadronic showers can also disturb BBN. The overall resulting constraint is [54] very strong allowing only limited regions of the parameter space of the CMSSM lying exclusively in the range where the NLSP is the $\tilde{\tau}_2$. Moreover, in these allowed regions, the NTP of gravitinos is not efficient enough to account for the observed CDM abundance for $M_{1/2} \lesssim 6$ TeV. However, we can compensate for the inefficiency of NTP by raising T_r to enhance the TP of $\tilde{G}$'s. The relic gravitino abundance from TP, for $m_{\tilde{G}} \ll m_{\tilde{g}}$, is [55]

$$
\Omega_{\tilde{G}}^{\mathrm{TP}} h^2 \approx 0.2 \left(\frac{T_r}{10^{10} \text{ GeV}} \right) \left(\frac{100 \text{ GeV}}{m_{\tilde{G}}} \right) \left(\frac{m_{\tilde{g}}(\mu)}{1 \text{ TeV}} \right),
\tag{1.30}
$$

where $m_{\tilde{g}}(\mu)$ is the running gluino mass (for the general formula, see [56]).

1.7 Yukawa Quasi-Unification

As already said in Sect. 1.3, exact YU in the framework of the CMSSM leads to wrong values for m_b and, thus, must be corrected. We will now present a model which naturally solves [39] (see also [57, 58]) this m_b problem and discuss the restrictions on its parameter space implied by CDM considerations and other phenomenological constraints. Exact YU can be achieved by embedding the MSSM in a SUSY GUT model with a gauge group containing SU(4)$_c$ and SU(2)$_R$. Indeed, assuming that the electroweak Higgs superfields H_1, H_2 and the third family right handed quark superfields t^c, b^c form SU(2)$_R$ doublets, we obtain [59] the asymptotic Yukawa coupling relation $h_t = h_b$ and, hence, large $\tan\beta \sim m_t/m_b$. Moreover, if the third generation quark and lepton $SU(2)_L$ doublets [singlets] q_3 and l_3 [b^c and τ^c] form a $SU(4)_c$ 4-plet [$\bar{4}$-plet] and the Higgs doublet H_1 which couples to them is a $SU(4)_c$ singlet, we get $h_b = h_\tau$ and the asymptotic relation $m_b = m_\tau$ follows. The simplest GUT gauge group which contains both $SU(4)_c$ and $SU(2)_R$ is the Pati-Salam (PS) group $G_{PS} = SU(4)_c \times SU(2)_L \times SU(2)_R$ and we will use it here.

As mentioned, applying YU in the context of the CMSSM and given the experimental values of the top-quark and tau-lepton masses (which naturally restrict $\tan\beta \sim 50$), the resulting value of the b-quark mass turns out to be unacceptable. This is due to the fact that, in the large $\tan\beta$ regime, the tree-level b-quark mass receives sizeable SUSY corrections [24, 60, 61, 62] (about 20%), which have the sign of μ (with the standard sign convention [63]) and drive, for $\mu > [<]\ 0$, the corrected b-quark mass at M_Z, $m_b(M_Z)$, well above [somewhat below] its 95% c.l. experimental range

$$2.684 \text{ GeV} \lesssim m_b(M_Z) \lesssim 3.092 \text{ GeV} \quad \text{with} \quad \alpha_s(M_Z) = 0.1185 . \qquad (1.31)$$

This is derived by appropriately [39] evolving the corresponding range of $m_b(m_b)$ in the $\overline{MS}$ scheme (i.e. $3.95 - 4.55$ GeV) up to M_Z in accordance with [64]. We see that, for both signs of μ, YU leads to an unacceptable b-quark mass with the $\mu < 0$ case being less disfavored.

A way out of this m_b problem can be found [39] (see also [57, 58]) without having to abandon the CMSSM (in contrast to the usual strategy [62, 65, 66, 67]) or YU altogether. We can rather modestly correct YU by including an extra $SU(4)_c$ non-singlet Higgs superfield with Yukawa couplings to the quarks and leptons. The Higgs $SU(2)_L$ doublets contained in this superfield can naturally develop [68] subdominant VEVs and mix with the main electroweak doublets, which are assumed to be $SU(4)_c$ singlets and form a $SU(2)_R$ doublet. This mixing can, in general, violate the $SU(2)_R$ symmetry. Consequently, the resulting electroweak Higgs doublets H_1, H_2 do not form a $SU(2)_R$ doublet and also break the $SU(4)_c$ symmetry. The required deviation from YU is expected to be more pronounced for $\mu > 0$. Despite this, we will study here this case, since the $\mu < 0$ case has been excluded [69] by combining the WMAP restrictions [1] on the CDM in the universe with the experimental results [70] on the inclusive branching ratio BR($b \to s\gamma$). The same SUSY GUT model which, for $\mu > 0$ and universal boundary conditions, remedies the m_b problem leads to a new version [71] of shifted hybrid inflation [72], which, as the older version [72], avoids monopole overproduction at the end of inflation, but, in contrast to that version, is based only on renormalizable interactions.

In Sect. 1.7.1, we review the construction of a SUSY GUT model which naturally and modestly violates YU, yielding an appropriate Yukawa quasi-unification condition (YQUC), which is derived in Sect. 1.7.2. We then outline the resulting CMSSM in Sect. 1.7.3 and introduce the various cosmological and phenomenological requirements which restrict its parameter space in Sect. 1.7.4. In Sect. 1.7.5, we delineate the allowed range of parameters.

Finally, in Sect. 1.7.6, we briefly comment on the new version of shifted hybrid inflation which is realized in this model.

1.7.1 The PS SUSY GUT Model

We will take the SUSY GUT model of shifted hybrid inflation [72] (see also [73]) as our starting point. It is based on G_{PS}, which is the simplest GUT gauge group that can lead to exact YU. The representations under G_{PS} and the global charges of the various matter and Higgs superfields contained in this model are presented in Table 1.2, which also contains the extra Higgs superfields required for accommodating an adequate violation of YU for $\mu > 0$ (see below). The matter superfields are F_i and F_i^c ($i = 1, 2, 3$), while the electroweak Higgs doublets belong to the superfield h. So, all the requirements for exact YU are fulfilled. The spontaneous breaking of G_{PS} down to G_{SM} is achieved by the superheavy VEVs ($\sim M_{GUT}$) of the right handed neutrino-type components of a conjugate pair of Higgs superfields H^c, $\bar{H}^c$. The model also contains a gauge singlet S which triggers the breaking of G_{PS}, a SU(4)$_c$ 6-plet G which gives [74] masses to the right handed down-quark-type components of H^c, $\bar{H}^c$, and a pair of gauge singlets N, $\bar{N}$ for solving [75] the μ problem of the MSSM via a PQ symmetry (for an alternative solution of the μ problem, see [10]). In addition to G_{PS}, the model possesses two global U(1) symmetries, namely a R and a PQ symmetry, as well as the discrete matter parity symmetry Z_2^{mp}. Note that global continuous symmetries such as our PQ and R symmetry can effectively arise [77] from the rich discrete symmetry groups encountered in many compactified string theories (see e.g. [78]). Note that, although the model contains baryon- and lepton-number-violating superpotential terms, the proton is [39, 72] practically stable. The baryon asymmetry of the universe is generated via the non-thermal realization [79] of the leptogenesis scenario [80] (for recent papers on non-thermal leptogenesis, see e.g. [81]).

A moderate violation of exact YU can be naturally accommodated in this model by adding a new Higgs superfield h' with Yukawa couplings $FF^c h'$. Actually, $(\mathbf{15}, \mathbf{2}, \mathbf{2})$ is the only representation of G_{PS}, besides $(\mathbf{1}, \mathbf{2}, \mathbf{2})$, which possesses such couplings to the matter superfields. In order to give superheavy masses to the color non-singlet components of h', we need to include one more Higgs superfield $\bar{h}'$ with the superpotential coupling $\bar{h}' h'$, whose coefficient is of the order of M_{GUT}.

After the breaking of G_{PS} to G_{SM}, the two color singlet SU(2)$_L$ doublets h_1', h_2' contained in h' can mix with the corresponding doublets h_1, h_2 in h. This is mainly due to the terms $\bar{h}' h$ and $H^c \bar{H}^c \bar{h}' h$. Actually, since

$$H^c \bar{H}^c = (\bar{\mathbf{4}}, \mathbf{1}, \mathbf{2})(\mathbf{4}, \mathbf{1}, \mathbf{2}) = (\mathbf{15}, \mathbf{1}, \mathbf{1} + \mathbf{3}) + \cdots ,$$
$$\bar{h}' h = (\mathbf{15}, \mathbf{2}, \mathbf{2})(\mathbf{1}, \mathbf{2}, \mathbf{2}) = (\mathbf{15}, \mathbf{1}, \mathbf{1} + \mathbf{3}) + \cdots , \tag{1.32}$$

there are two independent couplings of the type $H^c \bar{H}^c \bar{h}' h$ (both suppressed by the string scale $M_S \approx 5 \times 10^{17}$ GeV, as they are non-renormalizable).

Table 1.2. Superfield content of the model

Superfields	Representations under G_{PS}	Global Charges R PQ Z_2^{mp}		
Matter Superfields				
F_i	$(\mathbf{4},\mathbf{2},\mathbf{1})$	$1/2$	-1	1
F_i^c	$(\bar{\mathbf{4}},\mathbf{1},\mathbf{2})$	$1/2$	0	-1
Higgs Superfields				
h	$(\mathbf{1},\mathbf{2},\mathbf{2})$	0	1	0
H^c	$(\bar{\mathbf{4}},\mathbf{1},\mathbf{2})$	0	0	0
$\bar{H}^c$	$(\mathbf{4},\mathbf{1},\mathbf{2})$	0	0	0
S	$(\mathbf{1},\mathbf{1},\mathbf{1})$	1	0	0
G	$(\mathbf{6},\mathbf{1},\mathbf{1})$	1	0	0
N	$(\mathbf{1},\mathbf{1},\mathbf{1})$	$1/2$	-1	0
$\bar{N}$	$(\mathbf{1},\mathbf{1},\mathbf{1})$	0	1	0
Extra Higgs Superfields				
h'	$(\mathbf{15},\mathbf{2},\mathbf{2})$	0	1	0
$\bar{h}'$	$(\mathbf{15},\mathbf{2},\mathbf{2})$	1	-1	0
ϕ	$(\mathbf{15},\mathbf{1},\mathbf{3})$	0	0	0
$\bar{\phi}$	$(\mathbf{15},\mathbf{1},\mathbf{3})$	1	0	0

One of these couplings is between the $SU(2)_{\mathrm{R}}$ singlets in $H^c\bar{H}^c$ and $\bar{h}'h$ and the other between the $SU(2)_{\mathrm{R}}$ triplets in these combinations. So, we obtain two bilinear terms $\bar{h}'_1 h_1$ and $\bar{h}'_2 h_2$ with different coefficients, which are suppressed by $M_{\mathrm{GUT}}/M_{\mathrm{S}}$. These terms together with the terms $\bar{h}'_1 h'_1$ and $\bar{h}'_2 h'_2$ from $\bar{h}'h'$, which have equal coefficients, generate different mixings between h_1, h'_1 and h_2, h'_2. Consequently, the resulting electroweak doublets H_1, H_2 contain $SU(4)_c$ violating components suppressed by $M_{\mathrm{GUT}}/M_{\mathrm{S}}$ and fail to form a $SU(2)_{\mathrm{R}}$ doublet by an equally suppressed amount. So, YU is naturally and moderately violated. Unfortunately, as it turns out, this violation is not adequately large for correcting the bottom-quark mass within the framework of the CMSSM with $\mu > 0$.

In order to allow for a more sizable violation of YU, we further extend the model by including the superfield ϕ with the coupling $\phi\bar{h}'h$. To give super-heavy masses to the color non-singlets in ϕ, we introduce one more superfield $\bar{\phi}$ with the coupling $\bar{\phi}\phi$, whose coefficient is of order M_{GUT}.

The superpotential terms $\bar{\phi}\phi$ and $\bar{\phi}H^c\bar{H}^c$ imply that, after the breaking of G_{PS} to G_{SM}, ϕ acquires a VEV of order M_{GUT}. The coupling $\phi\bar{h}'h$ then generates $SU(2)_{\mathrm{R}}$ violating unsuppressed bilinear terms between the doublets in $\bar{h}'$ and h. These terms can overshadow the corresponding ones from the non-renormalizable term $H^c\bar{H}^c\bar{h}'h$. The resulting $SU(2)_{\mathrm{R}}$ violating mixing of the doublets in h and h' is then unsuppressed and we can obtain stronger violation of YU.

1.7.2 The YQUC

To further analyze the mixing of the doublets in h and h', observe that the part of the superpotential corresponding to the symbolic couplings $\bar{h}'h'$, $\phi\bar{h}'h$ is properly written as

$$m\,\text{tr}\left(\bar{h}'\epsilon h'^{\text{T}}\epsilon\right) + p\,\text{tr}\left(\bar{h}'\epsilon\phi h^{\text{T}}\epsilon\right) , \qquad (1.33)$$

where m is a mass parameter of order M_{GUT}, p is a dimensionless parameter of order unity, tr denotes trace taken with respect to the $\text{SU}(4)_{\text{c}}$ and $\text{SU}(2)_{\text{L}}$ indices, and the superscript T denotes the transpose of a matrix.

After the breaking of G_{PS} to G_{SM}, ϕ acquires a VEV $\langle\phi\rangle \sim M_{\text{GUT}}$. Substituting it by this VEV in the above couplings, we obtain

$$\text{tr}(\bar{h}'\epsilon h'^{\text{T}}\epsilon) = \bar{h}_1'^{\text{T}}\epsilon h_2' + h_1'^{\text{T}}\epsilon\bar{h}_2' + \cdots , \qquad (1.34)$$

$$\text{tr}(\bar{h}'\epsilon\phi h^{\text{T}}\epsilon) = \frac{\langle\phi\rangle}{\sqrt{2}}\text{tr}(\bar{h}'\epsilon\sigma_3 h^{\text{T}}\epsilon) = \frac{\langle\phi\rangle}{\sqrt{2}}(\bar{h}_1'^{\text{T}}\epsilon h_2 - h_1^{\text{T}}\epsilon\bar{h}_2') , \qquad (1.35)$$

where the ellipsis in (1.34) contains the colored components of $\bar{h}'$, h' and $\sigma_3 = \text{diag}(1, -1)$. Inserting (1.34) and (1.35) into (1.33), we obtain

$$m\bar{h}_1'^{\text{T}}\epsilon(h_2' - \alpha h_2) + m(h_1'^{\text{T}} + \alpha h_1^{\text{T}})\epsilon\bar{h}_2' \quad \text{with} \quad \alpha = -\frac{p\langle\phi\rangle}{\sqrt{2}m} . \qquad (1.36)$$

So, we get two pairs of superheavy doublets with mass m. They are predominantly given by

$$\bar{h}_1', \quad \frac{h_2' - \alpha h_2}{\sqrt{1 + |\alpha|^2}} \quad \text{and} \quad \frac{h_1' + \alpha h_1}{\sqrt{1 + |\alpha|^2}}, \quad \bar{h}_2' . \qquad (1.37)$$

The orthogonal combinations of h_1, h_1' and h_2, h_2' constitute the electroweak doublets

$$H_1 = \frac{h_1 - \alpha^* h_1'}{\sqrt{1 + |\alpha|^2}} \quad \text{and} \quad H_2 = \frac{h_2 + \alpha^* h_2'}{\sqrt{1 + |\alpha|^2}} . \qquad (1.38)$$

The superheavy doublets in (1.37) must have vanishing VEVs, which readily implies that $\langle h_1'\rangle = -\alpha\langle h_1\rangle$ and $\langle h_2'\rangle = \alpha\langle h_2\rangle$. Equation (1.38) then gives $\langle H_1\rangle = (1 + |\alpha|^2)^{1/2}\langle h_1\rangle$, $\langle H_2\rangle = (1 + |\alpha|^2)^{1/2}\langle h_2\rangle$. From the third generation Yukawa couplings $y_{33}F_3hF_3^c$, $2y_{33}'F_3h'F_3^c$, we obtain

$$m_t = |y_{33}\langle h_2\rangle + y_{33}'\langle h_2'\rangle| = \left|\frac{1 + \rho\alpha/\sqrt{3}}{\sqrt{1 + |\alpha|^2}}y_{33}\langle H_2\rangle\right| , \qquad (1.39)$$

$$m_b = \left|\frac{1 - \rho\alpha/\sqrt{3}}{\sqrt{1 + |\alpha|^2}}y_{33}\langle H_1\rangle\right| , \quad m_\tau = \left|\frac{1 + \sqrt{3}\rho\alpha}{\sqrt{1 + |\alpha|^2}}y_{33}\langle H_1\rangle\right| , \qquad (1.40)$$

where $\rho = y'_{33}/y_{33}$. From (1.39) and (1.40), we see that YU is now replaced by the YQUC

$$h_t : h_b : h_\tau = (1 + c) : (1 - c) : (1 + 3c) \quad \text{with} \quad 0 < c = \rho\alpha/\sqrt{3} < 1 . \quad (1.41)$$

For simplicity, we restricted ourselves here to real values of c only which lie between zero and unity, although c is, in general, an arbitrary complex quantity with $|c| \sim 1$.

1.7.3 The Resulting CMSSM

Below the GUT scale M_{GUT}, the particle content of our model reduces to this of MSSM (modulo SM singlets). We assume universal soft SUSY breaking scalar masses m_0, gaugino masses $M_{1/2}$, and trilinear scalar couplings A_0 at M_{GUT}. Therefore, the resulting MSSM is the so-called CMSSM [18] with $\mu > 0$ supplemented by the YQUC in (1.41). With these initial conditions, we run the MSSM RGEs [21] between M_{GUT} and a common variable SUSY threshold M_{SUSY} (see Sect. 1.3) determined in consistency with the SUSY spectrum of the model. At M_{SUSY}, we impose radiative electroweak symmetry breaking, evaluate the SUSY spectrum and incorporate the SUSY corrections [24, 61, 62] to the b-quark and τ-lepton masses. Note that the corrections to the τ-lepton mass (almost 4%) lead [69] to a small reduction of $\tan\beta$. From M_{SUSY} to M_Z, the running of gauge and Yukawa coupling constants is continued using the SM RGEs.

For presentation purposes, $M_{1/2}$ and m_0 can be replaced [21] by the LSP mass m_{LSP} and the relative mass splitting between this particle and the lightest stau $\Delta_{\tilde{\tau}_2} = (m_{\tilde{\tau}_2} - m_{\text{LSP}})/m_{\text{LSP}}$ (recall that $\tilde{\tau}_2$ is the NLSP in this case). For simplicity, we restrict this presentation to the $A_0 = 0$ case (for $A_0 \neq 0$ see [39, 82]). So, our input parameters are m_{LSP}, $\Delta_{\tilde{\tau}_2}$, c, and $\tan\beta$.

For any given $m_b(M_Z)$ in the range in (1.31) and with fixed $m_t(m_t) = 166$ GeV and $m_\tau(M_Z) = 1.746$ GeV, we can determine the parameters c and $\tan\beta$ at M_{SUSY} so that the YQUC in (1.41) is satisfied. We are, thus, left with m_{LSP} and $\Delta_{\tilde{\tau}_2}$ as free parameters.

1.7.4 Cosmological and Phenomenological Constraints

Restrictions on the parameters of our model can be derived by imposing a number of cosmological and phenomenological requirements (for similar recent analyses, see [66, 67, 83]). These constraints result from

• *CDM Considerations.* As discussed in Sect. 1.3, in the context of the CMSSM, the LSP can be the lightest neutralino which is an almost pure bino. It naturally arises [84] as a CDM candidate. We require its relic abundance, $\Omega_{\text{LSP}}h^2$, not to exceed the 95% c.l. upper bound on the CDM abundance derived [1] by WMAP:

$$\Omega_{\text{CDM}}h^2 \lesssim 0.13 . \quad (1.42)$$

We calculate $\Omega_{\mathrm{LSP}}h^2$ using micrOMEGAs [44], which is certainly one of the most complete publicly available codes. Among other things, it includes all possible coannihilation processes [33] and one-loop QCD corrections [43] to the Higgs decay widths and couplings to fermions.

• *Branching Ratio of $b \to s\gamma$.* Taking into account the experimental results of [70] on this ratio, $\mathrm{BR}(b \to s\gamma)$, and combining [39] appropriately the experimental and theoretical errors involved, we obtain the 95% c.l. range

$$1.9 \times 10^{-4} \lesssim \mathrm{BR}(b \to s\gamma) \lesssim 4.6 \times 10^{-4} . \tag{1.43}$$

Although there exist more recent experimental data [85] on the branching ratio of $b \to s\gamma$, we do not use them here. The reason is that these data do not separate the theoretical errors from the experimental ones and, thus, the derivation of the 95% c.l. range is quite ambiguous. In any case, the 95% c.l. limits obtained in [86] on the basis of these latest measurements are not terribly different from the ones quoted in (1.43). In view of this and the fact that, in our case, the restrictions from $\mathrm{BR}(b \to s\gamma)$ are overshadowed by other constraints (see Sect. 1.7.5), we limit ourselves to the older data. We compute $\mathrm{BR}(b \to s\gamma)$ by using an updated version of the relevant calculation contained in the micrOMEGAs package [44]. In this code, the SM contribution is calculated following [87]. The charged Higgs $(H^\pm)$ contribution is evaluated by including the next-to-leading order (NLO) QCD corrections [88] and $\tan\beta$ enhanced contributions [88]. The dominant SUSY contribution includes resummed NLO SUSY QCD corrections [88], which hold for large $\tan\beta$.

• *Muon Anomalous Magnetic Moment.* The deviation, δa_μ, of the measured value of a_μ from its predicted value in the SM, a_μ^{SM}, can be attributed to SUSY contributions, which are calculated by using the micrOMEGAs routine [89]. The calculation of a_μ^{SM} is not yet stabilized mainly because of the instability of the hadronic vacuum polarization contribution. According to recent calculations (see e.g. [90, 91]), there is still a considerable discrepancy between the findings based on the e^+e^- annihilation data and the ones based on the τ-decay data. Taking into account the results of [90] and the experimental measurement of a_μ reported in [92], we get the following 95% c.l. ranges:

$$-0.53 \times 10^{-10} \lesssim \delta a_\mu \lesssim 44.7 \times 10^{-10}, \quad e^+e^-\text{-based} ; \tag{1.44}$$

$$-13.6 \times 10^{-10} \lesssim \delta a_\mu \lesssim 28.4 \times 10^{-10}, \quad \tau\text{-based} . \tag{1.45}$$

Following the common practice [83], we adopt the restrictions to parameters induced by (1.44), since (1.45) is considered as quite oracular, due to poor τ-decay data. It is true that there exist more recent experimental data [93] on a_μ than the ones we considered and more updated estimates of δa_μ than the one in [90] (see e.g. [91]). However, only the 95% c.l. upper limit on δa_μ enters into our analysis here and its new values are not very different from the one in (1.44).

• *Collider Bounds.* Here, as it turns out, the only relevant collider bound is the 95% c.l. LEP lower bound [94] on the mass of the lightest CP-even neutral Higgs boson h:

$$m_h \gtrsim 114.4 \text{ GeV} . \qquad (1.46)$$

The SUSY corrections to the lightest CP-even Higgs boson mass m_h are calculated at two loops by using the `FeynHiggsFast` program [95] included in the `micrOMEGAs` code [44].

1.7.5 The Allowed Parameter Space

We will now try to delineate the parameter space of our model with $\mu > 0$ which is consistent with the constraints in Sect. 1.7.4. The restrictions on the $m_{\text{LSP}} - \Delta_{\tilde{\tau}_2}$ plane, for $A_0 = 0$ and the central values of $\alpha_s(M_Z) = 0.1185$ and $m_b(M_Z) = 2.888$ GeV, are indicated in Fig. 1.4 by solid lines, while the upper bound on m_{LSP} from (1.42), for $m_b(M_Z) = 2.684$ [3.092] GeV, is depicted by a dashed [dotted] line. We observe the following:

– The lower bounds on m_{LSP} are not so sensitive to the variations of $m_b(M_Z)$.
– The lower bound on m_{LSP} from (1.46) overshadows all the other lower bounds on this mass.

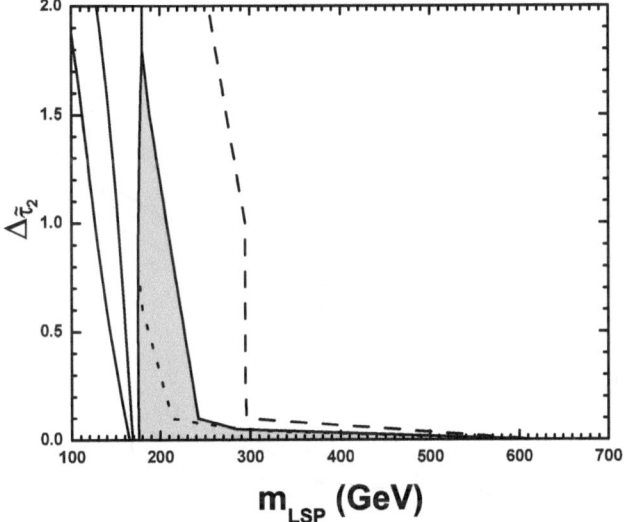

Fig. 1.4. The various restrictions on the $m_{\text{LSP}} - \Delta_{\tilde{\tau}_2}$ plane for $\mu > 0$, $A_0 = 0$, and $\alpha_s(M_Z) = 0.1185$. From left to right, the solid lines depict the lower bounds on m_{LSP} from $\delta a_\mu < 44.7 \times 10^{-10}$, BR($b \rightarrow s\gamma$) $> 1.9 \times 10^{-4}$, and $m_h > 114.4$ GeV and the upper bound on m_{LSP} from $\Omega_{\text{LSP}}h^2 < 0.13$ for $m_b(M_Z) = 2.888$ GeV. The dashed [*dotted*] line depicts the upper bound on m_{LSP} from $\Omega_{\text{LSP}}h^2 < 0.13$ for $m_b(M_Z) = 2.684$ [3.092] GeV. The allowed area for $m_b(M_Z) = 2.888$ GeV is shaded

- The upper bound on m_{LSP} from (1.42) is very sensitive to the variations of $m_b(M_Z)$. In particular, one notices the extreme sensitivity of the almost vertical part of the corresponding line, where the LSP annihilation via an A-boson exchange in the s-channel is [96] by far the dominant process, since m_A, which is smaller than $2m_{\mathrm{LSP}}$, is always very close to it as seen from Fig. 1.5. This sensitivity can be understood from Fig. 1.6, where m_A is depicted versus m_{LSP} for various $m_b(M_Z)$'s. We see that, as $m_b(M_Z)$ decreases, m_A increases and approaches $2m_{\mathrm{LSP}}$. The A-pole annihilation is then enhanced and $\Omega_{\mathrm{LSP}}h^2$ is drastically reduced causing an increase of the upper bound on m_{LSP}.
- For low $\Delta_{\tilde{\tau}_2}$'s, bino-stau coannihilations [33] take over leading to a very pronounced reduction of the LSP relic abundance $\Omega_{\mathrm{LSP}}h^2$, thereby enhancing the upper limit on m_{LSP}. So, we obtain the almost horizontal tail of the allowed region in Fig. 1.4.

For $\mu > 0$, $A_0 = 0$, $\alpha_s(M_Z) = 0.1185$ and $m_b(M_Z) = 2.888$ GeV, we find the following allowed ranges of parameters:

$$176 \text{ GeV} \lesssim m_{\mathrm{LSP}} \lesssim 615 \text{ GeV}, \quad 0 \lesssim \Delta_{\tilde{\tau}_2} \lesssim 1.8 \,,$$
$$58 \lesssim \tan\beta \lesssim 59, \quad 0.14 \lesssim c \lesssim 0.17 \,. \tag{1.47}$$

The splitting between the bottom (or tau) and top Yukawa coupling constants $\delta h \equiv -(h_b - h_t)/h_t = (h_\tau - h_t)/h_t = 2c/(1+c)$ ranges between 0.25 and 0.29.

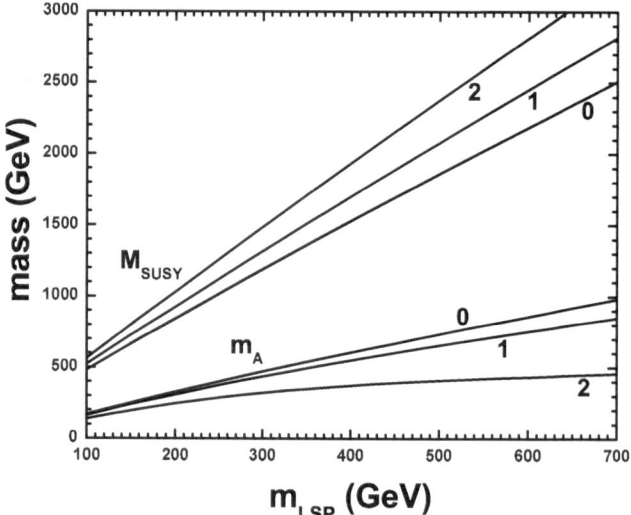

Fig. 1.5. The mass parameters m_A and M_{SUSY} versus m_{LSP} for various values of $\Delta_{\tilde{\tau}_2}$, which are indicated on the curves. We take $\mu > 0$, $A_0 = 0$, $m_b(M_Z) = 2.888$ GeV, and $\alpha_s(M_Z) = 0.1185$

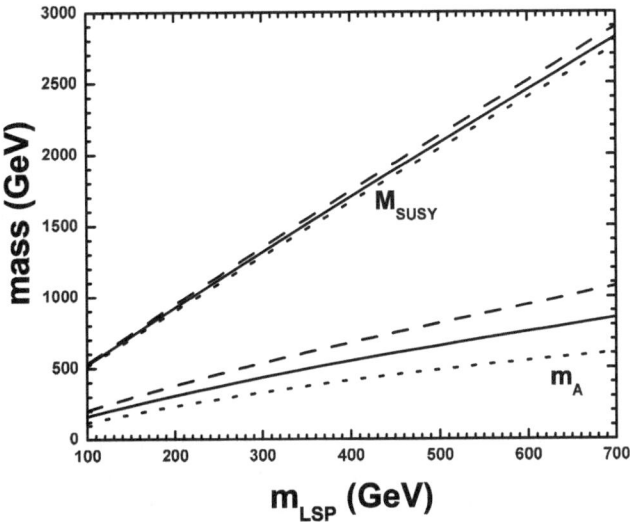

Fig. 1.6. The mass parameters m_A and M_{SUSY} as functions of m_{LSP} for $\mu > 0$, $A_0 = 0$, $\Delta_{\tilde{\tau}_2} = 1$, $\alpha_s(M_Z) = 0.1185$, and with $m_b(M_Z) = 2.684$ GeV (*dashed lines*), 3.092 GeV (*dotted lines*), or 2.888 GeV (*solid lines*)

1.7.6 The New Shifted Hybrid Inflation

It is interesting to note that our SUSY GUT model gives rise [71] naturally to a modified version of shifted hybrid inflation [72]. Hybrid inflation [97], which is certainly one of the most promising inflationary scenarios, uses two real scalars: one which provides the vacuum energy density for driving inflation and a second which is the slowly varying field during inflation. This scheme, which is naturally incorporated [98] in SUSY GUTs (for an updated review, see [99]), in its standard realization has the following property [100]: if the GUT gauge symmetry breaking predicts topological defects such as magnetic monopoles [101], cosmic strings [102], or domain walls [103], these defects are copiously produced at the end of inflation. In the case of monopoles or walls, this leads to a cosmological catastrophe [104]. The breaking of the G_{PS} symmetry predicts the existence of doubly charged monopoles [105]. So, any PS SUSY GUT model incorporating the standard realization of SUSY hybrid inflation would be unacceptable. One way to remedy this is to invoke [106] thermal inflation [107] to dilute the primordial monopoles well after their production. Alternatively, we can construct variants of the standard SUSY hybrid inflationary scenario such as smooth [100] or shifted [72] hybrid inflation which do not suffer from the monopole overproduction problem. In the latter scenario, we generate [72] a shifted inflationary trajectory so that G_{PS} is already broken during inflation. This could be achieved [72] in our SUSY GUT model even before the introduction of the extra Higgs superfields, but only by utilizing non-renormalizable terms. The inclusion of h' and $\bar{h}'$ does not change this situation. The inclusion of ϕ and $\bar{\phi}$, however, very naturally gives rise [71] to a shifted path, but now with renormalizable interactions alone.

1.8 Conclusions

We showed that particle physics provides us with a number of candidate particles out of which the CDM of the universe can be made. These particles are not invented solely for explaining the CDM, but they are naturally there in various particle physics models. We discussed in some detail the major candidates which are the axion, the lightest neutralino, the axino, and the gravitino. The last three particles exist only in SUSY theories and can be stable provided that they are the LSP.

The axion is a pseudo Nambu-Goldstone boson associated with the spontaneous breaking of a PQ symmetry. This is a global anomalous U(1) symmetry invoked to solve the strong CP problem. It is, actually, the most natural solution to this problem which is available at present. The axions are extremely light particles and are generated at the QCD phase transition carrying zero momentum. We argued that these particles can easily provide the CDM in the universe. However, if the PQ field emerges with non-zero value at the end of inflation, they lead to isocurvature perturbations, which, for superheavy inflationary scales, are too strong to be compatible with the recent results of the WMAP satellite on the CMBR anisotropies.

The most popular CDM candidate is, certainly, the lightest neutralino which is present in all SUSY models and can be the LSP for a wide range of parameters. We considered it within the simplest SUSY framework which is the MSSM whose salient properties were summarized. We used exclusively the constrained version of MSSM which is known as CMSSM and is based on universal boundary conditions. In this case, the lightest neutralino is an almost pure bino, whereas the NLSP is the lightest stau. We sketched the calculation of the neutralino relic abundance in the universe paying particular attention not only to the neutralino pair annihilations, but to the neutralino-stau coannihilations too. It is very important for the accuracy of the calculation to treat poles and final-state thresholds properly and include the one-loop QCD corrections to the Higgs boson decay widths and the fermion masses. We find that two effects help us reduce the neutralino relic abundance and satisfy the WMAP constraint on CDM: the resonantly enhanced neutralino pair annihilation via an A-pole exchange in the s-channel, which appears in the large $\tan\beta$ regime, and the strong neutralino-stau coannihilation, which is achieved when these particles are almost degenerate in mass.

The axino, which is the SUSY partner of the axion, can also be the LSP in many cases since its mass is a strongly model-dependent parameter in the CMSSM. It is produced thermally by 2-body scattering or decay processes in the thermal bath, or non-thermally by the decay of sparticles which are already frozen out of thermal equilibrium. For small axino masses, TP is more important yielding a very narrow favored region in the parameter space. For larger axino masses, however, NTP is more efficient and the favored region in the parameter space becomes considerably wider. One finds that, in the case of the CMSSM, almost any point on the $m_0 - M_{1/2}$ plane can be allowed by axino CDM considerations. The required reheat temperatures though are quite small ($\lesssim$ few $\times$ 100 GeV).

The mass of the gravitino is a practically free parameter in the CMSSM. So, the gravitino can easily be the LSP and, in principle, contribute to the CDM of the universe. It is produced thermally by 2-body scattering processes in the thermal bath as well as non-thermally by the decay of the NLSP, which can be either the neutralino or the stau. In contrast to the axino case, however, the NLSP can now have quite a long lifetime. The electromagnetic showers resulting from the NLSP decay can destroy the successful predictions of BBN. So, we obtain strong constraints which allow only very limited regions of the parameter space of the CMSSM. As it turns out, NTP in these regions is not efficient enough to account for CDM. We can, however, make these regions cosmologically favored by raising T_r to enhance TP of gravitinos.

We studied the CMSSM with $\mu > 0$ and $A_0 = 0$ applying a YQUC which originates from a PS SUSY GUT model. This condition yields an adequate deviation from YU which allows an acceptable $m_b(M_Z)$. We, also, imposed the constraints from the CDM in the universe, $b \to s\gamma$, $\delta\alpha_\mu$ and m_h. We found that there exists a wide and natural range of CMSSM parameters which is consistent with all the above constraints. The parameter $\tan \beta$ ranges between about 58 and 59 and the asymptotic splitting between the bottom (or tau) and the top Yukawa coupling constants varies in the range $25 - 29\%$ for central values of $m_b(M_Z)$ and $\alpha_s(M_Z)$. The predicted LSP mass can be as low as about 176 GeV. Moreover, the model resolves the μ problem of MSSM, predicts stable proton, generates the baryon asymmetry of the universe via primordial leptogenesis, and gives rise to a new version of shifted hybrid inflation which is based solely on renormalizable interactions.

Acknowledgements

We thank L. Roszkowski, P. Sikivie, and F.D. Steffen for useful suggestions. This work was supported by European Union under the contract MRTN-CT-2004-503369 as well as the Greek Ministry of Education and Religion and the EPEAK program Pythagoras.

References

1. C.L. Bennett et al., Astrophys. J. Suppl. **148**, 1 (2003); D.N. Spergel et al., Lect. Notes phys. **148**, 175 (2003).
2. W.L. Freedman et al., Astrophys. J. **553**, 47 (2001).

3. D.V. Ahluwalia-Khalilova, D. Grumiller, Phys. Rev. D **72**, 067701 (2005); J. Cosmol. Astropart. Phys. **07**, 012 (2005); M. Cirelli, N. Fornengo, A. Strumia, hep-ph/0512090.

4. R. Peccei, H. Quinn, Phys. Rev. Lett. **38**, 1440 (1977); S. Weinberg,: Lect. Notes phys. **40** 223 (1978); F. Wilczek, Lect. Notes phys. **40** 279 (1978).

5. D.S.P. Dearborn, D.N. Schramm, G. Steigman, Phys. Rev. Lett. **56**, 26 (1986).

6. J. Preskill, M.B. Wise, F. Wilczek, Phys. Lett. B **120**, 127 (1983); L.F. Abbott, P. Sikivie, Lect. Notes phys. **120** 133 (1983).

7. M. Dine, W. Fischler, Phys. Lett. B **120**, 137 (1983).

8. P.J. Steinhardt, M.S. Turner, Phys. Lett. B **129**, 51 (1983); K. Yamamoto, Lect. Notes phys. **161** 289 (1985).

9. G. Lazarides, C. Panagiotakopoulos, Q. Shafi, Phys. Lett. B **192**, 323 (1987); G. Lazarides, R.K. Schaefer, D. Seckel, Q. Shafi, Nucl. Phys. B **346**, 193 (1990).

10. K. Choi, E.J. Chun, J.E. Kim, Phys. Lett. B **403**, 209 (1997); K. Dimopoulos, G. Lazarides, Phys. Rev. D **73**, 023525 (2006).

11. A.H. Guth, Phys. Rev. D **23**, 347 (1981). For a recent review see G. Lazarides, Lect. Notes Phys. **592**, 351 (2002).

12. P. Sikivie, Phys. Rev. Lett. **48**, 1156 (1982); G. Lazarides, Q. Shafi, Phys. Lett. B **115**, 21 (1982); H. Georgi, M.B. Wise, Lect. Notes phys. **116** 122 (1882).

13. M.S. Turner, Phys. Rev. D **33**, 889 (1986).

14. L. Campanelli, M. Giannotti, Phys. Rev. Lett. **96**, 161303 (2006) astro-ph/0512458.

15. H.V. Peiris et al., Astrophys. J. Suppl. **148**, 213 (2003).

16. C. Gordon, A. Lewis, Phys. Rev. D **67**, 123513 (2003); P. Crotty, J. García-Bellido, J. Lesgourgues, A. Riazuelo: Phys. Rev. Lett. **91**, 171301 (2003); C. Gordon, K.A. Malik, Phys. Rev. D **69**, 063508 (2004); M. Bucher, J. Dunkley, P.G. Ferreira, K. Moodley, C. Skordis, Phys. Rev. Lett. **93**, 081301 (2004); K. Moodley, M. Bucher, J. Dunkley, P.G. Ferreira, C. Skordis, Phys. Rev. D **70**, 103520 (2004); M. Beltrán, J. García-Bellido, J. Lesgourgues, A. Riazuelo, Lect. Notes phys. **70** 103530 (2004).

17. K. Dimopoulos, G. Lazarides, D. Lyth, R. Ruiz de Austri, J. High Energy Phys. **05**, 057 (2003); G. Lazarides, R. Ruiz de Austri, R. Trotta, Phys. Rev. D **70**, 123527 (2004); G. Lazarides, Nucl. Phys. B (Proc. Sup.) **148**, 84 (2005).

18. G.L. Kane, C. Kolda, L. Roszkowski, J.D. Wells, Phys. Rev. D **49**, 6173 (1994).

19. J.R. Ellis, K.A. Olive, Y. Santoso and V.C. Spanos, Phys. Rev. D **70**, 055005 (2004).

20. B. Ananthanarayan, G. Lazarides and Q. Shafi, Phys. Rev. D **44**, 1613 (1991); Phys. Lett. B **300**, 245 (1993). For a more recent update see U. Sarid, hep-ph/9610341 (In: *Snowmass 1996, New Directions for High-Energy Physics*).

21. M.E. Gómez, G. Lazarides and C. Pallis, Phys. Rev. D **61**, 123512 (2000); Phys. Lett. B **487**, 313 (2000).

22. M. Drees and M.M. Nojiri, Phys. Rev. D **45**, 2482 (1992).

23. V. Barger, M.S. Berger, P. Ohmann, Phys. Rev. D **49**, 4908 (1994); M. Drees and M.M. Nojiri, Nucl. Phys. B **369**, 54 (1992); M. Olechowski and S. Pokorski, Lect. Notes phys. B **404** 590 (1993).

24. D. Pierce, J. Bagger, K. Matchev and R. Zhang, Nucl. Phys. B **491**, 3 (1997).

25. G. Gamberini, G. Ridolfi and F. Zwirner, Nucl. Phys. B **331**, 331 (1990).

26. H. Baer, C. Chen, M. Drees, F. Paige and X. Tata, Phys. Rev. Lett. **79**, 986 (1997).

27. F.M. Borzumati, M. Olechowski and S. Pokorski, Phys. Lett. B **349**, 311 (1995).

28. H. Baer, M. Brhlik, D. Castaño and X. Tata, Phys. Rev. D **58**, 015007 (1998).

29. K. Griest and D. Seckel, Phys. Rev. D **43**, 3191 (1991).

30. C. Pallis, Astropart. Phys. **21**, 689 (2004); J. Cosmol. Astropart. Phys. **10**, 015 (2005); hep-ph/0510234.

31. M. Drees and M.M. Nojiri, Phys. Rev. D **47**, 376 (1993); M. Drees, hep-ph/9703260.

32. J. Ellis, T. Falk and K.A. Olive, Phys. Lett. B **444**, 367 (1998); J. Ellis, T. Falk, G. Ganis, K.A. Olive, M. Schmitt, Phys. Rev. D **58**, 095002 (1998).

33. J. Ellis, T. Falk, K.A. Olive, M. Srednicki, Astropart. Phys. **13**, 181 (2000); (E) Lect. Notes phys. **15** 413 (2003).

34. E.W. Kolb and M.S. Turner, *The Early Universe* (Addison-Wesley, Redwood City CA 1990).

35. K. Griest, M. Kamionkowski and M.S. Turner, Phys. Rev. D **41**, 3565 (1990).

36. P. Gondolo and G. Gelmini, Nucl. Phys. B **360**, 145 (1991).

37. T. Falk, K.A. Olive and M. Srednicki, Phys. Lett. B **339**, 248 (1994); J. Edsjö and P. Gondolo, Phys. Rev. D **56**, 1879 (1997).

38. T. Nihei, L. Roszkowski and R. Ruiz de Austri, J. High Energy Phys. **03**, 031 (2002).

39. M.E. Gómez, G. Lazarides and C. Pallis, Nucl. Phys. B **638**, 165 (2002).

40. T. Nihei, L. Roszkowski and R. Ruiz de Austri, J. High Energy Phys. **07**, 024 (2002).

41. V.D. Barger and C. Kao, Phys. Rev. D **57**, 3131 (1998).

42. J.R. Ellis, T. Falk, G. Ganis, K.A. Olive and M. Srednicki, Phys. Lett. B **510**, 236 (2001); L. Roszkowski, R. Ruiz de Austri, T. Nihei, J. High Energy Phys. **08**, 024 (2001); A.B. Lahanas, D.V. Nanopoulos, V.C. Spanos, hep-ph/0211286 (In: *Oulu 2002, Beyond the Desert*).

43. A. Djouadi, J. Kalinowski and M. Spira, Comput. Phys. Commun. **108**, 56 (1998).

44. G. Bélanger, F. Boudjema, A. Pukhov and A. Semenov, Comput. Phys. Commun. **149**, 103 (2002).

45. P. Gondolo, J. Edsjö, L. Bergström, P. Ullio, E.A. Baltz, astro-ph/0012234 (In: *York 2000, The Identification of Dark Matter*).

46. L. Covi, J.E. Kim and L. Roszkowski, Phys. Rev. Lett. **82**, 4180 (1999); L. Covi, H.-B. Kim, J.E. Kim and L. Roszkowski, J. High Energy Phys. **05**, 033 (2001); A. Brandenburg and F.D. Steffen, J. Cosmol. Astropart. Phys. **08**, 008 (2004).

47. L. Roszkowski: hep-ph/0102325 (In: *York 2000, The Identification of Dark Matter*); J.E. Kim, astro-ph/0205146 (In: *Cape Town 2002, Dark Matter in Astro- and Particle Physics*).

48. K. Rajagopal, M.S. Turner and F. Wilczek, Nucl. Phys. B **358**, 447 (1991); E.J. Chun, J.E. Kim, H.P. Nilles, Phys. Lett. B **287**, 123 (1992).

49. L. Covi, L. Roszkowski, R. Ruiz de Austri, M. Small, J. High Energy Phys. **06**, 003 (2004); A. Brandenburg, L. Covi, K. Hamaguchi, L. Roszkowski, F.D. Steffen, Phys. Lett. B **617**, 99 (2005).

50. L. Covi, L. Roszkowski and M. Small, J. High Energy Phys. **07**, 023 (2002).

51. J. Ellis, J.E. Kim, D. Nanopoulos, Phys. Lett. B **145**, 181 (1984); T. Moroi, H. Murayama, M. Yamaguchi, Lect. Notes phys. **303** 289 (1993).

52. M. Bolz, W. Buchmüller and M. Plümacher, Phys. Lett. B **443**, 209 (1998).

53. E. Holtmann, M. Kawasaki, K. Kohri and T. Moroi, Phys. Rev. D **60**, 023506 (1999); T. Gherghetta, G.F. Giudice and A. Riotto, Phys. Lett. B **446**, 28 (1999); T. Asaka, K. Hamaguchi and K. Suzuki, Lect. Notes phys. **490** 136 (2000); J.L. Feng, A. Rajaraman, F. Takayama, Phys. Rev. Lett. **91**, 011302 (2003); Phys. Rev. D **68**, 063504 (2003); J.L. Feng, S. Su, F. Takayama, Lect. Notes phys. **70** 063514 (2004); Lect. Notes phys. **70** 075019 (2004).

54. L. Roszkowski, R. Ruiz de Austri and K.-Y. Choi, J. High Energy Phys. **08**, 080 (2005); D.G. Cerdeno, K.-Y. Choi, K. Jedamzik, L. Roszkowski, R. Ruiz de Austri, hep-ph/0509275.

55. M. Bolz, A. Brandenburg and W. Buchmüller, Nucl. Phys. B **606**, 518 (2001).

56. F.D. Steffen, hep-ph/0507003.

57. G. Lazarides and C. Pallis, hep-ph/0404266 (In: *Vrnjacka Banja 2003, Mathematical, Theoretical and Phenomenological Challenges beyond the Standard Model*).

58. G. Lazarides and C. Pallis, hep-ph/0406081.

59. G. Lazarides and C. Panagiotakopoulos, Phys. Lett. B **337**, 90 (1994); S. Khalil, G. Lazarides and C. Pallis, Lect. Notes phys. **508** 327 (2001).

60. L. Hall, R. Rattazzi and U. Sarid, Phys. Rev. D **50**, 7048 (1994); M. Carena, M. Olechowski, S. Pokorski and C.E.M. Wagner, Nucl. Phys. B **426**, 269 (1994).

61. M. Carena, D. Garcia, U. Nierste and C.E.M. Wagner, Nucl. Phys. B **577**, 88 (2000).

62. S.F. King and M. Oliveira, Phys. Rev. D **63**, 015010 (2001).

63. S. Abel et al. (SUGRA Working Group Collaboration), hep-ph/0003154.

64. H. Baer, J. Ferrandis, K. Melnikov and X. Tata, Phys. Rev. D **66**, 074007 (2002).

65. T. Blažek, R. Dermíšek, and S. Raby, Phys. Rev. Lett. **88**, 111804 (2002); Phys. Rev. D **65**, 115004 (2002).

66. D. Auto et al., J. High Energy Phys. **06**, 023 (2003).

67. U. Chattopadhyay, A. Corsetti and P. Nath, Phys. Rev. D **66**, 035003 (2002); C. Pallis, Nucl. Phys. B **678**, 398 (2004).

68. G. Lazarides, Q. Shafi and C. Wetterich, Nucl. Phys. B **181**, 287 (1981); G. Lazarides, Q. Shafi, Lect. Notes phys. B **350** 179 (1991).

69. M.E. Gómez, G. Lazarides and C. Pallis, Phys. Rev. D **67**, 097701 (2003); C. Pallis and M.E. Gómez, hep-ph/0303098.

70. R. Barate et al. (ALEPH Collaboration), Phys. Lett. B **429**, 169 (1998); K. Abe et al., (BELLE Collaboration), Lect. Notes phys. **511** 151 (2001); S. Chen et al., (CLEO Collaboration), Phys. Rev. Lett. **87**, 251807 (2001).

71. R. Jeannerot, S. Khalil and G. Lazarides, J. High Energy Phys. **07**, 069 (2002).

72. R. Jeannerot, S. Khalil, G. Lazarides and Q. Shafi, J. High Energy Phys. **10**, 012 (2000).

73. G. Lazarides, hep-ph/0011130 (In: *Cascais 2000, Recent Developments in Particle Physics and Cosmology*); R. Jeannerot, S. Khalil, G. Lazarides, hep-ph/0106035 (In: *Cairo 2001, High Energy Physics*).

74. I. Antoniadis and G.K. Leontaris, Phys. Lett. B **216**, 333 (1989).

75. G. Lazarides and Q. Shafi, Phys. Rev. D **58**, 071702 (1998).
76. G.R. Dvali, G. Lazarides and Q. Shafi, Phys. Lett. B **424**, 259 (1998).
77. G. Lazarides, C. Panagiotakopoulos and Q. Shafi, Phys. Rev. Lett. **56**, 432 (1986).
78. N. Ganoulis, G. Lazarides and Q. Shafi, Nucl. Phys. B **323**, 374 (1989); G. Lazarides, Q. Shafi, Lect. Notes phys. B **329** 182 (1990).
79. G. Lazarides and Q. Shafi, Phys. Lett. B **258**, 305 (1991).
80. M. Fukugita and T. Yanagida, Phys. Lett. B **174**, 45 (1986).
81. G. Lazarides, Phys. Lett. B **452**, 227 (1999); G. Lazarides and N.D. Vlachos, Lect. Notes phys. **459** 482 (1999); T. Dent, G. Lazarides and R. Ruiz de Austri Phys. Rev. D **69**, 075012 (2004); Lect. Notes phys. **72** 043502 (2005).
82. M.E. Gómez and C. Pallis hep-ph/0303094 (In: *Hamburg 2002, Supersymmetry and Unification of Fundamental Interactions*).
83. J. Ellis, K.A. Olive, Y. Santoso and V.C. Spanos, Phys. Lett. B **565**, 176 (2003); A.B. Lahanas and D.V. Nanopoulos, Lect. Notes phys. **568** 55 (2003); H. Baer and C. Balázs J. Cosmol. Astropart. Phys. **05**, 006 (2003); U. Chattopadhyay, A. Corsetti and P. Nath, Phys. Rev. D **68**, 035005 (2003).
84. H. Goldberg, Phys. Rev. Lett. **50**, 1419 (1983); J.R. Ellis, J.S. Hagelin, D.V. Nanopoulos, K.A. Olive and M. Srednicki, Nucl. Phys. B **238**, 453 (1984).
85. B. Aubert et al., (BABAR Collaboration), hep-ex/0207074; hep-ex/0207076.
86. J.R. Ellis, S. Heinemeyer, K.A. Olive and G. Weiglein, J. High Energy Phys. **02**, 013 (2005).
87. A.L. Kagan and M. Neubert, Eur. Phys. J. C **7**, 5 (1999); P. Gambino and M. Misiak, Nucl. Phys. B **611**, 338 (2001).
88. M. Ciuchini, G. Degrassi, P. Gambino and G. Giudice, Nucl. Phys. B **527**, 21 (1998); G. Degrassi, P. Gambino and G.F. Giudice, J. High Energy Phys. **12**, 009 (2000).
89. S. Martin and J. Wells, Phys. Rev. D **64**, 035003 (2001).
90. M. Davier, hep-ex/0312065 (In: *Pisa 2003, SIGHAD 03*).
91. A. Höcker, hep-ph/0410081 (In: *Beijing 2004, ICHEP 2004*); A. Vainshtein, Prog. Part. Nucl. Phys. **55**, 451 (2005).
92. G.W. Bennett et al., (Muon *g*-2 Collaboration), Phys. Rev. Lett. **89**, 101804 (2002), (E) Lect. Notes phys. **89** 129903 (2002).
93. G.W. Bennett et al., (Muon *g*-2 Collaboration), Phys. Rev. Lett. **92**, 161802 (2004); A. Aloisio et al., (KLOE Collaboration), Phys. Lett. B **606**, 12 (2005).
94. ALEPH, DELPHI, L3 and OPAL Collaborations, The LEP Higgs working group for Higgs boson searches hep-ex/0107029 (In: *Budapest 2001, High Energy Physics*); LHWG-NOTE/2002-01, http//lephiggs.web.cern.ch/LEPHIGGS/papers/July2002_SM/index.html.
95. S. Heinemeyer, W. Hollik and G. Weiglein, hep-ph/0002213.
96. A.B. Lahanas, D.V. Nanopoulos and V.C. Spanos, Phys. Rev. D **62**, 023515 (2000).
97. A.D. Linde, Phys. Rev. D **49**, 748 (1994).
98. E.J. Copeland, A.R. Liddle, D.H. Lyth, E.D. Stewart and D. Wands, Phys. Rev. D **49**, 6410 (1994); G.R. Dvali, Q. Shafi and R.K. Schaefer, Phys. Rev. Lett. **73**, 1886 (1994); G. Lazarides, R.K. Schaefer and Q. Shafi, Phys. Rev. D **56**, 1324 (1997).
99. V.N. Senoguz and Q. Shafi, Phys. Lett. B **567**, 79 (2003); Lect. Notes phys. **582**, 6 (2003).

100. G. Lazarides and C. Panagiotakopoulos, Phys. Rev. D **52**, 559 (1995).
101. G. 't Hooft, Nucl. Phys. B **79**, 276 (1974); A. Polyakov, JETP Lett. **20**, 194 (1974).
102. T.W.B. Kibble, G. Lazarides and Q. Shafi, Phys. Lett. B **113**, 237 (1982).
103. Ya.B. Zeldovich, I.Yu. Kobzarev and L.B. Okun, JETP (Sov. Phys.) **40**, 1 (1975).
104. T.W.B. Kibble, J. Phys. A **9**, 387 (1976).
105. G. Lazarides, M. Magg and Q. Shafi, Phys. Lett. B **97**, 87 (1980).
106. G. Lazarides and Q. Shafi, Phys. Lett. B **489**, 194 (2000).
107. G. Lazarides, C. Panagiotakopoulos and Q. Shafi, Phys. Rev. Lett. **56**, 557 (1986); D.H. Lyth and E.D. Stewart, Lect. Notes phys. **75** 201 (1995).

2

LSP as a Candidate for Dark Matter

Athanasios Lahanas

University of Athens, Physics Department, Nuclear and Particle Physics Section,
GR–15771 Athens, Greece
alahanas@phys.uoa.gr

Abstract. The most recent observations by the WMAP satellite provided us with
data of unprecedented accuracy regarding the parameters describing the Standard
Cosmological Model. The current matter-energy density of the Universe is close to
its critical value of which 73% is attributed to Dark Energy, 23% to Cold Dark
Matter and only 4% is ordinary matter of baryonic nature. The origins of the Dark
Energy (DE) and Dark Matter (DM) constitute the biggest challenge of Modern
Astroparticle Physics. Particle theories, which will be tested in the next round experiments in large accelerators, such as the LHC, provide candidates for DM while
at the same time can be consistent with the DE component. We give a pedagogical
account on the DM problem and the possibility that this has supersymmetric origin.

2.1 Introduction

The first evidence for Dark Matter (DM) stemmed from observations of clusters of Galaxies which are aggregates of a few hundred to a few thousand
galaxies otherwise isolated in space. In 1930 Smith and Zwicky examined
two nearby clusters, Virgo and Coma, and found that the velocities of the
galaxies making up the clusters were about ten times larger than they expected. This may be explained by assuming that there is more mass in the
clusters which accelerates the galaxies to higher velocities. In 1970 more reliable data, by observation of a larger number of clusters by Rubin, Freeman
and Peebles, confirmed that the velocities of the galaxies are indeed different
than one expects assuming that all matter comprising the galaxies is luminous. As you go to the edge of a spiral galaxy the amount of the light stars
emit falls off and if all matter were luminous the rotational speed would fall
off too. In fact from the distribution of luminous stars, the rotational velocity
at a distance r from the center of the galaxy turns out to be $v(r) \approx r^{-1/2}$
while observations showed instead that $v_{obs}(r) \approx const.$ This cannot be explained unless there is some sort of invisible matter or "Dark", not interacting

A. Lahanas: *LSP as a Candidate for Dark Matter*, Lect. Notes Phys. **720**, 35–68 (2007)
DOI 10.1007/978-3-540-71013-4_2   © Springer-Verlag Berlin Heidelberg 2007

electromagnetically therefore, which participates however in the gravitational dynamics[1].

Cosmologists usually measure the amount of mass and energy of the Universe in units of the critical density ρ_c by defining the fraction $\Omega \equiv \rho/\rho_c$, with ρ is the mass-energy density of the Universe. When $\Omega > 1$ Universe closes. Its value is accurately determined by WMAP [1] due to the high precision measurements of the Cosmic Microwave Background (CMB), $\Omega = 1.02 \pm 0.02$, confirming previous claims that our Universe is almost flat in concordance with inflation. After combining WMAP with other existing data and using the rescaled Hubble constant h we have for the directly measured quantities Ωh^2, which are Hubble parameter independent, $\Omega_{matter} h^2 = 0.134 \pm 0.006$, of which a small amount $\Omega_b h^2 = 0.023 \pm 0.001$ is of baryonic nature while the corresponding luminous mass density is smaller by an order of magnitude. The deficit $\Omega_{matter} h^2 - \Omega_b h^2$ is attributed to Dark Matter whose value at the 2σ level lies within the range

$$0.094 < \Omega_{DM} h^2 < 0.129 .$$

The value of the rescaled Hubble parameter is $h = 0.73 \pm 0.05$ from which one can infer the values of Ω's. The conclusion is that our Universe is dominated by a large amount of energy $\approx 73\%$, of unknown origin, the so called Dark Energy (DE) and a large amount of mass $\approx 23\%$, whose composition is also unknown the so called Dark Matter (DM). Only a small fraction, $\approx 4\%$, consists of ordinary matter. Therefore 96% of our Universe is a completely mystery!

Candidates for DM can be the neutrinos, axions, gravitinos or WIMPs or other more exotic particles such as cryptons, Kaluza-Klein excitations or branons existing in higher dimensional theories and so on. The energy loss limit from SN 1987A put upper limits on axion masses, $m_a \leq 10^{-2} eV$, and on these grounds they are considered as non-thermal relics with very small mass. From the requirement that axions do not overclose the Universe lower limits on axion mass are imposed if one follows standard scenarios. The allowed mass window in standard considerations is very narrow leaving little room to believe that the axion can explain the DM of the Universe. For other more exotic as yet undiscovered candidates, proposed in higher dimensional gravity theories, already mentioned above, the situation is more involved and we do not put them under consideration in these lectures.

Standard Model neutrinos and their antiparticles are existing particles and once believed that they could explain the Universe missing mass problem. However this possibility is rather ruled out in view of the latest data. If neutrinos are massless their density is $\Omega_\nu \approx 3.5 \times 10^{-5}$ and their contribution to the energy – matter density of the Universe is quite small. However

---

[1] Another explanation would be to assume that the gravitational force does not follow the simple inverse square law at galactic distances which although cannot be excluded it is rather ad-hoc lacking a firm theoretical foundation.

we now know that neutrinos are not completely massless and their relics are given by $\Omega_\nu h^2 = \sum m_i/92.5$. This assumes that their temperature today is $T_\nu = 1.95\ ^0K = 1.7 \times 10^{-4}eV$. Therefore they could offer as Hot Dark Matter (HDM) candidates. In general the possibility that the Universe is dominated by HDM seems to conflict with numerical simulations of structure formation. In fact relativistic matter streaming from overdense to underdense regions prevents structures from growing below the so called free-streaming scale [2]. In fact the combined results from WMAP and other data imply that $\Omega_\nu h^2 < 0.0076$, providing also limits on neutrino masses, and this is too low to account for the Universe's missing mass. Warm Dark Matter (WDM) neutrinos, or other low mass species, also seem unlikely. The reason is that star formation occurs relatively late in WDM models because small scale structure is suppressed. This is in conflict with the low-l CMB measurements by WMAP which indicate early re-ionization (at $z \approx 20$) and therefore early star formation.[2] Actually it is found that WDM candidates are inconsistent with WMAP data for masses $m_X \leq 10\ KeV$ [3] while masses larger than $100\ KeV$ are almost indistinguishable from Cold Dark Matter. Cold Dark Matter is however allowed and it is perhaps the most plausible possibility and candidates that may play this role naturally exist in some particle theories notably in Supersymmetric and Supergravity theories which are believed to be the low energy manifestation of String theories.

2.2 The Energy – Matter Content of the Universe

The Universe is homogeneous and isotropic at supergalactic scales being therefore described by the Friedmann-Robertson-Walker (FRW) geometry whose metric is read from the line element [4, 5]

$$ds^2 = -c^2\, dt^2 + a^2(t)\,(\frac{dr^2}{1 - kr^2} + r^2(d\theta^2 + \sin^2\theta d\phi^2))\,. \qquad (2.1)$$

In it $a(t)$ is the cosmic scale factor and depending on k we distinguish three types of Universe with the following characteristics

k	Type of Universe	3-d curvature	Spatial volume
1	closed	$k\,a^{-2} > 0$	$2\pi^2 a^3$
0	flat	0	∞
-1	open	$k\,a^{-2} < 0$	∞

The expansion rate of the Universe is defined by

$$H = \frac{\dot{a}}{a}\,. \qquad (2.2)$$

---

[2] We have assumed that structure formation is responsible for re-ionization.

Its value today, the well known "Hubble constant", is denoted by H_0 and can be written as $H_0 = 100\, h_0\ Km/sec/Mpc^3$ by defining the dimensionless quantity h_0 which is usually called "rescaled Hubble's constant". Its value is experimentally known with a fairly good accuracy by the WMAP data, $h_0 = 0.72 \pm 0.05$. The matter – energy density of the Universe is related to the expansion rate by

$$\varrho = \frac{3}{8\pi G_N}\left(H^2 + \frac{k}{a^2} \right), \tag{2.3}$$

while the "critical density" ϱ_c, is defined by,

$$\varrho_c = \frac{3}{8\pi G_N} H_0^2 . \tag{2.4}$$

Its value is $\rho_c = 1.88 \times 10^{-29}\ h_0^2\ gr/cm^3 = 8.1 \times 10^{-47}\ h_0^2\ GeV^4$. By (2.3) it is seen that if the value of the matter-energy density today is ϱ_0 then depending on whether this is larger, smaller or equal to the critical density our Universe is close, open or flat respectively,

$$
\begin{array}{lll}
\rho_0 > \rho_c & k > 0 & \text{closed Universe} \\
\rho_0 = \rho_c \implies & k = 0 & \text{flat Universe} \\
\rho_0 < \rho_c & k < 0 & \text{open Universe}
\end{array}
$$

Recent observations by WMAP and other sources point towards a flat Universe.

The total matter-energy density today can be expressed as fraction of the critical density by defining the ratio

$$\Omega = \frac{\varrho_0}{\rho_c} . \tag{2.5}$$

This can be written as a sum

$$\Omega = \sum_i \Omega_i \tag{2.6}$$

with Ω_i denoting the contribution of each particle species, including the contribution of the radiation $\Omega_{radiation}$ and that of the cosmological constant Ω_Λ. Note that Ω's as defined above refer to today's values.

From observations of distant Galaxies Hubble (1929) drew the conclusion that

$$d_L\, H_0 = c\,z \tag{2.7}$$

In (2.7) d_L is the luminosity distance and z the Doppler shift in the wavelength of the emitted radiation, $1 + z = \lambda_{observer}/\lambda_{source}$. For a moving

---

[3] 1 pc $\approx$ 3.26 light years.

source with velocity v, which radiates electromagnetic radiation, $v = cz$ and therefore this law states that Galaxies recede with velocities given by

$$v_r = d_L H_0 .$$

This is perhaps one of the most spectacular discoveries in the history of Astrophysics. *Our Universe is not static but it expands.* FRW geometry is actually encompassing Hubble's law since the distance d is proportional to the cosmic scale factor a and hence the receding velocities are $v = Hd$.

The luminosity distance, which is not actually the true geometric distance, between emitter and receiver, is defined by

$$\mathcal{F} = \frac{\mathcal{L}}{4\pi d_L^2} , \tag{2.8}$$

where the flux $\mathcal{F}$ is the energy received in the apparatus per unit area, per unit time and $\mathcal{L}$ is the "absolute luminosity". d_L would be the actual geometric distance to the Galaxy if our Universe were static. In an expanding Universe d_L is given by [4, 5]

$$d_L = a(t_0)\, r_1(t_1)\, (1+z) . \tag{2.9}$$

In this t_0 is observer's time and t_1 is the time the light signal was emitted from the Galaxy. r_1 is given by $\int_0^{r_1} \frac{dr}{\sqrt{1-kr^2}} = \int_{t_1}^{t_0} \frac{dt'}{a(t')}$. The $1+z$ in (2.9) effectively reduces the absolute luminosity $\mathcal{L}$ in (2.8) by a factor $(1+z)^{-2}$. One power of $(1+z)^{-1}$ accounts for the fact that photons are redshifted and their energy E' when they reach the observer is $1+z$ times smaller than their energy when the were emitted. The other $(1+z)^{-1}$ power duly takes care of the fact that the number of photons per unit length of the beam drops as the light beam increases with expansion and thus the observer's apparatus receives less photons per unit time. Due to these two effects the observed luminosity is $\mathcal{L}_o = \mathcal{L}/(1+z)^{-2}$ and (2.8) could have been equally well expressed as $\mathcal{F} = \mathcal{L}_0/4\pi d_G^2$ with $d_G = a(t_0)r_1(t_1)$ the geometric distance to the Galaxy.

The cosmological redshift is defined by

$$1 + z = \frac{\lambda(t_0)}{\lambda(t)} , \tag{2.10}$$

where $\lambda(t_0), \lambda(t)$ are respectively the wavelengths of the radiation today t_0 and at the time of emission t. If a signal is emitted from a source at t and it is received at t_0 it is easy to show that

$$\frac{\lambda(t_0)}{\lambda(t)} = \frac{a(t_0)}{a(t)} \tag{2.11}$$

which has the meaning that wavelengths also scale with the cosmic scale factor like distances. Using this we get from (2.10)

$$1 + z = \frac{a(t_0)}{a(t)} \, . \tag{2.12}$$

This relation between redshists and cosmic scale factor can be used to express the time t as function of the redshift z. Today $z = 0$ while $z = \infty$ at the beginning, $t = 0$. By using $H = \dot{a}/a$ it follows that

$$t(z) = \int_z^\infty \frac{dz}{H\,(1+z)} \, , \tag{2.13}$$

where H as function of z is given by

$$H(z) = H_0 \left[\sum_i \Omega_i\,(1+z)^{3(1+w_i)} + (\,1 - \sum_i \Omega_i\,)(1+z)^2\,\right]^{1/2} . \tag{2.14}$$

If the redshift of some astronomical source is known then from (2.12) we can estimate the relative size of the Universe when light was emitted from it. For instance quasars are characterized by $z = 5.82$ and from (2.12) we conclude that our Universe was 6.82 times smaller when light was emitted from these objects.

The luminosity distance as function of the redshift can be written as

$$d_L = c\,H_0^{-1}\,|\Omega_k|^{-1/2}(1+z)\,sinn[\,|\Omega_k|^{-1/2}H_0\int_0^z \frac{dz}{H}\,] . \tag{2.15}$$

where the function $sinn(\chi)$ is $sin(\chi), \chi$ or $sinh(\chi)$ for $k = 1, 0, -1$ respectively. Assuming only contribution of matter and cosmological constant Ω_M, Ω_Λ this can be expanded in powers of z with the result

$$d_L = cH_0^{-1}\,[\,z + \frac{z^2}{2}\,(\,1 + \Omega_\Lambda - \frac{\Omega_M}{2}\,) + ... \,] . \tag{2.16}$$

The ellipsis in (2.16) stand for terms of order $\sim z^3$ and higher. The deceleration parameter q_0 is defined by $q_0 \equiv -\ddot{a}_0 a_0/\dot{a}_0^2$ and assuming existence of only matter and cosmological constant it is given by

$$q_0 = -\Omega_\Lambda + \frac{\Omega_M}{2} \, . \tag{2.17}$$

On account of it one observes that the coefficient of the $z^2/2$ term in (2.16) is exactly equal to $1 - q_0$. In (2.16) we have kept terms up to $\sim z^2$ which is necessary for the study of distant supernovae, as is the case with the SNIa, which are characterized by redshifts in the range $z = 0.16 - 0.62$. Observations showed that these objects appear fainter and hence at larger distances than expected. The analyses by SCP and HZSS collaborations [6] showed that the data are consistent with theory if one assumes the existence of a non-vanishing contribution Ω_Λ from the cosmological constant which is correlated to Ω_M. The value of Ω_Λ turns out to be about 70% showing in a spectacular manner that the bulk of the matter-energy density of the Universe is void! From the values of $\Omega_{\Lambda,M}$ one can see that the deceleration parameter q_0 is negative or same the Universe is accelerating. The relevant Hubble diagram is shown in Fig. 2.1. Five years after the direct evidence for the existence of a nonvanishing cosmological constant which accelerates the Universe WMAP precision data confirmed previous analyses leaving little room for doubt that the bulk of the matter-energy of the Universe is vacuum energy while the majority of the matter is invisible or Dark Matter [1].

All discussion so far assumes that our space time is four-dimensional. The picture changes drastically if one assumes that there exist extra dimensions (ED) that are invisible at present energies. These ideas are very popular nowadays in particle physics community and they are inevitable in string theories which require the existence of extra dimensions. Especially the last five years there is a lot of activity in the direction of the Brane World physics stirring new interest to the field with a variety of theoretical predictions that are waiting for their experimental confirmation (or rejection!) in the next generation high energy accelerators. In such ED scenaria not only particle physics but also Cosmology is affected and the Standard Cosmological Model has to

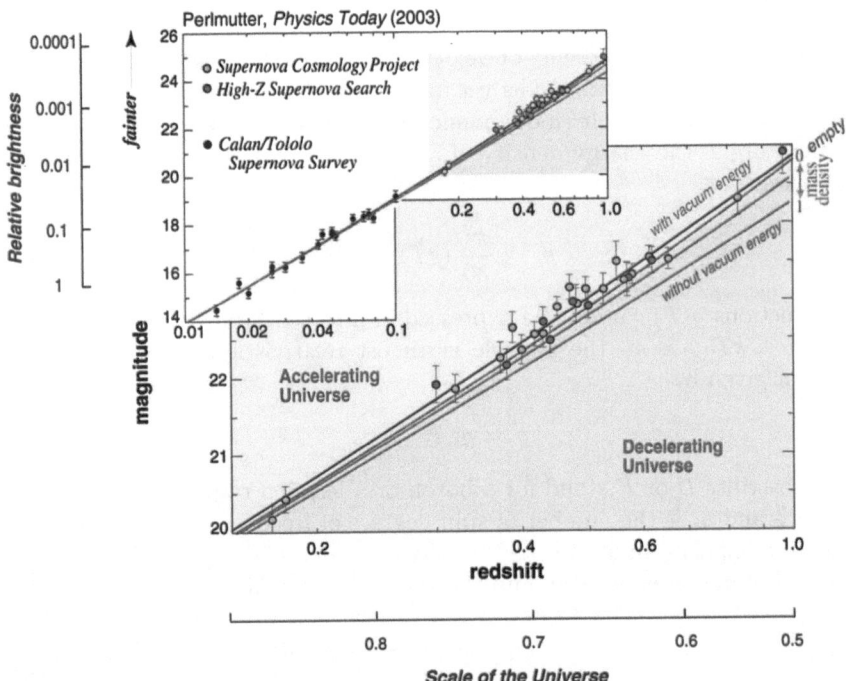

Fig. 2.1. The Hubble diagram for the high-z SNIa analyzed by the SCP and HZSS collaborations [6]

be reconsidered. For instance in a Universe confined to a 3-brane within a larger spacetime does not lead to Friedmann equations of the Standard Cosmology. In fact the Hubble expansion rate $H \equiv \dot{a}/a$ is not proportional to the density ϱ but rather to ϱ^2 [7]. Assuming the bulk contribution is negligible at Nucleosynthesis one has $a(t) \sim t^{1/4}$. Moreover $H \sim T^4$ unlike $H \sim T^2$ of the standard Cosmology. Therefore cooling is much slower and the freezing temperature T_D of the neutron to proton ratio changes from about $0.8\ MeV$ to about $3\ MeV$. Values of $T_D \sim 2\ MeV$ are then required by Big Bang Nucleosynthesis (BBN) and since $g_*^{-1/8}\ M_{(5)} \simeq 3\ MeV$, with $M_{(5)}$ the scale characterizing the 5-dimensional Gravity, this relation cannot be satisfied. Actually more dimensions are required. Thus it appears that although ED seems a quite intriguing idea new parameters enter the game, at least as far as Cosmology is concerned, which may upset the successes of the Standard Cosmological Model. Within the context of these ED scenaria various attempts to find a reasonable explanation for DE and DM have been put forward. In these lectures we shall be more conservative and pursue the idea that DM are non–Standard Model (SM) particles living in the standard four space-time dimensions. The issue of extra dimension will be covered by other speakers in this school.

2.3 The Thermal Universe

At early times the ingredients of our Universe were the Standard Model particles, and perhaps additional as yet undiscovered particle species, being in a state of hot plasma in thermodynamic and chemical equilibrium. At a given temperature T the energy density of a particular particle species of mass m is given by

$$\rho = \frac{\pi^2}{30}\, (kT)^4\, g(T) \, . \tag{2.18}$$

The functions $g(T)$ can not be expressed in a closed form. However for temperatures, $kT >> m$, the particle is almost relativistic and $g(T)$ is just a constant given by

$$g = g_s\, N'_{B,F} \, . \tag{2.19}$$

The subscripts B or F stand for a boson or a fermion respectively, $N'_B = 1$, $N'_F = \frac{7}{8}$ and g_s is the number of spin degrees of freedom.

In the opposite limit, $kT << m$, the functions $g(T)$ approaches its well-known Boltzmann expression and the energy density is

$$\rho \simeq g_s\, m \left(\frac{mkT}{2\pi}\right)^{\frac{3}{2}} \exp\left(-\frac{m}{kT}\right) \, . \tag{2.20}$$

In this limit $p << \rho$, that is, we deal with a pressureless gas or "dust".

Especially for the photons and neutrinos the situation is as follows:

Photons: The contribution of the photons to the total matter-energy density can be read from (2.18) with $g(T) = 2$ since the photon is a massless spin-1 particle, having therefore two helicity degrees of freedom. Thus employing the fact that today $T_\gamma \simeq 2.7 \; {}^0K$, one has,

$$\Omega_\gamma \equiv \frac{\rho_\gamma}{\rho_c} \simeq 2.5 \times 10^{-5}/h_0^2 \; . \tag{2.21}$$

With $h_0 \approx 0.7$ this yields $\Omega_\gamma \simeq 5.2 \times 10^{-5}$. That is, the energy density carried by photons is five orders of magnitude smaller than the critical density.

Neutrinos: If neutrinos or antineutrinos are massless then their energy density, for each species, is given by

$$\rho_\nu = \frac{7\pi^2}{240} (kT_\nu)^4 \; .$$

In deriving this we used the fact that neutrinos and antinieutrinos are massless and they are each characterized by one helicity state (see (2.18)). Their total contribution to Ω can be weighted relative to Ω_γ and is given by

$$\Omega_\nu = N_\nu \frac{7}{16} \left(\frac{T_\nu}{T_\gamma}\right)^4 \Omega_\gamma \; , \tag{2.22}$$

where N_ν counts the total number of neutrino and antineutrino species, i.e. $N_\nu = 6$ in the Standard Model. One should note that the neutrinos' temperature T_ν is different from that of photons T_γ and hence the subscripts in the equation above. The reason is that neutrinos decouple from the cosmic soup when the Universe's temperature is $kT_D \simeq$ MeV and then they expand freely. At a lower temperature $kT \simeq 2m_e$ photons reheat, due to the annihilation process $e^+ e^- \rightarrow 2\gamma$, and from entropy conservation it follows that $T_\nu/T_\gamma = \left(\frac{4}{11}\right)^{1/3}$ after reheating. Using this we have from (2.22),

$$\Omega_\nu \simeq 0.115 \, N_\nu \, \Omega_\gamma \; , \tag{2.23}$$

from which we deduce that the energy density of massless neutrinos is also much smaller than the critical density of the Universe.

The situation changes for massive neutrinos however. If we assume a massive neutrino (or antineutrino), of mass m_ν, then its energy density, at temperatures larger than its mass, is (see (2.20)), $\rho_\nu \simeq n_\nu m_\nu$ from which we have

$$\Omega_\nu \simeq 0.0053 \frac{\sum_\nu (m_\nu/eV)}{h_0^2} \; , \tag{2.24}$$

with m_ν in eV. In this expression the sum runs over all neutrino and antineutrino species. Employing a value $h_0 \simeq 0.7$ this yields $\Omega_\nu \simeq \sum_\nu m_\nu/(92.5 \text{ eV})$. The best evidence for neutrino masses comes from the

SuperKamionkande experiment [8] which detected oscillations $\nu_\mu \rightarrow \nu_\tau$. This result indicates a mass difference $m_{\nu_\tau} - m_{\nu_\mu} \simeq 0.05$ eV which points towards small neutrino masses of the order of a few eV. However experimental data from neutrino physics do not put an upper limit to neutrino masses able to exclude neutrinos as Dark Matter candidates. It is the WMAP data in combination with data from other sources that almost exclude the possibility that neutrinos can be either Hot or Warm Dark Matter as already mentioned in the introduction. In fact the limit $\Omega_\nu h^2 < 0.0076$ is extracted, from which an upper limit on $\sum_\nu m_\nu$ can be derived as is seen in Fig. 2.2. Therefore neutrinos contribute very little $\sim 10^{-2}$, or less, to Ω, as photons do, and hence unable to account for the observed amount of Dark Matter.

In the evolution of the Universe one distinguishes two main epochs during which radiation and matter dominated respectively.

Radiation Dominance: During radiation dominance, the energy density was dominated by relativistic particles whose thermal properties, apart from their spin content, is like that of photons. In this era $p = \rho/3$ and from energy conservation it follows that $d(\rho\, a^4)/dt = 0$ which is immediately solved to yield $\rho \simeq 1/a^4$. Then from (2.3) the density term dominates over k/a^2 and we have an equation for the cosmic scale factor which can be solved to yield $a \simeq t^{1/2}$. From these two we get

$$\rho = \frac{3}{32\,\pi\,G_N}\,t^{-2}\,, \tag{2.25}$$

which on account of (2.18) yields,

$$T = \left(\frac{16\,\pi^3\,G_N}{45}g(T)\right)^{-1/4}t^{-1/2}\,. \tag{2.26}$$

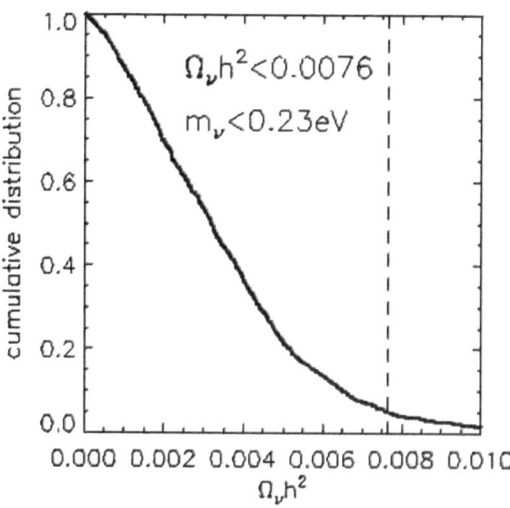

Fig. 2.2. WMAP bounds on Neutrino masses and their relic density, [1]

Putting it all together, during the radiation era, the cosmic scale factor, the energy density and the temperature behave as

$$a \sim t^{1/2} , \; \rho \sim t^{-2} , \; T \sim t^{-1/2} ,$$ (2.27)

and the expansion rate is,

$$H = 1/2t = 1.66 \sqrt{g(T)} \; \frac{(kT)^2}{M_{Planck}} .$$ (2.28)

The content of $g(T)$ entering (2.26) and (2.28) is different at different times. If a massive particle with mass m_i is in equilibrium with the cosmic soup, then for temperatures $kT >> m_i$ it is relativistic and contributes either $7/8 \, g_s$ or g_s, depending on whether it is a boson or a fermion, see (2.19). As the Universe expands, its temperature drops and eventually reaches a temperature for which $kT << m_i$. Therefore, its contribution is negligible because of Boltzmann suppression exp $(-m_i/kT)$, see (2.20), and can be ignored. In Table 2.1 the values of $g(T)$ at different temperatures are shown along with the radiation content at the temperature shown on the right. For temperatures above the top mass, $m_t \simeq 175.0 \, GeV$, the contributions of additional non Standard Model (SM) particles, if they exist, should be added. The contribution of the SM Higgs boson adds $+1$ to $g(T)$ for $kT > m_{Higgs}$ and is not shown. The LEP experimental limit put on the Higgs mass is $m_{Higgs} > 114 \, GeV$, [9]. We should remark that in Table 2.1, the effects of the neutrino decoupling and the photon reheat have not been counted for.

Matter Dominance: After radiation dominated era the Universe started entering the period in which matter dominated. During this period, the mass density was much larger than the pressure $\rho >> p$. Then in the equation

Table 2.1. Active degrees of freedom and their contribution to g(T) for the standard model particles

k T	Content of radiation	$g(T)$
$< m_e$	$\gamma + 3 \times (\nu + \bar{\nu})$	29 /4
$m_e - m_\mu$	$\cdots + e^+, e^-$	43/4
$m_\mu - m_\pi$	$\cdots + \mu^+, \mu^-$	57/4
$m_\pi - \Lambda_c$	$\cdots + \pi^+, \pi^-, \pi^0$	69/4
$\Lambda_c - m_s$	$\cdots + u, \bar{u}, d, \bar{d}, gluons$	205/4
$m_s - m_c$	$\cdots + s, \bar{s}$	247/4
$m_c - m_\tau$	$\cdots + c, \bar{c}$	289/4
$m_\tau - m_b$	$\cdots + \tau, \bar{\tau}$	303/4
$m_b - M_Z$	$\cdots + W^+, W^-, b, \bar{b}$	369/4
$M_Z - m_t$	$\cdots + Z$	381/4
$> m_t$	$\cdots + t, \bar{t}$	423/4
$\cdots$	$\cdots$	$\cdots$

for the energy conservation the pressure term can be neglected leading to $d(\rho\, a^3)/dt = 0$. This is solved to yield $\rho \simeq 1/a^3$. Then from (2.3) we have $a^{1/2}\, \dot{a} = const$ which is solved to yield $a \sim t^{2/3}$. During the matter dominated era the cosmic scale factor, the density and the temperature behave as

$$a \sim t^{2/3}\,,\ \rho \sim t^{-2}\,,\ T \sim t^{-2/3} \tag{2.29}$$

In this period the expansion rate is $H = 2/3t$. The temperature T_{EQ} at which the Universe entered the matter dominated era is estimated to be around $T_{EQ} \simeq 10^4\ {}^0K$ equivalent to 1 eV. This estimate follows by equating radiation ρ_r and matter density ρ_m.

Vacuum Dominated Universe: Since the Universe expands and its temperature drops, eventually the cosmological term within the density in (2.3), if it exists, will take over. Hence it is worth exploring this case too. If the Dark Energy of the Universe is attributed to a cosmological term with $\Lambda > 0$ this will dominate at some epoch. During this era from (2.3), by putting $\rho = \Lambda$, we get

$$\left(\frac{\dot{a}}{a}\right)^2 \simeq \frac{8\pi G_N}{3}\Lambda \tag{2.30}$$

while for the acceleration it is found

$$\frac{\ddot{a}}{a} = \frac{8\pi G_N}{3}\Lambda\,. \tag{2.31}$$

Note that in (2.30) the r.h.s is negative for $\Lambda < 0$ while the l.h.s. is positive. We therefore conclude that in the presence of a negative cosmological constant, the Universe never enters the regime in which the cosmological term is dominant. In fact the cosmic scale factor, in this case, attains a maximum value before reaching this regime. However this is not the case when the cosmological constant is positive as we have assumed. From (8.122) we get, in this case,

$$a \simeq \exp\sqrt{\frac{8\pi G_N \Lambda}{3}}\, t. \tag{2.32}$$

Since $\Lambda > 0$, (2.31) implies that $\ddot{a} > 0$, that is the Universe is accelerated in the vacuum dominated era. In other words gravitational forces are repulsive. In this period the pressure is negative, since $p_{vac} = -\rho_{vac} = -\Lambda$. The present cosmological data show that our Universe has already entered into this phase. In fact from the values of matter and energy densities we have already seen that the deceleration parameter, see (2.17), is negative.

2.4 Dark Matter

In a hot Universe which is filled with particles interacting with each other, these all are in thermal equilibrium, at some epoch. However as Universe cools and expands some of these may go out of thermal equilibrium and eventually decouple. The equilibrium criterion is that the mean free path $l_{m.f.p.}$ is smaller than the distance these particles travel since the beginning, i.e.

$$l_{m.f.p.} < v\,t\ . \tag{2.33}$$

If (2.33) holds, then particles will interact with the cosmic soup and there is no way of escaping. The mean free path is defined by $l_{m.f.p.} \equiv 1/(n\sigma)$ where σ is the interaction cross section of the particles under consideration and n their density. Since the expansion rate is inversely proportional to time $H \sim 1/t$ the equilibrium criterion (2.33) can be expressed as

$$\Gamma > H,\ \ (\,\Gamma \equiv \frac{1}{v\,n\,\sigma}\,)\ . \tag{2.34}$$

If $\Gamma > H$ at some epoch and $\Gamma < H$ at later times, then there is a temperature T_D for which $\Gamma = H$. T_D is called the decoupling or freeze-out temperature. For $T \leq T_D$ these particles do not interact any longer with the cosmic soup and they expand freely. Their total number, after decoupling, remains constant and thus their density decreases with the cube of the cosmic scale factor $n \sim 1/a^3$.

A notable example of this situation are neutrinos. They decoupled when the Universe was as hot as ten million Kelvin degrees and their relics today account for a small fraction of the total energy of the Universe. Neutrinos interact only weakly and they are nearly massless. For temperatures below the muon mass, that is $kT < m_\mu$, the active degrees of freedom in the hot Universe are the photons, the neutrinos and their antiparticles, the electrons and the positrons. The interaction cross section of neutrinos and antineutrinos with the electrons and positrons is $\sigma_\nu \sim (G_F/\sqrt{\pi})^2 (kT)^2$ where G_F is the Fermi coupling constant and kT is the energy neutrinos carry. The total density of neutrinos, antineutrinos, electrons and the positrons which interact weakly with each other is $n \simeq (kT)^3$, while all these are relativistic at temperatures $T >> m_e$. Therefore, their velocities are $v \simeq 1$ and the quantity Γ in (2.34) is

$$\Gamma \simeq \left(G_F/\sqrt{\pi}\right)^2 (k\,T)^5\ . \tag{2.35}$$

At this temperature the expansion rate is

$$H = 1.66\,\sqrt{g^*}\,\frac{(k\,T)^2}{M_{\text{Planck}}}\ . \tag{2.36}$$

where g^* is the value of the function $g(T)$ at this temperature which, as read from Table 2.1, is $43/4$. The neutrino freeze-out temperature is found, by equating (2.35) and (2.36), to be $k\,T_D \simeq 2$ MeV. For $T \leq T_D$ the neutrinos and their antiparticles decouple and they do not interact any longer with electrons and positrons. They can be conceived, as being in an isolated bath at temperature T_ν which equals to the photons temperature at the moment of decoupling. Since they do not interact any longer with the rest of the particles, namely γ, e^-, e^+, their total number is locked. When the temperature reached $T \sim 2\, m_e$ the electrons and positrons started being annihilated to two photons through the process $e^-\, e^+ \rightarrow 2\,\gamma$ but the photons did not have enough energy to produce back the electrons and the positrons. Because of that, the temperature of the photons is increased (photon reheat) but this is not felt by the neutrinos since the latter do not interact with the soup. The increase of the photon temperature can be calculated from the conservation of entropy. Since the neutrinos have the same temperature with the photons before photon reheat we finally get

$$ T_\nu \;=\; \left(\frac{4}{11} \right)^{1/3} T_\gamma \; . \tag{2.37} $$

Because of that the value of $g(T)$ in the first row of the Table 2.1 should be corrected. With the effect of neutrino decoupling and photon reheat taken into account the contributions of neutrinos and photons to the energy density after photon's reheat is proportional to $2\, T_\gamma^4 + (21/4)\, T_\nu^4$ which gives a value equal to $2 + (21/4)\,(T_\nu/T_\gamma)^4 \simeq 3.36$ for the function $g(T)$, which is almost half of $29/4$ appearing in Table 2.1. As already remarked neutrinos are rather ruled out as Hot or Warm Dark Matter candidates and hence we will drop them from the discussion in the following.

Various models of Particle Physics predict the existence of Weakly Interacting Massive Particles, called for short WIMPs, that have decoupled long ago and their densities at the present epoch may account for the "missing mass" or **Dark Matter** of the Universe [10, 11, 12, 13, 14]. In Supersymmetric theories in which R-parity is conserved the Lightest Supersymmetric particle (LSP) is, in most of the cases, the lightest of the neutralinos $\tilde{\chi}$, a massive stable and weakly interacting particle. However other options are available like for instance the gravitino, the axino or the sneutrino[4] This qualifies as Dark Matter candidate provided its relics is within the experimentally determined DM relic density that is $\Omega_{\tilde{\chi}}\, h_0^2 \sim 0.1$ (for a review see [15]). Its relic abundance can be calculated using the transport Boltzmann equation which will be the subject of the following section.

2.5 Calculating DM Relic Abundances

2.5.1 The Boltzmann Transport Equation

The number density of a decoupled particle can be calculated by use of the Boltzmann transport equation. Let us assume for definiteness that the LSP particle under consideration is the neutralino, $\tilde{\chi}$, although most of the discussion can be generalized to other sort of particles as well. In order to know its

---

[4] sneutrino is rather ruled out by accelerator and astrophysical data.

relic abundance and compare it with the current data we should compute its density today assuming that at some epoch $\tilde{\chi}$ s were in thermal equilibrium with the cosmic soup. Their density decreases because of annihilation only since they are the LSP s and hence stable. If the $\tilde{\chi}$ density at a time t is $n(t)$ then it satisfies the following equation known as *Boltzmann transport equation*

$$\frac{dn}{dt} = -3\frac{\dot{a}}{a}n - \langle v\,\sigma \rangle\,(\,n^2 - n_{eq}^2\,). \tag{2.38}$$

In (2.38), σ is the cross section of the annihilated $\tilde{\chi}$ s, and v is their relative velocity. The thermal average $\langle v\sigma \rangle$ is defined in the usual manner as any other thermodynamic quantity.

The first term on the r.h.s. of (2.38) is easy to understand. It expresses the fact that the density changes because of the expansion. If we momentarily ignore the interactions of the $\tilde{\chi}$ s with the rest of the particles then their total number remains constant. Therefore, $n\,a^3 = \text{const.}$ from which it follows, by taking the derivative with respect the time, that the density rate is given by the first term on the r.h.s. of (2.38). However the $\tilde{\chi}$ s do interact and their number decreases because of pair annihilation. In fact LSP's are stable Majorana particles and it is possible to be annihilated by pairs to Standard Model particles. Their number is therefore reduced until the freezing-out temperature below which they do not interact any further with the remaining particles and their total number is locked.[5]

Therefore, their density decreases as $dn/dt = -n\Gamma_{ann}$, where the annihilation rate Γ_{ann} is given by $\Gamma_{ann} = v\,\sigma\,n$. This explains the second term on the r.h.s. of (2.38). However the $\tilde{\chi}$ s do not only annihilate but are also produced through the inverse process. The last term on the r.h.s of (2.38) expresses exactly this fact. Note that when $\tilde{\chi}$ s were in thermal equilibrium with the rest of the particles and the environment was hot enough, the annihilated products had enough energy to produce back the $\tilde{\chi}$ s at equal rates. During this period $n = n_{eq}$ and the last two terms on the r.h.s. of (2.38) cancel each other, as they should.

Thus the picture is the following. The $\tilde{\chi}$ s are in thermal and chemical equilibrium at early times. During this period $\Gamma >> H$, see (2.34), and $n = n_{eq}$. However as the temperature drops and eventually passes $k\,T \sim m_{\tilde{\chi}}$ the $\tilde{\chi}$ s annihilate but their products do not have enough thermal energy to produce back the annihilated $\tilde{\chi}$ s. The $\tilde{\chi}$ s are in thermal but not in chemical equilibrium any more. In addition their density drops exponentially $\exp\,(-m_{\tilde{\chi}}/kT)$ following the Boltzmann distribution law and $\Gamma \equiv nv\sigma$ decreases so that eventually at a temperature T_f, the freeze – out temperature, Γ equals to the expansion rate H. Below this temperature $\Gamma < H$ and the $\tilde{\chi}$ s are out

---

[5] If other unstable supersymmetric particles are almost degenerate in mass with the $\tilde{\chi}$ s they are in thermal equilibrium with these almost until the decoupling temperature affecting the relic abundance of $\tilde{\chi}$ s through the mechanism of the *co-annihilation*. We drop momentarily this very interesting case from the discussion.

of thermal equilibrium. They decouple not interacting any longer with the cosmic soup and they expand freely. Their total number is locked to a constant value and their density changes because of the expansion. Actually for $T < T_{\tilde{\chi}}$ their density is much larger than the equilibrium density $n >> n_{eq}$ and since $\Gamma >> H$, the first term dominates in (2.38) so that $n\, a^3$ is indeed a constant.

Concerning the cross section thermal average, in general for two annihilating particles $1, 2$, under the assumption that they obey Boltzmann statistics, which is valid for $T \leq m_{1,2}$,[6] one finds [16, 17]

$$\langle v\sigma_{12} \rangle \;=\; \frac{\int_{(m_1+m_2)^2}^{\infty} ds\, K_1(\sqrt{s}/T)\, p_{cm}\, W(s)}{2\, m_1^2\, m_2^2\, T\, K_2(m_1/T)\, K_2(m_2/T)} \tag{2.39}$$

where p_{cm} is the magnitude of the momentum of each incoming particle in their CM frame and $K_{1,2}$ are Bessel functions. The quantity W within the integral is related to the total cross section $\sigma(s)$ through $\sigma = 4\, p_{cm}\, \sqrt{s}\, W(s)/\lambda(s, m_1^2, m_2^2)$.[7] This expression can be considerably simplified if one follows a non-relativistic treatment expanding the cross section in powers of their relative velocity v,

$$v\,\sigma \;=\; a \,+\, \frac{b}{6}\, v^2\,. \tag{2.40}$$

With this approximation the LSP s annihilation thermal average is

$$\langle v\,\sigma \rangle \;=\; a \,+\, \left(b - \frac{3}{2}\, a\right) \frac{kT}{m_{\tilde{\chi}}}\,. \tag{2.41}$$

The non-relativistic expansion is legitime in energy regions away from poles, some of which may be of particular physical interest, and thresholds as well. However near such points this approximation behaves badly invalidating physical results. Thus we had better used the thermal average in the form given by the (2.39).

The goal is to solve (2.38), in order to know the density at today's temperature $T_0 \simeq 2.7\ ^0K$, provided that $n = n_{eq}$ long before decoupling time. By defining $Y \equiv n/s$, where s is the entropy density, and using a new variable $x = T/m_{\tilde{\chi}}$,[8] one arrives at a simpler looking equation

$$\frac{dY}{dx} \;=\; m_{\tilde{\chi}}\, \langle v\,\sigma \rangle \left(\frac{45 G_N}{\pi}\right)^{-1/2} g \left(h + \frac{x}{3}\frac{dh}{dx}\right)\left(Y^2 - Y_{eq}^2\right). \tag{2.42}$$

---

[6] The initial condition in solving Boltzmann's equation should lie in this regime and at a point above the decoupling temperature. This is perfectly legitimate for the case of neutralinos since their decoupling temperature is well below $m_{\tilde{\chi}}$, $T_f \sim m_{\tilde{\chi}}/20$.

[7] $\lambda(x, y, z) \equiv x^2 + y^2 + z^2 - 2\,(xy + yz + zx)$.

[8] We use now units in which the Boltzmann constant k is unity, or same we absorb it within the temperature T.

The function h that appears in this equation counts the effective entropy degrees of freedom related to the entropy density through $s = k \frac{2\pi^2}{45} (kT)^3 h(T)$. The prefactor of $Y^2 - Y_{eq}^2$ is usually a large number, due to the appearance of the gravitational constant $G_N^{-1/2}$, and this can be exploited in using numerical approximations reminiscent of the WKB in Quantum Mechanics [18, 19]. One can solve this to find Y today T_γ which is very close to $Y_0 = Y(0)$, and from this the matter density of $\tilde\chi$ s. The latter is

$$\varrho_{\tilde\chi} = n_{\tilde\chi} m_{\tilde\chi} = m_{\tilde\chi} s_0 Y_0 = \frac{2\pi^2}{45} m_{\tilde\chi} h_{eff}^0 Y_0 T_\gamma^3 , \qquad (2.43)$$

where $h_{eff}^0 \equiv h(T_\gamma)$ is today's value for the effective entropy degrees of freedom which is $h_{eff}^0 \simeq 3.918$[9]. The relic density is then

$$\Omega_{\tilde\chi} h_0^2 = h_0^2 \frac{\varrho_{\tilde\chi}}{\varrho_c} = 0.6827 \times 10^8 \frac{m_{\tilde\chi}}{GeV} h_{eff}^0 Y_0 \left(\frac{T_\gamma}{T_0}\right)^3 . \qquad (2.44)$$

In writing (2.44) we use the fact that $\varrho_c = 8.1 \times 10^{-47} h_0^2 \, GeV^4$. We have also expressed the dependence on temperature through the ratio T_γ/T_0 by using a reference temperature $T_0 = 2.7\,^0K$. The CMB temperature has been determined by measurements of the CMB to be $T_\gamma = 2.752 \pm 0.001\,^0K$. The value of Y_0 required can be found by solving (2.42) numerically with the boundary condition that $Y \to Y_{eq}$ at temperatures well above the freeze-out temperature. The numerical solution proves to be rather time consuming and for this reason many authors use approximate solutions which are less accurate, by only 5–10 %, having the advantage that the calculation performs fast and the physical content is more transparent. There are good packages in the literature, like DarkSUSY [20] and microOMEGAs [21], which can be used to handle numerically the Boltzmann equation and find the LSP relic density in supersymmetric theories.

2.5.2 Approximate Solutions to Boltzmann Equation

Approximate solutions can be found under the assumption that $Y \ll Y_{eq}$ below the freezing point $x_f = T_f/m_{\tilde\chi}$ while $Y \simeq Y_{eq}$ above it. Omitting then the Y_{eq} term in (2.42), which is valid for x between x_f and $x_0 \simeq 0$, and putting $x = x_f$ in (2.42) an equation for x_f is derived

$$x_f^{-1} = \ln \left[0.03824 \, g_s \frac{M_{Planck} \, m_{\tilde\chi}}{\sqrt{g^*}} \langle v \sigma \rangle \, c(c+2) \, x_f^{1/2} \right], \qquad (2.45)$$

---

[9] At today's temperatures only photons and neutrinos contribute to h_{eff}. Its value is almost half of $2+6\times\frac{7}{8} = 7.25$ one naively expects by merely counting the spin degrees of freedom of the photons and neutrinos due mainly to the decoupling of the neutrinos.

which can be solved numerically to obtain x_f. In this g^* stands for the effective energy degrees of freedom at the freeze-out temperature $g^* = g(x_f)$ and the derivative term dh/dx in (2.42) has been ignored in this approximation. g_s are the spin degrees of freedom and the $c(c+2)$ within this expression equals to one. The reason we present it in the above equation is that empirically it is found that very good approximation for the freeze-out point temperature is obtained with values $c \simeq 1/2$. The freeze out temperature for a WIMP is close to $T_f \simeq m_{\tilde{\chi}}/20$. Now that x_f has been determined one can integrate (2.42) from x_f to zero, under the same assumptions, to obtain $Y(0)$ and from this the relic abundance. The solution for $Y(0)$ entails to a density

$$\varrho_{\tilde{\chi}} = (\frac{4\pi^3}{45})^{1/2} (\frac{T_{\tilde{\chi}}}{T_\gamma})^3 \frac{T_\gamma^3}{M_{Planck}} \frac{\sqrt{g^*}}{J} . \tag{2.46}$$

In (2.46) the quantity J is given by the integral $J \equiv \int_0^{x_f} \langle v\sigma \rangle \, dx$. Recall that $g^* = g(x_f)$ in the notation we follow here. In this expression the $\tilde{\chi}$'s temperature $T_{\tilde{\chi}}$ appears explicitly which is different from that of photons. In fact it is found [22] that due to the decoupling of both $\tilde{\chi}$ s and neutrinos that $(T_{\tilde{\chi}}/T_\gamma)^3 = 4/11 \cdot g(T_\nu)/g(T_f) = 3.91/g(T_f)$ with T_ν the neutrino decoupling temperature and T_f that of $\tilde{\chi}$ s. Using this and (2.46) we obtain for the relic density $\Omega_{\tilde{\chi}} h_0^2 = \varrho_{\tilde{\chi}}/(8.1 \times 10^{-47} \, GeV^4)$ the following result

$$\Omega_{\tilde{\chi}} h_0^2 = \frac{1.066 \times 10^9 \, GeV^{-1}}{M_{Planck} \sqrt{g^*} \, J} , \tag{2.47}$$

which is the expression quoted in many articles. The quantity J, which has already been defined, is given in GeV^{-2} units. Note that the relic density is roughly inverse proportional to the total cross section. This means that the larger the cross section the smaller the relic density is and vice versa.

To have an estimate of the predicted relic density for the case of a neutralino LSP we further approximate $J \approx x_f \langle v \sigma \rangle_f$ and $\langle v\sigma \rangle \sim \alpha/m_{\tilde{\chi}}^2$ where α is a typical electroweak coupling. Since x_f is of the order of ~ 0.1 and g^* is $\simeq 100$ for masses in the range $m_{\tilde{\chi}} \simeq 20 \, GeV - 1 \, TeV$, the relic density (2.47) above turns out to be $\Omega_{\tilde{\chi}} h_0^2 \sim 0.1$ in the physically interesting region $m_{\tilde{\chi}} \simeq 100 \, GeV$. Therefore we conclude that relic densities of the right order of magnitude can naturally arise in supersymmetric theories if one interprets the Dark Matter as due to a stable neutralino.

2.5.3 Co-annihilations

All discussion so far concerned cases in which the stable WIMP is not degenerate in mass with other heavier species that can decay to it. Therefore there is an epoch where the Universe is filled by SM particles and the LSPs, whose density decreases because of pair annihilations when the temperature starts passing the point where SM particles do not have enough energies to produce back LSPs. However it may happen that although lighter the LSP's mass $m_{\tilde{\chi}}$ is not very different from other particles's masses m_i that they decay to it. In fact when $\delta m_i \equiv m_i - m_{\tilde{\chi}} \sim T_f$ these particles are thermally accessible and this implies that they are as abundant as the relic species. This drastically affects the calculation of the LSP relic density. This effect is known as co-annihilation [17, 23, 24, 25, 26] and it is not of academic interest. Actually in the most popular supersymmetric schemes, advocating the existence of good CDM candidates, there are regions having this characteristic so it is worth discussing this case.

Since all nearly degenerate particles with the LSP will eventually decay to it the relevant quantity to calculate for the relic abundance of the LSP is the density $n = \sum_i n_i$. In it n_i is the density of the particle i and the sum runs from $i = 1, ...N$. With $i = 1$ we label the LSP and with $i = 2, ...N$ the rest of the particles that are almost degenerate in mass with it. Following [24] the Boltzmann transport (2.38) is generalized to,

$$\frac{dn}{dt} = -3\frac{\dot{a}}{a}n - \sum_{i,j}^{N} \langle v_{ij}\,\sigma_{ij}\rangle\,(\,n_i n_j - n_i^{eq} n_j^{eq}\,). \qquad (2.48)$$

The notation in (2.48) is obvious. Since the criterion for which particles co-annihilate with the LSP is roughly given by $\delta m \sim T_f$ and $T_f \simeq m_{\tilde{\chi}}/20$ the particles participating in the co-annihilation process are those for which the mass differences are $\delta m_i \simeq 5\%\ m_{\tilde{\chi}}$. Their equilibrium densities are given by $n_i^{eq} = g_i \frac{T}{2\pi} m_i^2 K_2(m_i/T)$ where g_i denotes the spin degrees of freedom. Approximating $n_i/n \simeq n_i^{eq}/n^{eq}$, (2.48) can be cast in the following form

$$\frac{dn}{dt} = -3\frac{\dot{a}}{a}n - \sum_{i,j}^{N} \langle v\,\sigma_{eff}\rangle\,(\,n^2 - n_{eq}^2\,). \qquad (2.49)$$

where the effective thermal average appearing in this equation is a generalization of (2.39) given by

$$\langle v\sigma_{eff}\rangle = \sum_{i}^{N} \frac{n_i^{eq}\,n_j^{eq}}{n_{eq}^2}\,\langle v_{ij}\sigma_{ij}\rangle\,. \qquad (2.50)$$

This can be written as a single integral generalizing the results of [16]

$$\langle v\sigma_{eff}\rangle = \frac{\int_2^\infty da\,K_1(a/x)\sum_{i,j}\lambda(a^2,b_i^2,b_j^2)\,g_i g_j\,\sigma_{ij}(a)}{4x\,(\,\sum_i g_i b_i^2 K_2(b_i/x))^2}, \qquad (2.51)$$

where $b_i \equiv m_i/m_{\tilde{\chi}}$ with $\tilde{\chi}$ denoting the LSP labelled by $i = 1$. If we seek for an approximate solution, as was done in the no co-annihilation case, then the freeze-out point is

$$x_f^{-1} = \ln\left[\,0.03824\,g_{eff}\,\frac{M_{Planck}\,m_{\tilde{\chi}}}{\sqrt{g^*}}\,\langle v\,\sigma_{eff}\rangle\,c(c+2)\,x_f^{1/2}\,\right]. \qquad (2.52)$$

In this g_{eff} is defined as

$$g_{eff} \equiv \sum_i g_i \left(1 + \Delta_i \right)^{3/2} \exp \left(-\Delta_i / x_f \right) \tag{2.53}$$

where $\Delta_i \equiv (m_i - m_{\tilde{\chi}})/m_{\tilde{\chi}}$. As for the relic abundance, this is given by (2.47) with the quantity J defined now as $J \equiv \int_0^{x_f} \langle v\sigma_{eff} \rangle \, dx$. In calculating this we have made use of the fact that all nearly degenerate particles with the LSP will eventually decay and therefore the final LSP abundance will be extracted from $n = \sum_i n_i$. In supersymmetric models the relevant co-annihilation channels are between neutralinos, charginos and sfermions.

As a preview of the importance of the co-annihilation process we mention that the cosmological bound on the LSP neutralino is pushed to $\simeq 600 \, GeV$, from about $200 \, GeV$ in the stau co-annihilation region and to $1.5 \, TeV$ in the chargino co-annihilation, increasing upper bounds and weakening the prospects of discovering supersymmetry in high energy accelerators.

2.6 Supersymmetry and its Cosmological Implications

Supersymmetry, or SUSY for short, is a fermion-boson symmetry and it is an indispensable ingredient of Superstring Theories. Fermions and bosons go in pairs (partners) having similar couplings and same mass!. Spontaneous Symmetry Breaking (SSB) of local SUSY at Planckian energies makes the partner of the graviton, named gravitino, massive $m_{3/2} \neq 0$ (**SuperHiggs effect**) and lifts the mass degeneracy between fermions and bosons by amounts M_S, a parameter which depends on the particular supersymmetry breaking mechanism. At much lower energy scales $E \ll M_{Planck}$, accessible to LHC if $M_S \simeq \mathcal{O}(TeV)$, the theory is supersymmetric but there appear terms that break SUSY softly. These are scalar mass terms, gaugino mass terms or scalar trilinear couplings m_0, $M_{1/2}$, A_0 (for a review see [27]). Supersymmetric extensions of the SM naturally predict the existence of Dark Matter candidates which is the LSP. This may be the "Neutralino" or the "Gravitino", whichever is the lightest, or other non-SM particle provided its relic abundance is within the cosmological limits while all accelerator bounds are respected.

In supersymmetric theories the generators of the fermion-boson symmetry are spinorial operators Q, $Q^\dagger$ which turn a boson state to a fermion and vice versa,

$$Q^\dagger \mid boson \; \rangle = |fermion\rangle \tag{2.54}$$

$$Q \; |fermion\rangle = \mid boson \rangle \, . \tag{2.55}$$

These commute with the supersymmetric Hamiltonian H_S,

$$[H_S, Q] = [H_S, Q^\dagger] = 0 \tag{2.56}$$

resulting to a degenerate mass spectrum between fermions and bosons. The field content of a supersymmetry theory is larger than in ordinary theories in the sense that additional degrees of freedom are needed, the so called "sparticles", that are superpartners of the known particles. The most economic supersymmetric extension of the Standard Model, known as MSSM (Minimal Supersymmetric Standard Model) has the physical content appearing in Table 2.2. In it the SM particles are shown on the left with their corresponding spins and their superpartners on the right. Note that unlike the SM the Higgses $H_{1,2}$ are not independent and thus five Higgses survive the Electroweak Symmetry Breaking. In addition MSSM can posses a symmetry known as R-parity under which each particle bears a quantum number

$$PR = (-1)^{3(B-L)+2s},   (2.57)$$

with B, L the baryon and lepton number of the particle and s its spin. P_R is +1 for particles and −1 for the superparticles. This quantum number is multiplicatively conserved in theories possessing R-parity and prohibits Baryon and Lepton number violations. Another virtue of this symmetry is the fact that the lightest supersymmetric particle (LSP) is stable. The reason for this is that, in R − parity conserving theories, the vertices have an even number of sparticles. Because of this, a sparticle can only decay to an odd number of sparticles and an even or odd number of SM particles. For the LSP such a decay is however energetically forbidden since it is the lightest sparticle and hence it is stable. If, in addition, it is electrically neutral and does not interact strongly the LSP qualifies as a WIMP.

The CMSSM (constrained MSSM) is the most popular and extensively studied supersymmetric model encompassing SM. It is motivated by the minimal supergavity theories (mSUGRA) and differs from MSSM in that universal boundary conditions are imposed for the scalar, gaugino and trilinear couplings at a unification scale so that there is a single m_0, $M_{1/2}$ and A_0. These along with the ratio $\langle H_2 \rangle / \langle H_1 \rangle$ can be chosen to be the only arbitrary parameters of the theory. Other parameters μ, m_3^2, inducing mixing between the Higgs multiplets are determined from the electrowek symmetry breaking conditions[10].

Table 2.2. Particle content of the minimal supersymmetric standard model

	SM particles				SUSY particles	
Particle	Name	Spin		Sparticle	Name	Spin
q	quarks	1/2		$\tilde{q}$	squarks	0
l	leptons	1/2	$Q, Q^\dagger$	$\tilde{l}$	sleptons	0
$W^\pm, W^0$	W-bosons	1	$\Longleftrightarrow$	$\tilde{W}^\pm, \tilde{W}^0$	winos	1/2
B	B-boson	1		$\tilde{B}$	bino	1/2
G	gluons	1		$\tilde{G}$	gluinos	1/2
$H_{1,2}$	Higgses	0		$\tilde{H}_{1,2}$	Higgsinos	1/2

---

[10] The sign of μ is also a free parameter.

2.6.1 The Neutralino DM

The neutral components of the Higgsinos $\tilde{H}^0_{1,2}$, the neutral "wino" $\tilde{W}^0$ and the "Bino" $\tilde{B}$ (see Table 2.2) interact weakly but they are not mass eigenstates. The four mass eigenstates $\tilde{\chi}^0_i$ $i = 1,...4$, named "Neutralinos", are linear combinations of these. These are Majorana fermions which means that particles are same as their antiparticles and thus they possess half the degrees of freedom of a "charged" Dirac particle like the electron for instance.

In the MSSM, briefly discussed in the previous section, and depending on the inputs for the SUSY breaking parameters m_0, $M_{1/2}$, A_0 and $tan\beta$, the lightest of the neutralinos, which we shall denote by $\tilde{\chi}$, may be the LSP. Being a linear combination of the Higgsinos, Wino and Bino fields this can be written as

$$\tilde{\chi} = a_1 \tilde{H}^0_1 + a_2 \tilde{H}^0_2 + a_W \tilde{W}^0 + a_B \tilde{B} . \qquad (2.58)$$

where normalization requires that $\sum_i |a_i|^2 = 1$. Depending on the magnitudes of a_i appearing in (2.58) we can distinguish the following two cases

$$|a_1|^2 + |a_2|^2 >> |a_W|^2 + |a_B|^2, \quad \text{(Higgsino - like)}$$
$$|a_1|^2 + |a_2|^2 << |a_W|^2 + |a_B|^2, \quad \text{(gaugino - like)} . \qquad (2.59)$$

In the upper case the LSP is mostly Higgsino and in the lower mostly Gaugino. Other cases encountered in model studies are somewhere in between. In the CMSSM, the $\tilde{\chi}$ is gaugino – like, actually a bino, in the major portion of the parameter space[11]. Absence of its detection in accelerator experiments puts a lower bound on its mass, $m_{\tilde{\chi}} > 46$ GeV.

If $\tilde{\chi}$ is the LSP, and thus stable in R - parity conserving theories, then at some epoch, the cosmic soup contains the $\tilde{\chi}$'s and Standard Model particles. All other supersymmetric particles have already decayed to $\tilde{\chi}$ and SM particles. Then the number of the LSP particles can only decrease through pair annihilations to SM particles through the reactions $\tilde{\chi} + \tilde{\chi} \rightarrow A + B + ...$ In the MSSM and in leading order in the coupling constants involved, only two body pair annihilations take place and the SM particles in the final state occur in the combinations displayed in Table 2.3. As already remarked, in the minimal supersymmetric extension of the Standard Model, there exist five Higgs mass eigenstates $H^\pm, H, h, A$, the last three being neutral, in contrast to the Standard Model where only one neutral Higgs survives after Electroweak Symmetry breaking. We should mention that an upper theoretical bound on the lightest of the neutral Higgses mass m_h exists which is ≈ 138 GeV.

Accelerator experiments impose various constraints on sparticle masses which along with the cosmological bound on the DM relic abundance restricts

---

[11] There are narrow regions in the parameter space, like for instance the so-called Hyperbolic Branch, which are of particular phenomenological interest and in which the LSP may be a Higgsino.

Table 2.3. Neutralino pair annihilations $\tilde{\chi}\tilde{\chi} \rightarrow A + B$

Particles in the final state A , B

Fermion − Antifermion:	$q \, \bar{q}$
	$l \, \bar{l}$
Gauge Bosons:	$W^+ \, W^-$
	$Z \, Z$
Gauge Bososn + 1Higgs:	$W^\pm \, H^\mp$
	$Z \, A$
	$Z \, H$
	$Z \, h$
Higgses:	$H^+ \, H^-$
	$H \, H$
	$h \, h$
	$H \, h$
	$A \, H, \; A \, h$
	$A \, A$

the allowed parameter space describing the model. The potential of discovering supersymmetric particles in future experiments depends on the bounds put on sparticle masses and these are constrained by the cosmological data. LEP and Tevatron colliders give a lower bound of 104 GeV for the chargino mass while sleptons should not weigh less than about 99 GeV. The bounds imposed on all sparticles can be traced in [28]. LEP has also provided us with the important Higgs mass bound $m_h > 114 \, GeV$ [9]. Other important constraints stem from the decay $b \rightarrow s \, \gamma$ whose branching ratio should lie in the range $1.8 \times 10^{-4} < BR(b \rightarrow s \, \gamma) < 4.5 \times 10^{-4}$ at the 2 σ level [29, 30]. The BNL E821 experiment derived a very precise value for the anomalous magnetic moment of the muon $\alpha_\mu \equiv (g_\mu - 2)/2 = 11659203(8) \times 10^{-10}$ [31] pointing to a discrepancy between the SM theoretical prediction and the experimental value given by $\delta\alpha_\mu = (361 \pm 106) \times 10^{-11}$ which shows a 3.3σ deviation. This can put severe upper bounds on sparticle masses. However theoretical uncertainties due to disagreement between the e^+e^- and τ decay data used to calculate the contributions to $g_\mu - 2$ forces us to consider these data with a grain of salt until the disagreement between the two theoretical approaches is finally resolved [32].

There are numerous phenomenological analyses by various groups and regions in the parameter space which are compatible with cosmological and accelerator data have been delineated. Three main regions have been identified as conforming with all data. The "funel" region $m_0 \sim M_{1/2}$ in which neutralinos rapidly annihilate via direct s-channel pseudoscalar Higgs poles, which opens up for large values of $tan\beta$, $\tilde{\chi}, \tilde{\tau}$ (stau) co-annihilation region which extends to large $M_{1/2} >> m_0$ and the hyperbolic branch (HB), which includes the "focus point" region, in which $m_0 \sim$ few TeV $>> M_{1/2}$.

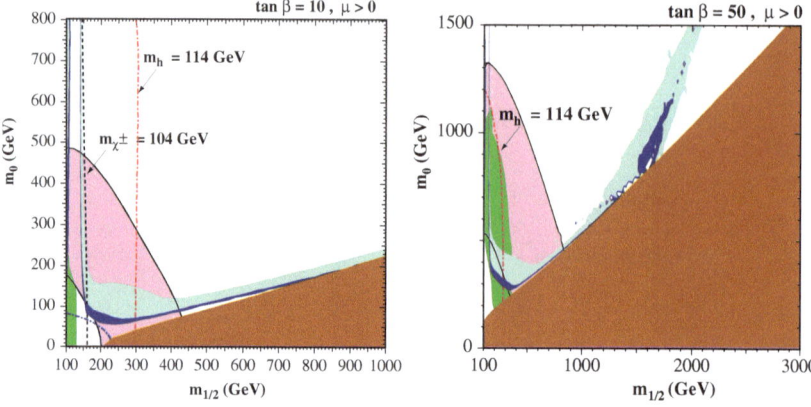

Fig. 2.3. mSUGRA/CMSSM constraints after WMAP. The very dark shaded region (*dark blue*) is favoured by WMAP ($0.094 \leq \Omega_\chi h^2 \leq 0.129$). In the medium shaded region (*turqoise*) $0.1 \leq \Omega_\chi h^2 \leq 0.3$. The shaded region at the bottom (*brick red*) is excluded because LSP is charged. Dark regions (*green*) on the left are excluded by $b \rightarrow s\gamma$. The shaded stripes (*pink*) on the left are favoured by $g_\mu - 2$ at the $2 - \sigma$ level. The LEP bounds on the chargino mass 104 GeV and the Higgs mass 114 GeV are also shown. (from [33])

In Fig. 2.3 the allowed regions in the $m_0, m_{1/2}$ plane are displayed showing clearly the tight constraints imposed by the WMAP data in conjunction with accelerator bounds. In Table 2.4, upper bounds on sparticle masses are shown which are derived if the WMAP value for $\Omega_{CDM}\ h_0^2$ is imposed and the 2σ E821 bound $149 \times 10^{-11} < \alpha_\mu^{SUSY} < 573 \times 10^{-11}$ is observed satisfying at the same time all other experimental constraints. From this table it can be seen that Supersymmetry (CMSSM) will be accessible in the LHC and to any or other linear e^+e^- collider with center of mass energy $\geq 1.1\ TeV$ [34].

Table 2.4. Upper bounds, in GeV, on the masses of the lightest of the neutralinos, charginos, staus, stops and Higgs bosons for various values of $\tan\beta$ if the new WMAP value [1] for $\Omega_{CDM}h^2$ and the 2σ E821 bound, $149 \times 10^{-11} < \alpha_\mu^{SUSY} < 573 \times 10^{-11}$, is imposed (from [34])

$\tan\beta$	$\tilde{\chi}$	χ^+	$\tilde{\tau}$	$\tilde{t}$	$Higgs$
10	155	280	170	580	116
15	168	300	185	640	116
20	220	400	236	812	118
30	260	470	280	990	118
40	290	520	310	1080	119
50	305	553	355	1120	119
55	250	450	585	970	117

From this it is apparent the importance of the E821 results and the need of further theoretical work lifting the discrepancies concerning the muon's anomalous magnetic moment calculations. Detailed studies have shown that the LHC will probe the region of the parameter space allowed by Cosmology and present accelerator data even if the data by E821 are not taken into account [36] as is seen in Fig. 2.4. In a χ^2 analysis performed in [37] it is shown that particular regions of the parameter space are favoured including the focus point region and the fast s – channel Higgs resonance annihilation as shown in Fig. 2.5.

Supersymmetric direct DM searches through elastic neutralino-Nucleon scattering $\tilde{\chi}+N \to \tilde{\chi}+N$ are very important for DM detection. The rates for the spin-independent cross sections as calculated in the CMSSM are of the order $\sigma_{s.i.} \sim 10^{-7}-10^{-8}$ pb at the maximum, still far from the sensitivity limits of CDMS II experiment as shown in Fig. 2.6. The spin independent cross section will be of interest for future experiments, CDMS, EDELWEISS, ZEPLIN and GENIUS, which will search for DM and will access the as yet unexplored

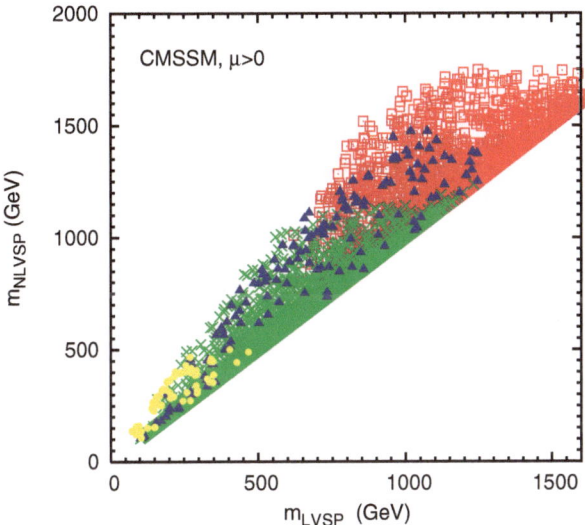

Fig. 2.4. Scatter plot of the masses of the lightest visible supersymmetric particle (LVSP) and the next-to-lightest visible supersymmetric particle (NLVSP) in the CMSSM. The darker (*blue*) triangles satisfy all the laboratory, astrophysical and cosmological constraints. For comparison, the dark (*red*) squares and medium-shaded (*green*) crosses respect the laboratory constraints, but not those imposed by astrophysics and cosmology. In addition, the (*green*) crosses represent models which are expected to be visible at the LHC. The very light (*yellow*) points are those for which direct detection of supersymmetric dark matter might be possible (from [36])

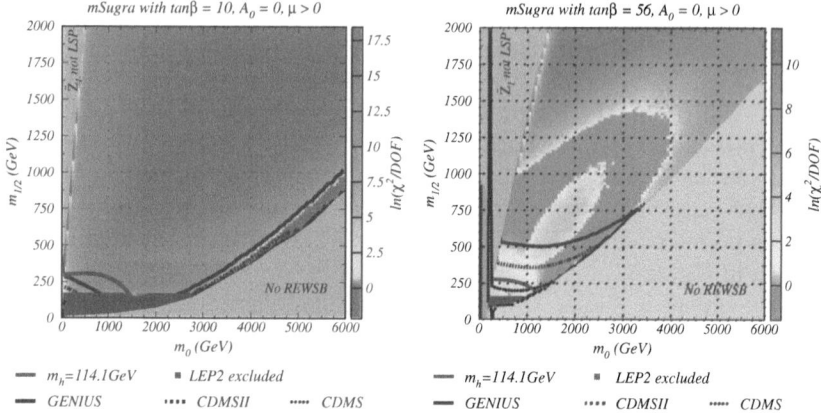

Fig. 2.5. WMAP data seem to favour ($\frac{\chi^2}{dof} < 4/3$) the HB/focus point region (moderate to large values of μ, large m_0 scalar masses) for almost all tanβ (*narrow stripe on the right of the Left panel*), as well as s - channel Higgs resonance annihilation (*ring-like stripe on the Right panel*) for $\mu > 0$ and large tanβ (from [37])

CMSSM region [38]. The present status and the sensitivity of future experiments, in conjuction with the theoretical predictions, is displayed in Fig. 2.7.

The annihilations of relic particles in the Galactic halo, $\tilde{\chi}\tilde{\chi} \rightarrow \bar{p}, e^+ + ...$, the Galactic center $\tilde{\chi}\tilde{\chi} \rightarrow \gamma + ...$ or the core of the Sun $\tilde{\chi}\tilde{\chi} \rightarrow \nu + ... \rightarrow \mu + ...$ are of great interest too. The annihilation positrons in the CMSSM seem to fall below the cosmic-ray background. The annihilated photons may be detected in

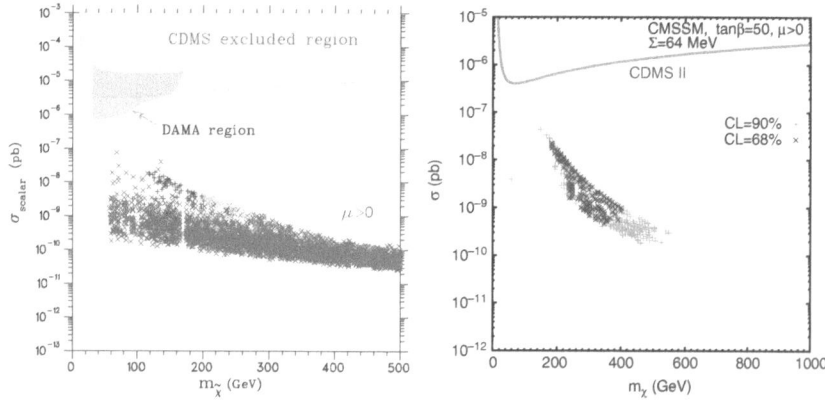

Fig. 2.6. (**a**) *Left panel.* Scatter plot of the spin independent neutralino-nucleon elastic cross section vs. m_χ predicted in the CMSSM. + signs are compatible with the E821 experiment and also cosmologically acceptable. The sample consists of 40,000 random points (for details see [34]). (**b**) *Right panel.* The same as in panel a) for tan$\beta = 50, \mu > 0$, with $\sigma_{\pi N} = 64$ MeV. The predictions for models allowed at the 68% (90%) confidence are shown by $\times$ signs (from [35])

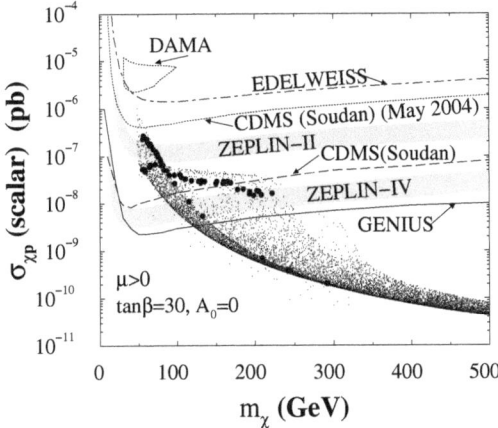

Fig. 2.7. Spin independent cross sections vs. $m_{\tilde{\chi}}$. The black circles are compatible with WMAP data. The present and future DM detection limits are shown (from [39])

GLAST experiment, while the annihilations inside the Sun may be detectable in the experiments AMANDA, ANTARES, NESTOR and IceCUBE [40].

2.6.2 The Gravitino

The gravitino, $g_{3/2}$, spin-3/2 superpartner of the graviton in Supergravity theories, gets non-vanishing mass after spontaneous symmetry breaking of local supersymmetry. The couplings of the gravitino are suppressed by the Planck scale. The dominant gravitino couplings are discussed in Sect. 1.6 in the Lazarides' article (see in particular (1.28)). Gravitino couples to the supersymmetric matter and if its mass $m_{3/2}$ is larger than that of the lightest of the neutralinos $\tilde{\chi}$ it can decay gravitationally to it through $g_{3/2} \to \tilde{\chi} + \gamma$. If it is lighter than $\tilde{\chi}$ then it is the LSP and $\tilde{\chi}$ decays to it via $\tilde{\chi} \to g_{3/2} + \gamma$. Such decays produce electromagnetic radiation and may be upset the Big Bang Nucleosynthesis (BBN) predictions for the light element abundances. In fact the emitted Electromagnetic radiation can destroy $D, {}^4He, {}^7Li$ and/or overproduse ${}^6Li$.

The most direct and accurate estimate of the baryon to photon ratio $\eta = n_B/n_\gamma$ is provided by the acoustic structures of the CMB perturbations [1] and its value, $\eta = 6.14 \pm 0.25 \times 10^{-10}$, controls the BBN calculations yielding very definite predictions for the abundances of the light elements as shown in Table 2.5. The predicted values quoted in the table are impressively close to the observed values with the exception of the Lithium cases. In fact the prediction for ${}^7Li$ is predicted higher, by almost a factor of three, than the values of the astrophysical data and ${}^6Li$ is predicted much smaller by a factor of 10^{-3}. However these discrepancies are not disturbing due to the large systematic errors existing in the astrophysical data. The decays of massive

Table 2.5. Predictions of the light element abundances

Element	Predicted[1]	Observed
Y_p	0.2485 ± 0.0005	0.232 to 0.258
D/H	$2.55^{+0.21}_{-0.20} \times 10^{-5}$	$2.78 \pm 0.29 \times 10^{-5}$
^{3}He/H	$1.01 \pm 0.07 \times 10^{-5}$	$1.5 \pm 0.5 \times 10^{-5}$
^{7}Li/H	$4.26^{+0.73}_{-0.60} \times 10^{-10}$	$1.23^{+0.68}_{-0.32} \times 10^{-10}$
^{6}Li/H	$1.3 \pm 0.1 \times 10^{-14}$	$6^{+7}_{-3} \times 10^{-12}$

[1]R. H. Cyburt, Phys. Rev. D70 (2004) 023505

unstable products with lifetimes $\tau > 10^2$ s produce Electromagnetic radiation and or hadronic showers in the Early Universe which may destroy or create nuclei spoiling the successful BBN predictions. The agreement with BBN imposes limits to the density of the decaying particles which depend on their lifetimes and on the value of η. These limits are shown in Fig. 2.8 [41] where the quantity ζ_X is defined as $\zeta_X \equiv m_X \, n_X/n_\gamma$. For instance if a decaying particle X has a lifetime $\tau_X = 10^8$ s, the bound on the ζ_X extracted from this figure is $m_X \, n_X/n_\gamma < 5 \times 10^{-12}$, with m_X in GeV units (from [41]). These constraints were updated in [42] where the effects of unstable heavy particles

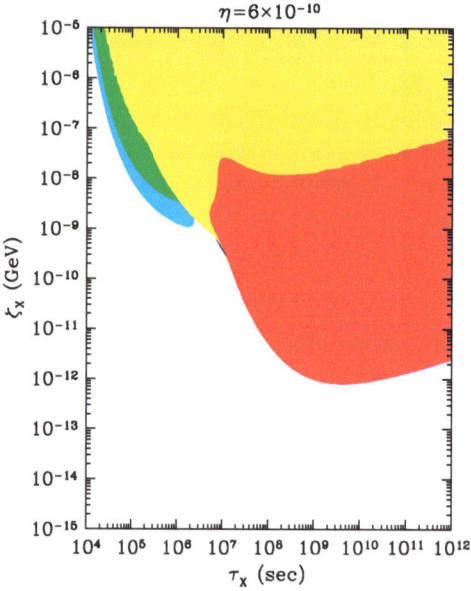

Fig. 2.8. Limits on ζ_X, τ_X imposed by the abundances of the light elements. Shaded (*Colored*) areas are excluded. Dark(*Red*)=^6Li, Heavy Grey(*Green*)=^7Li, Light Grey(*Yellow*)=$^6Li/^7Li$, Medium Grey(*Light Blue*)=D/H (from [41])

were reconsidered in an attempt to reconcile the high primordial ^7Li abundance, as implied by the baryon-to-photon ratio, with the lower ^7Li observed in halo stars. In Supergravity the gravitino can either decay to $g_{3/2} \to \tilde{\chi} + \gamma$ or, if it is the LSP, $\tilde{\chi} \to g_{3/2} + \gamma$ as mentioned earlier. These late decays can affect the light element abundances and the previously discussed limits apply.

a) Unstable gravitino $(m_{3/2} > m_{\tilde{\chi}})$:
The gravitino's decay width is

$$\Gamma\left(g_{3/2} \to \tilde{\chi} + \gamma\right) = \frac{1}{4} \frac{m_{3/2}^3}{M_{Planck}} O_{\tilde{\chi}\tilde{\gamma}}^2, \qquad (2.60)$$

Where $O_{\tilde{\chi}\tilde{\gamma}}$ is the matrix element relating the mass eigenstate $\tilde{\chi}$ to the photino $\tilde{\gamma}$. Assuming for the sake of the argument that $\tilde{\chi}$ is a bino, $\tilde{\chi} = \tilde{B}$, this matrix element is $O_{\tilde{\chi}\tilde{\gamma}} = \cos\theta_w$. Then if no other decay is significant

$$\tau_{3/2} = 2.9 \times 10^8 \left(\frac{100 \, GeV}{m_{3/2}}\right)^3 s. \qquad (2.61)$$

With a gravitino mass in the range $100 \, GeV - 10 \, TeV$ the gravitino's lifetime is $\tau_{3/2} = 10^2 - 10^8 \, s$, so it is indeed a late decaying particle. On the other hand the thermal production of the gravitinos is estimated to be [43]

$$Y_{3/2} \equiv \frac{n_{3/2}}{n_\gamma} = 1.2 \times 10^{-11} \left(1 + \frac{m_{\tilde{g}}^2}{12 \, m_{3/2}^2}\right) \frac{T_R}{10^{10} \, GeV}, \qquad (2.62)$$

with T_R the maximum temperature reached in the Universe. With reasonable values of the gluino and gravitino masses $m_{\tilde{g}}, m_{3/2}$

$$Y_{3/2} \simeq (0.7 - 2.7) \times 10^{-11} \frac{T_R}{10^{10} \, GeV}. \qquad (2.63)$$

From the BBN limits which we discussed previously we can infer limits on $Y_{3/2}$. For instance for a gravitino mass $100 \, GeV$ with lifetime $\tau_{3/2} = 10^8 \, s$ we get

$$Y_{3/2} \leq 5 \times 10^{-14}, \qquad (2.64)$$

and relations (2.63) and (2.64) are combined to yield an upper limit on T_R

$$T_R < 7 \times 10^7 \, GeV. \qquad (2.65)$$

This limit is much smaler than the reheating temperature expected in inflationary models $T_R \simeq 10^{12} \, GeV$. Therefore with unstable gravitinos we need an explanation to resolve this problem.

a) Stable gravitino $(m_{3/2} < m_{\tilde{\chi}})$:
With a stable gravitino the next to it NSP (Next Supersymmetric Particle),

which can be the $\tilde{\tau}$ or $\tilde{\chi}$, can decay to it. For the neutralino $\tilde{\chi}$ decay we have for instance

$$\Gamma \left(\tilde{\chi} \rightarrow g_{3/2} + \tilde{\gamma} \right) = \frac{1}{16\,\pi} \frac{O^2_{\tilde{\chi}\gamma}}{M^2_{Planck}} \frac{m^5_{\tilde{\chi}}}{m^2_{3/2}} \left(1 - \frac{m^2_{3/2}}{m^2_{\tilde{\chi}}} \right)^3 \left(\frac{1}{3} + \frac{m^2_{3/2}}{m^2_{\tilde{\chi}}} \right)$$

(2.66)

with $O_{\tilde{\chi}\tilde{\gamma}} = O_{1\tilde{\chi}} \cos\theta_w + O_{2\tilde{\chi}} \sin\theta_w$. The DM gravitino is produced via the decay $NSP \rightarrow g_{3/2} + \gamma$ and its relic density is

$$\Omega_{3/2}\, h_0^2 = \frac{m_{3/2}}{m_{NSP}}\, \Omega_{NSP}\, h_0^2 < \Omega_{NSP}\, h_0^2 .$$

(2.67)

From the BBN constraints and especially from ^6Li abundance

$$\frac{n_{NSP}}{n_\gamma} < 5 \times 10^{-14} \left(\frac{100\ GeV}{m_{NSP}} \right) ,$$

(2.68)

for $\tau_{NSP} = 10^8\ s$, before the decay of the NSP. Using $n_B/n_\gamma = 6 \times 10^{-10}$ we have the bounds

$$\Omega_{NSP}\, h_0^2 < \frac{10^{-2}}{m_{NSP}}\, \Omega_B\, h_0^2 < 10^{-2}\, \Omega_B\, h_0^2 \sim 2 \times 10^{-4} .$$

(2.69)

Plugging (2.69) into (2.67) yields the bound

$$\Omega_{3/2}\, h_0^2 < 2 \times 10^{-4}$$

(2.70)

This is too small far away from ~ 0.1 required by the cosmological data. At this point recall that values in the range 0.1 are generic in supersymmetric models having the lightest of the neutralinos as the LSP. However in deriving this estimate we assumed a lifetime for the NSP of the order of $\sim 10^8\ s$. For shorter lived NSPs this tight constraint is relaxed. However still $\Omega_{3/2}\, h_0^2 < \Omega_{NSP}\, h_0^2$ and on account of the fact that $\Omega_{NSP}\, h_0^2 \sim 0.1$ we may need some supplementary mechanisms to produce gravitinos as for instance reheating after inflation in addition to NSP decays.

In the analysis of [44], whose arguments we closely followed in the previous discussion, the authors calculate the relic density the NSP would have today if it had not decayed, $\Omega_{NSP}\, h_0^2 = 3.9 \times 10^7\ \zeta_{NSP}\ GeV^{-1}$, compute the lifetime τ_{NSP} and impose the detailed bounds from BBN on ζ_{NSP}. They delineate regions of the parameter space where the gravitino relic density $\Omega_{3/2}\, h_0^2 = \frac{m_{3/2}}{m_{NSP}}\, \Omega_{NSP}\, h_0^2$ is less than 0.129 the highest CDM matter density allowed by WMAP data at 2σ. In their analysis they do not consider regions in which $\tau_{NSP} < 10^4\ s$. For such short lifetimes the hadronic decays should be considered too which would give additional constraints strengthening the limits on gravitino DM derived on the basis of the Electromagnetic showers only. Some sample outputs of the limits imposed are shown in Fig. 2.9. In the light grey (yellow) shaded areas, designated by $r < 1$, $\Omega_{3/2}\, h_0^2 < 0.129$

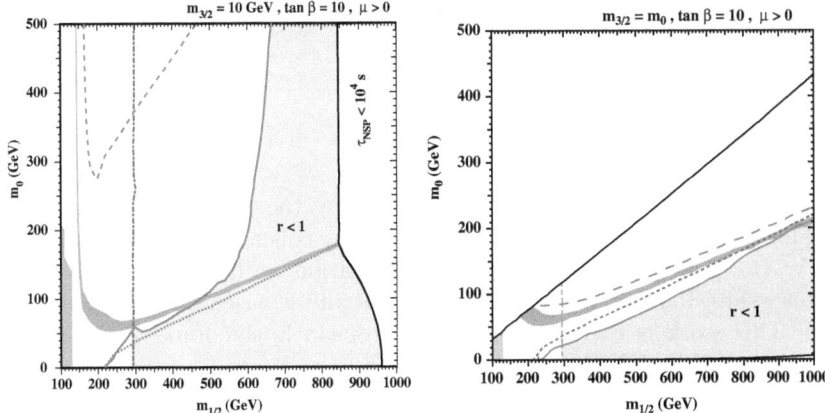

Fig. 2.9. The $(m_0, M_{1/2})$ planes for $\tan\beta = 10$ and values of the gravitino mass equal to 10 GeV (*left panel*) and $m_{3/2} = m_0$ (*right panel*). In the shaded area designated by $r < 1$ we have $\Omega_{3/2}\, h_0^2 < 0.129$ and the BBN limits are observed (from [44])

and the BBN limits, as well as all accelerator data, are observed. In the allowed domain the relic density is typically less than that favoured by astrophysics and cosmological data and as already remarked supplementary mechanisms for the production of gravitinos may be needed for these to constitute the DM of the Universe.

2.7 Conclusions

Supersymmetric Cold Dark Matter is offered as a plausible explanation for explaining the missing mass of the Universe. Supersymmetric Theories will be tested in the near future at LHC or other high energy machines and will confirm or refute the existence of supersymmetry with consequences of paramount importance for the future of the small scale physics. The confirmation of the existence of supersymmetric matter in the laboratory will have important consequences for Astrophysics too. The abundances of the LSP neutralinos in supersymmetric models come out to be in the right amount as required by astrophysical observations and it is the ideal candidate for explaining the DM problem. The gravitinos are also good CDM candidates but their case needs further investigation. In the most of the parameter space describing supersymmetric theories the relic density of the gravitino is less than that required by observations and significant thermal gravitino production is needed in addition to the NSP decay mechanism. The precision of the cosmological data is a valuable input for particle physics. In conjunction with the laboratory experiments they put severe phenomenological constraints which particle physicists should observe enabling them to learn more on fundamental scale physics from the Universe.

Acknowledgements

I thank the organizers for inviting me to enjoy the pleasant atmosphere of the Third Aegean School on Cosmology and my colleagues D. V. Nanopoulos and V. C. Spanos for the enjoyable collaboration we have had on the Dark Matter issues. Part of the material of this lecture is based on work done with them. This work is co-funded by the European Social Fund and National Resources – EPEAEK B – PYTHAGORAS.

References

1. C. L. Bennett et al., Astrophys. J. Suppl. **148**, 1 (2003) [arXiv:astro-ph/0302207]; D.N. Spergel et al., Astrophys. J. Suppl. 148, 175, (2003) [arXiv:astro-ph/0302209].
2. P. Colin et al., Astrophys. J **542**, 622 (2000).
3. N. Yoshida, A. Sokasian, L. Hernquist and V. Springel, Astrophys. J. **591**, L1 (2003) [arXiv: astro-ph/0312194].
4. S. Weinberg,'Gravitation and Cosmology', Principles and Applications of the General Theory of Relativity, (John Wiley & Sons, New York 1972).
5. E.W. Kolb and M.S. Turner, 'The Early Universe', (Addisson – Wesley, Redwood City, CA, 1990).
6. S. Perlmutter et al., Astrophys. J **483**, 565 (1997); B.P. Schmidt et al., Astrophys. J **507**, 46 (1998); A. G. Riess et al., Astron. J. **116**, 1009 (1998); P.M. Garnavich et al., Astrophys. J **509**, 74 (1998); S. Perlmutter et al., Astrophys. J **517**, 565 (1999).
7. P. Binetruy, C. Deffayet and D. Langlois, Nucl.Phys. B **565**, 269 (2000) [arXiv: hep-th/9905012]; P. Binetruy, C. Deffayet, U. Ellwanger and D. Langlois, Physics Letters B **477**, 285 (2000) [arXiv: hep-th/9910219].
8. Y. Fukuda et al., Phys. Rev. Lett. **81**, 1562 (1998).
9. "Searches for the Neutral Higgs Bosons of the MSSM, Preliminary Combined Results Using LEP Data Collected at Energies up to 209 GeV", The ALEPH, DELPHI, L3 and OPAL coll. and the LEP Higgs Working Group, http,//lephiggs.web.cern.ch/LEPHIGGS.
10. B. Lee and S. Weinberg, Phys. Rev. Lett. **39**, 165 (1977).
11. G. Steigman et al., Astron. J **83**, 1050 (1978); J. E. Gunn et al., Astrophys. J **223**, 1015 (1978).
12. H. Goldberg, Phys. Rev. Lett. **50**, 1419 (1983).
13. J. Ellis, J. Hagelin, D.V. Nanopoulos, K. Olive and M. Srednicki, Nucl. Phys. B **238**, 453 (1984).
14. G. Jungman, M. Kamionkowski and K. Griest, Phys. Rept. **267**, 195 (1996) [arXiv:hep-ph/9506380].
15. A.B. Lahanas, N.E. Mavromatos and D.V. Nanopoulos, Int. J. Mod. Phys. D **12**, 1529 (2003) [arXiv:hep-ph/0308251].

16. P. Gondolo and G. Gelmini, Nucl. Phys B **360**, 145 (1991).
17. J. Edsjo and P. Gondolo, Phys. Rev D **56**, 1879 (1997), [arXiv:hep-ph/9704361].
18. J.L. Lopez, D.V. Nanopoulos and K.J. Yuan, Nucl. Phys. B **370**, 445 (1992).
19. A.B. Lahanas, D.V. Nanopoulos and V.C. Spanos, Phys. Rev. D **62**, 023515 (2000) [arXiv:hep-ph/9909497].
20. P. Gondolo, L. Edsjö, L. Bergström, P. Ullio and E. Baltz, arXiv:astro-ph/0012234; P. Gondolo, L. Edsjö, L. Bergström, P. Ullio, M. Schelke and E. Baltz, arXiv:astro-ph/0211238, http://www.physto.se/ edsjo/darksusy.
21. G. Belanger, F. Boudjema, A. Pukhov and A. Semenov, arXiv:hep-ph/0405253, **micrOMEGAS: Version 1.3**.
22. K.A. Olive, D.N. Schramm and G. Steigman, Nucl. Phys. B **180**, 497 (1981).
23. P. Binetruy, G. Girardi and P. Salati, Nucl. Phys. B **237**, 285 (1984).
24. K. Griest and D. Seckel, Phys. Rev D **43**, 3191 (1991).
25. J. Edsjö, M. Schelke, P. Ullio and P. Gondolo, JCAP **04**, 001 (2003).
26. J.R. Ellis, T. Falk and K.A. Olive, Phys. Lett. B **444**, 367 (1998); [arXiv:hep-ph/9810360];
    J.R. Ellis, T. Falk, K.A. Olive and M. Srednicki, Astropart. Phys. **13**, 181 (2000)[Erratum-lect. Notes phys. **15**, 413 (2001)]; [arXiv:hep-ph/9905481];
    M. Gomez, G. Lazarides and C. Pallis, Phys. Rev. D **61**, 123512 (2000); Phys. Lett. B **487**, 313 (2000);
    J.R. Ellis, K.A. Olive and Y. Santoso, Asrtpart. Phys. **18**, 395 (2003) [arXiv: hep-ph/0112113].
    A. Djouadi, M. Drees and J. Kneur, JHEP **0108**, 055 (2001);
    S. Mizuta and M. Yamaguchi, Phys. Lett. B **298**, 120 (1993) [arXiv:hep-ph/9208251];
    R. Arnowitt, B. Dutta and Y. Santoso, Nucl. Phys. B **606**, 59 (2001) [arXiv:hep-ph/0102181];
    T. Nihei, L. Roszkowski and R. Ruiz de Austri, [arXiv:hep-ph/0206266]; JHEP **0108**, 024 (2001); JHEP **0207**, 024 (2002); [arXiv:hep-ph/0206266];
    R. Arnowitt, B. Dutta and Y. Santoso, Nucl. Phys. B **606**, 59 (2001); [arXiv:hep-ph/0102181];
    V.A. Bednyakov, H.V. Klapdor-Kleingrothaus and V. Gronewold, Phys. Rev. D **66**, 115005 (2002) [arXiv:hep-ph/0208178].
27. H.P. Nilles, Phys. Rep. **110**, 1 (1984); H.E. Haber and G.L. Kane, Phys. Rep. **117**, 75 (1985); A.B. Lahanas and D.V. Nanopoulos: **147**, 1 (1987).
28. The Joint LEP2 Supersymmetry Working Group, http://lepsusy.web.cern.ch/lepsusy/.
29. M.S. Alam et al. [CLEO Collaboration], Phys. Rev. Lett. **74**, 2885 (1995), [arXiv:hep-ex/9908023]; K. Abe et al., [Belle Collaboration], Phys. Lett. B511, 151 (2001) [arXiv:hep-ex/0103042 and hep-ex/0107065]; L. Lista [BaBar Collaboration],[arXiv:hep-ex/0110010].
30. C. Degrassi, P. Gambino and G.F. Giudice, JHEP **0012**, 009 (2000) [arXiv:hep-ph/0009337]; M. Carena, D. Garcia, U. Nierste and C.E. Wagner, Phys. Lett. B **499**, 141 (2001) [arXiv:hep-ph/0010003]; P. Gambino and M. Misiak, Nucl. Phys. B **611**, 338 (2001); D.A. Demir and K.A. Olive, Phys. Rev. D **65** 034007 (2002) [arXiv:hep-ph/0107329]; T. Hurth, Rev. Mod. Phys. **75**, 1159 (2003) [arXiv:hep-ph/0212304].
31. G.W. Bennett et al. [Muon g-2 Collaboration], Phys. Rev. Lett. **92**, 161802 (2004) [arXiv:hep-ex/0401008].

32. M. Davier, S. Eidelman, A. Hocker and Z. Zhang, Eur. Phys. J. C **31**, 503 (2003) [arXiv:hep-ph/0308213]; K. Hagiwara, A.D. Martin, D. Nomura and T. Teubner, [arXiv:hep-ph/0312250]; J.F. de Trocóniz and F.J. Ynduráin, [arXiv:hep-ph/0402285]; K. Melnikov and A. Vainshtein, [arXiv:hep-ph/0312226]; M. Passera, [arXiv:hep-ph/0411168].

33. J. Ellis, K.A. Olive, Y. Santoso and V.C. Spanos, Phys.Lett. B **565** (2003) 176–182.

34. A.B. Lahanas and D.V. Nanopoulos, Phys. Lett. B **568** (2003) 55 [arXiv:hep-ph/0303130].

35. J.R. Ellis, K.A. Olive, Y. Santoso and V.C. Spanos, Phys. Rev. D **71**, 095007 (2005) [arXiv:hep-ph/0502001].

36. J.R. Ellis, K.A. Olive, Y. Santoso and V.C. Spanos, Phys. Lett. B **603**, 51 (2004) [arXiv:hep-ph/0408118].

37. H. Baer and C. Balazs, JCAP **0305**, 006 (2003) [arXiv:hep-ph/0303114].

38. CDMS Collaboration, arXiv:astro-ph/0405033; G. Chardin et al., [EDELWEISS Collaboration], Nucl. Instrum. Meth. A **520**, 101 (2004); H.V. Klapdor-Kleingrothaus, Nucl. Phys. Proc. Suppl. **110**, 58 (2002), [arXiv:hep-ph/0206250]; D. Cline, arXiv:astro-ph/0306124; D.R. Smith and N. Weiner, Nucl. Phys. Proc. Suppl. **124**, 197 (2003), [arXiv:astro-ph/0208403].

39. U. Chattopadhyay, PASCOS 04 talk, [arXiv:hep-ph/0412168].

40. J. Ellis, J. Phys. Conf. Ser. **50**, 8 (2006) [arXiv:astro-ph/0504501].

41. R. Cyburt, J. Ellis, B. Fields and K.A.Olive, Phys. Rev. D **67**, 103521 (2003) [arXiv:astro-ph/0211258].

42. J.R. Ellis, K.A. Olive and E. Vangioni, Phys. Lett. B **619**, 30 (2005) [arXiv:astro-ph/0503023].

43. M. Bolz, A. Brandenburg and W. Buchmuller, Nucl. Phys. **606**, 218 (2001) [arXiv:hep-ph/0012051].

44. J. Ellis, K.A. Olive, Y. Santoso and V.C. Spanos, Phys. Lett. B **588**, 7 (2004) [arXiv:hep-ph/0312262].

3

On the Direct Detection of Dark Matter

John Vergados

Physics Department, University of Ioannina, Gr 451 10, Ioannina, Greece
vergados@cc.uoi.gr

Abstract. Various issues related to the direct detection of supersymmetric dark matter are reviewed. Such are: 1) Construction of supersymmetric models with a number of parameters, which are constrained from the data at low energies as well as cosmological observations. 2) A model for the nucleon, in particular the dependence on the nucleon cross section on quarks other than u and d. 3) A nuclear model, i.e. the nuclear form factor for the scalar interaction and the spin response function for the axial current. 4) Information about the density and the velocity distribution of the neutralino (halo model). Using the present experimental limits on the rates and proper inputs in 3)-4) we derive constraints in the nucleon cross section, which involves 1)-2). Since the expected event rates are extremely low we consider some additional signatures of the neutralino nucleus interaction, such as the periodic behavior of the rates due to the motion of Earth (modulation effect), which, unfortunately, is characterized by a small amplitude. This leads us to examine the possibility of suggesting directional experiments, which measure not only the energy of the recoiling nuclei but their direction as well. In these, albeit hard, experiments one can exploit two very characteristic signatures: a)large asymmetries and b) interesting modulation patterns. Furthermore we extended our study to include evaluation of the rates for other than recoil searches such as: i) Transitions to excited states, ii) Detection of recoiling electrons produced during the neutralino-nucleus interaction and iii) Observation of hard X-rays following the de-excitation of the ionized atom.

3.1 Introduction

The combined MAXIMA-1 [1], BOOMERANG [2], DASI [3], COBE/DMR Cosmic Microwave Background (CMB) observations [4], the recent WMAP data [6] and SDSS [7] imply that the Universe is flat [5] and and that most of the matter in the Universe is dark, i.e. exotic.

$$\Omega_b = 0.044 \pm 0.04, \, \Omega_m = 0.27 \pm 0.04, \, \Omega_\Lambda = 0.69 \pm 0.08$$

for baryonic matter, cold dark matter and dark energy respectively. An analysis of a combination of SDSS and WMAP data yields [7] $\Omega_m \approx 0.30 \pm 0.04 (1\sigma)$.

J. Vergados: *On the Direct Detection of Dark Matter*, Lect. Notes Phys. **720**, 69–100 (2007)
DOI 10.1007/978-3-540-71013-4_3   © Springer-Verlag Berlin Heidelberg 2007

Crudely speaking and easy to remember

$$\Omega_b \approx 0.05, \Omega_{CDM} \approx 0.30, \Omega_\Lambda \approx 0.65 .$$

Additional indirect information comes from the rotational curves [8]. The rotational velocity of an object increases so long is surrounded but matter. Once outside matter the velocity of rotation drops as the square root of the distance. Such observations are not possible in our own galaxy. The observations of other galaxies, similar to our own, indicate that the rotational velocities of objects outside the luminous matter do not drop. So there must be a halo of dark matter out there. Since the non exotic component cannot exceed 40% of the CDM [9], there is room for exotic WIMP's (Weakly Interacting Massive Particles).

In fact the DAMA experiment [10] has claimed the observation of one signal in direct detection of a WIMP, which with better statistics has subsequently been interpreted as a modulation signal [11]. These data, however, if they are due to the coherent process, are not consistent with other recent experiments, see e.g. EDELWEISS and CDMS [12]. It could still be interpreted as due to the spin cross section, but with a new interpretation of the extracted nucleon cross section. The above developments are in line with particle physics considerations. Thus, in the currently favored supersymmetric (SUSY) extensions of the standard model, the most natural WIMP candidate is the LSP, i.e. the lightest supersymmetric particle. In the most favored scenarios the LSP can be simply described as a Majorana fermion, a linear combination of the neutral components of the gauginos and Higgsinos [8, 9, 10, 11, 12, 13].

Since this particle is expected to be very massive, $m_\chi \geq 30 GeV$, and extremely non relativistic with average kinetic energy $T \leq 100 KeV$, it can be directly detected mainly via the recoiling of a nucleus (A,Z) in the elastic scattering process:

$$\chi + (A, Z) \rightarrow \chi + (A, Z)^* \tag{3.1}$$

(χ denotes the LSP). In order to compute the event rate one needs the following ingredients:

1. An effective Lagrangian at the elementary particle (quark) level obtained in the framework of supersymmetry [8, 13] and [15]. One starts with representative input in the restricted SUSY parameter space as described in the literature, e.g. Ellis et al. [14], Bottino et al., Kane et al., Castano et al. and Arnowitt et al. [15] as well as elsewhere [18, 19, 20, 21, 22, 23, 24, 25, 26, 27, 28, 29, 30, 31, 32, 33, 34, 35, 36, 37, 38, 39, 40]. We will not, however, elaborate on how one gets the needed parameters from supersymmetry, since this topic will be covered by another lecture in this school by professor Lahanas. For the reader's convenience, however, we will give a description in Sect. 3.3 of the basic SUSY ingredients needed

to calculate LSP-nucleus scattering cross section. Our own SUSY input parameters can be found elsewhere [16, 17].

2. A procedure in going from the quark to the nucleon level, i.e. a quark model for the nucleon. The results depend crucially on the content of the nucleon in quarks other than u and d. This is particularly true for the scalar couplings as well as the isoscalar axial coupling [41, 42, 43]. Such topics will be discussed in Sect. 3.4.

3. computation of the relevant nuclear matrix elements [44, 45, 46, 47, 48, 49, 50, 51] using as reliable as possible many body nuclear wave functions. By putting as accurate nuclear physics input as possible, one will be able to constrain the SUSY parameters as much as possible. The situation is a bit simpler in the case of the scalar coupling, in which case one only needs the nuclear form factor.

4. Convolution with the LSP velocity Distribution. To this end we will consider here Maxwell-Boltzmann [8] (MB) velocity distributions, with an upper velocity cut off put in by hand. Other distributions are possible, such as non symmetric ones, like those of Drukier [53] and Green [54], or non isothermal ones, e.g. those arising from late in-fall of dark matter into our galaxy, like Sikivie's caustic rings [55]. In any event in a proper treatment the velocity distribution ought to be consistent with the dark matter density in the context of the Eddington theory [56].

After this we will specialize our study in the case of the nucleus ^{127}I, which is one of the most popular targets [10, 45].

Since the expected rates are extremely low or even undetectable with present techniques, one would like to exploit the characteristic signatures provided by the reaction. Such are:

1. The modulation effect, i.e. the dependence of the event rate on the velocity of the Earth

2. The directional event rate, which depends on the velocity of the sun around the galaxy as well as the the velocity of the Earth. has recently begun to appear feasible by the planned experiments [47, 48].

3. Detection of other than nuclear recoils, such as
   - Detection of γ rays following nuclear de-excitation, whenever possible [49, 50].
   - Detection of ionization electrons produced directly in the LSP-nucleus collisions [57, 58].
   - Observations of hard X-rays produced [59], when the inner shell electron holes produced as above are filled.

In all calculations we will, of course, include an appropriate nuclear form factor and take into account the influence on the rates of the detector energy cut off. We will present our results a function of the LSP mass, m_χ, in a way which can be easily understood by the experimentalists.

3.2 The Nature of the LSP

Before proceeding with the construction of the effective Lagrangian we will briefly discuss the nature of the lightest supersymmetric particle (LSP) focusing on those ingredients which are of interest to dark matter.

In currently favorable supergravity models the LSP is a linear combination [8] of the neutral four fermions $\tilde{B}, \tilde{W}_3, \tilde{H}_1$ and $\tilde{H}_2$ which are the supersymmetric partners of the gauge bosons B_μ and W_μ^3 and the Higgs scalars H_1 and H_2. Admixtures of s-neutrinos are expected to be negligible.

In the above basis the mass-matrix is given in Lazarides' article relation (1.7) [8, 13] where $tan\beta = \langle v_2 \rangle / \langle v_1 \rangle$ is the ratio of the vacuum expectation values of the Higgs scalars H_2 and H_1 and μ is a dimensionful coupling constant which is not specified by the theory (not even its sign).

By diagonalizing the above matrix we obtain a set of eigenvalues m_j and the diagonalizing matrix C_{ij} as follows

$$\begin{pmatrix} \tilde{B}_R \\ \tilde{W}_{3R} \\ \tilde{H}_{1R} \\ \tilde{H}_{2R} \end{pmatrix} = (C_{ij}^R) \begin{pmatrix} \chi_{1R} \\ \chi_{2R} \\ \chi_{3R} \\ \chi_{4R} \end{pmatrix} \qquad \begin{pmatrix} \tilde{B}_L \\ \tilde{W}_{2L} \\ \tilde{H}_{1L} \\ \tilde{H}_{2L} \end{pmatrix} = (C_{ij}) \begin{pmatrix} \chi_{1L} \\ \chi_{2L} \\ \chi_{3L} \\ \chi_{4L} \end{pmatrix} \qquad (3.2)$$

with $C_{ij}^R = C_{ij}^* e^{i\lambda_j}$ The phases are $\lambda_i = 0, \pi$ depending on the sign of the eigenmass.

Another possibility to express the above results in photino-zino basis $\tilde{\gamma}, \tilde{Z}$ via

$$\tilde{W}_3 = sin\theta_W \tilde{\gamma} - cos\theta_W \tilde{Z} , \quad \tilde{B}_0 = cos\theta_W \tilde{\gamma} + sin\theta_W \tilde{Z} . \qquad (3.3)$$

In the absence of supersymmetry breaking ($M_1 = M_2 = M$ and $\mu = 0$) the photino is one of the eigenstates with mass M. One of the remaining eigenstates has a zero eigenvalue and is a linear combination of $\tilde{H}_1$ and $\tilde{H}_2$ with mixing $sin\beta$. In the presence of SUSY breaking terms the $\tilde{B}, \tilde{W}_3$ basis is superior since the lowest eigenstate χ_1 or LSP is primarily $\tilde{B}$. From our point of view the most important parameters are the mass m_x of LSP and the mixing $C_{j1}, j = 1, 2, 3, 4$ which yield the χ_1 content of the initial basis states.

We are now in a position to find the interaction of χ_1 with matter. We distinguish three possibilities involving Z-exchange, s-quark exchange and Higgs exchange.

3.3 The Feynman Diagrams Entering the Direct Detection of LSP

The diagrams involve Z-exchange, s-quark exchange and Higgs exchange.

3.3.1 The Z-exchange Contribution

This can arise from the interaction of Higgsinos with Z (see Fig. 3.1) which can be read from C86 of [13]

$$L = \frac{g}{cos\theta_W}\frac{1}{4}[\tilde{H}_{1R}\gamma_\mu\tilde{H}_{1R} - \tilde{H}_{1L}\gamma_\mu\tilde{H}_{1L} - (\tilde{H}_{2R}\gamma_\mu\tilde{H}_{2R} - \tilde{H}_{2L}\gamma_\mu\tilde{H}_{2L})]Z^\mu \quad (3.4)$$

Using (3.2) and the fact that for Majorana particles $\bar{\chi}\gamma_\mu\chi = 0$, we obtain

$$L = \frac{g}{cos\theta_W}\frac{1}{4}(|C_{31}|^2 - |C_{41}|^2)\bar{\chi}_1\gamma_\mu\gamma_5\chi_1 Z^\mu , \quad (3.5)$$

which leads to the effective 4-fermion interaction

$$L_{eff} = \frac{g}{cos\theta_W}\frac{1}{4}2(|C_{31}|^2 - |C_{41}|^2)(-\frac{g}{2cos\theta_W}\frac{1}{q^2 - m_Z^2}\bar{\chi}_1\gamma^\mu\gamma_5\chi_1)J_\mu^Z , \quad (3.6)$$

where the extra factor of 2 comes from the Majorana nature of χ_1. The neutral hadronic current J_λ^Z is given by

$$J_\lambda^Z = -\bar{q}\gamma_\lambda\{\frac{1}{3}sin^2\theta_W - \left[\frac{1}{2}(1 - \gamma_5) - sin^2\theta_W\right]\tau_3\}q \quad (3.7)$$

at the nucleon level it can be written as

$$\tilde{J}_\lambda^Z = -\bar{N}\gamma_\lambda\{ sin^2\theta_W - g_V(\frac{1}{2} - sin^2\theta_W)\tau_3 + \frac{1}{2}g_A\gamma_5\tau_3\}N . \quad (3.8)$$

Thus we can write

$$L_{eff} = -\frac{G_F}{\sqrt{2}}(\bar{\chi}_1\gamma^\lambda\gamma^5\chi_1)J_\lambda(Z) , \quad (3.9)$$

where

$$J_\lambda(Z) = \bar{N}\gamma_\lambda[f_V^0(Z) + f_V^1(Z)\tau_3 + f_A^0(Z)\gamma_5 + f_A^1(Z)\gamma_5\tau_3]N \quad (3.10)$$

and

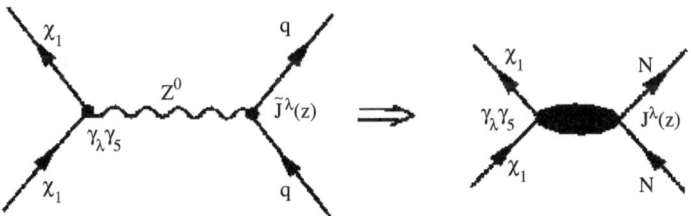

Fig. 3.1. The LSP-quark interaction mediated by Z-exchange

$$f_V^0(Z) = 2(|C_{31}|^2 - |C_{41}|^2)\frac{m_Z^2}{m_Z^2 - q^2}sin^2\theta_W \ , \qquad (3.11)$$

$$f_V^1(Z) = -2(|C_{31}|^2 - |C_{41}|^2)\frac{m_Z^2}{m_Z^2 - q^2}g_V(\frac{1}{2} - sin^2\theta_W) \ , \qquad (3.12)$$

$$f_A^0(Z) = 0 \ , \quad f_A^1(Z) = 2(|C_{31}|^2 - |C_{41}|^2)\frac{m_Z^2}{m_Z^2 - q^2}\frac{1}{2}g_A \ , \qquad (3.13)$$

with $g_V = 1.0$ and $g_A = 1.24$. We can easily see that

$$f_V^1(Z)/f_V^0(Z) = -g_V(\frac{1}{2sin^2\theta_W} - 1) \simeq -1.15 \ . \qquad (3.14)$$

Note that the suppression of this Z-exchange interaction compared to the ordinary neutral current interactions arises from the smallness of the mixing C_{31} and C_{41}, a consequence of the fact that the Higgsinos are normally quite a bit heavier than the gauginos. Furthermore, the two Higgsinos tend to cancel each other.

We should also mention that the vector contribution, the time component of which can lead to coherence, contributes only to order $v/c \approx 10^{-3}$ due to the Majorana nature of the LSP. Thus to leading order only the axial current can contribute to the direct detection of the neutralino.

3.3.2 The S-quark Mediated Interaction

The other interesting possibility arises from the other two components of χ_1, namely $\tilde{B}$ and $\tilde{W}_3$. Their corresponding couplings to s-quarks (see Fig. 3.2) can be read from the appendix C4 of [13] They are

$$L_{eff} = -g\sqrt{2}\{\bar{q}_L[T_3\tilde{W}_{3R} - tan\theta_W(T_3 - Q)\tilde{B}_R]\tilde{q}_L$$
$$-tan\theta_W\bar{q}_RQ\tilde{B}_L\tilde{q}_R\} + HC \ , \qquad (3.15)$$

where $\tilde{q}$ are the scalar quarks (SUSY partners of quarks). A summation over all quark flavors is understood. Using (3.2) we can write the above equation in the χ_i basis. Of interest to us here is the part

$$L_{eff} = g\sqrt{2}\{(tan\theta_W(T_3 - Q)C_{11}^R - T_3C_{21}^R)\tilde{q}_L\chi_{1R}\tilde{q}_L$$
$$+tan\theta_W C_{11}Q\bar{q}_R\chi_{1L}\tilde{q}_R\} \ . \qquad (3.16)$$

The above interaction is almost diagonal in the quark flavor. There exists, however, mixing between the s-quarks $\tilde{q}_L$ and $\tilde{q}_R$ (of the same flavor) i.e.

$$\tilde{q}_L = cos\theta_{\tilde{q}}\tilde{q}_1 + sin\theta_{\tilde{q}}\tilde{q}_2, \ \tilde{q}_R = -sin\theta_{\tilde{q}}\tilde{q}_1 + cos\theta_{\tilde{q}}\tilde{q}_2 \qquad (3.17)$$

with

$$tan2\theta_{\tilde{u}} = \frac{m_u(A + \mu cot\beta)}{m_{\tilde{u}_L}^2 - m_{\tilde{u}_R}^2 + m_z^2 cos2\beta/2}, \ tan2\theta_{\tilde{d}} = \frac{m_d(A + \mu tan\beta)}{m_{\tilde{d}_R}^2 - m_{\tilde{d}_R}^2 + m_Z^2 cos2\beta/2} \ . \qquad (3.18)$$

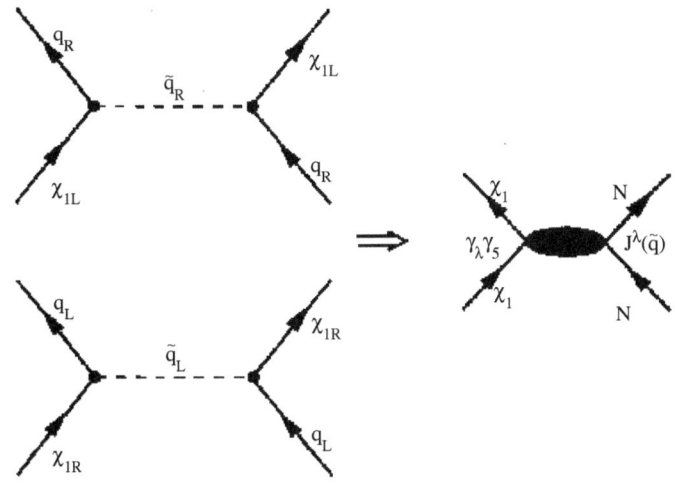

Fig. 3.2. The LSP-quark interaction mediated by s-quark exchange

Thus (3.16) becomes

$$L_{eff} = g\sqrt{2}\left\{[B_L cos\theta_{\tilde{q}}\ \bar{q}_L\chi_{1R} - B_R sin\theta_{\tilde{q}}\bar{q}_R\chi_{1L}]\tilde{q}_1 \right.$$
$$\left. + [B_L sin\theta_{\tilde{q}}\bar{q}_L\chi_{1R} + B_R cos\theta_{\tilde{q}}\bar{q}_R\chi_{1L}]\ \tilde{q}_2\right\}$$

with

$$B_L(q) = -\frac{1}{6}C_{11}^R tan\theta_w - \frac{1}{2}C_{21}^R,\quad q = u\ \ (charge\ 2/3)\,,$$
$$B_L(q) = -\frac{1}{6}C_{11}^R tan\theta_w + \frac{1}{2}C_{21}^R,\quad q = d\ \ (charge\ -1/3)\,,$$
$$B_R(q) = \frac{2}{3}tan\theta_w C_{11},\quad q = u\ \ (charge\ 2/3)\,,$$
$$B_R(q) = -\frac{1}{3}tan\theta_w C_{11},\quad q = d\ \ (charge\ -1/3)\,.$$

The effective four fermion interaction takes the form

$$L_{eff} = (g\sqrt{2})^2\{(B_L cos\theta_{\tilde{q}}\bar{q}_L\chi_{1R} - B_R sin\theta_{\tilde{q}}\bar{q}_R\chi_{1L})$$
$$\frac{1}{q^2 - m_{\tilde{q}_1^2}}(B_L cos\theta_q\bar{\chi}_{1R}q_L - B_R sin\theta_{\tilde{q}}\bar{\chi}_{1L}q_R)$$
$$+(B_L sin\theta_q q_L\chi_{1R} + cos\theta_{\tilde{q}}\bar{q}_R\chi_{1L})\frac{1}{q^2 - m_{\tilde{q}_2^2}}$$
$$(B_L sin\theta_q\bar{\chi}_{1R}q_L + B_R cos\theta_{\tilde{q}}\bar{\chi}_{1L}q_R)\}\,. \tag{3.19}$$

The above effective interaction can be written as

$$L_{eff} = L_{eff}^{LL+RR} + L_{eff}^{LR}\,. \tag{3.20}$$

The first term involves quarks of the same chirality and is not much effected by the mixing (provided that it is small). The second term involves quarks of opposite chirality and is proportional to the s-quark mixing.

The Part L_{eff}^{LL+RR}

Employing a Fierz transformation L_{eff}^{LL+RR} can be cast in the more convenient form

$$
\begin{aligned}
L_{eff}^{LL+RR} = {} & (g\sqrt{2})^2 2(-\frac{1}{2})\{|B_L|^2 \\
& (\frac{cos^2\theta_{\tilde{q}}}{q^2 - m_{\tilde{q}_1}^2} + \frac{sin^2\theta_{\tilde{q}}}{q^2 - m_{\tilde{q}_2}^2})\bar{q}_L\gamma_\lambda q_L\chi_{1R}\gamma^\lambda\chi_{1R} \\
& + |B_R|^2 (\frac{sin^2\theta_{\tilde{q}}}{q^2 - m_{\tilde{q}_1}^2} + \frac{cos^2\theta_{\tilde{q}}}{q^2 - m_{\tilde{q}_2}^2})\bar{q}_R\gamma_\lambda q_R\chi_{1L}\gamma^\lambda\chi_{1L}\} \, . \quad (3.21)
\end{aligned}
$$

The factor of 2 comes from the Majorana nature of LSP and the (-1/2) comes from the Fierz transformation. Equation (3.21) can be written more compactly as

$$
\begin{aligned}
L_{eff} = {} & -\frac{G_F}{\sqrt{2}}2\{\bar{q}\gamma_\lambda(\beta_{0R} + \beta_{3R}\tau_3)(1 + \gamma_5)q \\
& - \bar{q}\gamma_\lambda(\beta_{0L} + \beta_{3L}\tau_3)(1 - \gamma_5)q\}(\bar{\chi}_1\gamma^\lambda\gamma^5\chi_1) \, , \quad (3.22)
\end{aligned}
$$

with

$$
\begin{aligned}
\beta_{0R} &= \left(\frac{4}{9}\chi_{\tilde{u}_R}^2 + \frac{1}{9}\chi_{\tilde{d}_R}^2\right)|C_{11}tan\theta_W|^2 \, , \\
\beta_{3R} &= \left(\frac{4}{9}\chi_{\tilde{u}_R}^2 - \frac{1}{9}\chi_{\tilde{d}_R}^2\right)|C_{11}tan\theta_W|^2 \, , \quad (3.23) \\
\beta_{0L} &= |\frac{1}{6}C_{11}^R tan\theta_W + \frac{1}{2}C_{21}^R|^2\chi_{\tilde{u}_L}^2 + |\frac{1}{6}C_{11}^R tan\theta_W - \frac{1}{2}C_{21}^R|^2\chi_{\tilde{d}_L}^2 \, , \\
\beta_{3L} &= |\frac{1}{6}C_{11}^R tan\theta_W + \frac{1}{2}C_{21}^R|^2\chi_{\tilde{u}_L}^2 - |\frac{1}{6}C_{11}^R tan\theta_W - \frac{1}{2}C_{21}^R|^2\chi_{\tilde{d}_L}^2 \, ,
\end{aligned}
$$

with

$$
\chi_{qL}^2 = c_{\tilde{q}}^2\frac{m_W^2}{m_{\tilde{q}_1}^2 - q^2} + s_{\tilde{q}}^2\frac{m_W^2}{m_{\tilde{q}_2}^2 - q^2} \, , \quad \chi_{qR}^2 = s_{\tilde{q}}^2\frac{m_W^2}{m_{\tilde{q}_1}^2 - q^2} + c_{\tilde{q}}^2\frac{m_W^2}{m_{\tilde{q}_2}^2 - q^2} \, , \quad (3.24)
$$

where $c_{\tilde{q}} = cos\theta_{\tilde{q}}$, $s_{\tilde{q}} = sin\theta_{\tilde{q}}$. The above parameters are functions of the four-momentum transfer which in our case is negligible.
Equation (3.22) can be explicitly rewritten [30] as:

$$
L_{eff} = -\frac{G_F}{\sqrt{2}}2 \left[\bar{u}\gamma_\lambda(d^0(u) + \gamma_5 d(u))u + \bar{d}\gamma_\lambda(d^0(d) + \gamma_5 d(d))d\right] (\bar{\chi}_1\gamma^\lambda\gamma^5\chi_1) \quad (3.25)
$$

where

$$
d^0(u) = \beta_{0R} + \beta_{3R} - \beta_{0L} - \beta_{3L}, d(u) = \beta_{0R} + \beta_{3R} + \beta_{0L} + \beta_{3L} \, , \quad (3.26)
$$

$$d^0(d) = \beta_{0R} - \beta_{3R} - \beta_{0L} + \beta_{3L}, d(u) = \beta_{0R} - \beta_{3R} + \beta_{0L} - \beta_{3L} . \qquad (3.27)$$

Proceeding as in Sect. 3.3.1 we can obtain the effective Lagrangian at the nucleon level as

$$L_{eff}^{LL+RR} = -\frac{G_F}{\sqrt{2}}(\bar{\chi}_1\gamma^\lambda\gamma^5\chi_1)J_\lambda(\tilde{q}) , \qquad (3.28)$$

$$J_\lambda(\tilde{q}) = \bar{N}\gamma_\lambda\{f_V^0(\tilde{q}) + f_V^1(\tilde{q})\tau_3 + f_A^0(\tilde{q})\gamma_5 + f_A^1(\tilde{q})\gamma_5\tau_3\}N , \qquad (3.29)$$

with

$$f_V^0 = 6(\beta_{0R} - \beta_{0L}) , \qquad f_V^1 = 2g_V(\beta_{3R} - \beta_{3L})$$
$$f_A^0 = 2g_A^0(\beta_{0R} + \beta_{0L}) , \qquad f_A^1 = 2g_A(\beta_{3R} + \beta_{3L}) \qquad (3.30)$$

with $g_v = 1.0$ and $g_A = 1.25$. The quantity g_A^0 depends on the quark model for the nucleon. It can be anywhere between 0.12 and 1.00 (see below 3.4.2).

We should note that this interaction is more suppressed than the ordinary weak interaction by the fact that the masses of the s-quarks are usually larger than that of the gauge boson Z^0. In the limit in which the LSP is a pure bino ($C_{11} = 1, C_{21} = 0$) we obtain

$$\beta_{0R} = tan^2\theta_W\left(\frac{4}{9}\chi_{\tilde{u}_R}^2 + \frac{1}{9}\chi_{\tilde{d}_R}^2\right) , \ \beta_{3R} = tan^2\theta_W\left(\frac{4}{9}\chi_{\tilde{u}_R}^2 - \frac{1}{9}\chi_{\tilde{d}_R}^2\right) , \quad (3.31)$$

$$\beta_{0L} = \frac{tan^2\theta_W}{36}(\chi_{\tilde{u}_L}^2 + \chi_{\tilde{d}_L}^2) , \ \beta_{3L} = \frac{tan^2\theta_W}{36}(\chi_{\tilde{u}_L}^2 - \chi_{\tilde{d}_L}^2) . \qquad (3.32)$$

Assuming further that $\chi_{\tilde{u}_R} = \chi_{\tilde{d}_R} = \chi_{\tilde{u}_L} = \chi_{\tilde{d}_L}$ we obtain

$$f_V^1(\tilde{q})/f_V^0(\tilde{q}) \simeq +\frac{2}{9}, \ f_A^1(\tilde{q})/f_A^0(\tilde{q}) \simeq +\frac{6}{11} . \qquad (3.33)$$

If, on the other hand, the LSP were the photino ($C_{11} = cos\theta_W$, $C_{21} = sin\theta_W$, $C_{31} = C_{41} = 0$) and the s-quarks were degenerate there would be no coherent contribution ($f_V^0 = 0$ if $\beta_{0L} = \beta_{0R}$).

As we have mentioned in the previous section, to leading order, only the axial current contributes to the direct detection of the neutralino.

The Part L_{eff}^{LR}

From (3.19) we obtain

$$L_{eff}^{LR} = -(g\sqrt{2})^2 sin2\theta_{\tilde{q}}B_L(q)B_R(q)\frac{1}{2}[\frac{1}{q^2 - m_{\tilde{q}_1^2}} - \frac{1}{q^2 - m_{\tilde{q}_2^2}}] \qquad (3.34)$$

$$(\bar{q}_L\chi_{1R}\bar{\chi}_{1L}q_R + \bar{q}_R\chi_{1L}\bar{\chi}_{1R}q_L) .$$

Employing a Fierz transformation we can cast it in the form

$$L_{eff} = -\frac{G_F}{\sqrt{2}} \sum_q \beta(q) \left[(\bar{q}q\bar{\chi}_1\chi_1 + \bar{q}\gamma_5 q\bar{\chi}_1\gamma_5\chi_1 - (\bar{q}\sigma_{\mu\nu}q)(\bar{\chi}_1\sigma^{\mu\nu}\chi_1)) \right] ,$$

$$(3.35)$$

where

$$\beta(u) = \frac{2}{3} tan\theta_W C_{11} \{ 2sin2\theta_{\tilde{u}} [\frac{1}{6} CR_{11} tan\theta_W + \frac{1}{2} CR_{21}] \Delta_{\tilde{u}} , \qquad (3.36)$$

$$\beta(d) = sin2\theta_{\tilde{d}} [\frac{1}{6} CR_{11} tan\theta_W - \frac{1}{2} C_{21}^R] \Delta_{\tilde{d}} \} . \qquad (3.37)$$

Where in the last expressions u indicates quarks with charge 2/3 and d quarks with charge -1/3. In all cases

$$\Delta_{\tilde{u}} = \frac{(m_{\tilde{u}_1}^2 - m_{\tilde{u}_2}^2) M_W^2}{(m_{\tilde{u}_1}^2 - q^2)(m_{\tilde{u}_2}^2 - q^2)}$$

and an analogous equation for $\Delta_{\tilde{d}}$.

The appearance of scalar terms in s-quark exchange [22] has been first noticed by Griest [19]. It has also been noticed there that one should consider explicitly the effects of quarks other than u and d [41] in going from the quark to the nucleon level. We first notice that with the exception of t s-quark the $\tilde{q}_L - \tilde{q}_R$ mixing small. Thus

$$sin2\theta_{\tilde{u}}\Delta\tilde{u} \simeq \frac{2m_u(A + \mu cot\beta)m_W^2}{(m_{\tilde{u}_L}^2 - q^2)(m_{\tilde{u}_R}^2 - q^2)} , \quad sin2\theta_{\tilde{d}}\Delta\tilde{d} \simeq \frac{2m_d(A + \mu tan\beta)m_W^2}{(m_{\tilde{d}_L}^2 - q^2)(m_{\tilde{d}_R}^2 - q^2)} .$$

$$(3.38)$$

In going to the nucleon level and ignoring the negligible pseudoscalar and tensor components we only need modify the above expressions for all quarks, with the possible exception of the t quarks, by the substitution $m_q \rightarrow f_q m_N$ (see Sect. 3.4.1). For the t s-quark the mixing is complete, which implies that the amplitude is independent of the top quark mass. Hence in the case of the top quark we may not get an extra enhancement in going from the quark to the nucleon level. In any case this way we get

$$L_{eff} = \frac{G_F}{\sqrt{2}} [f_S^0(\tilde{q})\bar{N}N + f_S^1(\tilde{q})\bar{N}\tau_3 N]\bar{\chi}_1\chi_1 \qquad (3.39)$$

with

$$f_S^0(\tilde{q}) = \frac{f_u\beta(u) + f_d\beta(d)}{2} + \sum_{q=s,c,b,t} f_q\beta(q) , \qquad (3.40)$$

$$f_S^1(\tilde{q}) = \frac{f_u\beta(u) - f_d\beta(d)}{2} . \qquad (3.41)$$

(see Sect. 3.4.1 for details). In the allowed SUSY parameter space considered in this work this contribution can be neglected in front of the Higgs exchange contribution. This happens because for quarks other than t the s-quark mixing is small. For the t-quark, as it has already been mentioned, we have large mixing, but we do not get the advantage of the mass enhancement.

3.3.3 The Intermediate Higgs Contribution

The coherent scattering can be mediated via the intermediate Higgs particles which survive as physical particles (see Fig. 3.3). The relevant interaction can arise out of the Higgs-Higgsino-gaugino interaction which takes the form

$$L_{H\chi\chi} = \frac{g}{\sqrt{2}}\left(\tilde{\bar{W}}_R^3\tilde{H}_{2L}H_2^{0*} - \tilde{\bar{W}}_R^3\tilde{H}_{1L}H_1^{0*}\right.$$

$$\left. -tan\theta_w(\tilde{\bar{B}}_R\tilde{H}_{2L}H_2^{0*} - \tilde{\bar{B}}_R\tilde{H}_{1L}H_1^{0*})\right) + H.C. \qquad (3.42)$$

Proceeding as above we can express $\tilde{W}$ an $\tilde{B}$ in terms of the appropriate eigenstates and retain the LSP to obtain

$$L = \frac{g}{\sqrt{2}}\left((C_{21}^R - tan\theta_w C_{11}^R)C_{41}\bar{\chi}_{1R}\chi_{1L}H_2^{o*}\right.$$

$$\left. -(C_{21}^R - tan\theta_w C_{11}^R)C_{31}\bar{\chi}_{1R}\chi_{1L}H_1^{o*}\right) + H.C. \qquad (3.43)$$

We can now proceed further and express the fields H_1^{0*}, H_2^{0*} in terms of the physical fields h, H and A. The term which contains A will be neglected, since it yields only a pseudoscalar coupling which does not lead to coherence.

Thus we can write

$$\mathcal{L}_{eff} = -\frac{G_F}{\sqrt{2}}\bar{\chi}\chi\,\bar{N}[f_s^0(H) + f_s^1(H)\tau_3]N \qquad (3.44)$$

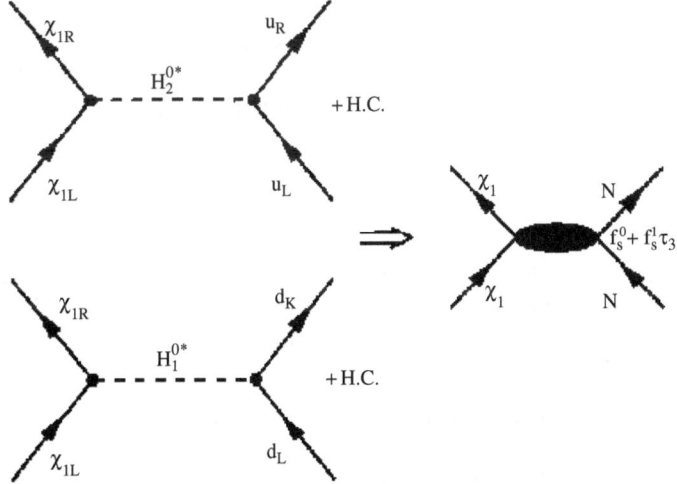

Fig. 3.3. The LSP-quark interaction mediated by Higgs exchange

where

$$f_S^0(H) = \frac{1}{2}(g_u + g_d) + g_s + g_c + g_b + g_t , \qquad (3.45)$$

$$f_S^1(H) = \frac{1}{2}(g_u - g_d) , \qquad (3.46)$$

with

$$g_{a_i} = \left[g_1(h)\frac{cos\alpha}{sin\beta} + g_2(H)\frac{sin\alpha}{sin\beta}\right]\frac{m_{a_i}}{m_N} , \quad a_i = u, c, t \qquad (3.47)$$

$$g_{\kappa_i} = \left[- g_1(h)\frac{sin\alpha}{cos\beta} + g_2(H)\frac{cos\alpha}{cos\beta}\right]\frac{m_{\kappa_i}}{m_N} , \quad \kappa_i = d, s, b \qquad (3.48)$$

$$g_1(h) = 4(C_{11}^R tan\theta_W - C_{21}^R)(C_{41}cos\alpha + C_{31}sin\alpha)\frac{m_N m_W}{m_h^2 - q^2} , \qquad (3.49)$$

$$g_2(H) = 4(C_{11}^R tan\theta_W - C_{21}^R)(C_{41}sin\alpha - C_{31}cos\alpha)\frac{m_N m_W}{m_H^2 - q^2} , \qquad (3.50)$$

where m_N is the nucleon mass, and the parameters m_h, m_H and α depend on the SUSY parameter space (see Table 1).

3.4 Going from the Quark to the Nucleon Level

As we have already mentioned, one has to be a bit more careful in handling quarks other than u and d.

3.4.1 The Scalar Interaction

As we have seen the scalar couplings of the LSP to the quarks are proportional to their mass [41]. One encounters in the nucleon not only sea quarks ($u\bar{u}, d\bar{d}$ and $s\bar{s}$) but the heavier quarks also due to QCD effects [42]. This way one obtains the scalar Higgs-nucleon coupling by using effective parameters f_q defined as follows:

$$\left\langle N|m_q\bar{q}q|N\right\rangle = f_q m_N , \qquad (3.51)$$

where m_N is the nucleon mass. The parameters f_q, $q = u, d, s$ can be obtained by chiral symmetry breaking terms in relation to phase shift and dispersion analysis. The isoscalar component can be obtained by considering the following quantities:

1. The phenomenologically determined mass ratios:

$$\frac{m_u}{m_d} = 0.553 \pm 0.043, \quad \frac{m_s}{m_d} = 18.9 \pm 0.08 . \qquad (3.52)$$

2. The quantities:

$$z = \frac{B_u - B_s}{B_d - B_s} \approx 1.49, \quad y = \frac{2B_s}{B_d + B_u} \,. \tag{3.53}$$

One then finds that:

$$\frac{B_u}{B_d} = \frac{2z - (z-1)y}{2 + (z-1)y} \text{ for protons}, \quad \frac{B_u}{B_d} = \frac{2 + (z-1)y}{2z - (z-1)y} \text{ for neutrons}$$
$$\tag{3.54}$$

with $B_q = <N|\bar{q}q|N>$.

3. The pion-nucleon sigma-term, $\sigma_{\pi N}$:
   this term is obtained from the isospin even π-N scattering amplitude with vanishing external momenta. It is defined by the scalar quark density operator averaged over the nucleon or equivalently by $\sigma_{\pi N}(t = 0)$, the scalar form factor of the nucleon at zero momentum transfer squared. The value of the sigma term is deduced from the analysis of two quantities: $\sigma_{\pi N}(t = 2M_\pi^2)$ the scalar form factor at the Cheng-Dashen point, which is experimentally accessible, and the difference $\Delta_\sigma = \sigma_{\pi N}(2M_\pi^2) - \sigma_{\pi N}(0) = 15.2 \pm 0.4$ MeV [60, 61] as induced by explicit chiral symmetry breaking. Experimentally, after efforts of many years, the value of the sigma-tern is still quite uncertain [61]. The canonical value of the πN sigma term with

$$\sigma_{\pi N} = \frac{m_u + m_d}{2}(B_u + B_d) = (45 \pm 8) \; MeV \tag{3.55}$$

is deduced from an earlier analyses with $\sigma_{\pi N}(t = 2M_\pi^2) = 60 \pm 8$ [60]. During the last few years analyses of also more recent pion-nucleon scattering data lead to an increase in the value of scalar form factor at the Cheng-Dashen point $\sigma_{\pi N}(M_\pi^2)$ with 88 ± 15 MeV [62], 71 ± 9 MeV [63], 79 ± 7 MeV [64] and $(80 - 90)$ MeV [65]. Thus the recent analyses suggest a value for the pion-nucleon sigma term of about

$$\sigma_{\pi N} = \frac{m_u + m_d}{2}(B_u + B_d) = (56 - 75) \; MeV \,. \tag{3.56}$$

4. Theoretical analysis of the $\sigma_{\pi N}$ term:
   In the context of chiral perturbation one can show that:

$$\sigma_{\pi N} = \frac{\sigma_0}{1 - y} \,, \quad \sigma_0 = (35 \pm 5) \; MeV \,. \tag{3.57}$$

Equations (3.55) and (3.56) together with (3.57) will provide the range of variation in the parameter y. Taking:

$$m_u = 5.1 \; MeV, \; m_d = 9.3 \; MeV \tag{3.58}$$

together with y will in turn provide by (3.54) the range of variation of the ratio B_u/B_d. The uncertainties in (3.55, 3.56, 3.57) provide a wide range in which the parameter y can vary. In other words the experimental and theoretical uncertainties permit us, we will exploit the possible consequences of variation in y to SUSY dark matter detection. For $\sigma_{\pi N}$ we choose 45, 55, 65 and 75 MeV to reflect the range of values set by (3.55) and (3.56). Thus from (3.57) we extract the corresponding y parameters with 0.22 ± 0.11, 0.36 ± 0.09, 046 ± 0.08 and 0.53 ± 0.07, respectively. Then we will combine these values with (3.54) to get the desired values of f_q given below.

From the above analysis we get in the case of the proton:

$$f_d^p = \frac{\Sigma_{\pi N}}{0.756\, m_p}[1 + \frac{2z - (z-1)y}{2 + (z-1)y}]^{-1} , \qquad (3.59)$$

$$f_u^p = 0.553\, f_d^p\, [\frac{2z - (z-1)y}{2 + (z-1)y}], \quad f_s^p = \frac{\Sigma_{\pi N}}{0.756\, m_p}\frac{19}{2}\, y , \qquad (3.60)$$

$$f_s^p = \frac{\Sigma_{\pi N}}{0.756\, m_p}\frac{19}{2}\, y . \qquad (3.61)$$

In the case of the neutron our expressions are analogous, the ratio B_u/B_d getting the inverse value.

For the heavy quarks, to leading order via quark loops and gluon exchange with the nucleon, one finds:

$$f_Q = 2/27(1 - \sum_q f_q).$$

There is a correction to the above parameters coming from loops involving s-quarks [42]. The leading contribution can be absorbed into the definition, if the functions $g_1(h)$ and $g_2(H)$ as follows:

$$g_1(h) \rightarrow g_1(h)[1 + \tfrac{1}{8}(2\frac{m_Q^2}{m_W^2} - \frac{sin(\alpha+\beta)}{cos^2\theta_W}\frac{sin\beta}{cos\alpha})] ,$$

$$g_2(H) \rightarrow g_1(h)[1 + \tfrac{1}{8}(2\frac{m_Q^2}{m_W^2} + \frac{cos(\alpha+\beta)}{cos^2\theta_W}\frac{sin\beta}{sin\alpha})] ,$$

for $Q = c$ and t For the b-quark we get:

$$g_1(h) \rightarrow g_1(h)[1 + \tfrac{1}{8}(2\frac{m_b^2}{m_W^2} - \frac{sin(\alpha+\beta)}{cos^2\theta_W}\frac{cos\beta}{cos\alpha})] ,$$

$$g_2(H) \rightarrow g_1(h)[1 + \tfrac{1}{8}(2\frac{m_b^2}{m_W^2} - \frac{cos(\alpha+\beta)}{cos^2\theta_W}\frac{cos\beta}{sin\alpha})] .$$

In addition to the above effects one has to consider QCD effects. These effects renormalize the contribution of the quark loops as follows [42]:

$$f_{QCD}(q) = \tfrac{1}{4}\frac{\beta(\alpha_s)}{1+\gamma_m(\alpha_s)}$$

with

$$\beta(\alpha_s) = \tfrac{\alpha_s}{3\pi}[1 + \tfrac{19}{4}\alpha_s\pi] , \; \gamma_m(\alpha_s) = 2\tfrac{\alpha_s}{\pi} .$$

Thus

$$f_{QCD}(q) = 1 + \tfrac{11}{4}\tfrac{\alpha_s}{\pi} \ .$$

The QCD correction associated with the s-quark loops is:

$$f_{QCD}(\tilde{q}) = 1 + \tfrac{25}{6}\tfrac{\alpha_s}{\pi} \ .$$

The above corrections depend on Q since α_s must be evaluated at the scale of m_Q.

It convenient to introduce the factor $f_{QCD}(\tilde{q})/f_{QCD}(q)$ into the factors $g_1(h)$ and $g_2(H)$ and the factor of $f_{QCD}(q)$ into the the quantities f_Q. If, however, one restricts oneself to the large $tan\beta$ regime, the corrections due to the s-quark loops is independent of the parameters α and β and significant only for the t-quark.

For a more detailed discussion we refer the reader to [41, 42]. We thus obtain the results presented in Table 3.1.

We notice that there exist differences between the proton and neutron components. These, however, cannot be taken as the sole contribution to isovector contribution, since all quantities were derived with isoscalar operators. So the isovector contribution will be discussed elsewhere. Here we will limit ourselves to the isoscalar component $f_q = (f_q^p + f_q^n)/2$.

3.4.2 The Axial Current Contribution

The amplitudes $a_p = f_A^0 + f_A^1$ and $a_n = f_A^0 - f_A^1$ are defined by [66]:

$$a_N = \sum_{q=u,d,s} d_q \Delta q_N \ , \tag{3.62}$$

$$2s_\mu \Delta q_N = \langle N|\bar{q}\gamma_\mu\gamma_5 q|N\rangle \ , \tag{3.63}$$

Table 3.1. The parameters f_q^p and f_Q^p (upper part) as well as f_q^n and f_Q^n (lower part) for the twelve cases discussed in the text

#	f_d^p	f_u^p	f_s^p	f_c^p	f_b^p	f_t^p	f_d^n	f_u^n	f_s^n	f_c^n	f_b^n	f_t^n
1	0.026	0.021	0.067	0.098	0.104	0.161	0.037	0.014	0.066	0.098	0.104	0.161
2	0.027	0.020	0.133	0.087	0.092	0.144	0.037	0.015	0.133	0.086	0.092	0.143
3	0.028	0.020	0.199	0.075	0.080	0.126	0.036	0.015	0.199	0.075	0.080	0.126
4	0.033	0.025	0.199	0.078	0.083	0.132	0.044	0.018	0.199	0.077	0.083	0.122
5	0.034	0.024	0.265	0.068	0.072	0.117	0.044	0.019	0.265	0.067	0.172	0.117
6	0.031	0.025	0.332	0.057	0.061	0.106	0.043	0.017	0.332	0.057	0.062	0.102
7	0.040	0.028	0.331	0.061	0.065	0.109	0.051	0.022	0.331	0.060	0.065	0.109
8	0.041	0.028	0.400	0.051	0.055	0.095	0.051	0.023	0.400	0.050	0.055	0.095
9	0.047	0.028	0.470	0.041	0.047	0.081	0.051	0.023	0.400	0.050	0.055	0.095
10	0.047	0.027	0.462	0.045	0.050	0.090	0.050	0.023	0.470	0.040	0.044	0.060
11	0.048	0.032	0.532	0.036	0.040	0.076	0.058	0.027	0.532	0.035	0.040	0.076
12	0.049	0.032	0.603	0.026	0.030	0.063	0.057	0.027	0.603	0.026	0.030	0.063

where s_μ is the nucleon spin and d_q the relevant spin amplitudes at the quark level obtained in a given SUSY model.

The isoscalar and the isovector axial current couplings at the nucleon level, f_A^0, f_A^1, are obtained from the corresponding ones given by the SUSY models at the quark level, $f_A^0(q)$, $f_A^1(q)$, via renormalization coefficients g_A^0, g_A^1, i.e. $f_A^0 = g_A^0 f_A^0(q)$, $f_A^1 = g_A^1 f_A^1(q)$. The renormalization coefficients are given terms of Δq defined above [66], via the relations

$$g_A^0 = \Delta u + \Delta d + \Delta s = 0.77 - 0.49 - 0.15 = 0.13 \ , \ g_A^1 = \Delta u - \Delta d = 1.26.$$

We see that, barring very unusual circumstances at the quark level, the isoscalar contribution is negligible. It is for this reason that one might prefer to work in the isospin basis.

3.5 The Nucleon Cross Sections

With the above ingredients we are in a position to evaluate the nucleon cross sections.

– The scalar cross section. As we have mentioned this is primarily due to the Higgs exchange.

$$\sigma_{p,\chi^0}^S = \sigma_0 |f_S^0 + f_S^1|^2 \ , \ \sigma_{n,\chi^0}^S = \sigma_0 |f_S^0 - f_S^1|^2 \tag{3.64}$$

with

$$\sigma_0 = \frac{1}{2\pi}(G_F m_p)^2 = 0.77 \times 10^{-38} cm^2 = 0.77 \times 10^{-2} pb \ . \tag{3.65}$$

Since, however, the process is dominated by quarks other than u and d, the isovector contribution is negligible. So we can talk about the nucleon cross section.

– The proton spin cross section is given by:

$$\sigma_{p,\chi^0}^{spin} = 3\sigma_0 |f_A^0 + f_A^1|^2 = 3\sigma_0 |a_p|^2 \ . \tag{3.66}$$

3.6 The Allowed SUSY Parameter Space

It is clear from the above discussion that the nucleon cross section depends:

– The the quark structure of the nucleon
   The allowed range of the parameters f_q may induce variations in the nucleon cross section as large as an order of magnitude.
– The parameters of supersymmetry.
   This is the most crucial input. One starts with a set of parameters at the GUT scale and predicts the low energy observable via the renormalization group equations (RGE). Conversely starting from the low energy phenomenology one can constrain the input parameters at the GUT scale.

The parameter space is the most crucial. In SUSY models derived from minimal SUGRA the allowed parameter space is characterized at the GUT scale by five parameters:

- two universal mass parameters, one for the scalars, m_0, and one for the fermions, $m_{1/2}$.
- $tan\beta$.
- The trilinear coupling A_0 (or m_t^{pole}) and
- The sign of μ in the Higgs self-coupling $\mu H_1 H2$.

The experimental constraints are

1. The LSP relic abundance (including co-annihilations):

$$0.09 \leq \Omega_{LSP}h^2 \leq 0.22 \text{ (previous)}, \ 0.09 \leq \Omega_{LSP}h^2 \leq 0.124 \text{ (WMAP)}.$$

2. the $b \rightarrow s\gamma$ constraint (CLEO, BELLE)

$$2 \times 10^{-4} \leq BR \leq 4.1 \times 10^{-4}.$$

3. The Higgs mass: $\geq 114.1 \ GeV$. This applies on the standard model Higgs. So For SUSY one must correct for factor $\sin^2(\alpha - \beta)$ where α is the Higgs mixing angle. So this imposes limits on $\tan\beta$

4. Limits on $g_s - 2$ (e^-, e^+) experiments (E821 at BNL)

$$a_\mu = (g_\mu - 2)/2 = (33.7 \pm 11.2) \times 10^{-10}$$

yields (2σ level):

$$11.3 \times 10^{-10} \leq \delta a_\mu(SUGRA) \leq 56.1 \times 10^{-10}$$

5. The fermion masses:
$m_t(pole) = 175 \ GeV$, $m_b(m_b) = 4.25 \ GeV \Rightarrow$
$m_b(m_Z) = 2.888 \ GeV$, $m_\tau(M_Z) = 1.7463 \ GeV$.

We are not going to elaborate further on this interesting aspect, since this is covered by A. Lahanas' contribution.

3.7 Rates

The differential non directional rate can be written as

$$dR_{undir} = \frac{\rho(0)}{m_\chi} \frac{m}{Am_N} d\sigma(u, v)|v| , \qquad (3.67)$$

where A is the nuclear mass number, $\rho(0) \approx 0.3 GeV/cm^3$ is the LSP density in our vicinity, m is the detector mass, m_χ is the LSP mass and $d\sigma(u, v)$ is the differential cross section.

The directional differential rate, i.e. that obtained, if nuclei recoiling in the direction $\hat{e}$ are observed, is given by [26]:

$$dR_{dir} = \frac{\rho(0)}{m_\chi} \frac{m}{Am_N} |v| \hat{v}.\hat{e} \; \Theta(\hat{v}.\hat{e}) \; \frac{1}{2\pi} \; d\sigma(u,v) \; \delta(\frac{\sqrt{u}}{\mu_r v \sqrt{2}} - \hat{v}.\hat{e}) , \qquad (3.68)$$

where $\Theta(x)$ is the Heaviside function.

The differential cross section is given by:

$$d\sigma(u,v) = \frac{du}{2(\mu_r b v)^2} [(\bar{\Sigma}_S F(u)^2 + \bar{\Sigma}_{spin} F_{11}(u)] , \qquad (3.69)$$

where u the energy transfer Q in dimensionless units given by

$$u = \frac{Q}{Q_0} , \quad Q_0 = [m_p A b]^{-2} = 40 A^{-4/3} \; MeV \qquad (3.70)$$

with b is the nuclear (harmonic oscillator) size parameter. $F(u)$ is the nuclear form factor and $F_{11}(u)$ is the spin response function associated with the isovector channel.

The scalar cross section is given by:

$$\bar{\Sigma}_S = (\frac{\mu_r}{\mu_r(p)})^2 \sigma^S_{p,\chi^0} A^2 \left[\frac{1 + \frac{f_S^1}{f_S^0} \frac{2Z-A}{A}}{1 + \frac{f_S^1}{f_S^0}} \right]^2 \approx \sigma^S_{N,\chi^0} (\frac{\mu_r}{\mu_r(p)})^2 A^2 \qquad (3.71)$$

(since the heavy quarks dominate the isovector contribution is negligible). σ^S_{N,χ^0} is the LSP-nucleon scalar cross section. The spin Cross section is given by:

$$\bar{\Sigma}_{spin} = (\frac{\mu_r}{\mu_r(p)})^2 \sigma^{spin}_{p,\chi^0} \zeta_{spin}, \zeta_{spin} = \frac{1}{3(1 + \frac{f_A^0}{f_A^1})^2} S(u) , \qquad (3.72)$$

$$S(u) \approx S(0) = [(\frac{f_A^0}{f_A^1} \Omega_0(0))^2 + 2 \frac{f_A^0}{f_A^1} \Omega_0(0) \Omega_1(0) + \Omega_1(0))^2] . \qquad (3.73)$$

The couplings f_A^1 (f_A^0) and the nuclear matrix elements $\Omega_1(0)$ ($\Omega_0(0)$) associated with the isovector (isoscalar) components are normalized so that, in the case of the proton at $u = 0$, they yield $\zeta_{spin} = 1$.

With these definitions in the proton neutron representation we get:

$$\zeta_{spin} = \frac{1}{3} S'(0) , \qquad (3.74)$$

$$S'(0) = \left[(\frac{a_n}{a_p} \Omega_n(0))^2 + 2 \frac{a_n}{a_p} \Omega_n(0) \Omega_p(0) + \Omega_p^2(0) \right] , \qquad (3.75)$$

where $\Omega_p(0)$ and $\Omega_n(0)$ are the proton and neutron components of the static spin nuclear matrix elements. In extracting limits on the nucleon cross sections from the data we will find it convenient to write:

$$\sigma_{p,\chi^0}^{spin} \, \zeta_{spin} = \frac{\Omega_p^2(0)}{3}|\sqrt{\sigma_p} + \frac{\Omega_n}{\Omega_p}\sqrt{\sigma_n}e^{i\delta}|^2 \,. \tag{3.76}$$

In (3.76) δ the relative phase between the two amplitudes a_p and a_n. The static spin matrix elements are obtained in the context of a given nuclear model. Some such matrix elements of interest to the planned experiments are given in Table 3.2. The shown results are obtained from Divari [46], Ressel et al. (*) [44], the Finish group (**) [51] and the Ioannina team (+) [22, 52].

The spin ME are defined as follows:

$$\Omega_p(0) = \sqrt{\frac{J+1}{J}} \prec J \, J|\sigma_z(p)|J \, J \succ \,, \quad \Omega_n(0) = \sqrt{\frac{J+1}{J}} \prec J \, J|\sigma_z(n)|J \, J \succ \,. \tag{3.77}$$

where J is the total angular momentum of the nucleus and $\sigma_z = 2S_z$. The spin operator is defined by $S_z(p) = \sum_{i=1}^{Z} S_z(i)$, i.e. a sum over all protons in the nucleus, and $S_z(n) = \sum_{i=1}^{N} S_z(i)$, i.e. a sum over all neutrons. Furthermore

$$\Omega_0(0) = \Omega_p(0) + \Omega_n(0) \,, \quad \Omega_1(0) = \Omega_p(0) - \Omega_n(0) \,. \tag{3.78}$$

3.8 Expressions for the Rates

To obtain the total rates one must fold with LSP velocity distribution and integrate the above expressions over the energy transfer from Q_{min} determined by the detector energy cutoff to Q_{max} determined by the maximum LSP velocity (escape velocity, put in by hand in the Maxwellian distribution), i.e. $v_{esc} = 2.84 \, v_0$ with v_0 the velocity of the sun around the center of the galaxy (229 Km/s).

For a given velocity distribution $f(v')$, with respect to the center of the galaxy, one can find the velocity distribution in the Lab $f(v, v_E)$ by writing $v' = v + v_E$, $v_E = v_0 + v_1$, with v_1 the Earth's velocity around the sun.

Table 3.2. The static spin matrix elements for various nuclei. For ^{3}He see Moulin, Mayet and Santos [67, 68]. For the other light nuclei the calculations are from DIVARI [46]. For ^{73}Ge and ^{127}I the results presented are from Ressel et al. [44] (*) and the Finish group et al. [51] (**). For ^{207}Pb they were obtained by the Ioannina team (+). [22, 52]

	^{3}He	^{19}F	^{29}Si	^{23}Na	^{73}Ge	^{127}I*	^{127}I**	^{207}Pb+
$\Omega_0(0)$	1.244	1.616	0.455	0.691	1.075	1.815	1.220	0.552
$\Omega_1(0)$	−1.527	1.675	−0.461	0.588	−1.003	1.105	1.230	−0.480
$\Omega_p(0)$	−0.141	1.646	−0.003	0.640	0.036	1.460	1.225	0.036
$\Omega_n(0)$	1.386	−0.030	0.459	0.051	1.040	0.355	−0.005	0.516
μ_{th}		2.91	−0.50	2.22				
μ_{exp}		2.62	−0.56	2.22				
$\frac{\mu_{th}(spin)}{\mu_{exp}}$		0.91	0.99	0.57				

It is convenient to choose a coordinate system so that $\hat{x}$ is radially out in the plane of the galaxy, $\hat{z}$ in the sun's direction of motion and $\hat{y} = \hat{x} \times \hat{z}$.

Since the axis of the ecliptic lies very close to the x, y plane ($\omega = 186.3^0$) only the angle $\gamma = 29.8^0$ (see Fig. 3.4) becomes relevant. Thus the velocity of the earth around the sun is given by

$$\boldsymbol{v}_E = \boldsymbol{v}_0\hat{z} + \boldsymbol{v}_1(\, sin\alpha \, \hat{\mathbf{x}} - cos\alpha \, cos\gamma \, \hat{\mathbf{y}} + cos\alpha \, sin\gamma \, \hat{\mathbf{z}}) \,, \tag{3.79}$$

where α is phase of the earth's orbital motion. The LSP velocity distribution $f(\boldsymbol{v}')$ is not known. Many velocity distributions are employed. In the present work we will adopt the standard practice and assume it to be Gaussian:

$$f(\boldsymbol{v}') = \frac{1}{(\sqrt{\pi}\boldsymbol{v}_0)^3}e^{-(\boldsymbol{v}'/\boldsymbol{v}_0)^2} \,. \tag{3.80}$$

Since $\boldsymbol{v}_1 \ll \boldsymbol{v}_0$ we will ignore, for the moment, the motion of the Earth. Then the total (non directional) rate is given by

$$R = \bar{R}t(a, Q_{min}) \,, \tag{3.81}$$

$$\bar{R} = \frac{\rho(0)}{m_{\chi^0}} \frac{m}{Am_p} \left(\frac{\mu_r}{\mu_r(p)} \right)^2 \sqrt{\langle v^2\rangle}[\sigma_{p,\chi^0}^S \, A^2 + \sigma_{p,\chi^0}^{spin} \, \zeta_{spin}] \,.$$

The SUSY parameters have been absorbed in $\bar{R}$. The parameter t takes care of the nuclear form factor and the folding with LSP velocity distribution

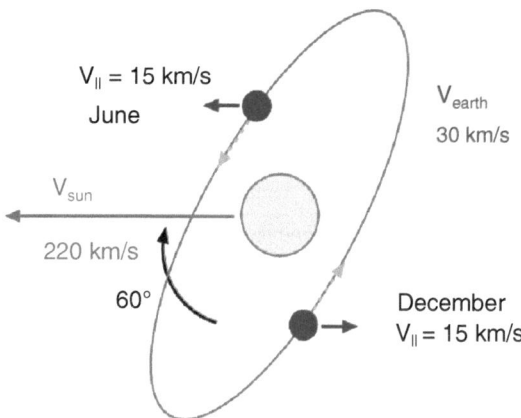

Fig. 3.4. The galactic plane is perpendicular to the paper containing the sun's velocity. The normal to the two planes form an angle $\gamma^{'} = \pi/2 - \gamma \approx \pi/6$. The modulation is affected by the projection of the Earth's velocity along the sun's velocity. Thus the velocity of the detector relative to the center of the galaxy is $220 + 15 = 235$ km/s around June 3nd (when the maximum of the event rate is expected) and $220 - 15 = 205$ km/s around December 3 (minimum of the event rate)

[26, 37, 38, 69]. It depends on Q_{min}, i.e. the energy transfer cutoff imposed by the detector and $a = [\mu_r b v_0 \sqrt{2}]^{-1}$.

In the present work we find it convenient to re-write it as:

$$R = \bar{K} \left[c_{coh}(A, \mu_r(A)) \sigma^S_{p,\chi^0} + c_{spin}(A, \mu_r(A)) \sigma^{spin}_{p,\chi^0} \zeta_{spin} \right] , \qquad (3.82)$$

where

$$\bar{K} = \frac{\rho(0)}{100 \text{ GeV}} \frac{m}{m_p} \sqrt{\langle v^2 \rangle} \simeq 160 \ 10^{-4} \ (pb)^{-1} y^{-1} \frac{\rho(0)}{0.3 GeV cm^{-3}} \frac{m}{1 Kg} \frac{\sqrt{\langle v^2 \rangle}}{280 kms^{-1}} \qquad (3.83)$$

and

$$c_{coh}(A, \mu_r(A)) = \frac{100 \text{ GeV}}{m_{\chi^0}} \left[\frac{\mu_r(A)}{\mu_r(p)} \right]^2 A \ t_{coh}(A) , \qquad (3.84)$$

$$c_{spin}(A, \mu_r(A)) = \frac{100 GeV}{m_{\chi^0}} \left[\frac{\mu_r(A)}{\mu_r(p)} \right]^2 \frac{t_{spin}(A)}{A} . \qquad (3.85)$$

The parameters $c_{coh}(A, \mu_r(A))$, $c_{spin}(A, \mu_r(A))$, which give the relative merit for the coherent and the spin contributions in the case of a nuclear target compared to those of the proton, have already been tabulated [69] for energy cutoff $Q_{min} = 0$, 10 keV.

Via (3.82) we can extract the nucleon cross section from the data (see below).

Neglecting the isoscalar contribution and using $\Omega_1^2 = 1.22$ and $\Omega_1^2 = 2.8$ for ^{127}I and ^{19}F respectively the extracted nucleon cross sections satisfy:

$$\frac{\sigma^{spin}_{p,\chi^0}}{\sigma^S_{p,\chi^0}} = \left[\frac{c_{coh}(A, \mu_r(A))}{c_{spin}(A, \mu_r(A))} \right] \frac{3}{\Omega_1^2} \Rightarrow \approx \times 10^4 \ (A = 127), \ \approx \times 10^2 \ (A = 19)$$

$$(3.86)$$

It is for this reason that the limit on the spin nucleon cross section extracted from both targets is much poorer.

The factors $c19 = c_{coh}(19, \mu_r(19))$, $s19 = c_{spin}(19, \mu_r(19))$, $c19 = c_{coh}(73, \mu_r(73))$, $s73 = c_{spin}(73, \mu_r(73))$ and $c127 = c_{coh}(127, \mu_r(127))$, $s127 = c_{spin}(127, \mu_r(127))$ for two values of Q_{min} and $s3 = c_{spin}(3, \mu_r(3))$ for $Q_{min} = 0$ can be found elsewhere [69].

3.9 Bounds on the Scalar Proton Cross Section

Before proceeding with the analysis of the spin contribution we would like to discuss the limits on the scalar proton cross section. In what follows we will employ for all targets [70, 71, 72, 73, 74, 75, 76, 77] the limit of CDMS II for the Ge target [73], i.e. < 2.3 events for an exposure of 52.5 Kg-d with a threshold of 10 keV. This event rate is similar to that for other systems [71]. The thus obtained limits are exhibited in Fig. 3.5.

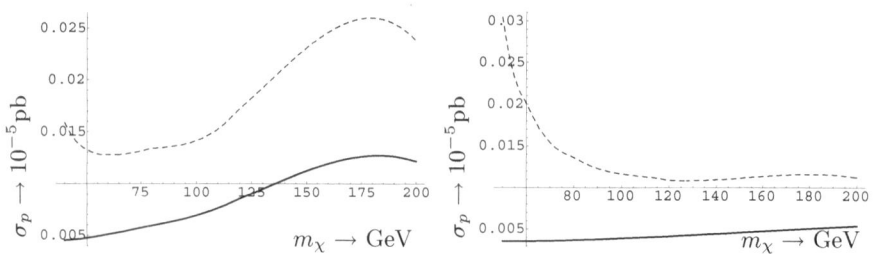

Fig. 3.5. The limits on the scalar proton cross section for A= 127 on the left and A= 73 on the right as functions of m_χ. The continuous (*dashed*) curves correspond to $Q_{min} = 0$ (10) keV respectively. Note that the advantage of the larger nuclear mass number of the A= 127 system is counterbalanced by the favorable form factor dependence of the A= 73 system

3.10 Exclusion Plots in the a_p, a_n and σ_p, σ_n Planes

From the data one can extract a restricted region in the σ_p, σ_n plane, which depends on the event rate and the LSP mass. Some such exclusion plots have already appeared [71, 72]. One can plot the constraint imposed on the quantities $|a_p + \frac{\Omega_n}{\Omega_p} a_n|$ and $|\sqrt{\sigma_p} + \frac{\Omega_n}{\Omega_p} \sqrt{\sigma_n} e^{i\delta}|^2$ derived from the experimental limits via relations:

$$| \sqrt{\sigma_p} + \frac{\Omega_n}{\Omega_p} \sqrt{\sigma_n})e^{i\delta}|^2 \preceq \sigma_{bound}(A) \, r(m_\chi, A) , \qquad (3.87)$$

$$\sigma_{bound}(A) = \frac{R}{\bar{K}} \frac{3}{\Omega_p^2} \frac{10^{-5}pb}{c_{spin}^{100}(A, \mu_r(A))}, \quad r(m_\chi, A) = \frac{c_{spin}^{100}(A, \mu_r(A))}{c_{spin}(A, \mu_r(A))} ,$$

where δ is the phase difference between the two amplitudes and $c_{spin}^{100}(A, \mu_r(A))$ is the value of $c_{spin}(A, \mu_r(A))$ evaluated for the LSP mass of 100 GeV. Furthermore

$$|a_p + \frac{\Omega_n}{\Omega_p} a_n| \preceq a_{bound}(A) \, [r(m_\chi, A)]^{1/2} , \quad a_{bound}(A) = \left[\frac{\sigma_{bound}(A)}{3\sigma_0} \right]^{1/2} .$$
$$(3.88)$$

The constraints will be obtained using the functions $c_{spin}^{100}(A, \mu_r(A))$, obtained without energy cut off, $Q_{min} = 0$, even though the experiments have energy cut offs greater than zero. Furthermore even though we know of no model such that $e^{i\delta}$ is complex, for completeness we will examine below this case as well. Such plots depend on the relative magnitude of the spin matrix elements. They will be given in units of the A-dependent quantity $\sigma_{bound}(A)$ for the nucleon cross sections and the dimensionless quantity a_{bound} for the amplitudes respectively. Before we proceed further we should mention that, if both protons and neutrons contribute, the standard exclusion plot, must be replaced by a sequence of plots, one for each LSP mass or via three dimensional plots. We found it is adequate to provide one such plot for a standard LSP mass,

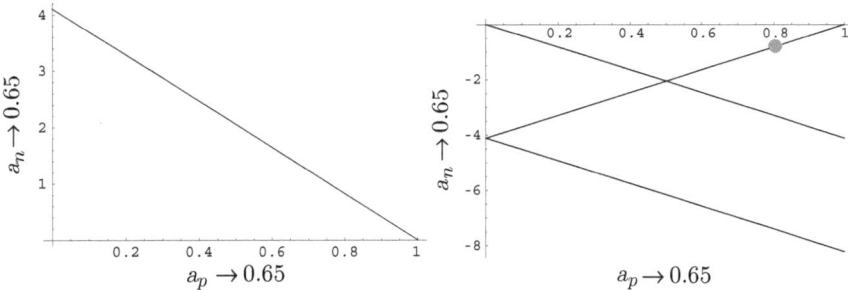

Fig. 3.6. The boundary in the a_p, a_n plane extracted from the data for the target ^{127}I is shown assuming that the amplitudes are relatively real. The scale depends on the event rate and the LSP mass. Shown here is the scale for $m_\chi = 100$ GeV. Note that the allowed region is confined when the amplitudes are of the same sign (*left plot*), but they are not confined when the amplitudes are of opposite sign. The allowed space now is i) The small triangle and ii) The space between the two parallel lines and on the right of the line that intercepts them. We also indicate by a dot the point $a_p = -a_n$ favored by the spin structure of the nucleon. The nuclear ME employed were those of Ressel and Dean (see Table 3.2)

e.g. 100 GeV, and zero energy threshold. The interested reader can find the scale for any other case in work already published [69]. The situation is exhibited in Figs. 3.6–3.8 in the interesting case of the A=127 system using the nuclear matrix elements of Ressel et al. given in Table 3.2. For other targets we refer to the literature [69].

One can understand the asymmetry in the plot due to the fact that Ω_p is much larger than Ω_n. In other words if σ_p happens to be very small a large σ_n will be required to accommodate the data. In the example considered here,

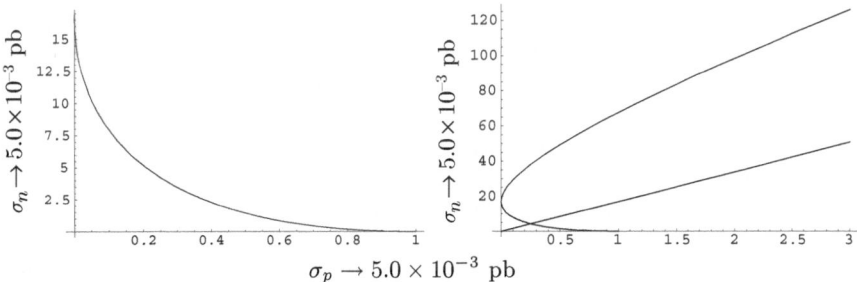

Fig. 3.7. The same as in Fig. 3.6 for the σ_p, σ_n plane. On the left the allowed region is that below the curve (the amplitudes are relatively real and have the same sign). In the plots on the right the amplitudes are relatively real and of opposite sign. The allowed region is i) between the higher segment of the hyperbola and the straight line and ii) Between the straight line and the lower segment of the curve are The nuclear ME employed were those of Ressel and Dean (see Table 3.2)

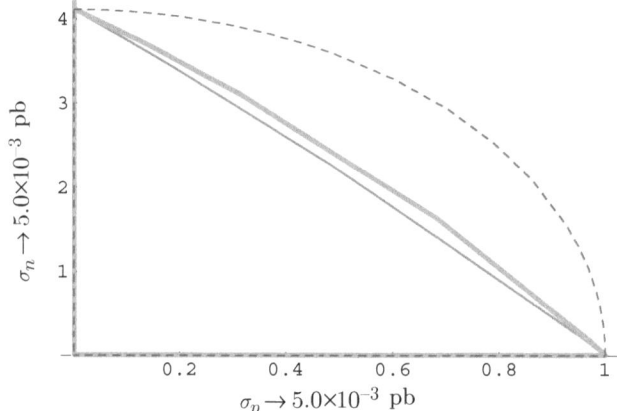

Fig. 3.8. The same as in Fig. 3.7 assuming that the amplitudes are not relatively real, but are characterized by a phase difference δ. The allowed space is now confined. The results shown for the thin solid, thick solid and dashed curves correspond to $\delta = \pi/3$, $\pi/6$ and $\pi/2$ respectively

however, the extreme values differ only by 20% from the values on the axes, which arise, if one assumes that one mechanism at a time (proton or neutron) dominates.

3.11 The Modulation Effect

As we have mentioned the expected event rate is so low that, even if one goes underground, the background is formidable. Especially since the signal coming from the detection of the energy energy of the recoiling nucleus has the same shape as that of the background. One, therefore, looks for specific signatures associated with the reaction. Since the event rate depends on the relative velocity between the LSP and the target, a periodic seasonal dependence is expected due to the motion of the Earth around the sun. What counts is the the is the projection of the velocity of the earth on the sun's velocity (see Fig. 3.4).

If the effects of the motion of the Earth around the sun are included, the total non directional rate is given by

$$R = \bar{K} \left[c_{coh}(A, \mu_r(A)) \sigma^S_{p,\chi^0} (1 + h(a, Q_{min}) cos\alpha) \right] \tag{3.89}$$

and an analogous one for the spin contribution. h is the modulation amplitude, which is quite small, less than 2% and it depends on the velocity distribution, the nuclear form factor and, for a given target, on the LSP mass. α is the phase of the Earth, which is zero around June 2nd. In the case of the target ^{127}I the modulation amplitude is shown in Fig. 3.9. We see that the modulation amplitude is small, especially for $Q_{min} = 0$. Furthermore its sign is uncertain,

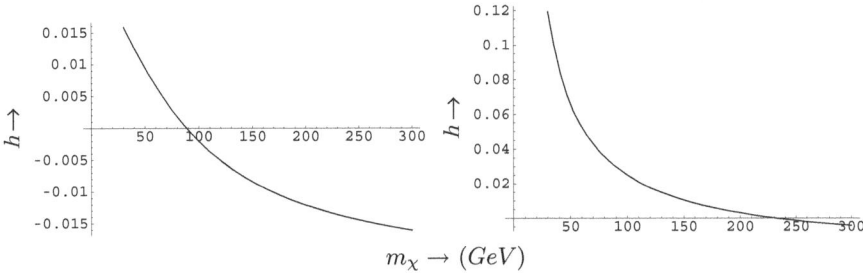

$$m_\chi \to (GeV)$$

Fig. 3.9. The modulation amplitude h as a function of the LSP mass in the case of 127I for $Q_{min} = 0$ on the left and $Q_{min} = 10$ keV on the right. We should mention that the average LSP energy for an LSP mass $m_\chi = 100$ GeV is $\simeq 40$ keV. For the definitions see text

since it depends on the LSP mass. The modulation amplitude increases as the threshold cut off energy increases, but, unfortunately, this occurs at the expense of the total number of counts. Furthermore many experimentalists worry that there are may be seasonal variations in the relevant backgrounds as well.

3.12 Transitions to Excited States

As we have mentioned the average neutralino energy scales with its mass. It is $\simeq 40$ keV for $m_\chi = 100$ GeV. Thus the neutralino energy is not high enough to excite the nucleus. In some rare cases involving odd mass nuclei there exist excited states at low energies, which can be populated in the LSP-nucleus collision due to the high velocity tail of the neutralino velocity distribution (see Fig. 3.10). From an experimental point of view this is very interesting [49], since the signature of the γ−ray emission following the nuclear de-excitation is much easier than nuclear recoils. An interesting target is ^{127}I, which has an excited state at $\simeq 50$ keV. It has recently been found [50] that the branching ratio to this excited state is appreciable from an experimental point of view.

3.13 The Directional Rates

As we have already mentioned one may attempt to measure not only the energy of the recoiling nucleus, but observe its direction of recoil. Admittedly such experiments are quite hard [48], but they are expected to provide unambiguous signature against background rejection. Since the sun is moving around the galaxy in a directional experiment, i.e. one in which the direction of the recoiling nucleus is observed, one expects a strong correlation of the event rate with the motion of the sun [26]. In fact the directional rate can be written as:

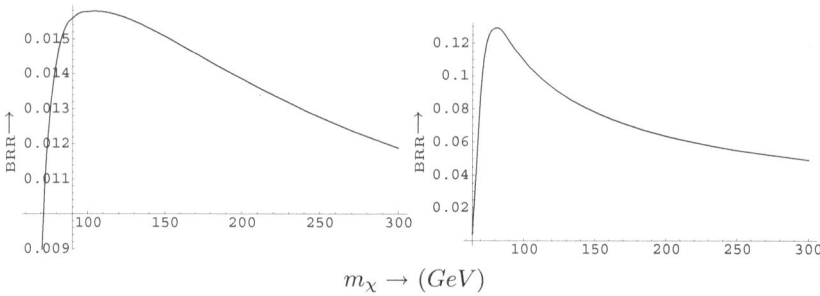

$$m_\chi \rightarrow (GeV)$$

Fig. 3.10. The ratio of the rate to the excited state divided by that of the ground state as a function of the LSP mass (in GeV) for ^{127}I. It was found that the static spin matrix element of the transition from the ground to the excited state is a factor of 1.9 larger than that involving the ground state. The spin response functions $F_{11}(u)$ were assumed to be the same. On the left we show the results for $Q_{min} = 0$ and on the right for $Q_{min} = 10\ KeV$. In the last case, due to the detector energy cut, off the denominator (recoil rate) is reduced, while the numerator (the rate to the excited state) is not affected

$$R_{dir} = \frac{\kappa}{2\pi} \bar{R}\, t \left[1 + h_m cos(\alpha - \alpha_m\ \pi)\right], \qquad (3.90)$$

where h_m is the modulation and α_m is the "shift" in the phase of the Earth α, since now the maximum occurs at $\alpha = \alpha_m \pi$. $\kappa/(2\pi)$ is the reduction factor of the unmodulated directional rate relative to the non-directional one. The parameters κ, h_m, α_m depend on the direction of observation:

$$\hat{e} = (\sin\Theta\cos\Phi,\ \sin\Theta\sin\Phi,\ \cos\Theta).$$

The parameter κt for a typical LSP mass 100 GeV is shown in Fig. 3.11 as a function of the angle Θ for the targets $A = 19$ and $A = 127$. We see that the change of the rate as a function of the angle Θ for the Maxwellian LSP velocity distribution is quite dramatic. This figure is important in the analysis of the angular correlations, since, among other things, there is always un uncertainty in the determination of the angle in a directional experiment. We prefer to use the parameters κ and h_m, since, being ratios, are expected to be less dependent on the parameters of the theory. We exhibit the dependence of the parameters t, h, κ, h_m, and α_m, which are essentially independent of the LSP mass for target $A = 19$, in Table 3.3 (for the other light systems the results are almost identical).

The asymmetry is quite large. For a Gaussian velocity distribution we find:

$$As = \frac{R(-z) - R(+z)}{R(-z) + R(+z)} \approx 0.97\ .$$

In the other directions it depends on the phase of the Earth and is equal to almost twice the modulation. For a heavier nucleus the situation is a bit

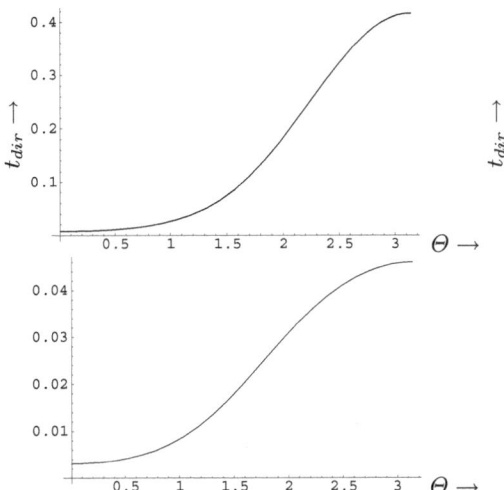

Fig. 3.11. The quantity κt as a function of the angle Θ, the polar angle from the sun's direction of motion, for $A = 19$ on the left and $A = 127$ on the right. The results presented correspond to an LSP mass of $100\ GeV$

complicated. Now the parameters κ and h_m depend on the LSP mass [26]. It is clear that, if such experiments will ever be performed, such signatures cannot be mimicked by background events.

Table 3.3. The parameters t, h, κ, h_m and α_m for the isotropic Gaussian velocity distribution and $Q_{min} = 0$. The results presented are associated with the spin contribution, but those for the coherent mode are similar. The results shown are for the light systems. For intermediate and heavy nuclei there is a dependence on the LSP mass. $+x$ is radially out of the galaxy ($\Theta = \pi/2, \Phi = 0$), $+z$ is in the sun's direction of motion ($\Theta = 0$) and $+y$ is vertical to the plane of the galaxy ($\Theta = \pi/2, \Phi = \pi/2$) so that (x, y, z) is right-handed. $\alpha_m = 0, 1/2, 1, 3/2$ means that the maximum occurs on the 2nd of June, September, December and March respectively

type	t	h	dir	κ	h_m	α_m
			+z	0.0068	0.227	1
dir			+(-)x	0.080	0.272	3/2(1/2)
			+(-)y	0.080	0.210	0 (1)
			-z	0.395	0.060	0
all	1.00					
all		0.02				

3.14 Observation of Electrons Produced During the LSP-nucleus Collisions

Since the detection of recoiling nuclei is quite hard one may look for other events. One such possibility is the observation of ionization electrons produced directly during the LSP nuclear collisions [57, 58]. Due to the properties of the bound electron wf, the event rate peaks at very low electron energies. One therefore must be able to achieve very low energy thresholds. In order to avoid uncertainties arising from the constraint SUSY parameter space we have opted to present the ratio of the event rate for producing electrons divided by the standard coherent recoil rate. This ratio is exhibited as a function of the electron threshold energy in Fig. 3.12. We see that for large atomic number Z and sufficiently low threshold energy this ratio may exceed unity.

It has also been found that inner 1s electrons can be ejected with a non negligible probability [59]. The produced electron holes can be filled via the Auger process or a sizable fraction can proceed via very hard (32 keV) X-ray emission. The detection of such X-rays, in or without coincidence with

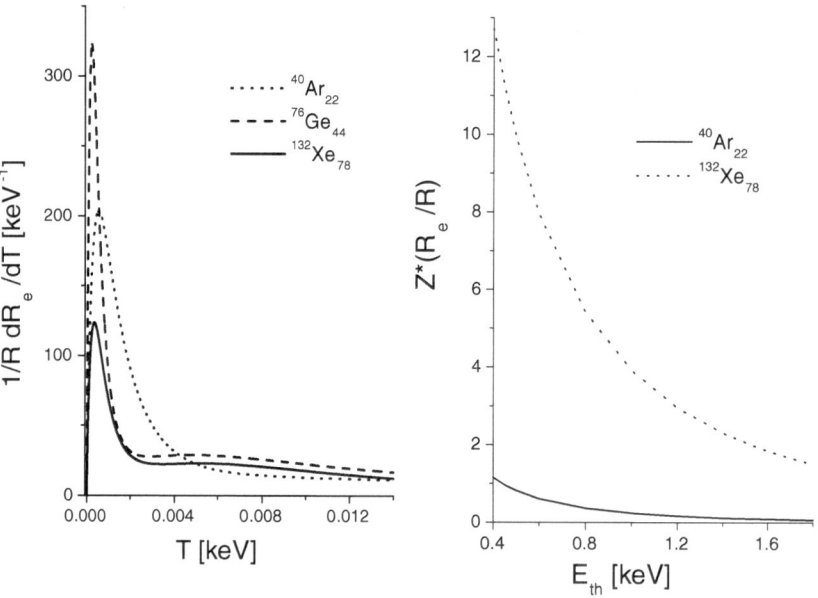

Fig. 3.12. On the left we show the differential rate for ionization electrons, divided by the total rate associated with the nuclear recoils, as a function of the electron energy T (in keV) for various atoms. On the right we show the total rate for producing electrons divided by the corresponding rate for nuclear recoil as a function of the threshold energy. The event rate is per atom, i.e. all electrons in the atom have been considered. The results exhibited were obtained for a typical LSP mass $m_\chi = 100$ GeV

nuclear recoils, will provide a signature very hard to miss, if SUSY allows for detectable recoil rates.

3.15 Conclusions

In this review we have dealt with various issues involving the direct detection of supersymmetric dark matter. The standard experiments employ various techniques of measuring the energy of the recoiling nuclei after their elastic scattering with the dark matter candidates. We have seen that the evaluation of the event rates involves a number of issues: 1) A supersymmetric model with a number of parameters, which at present can only be constrained from laboratory data at low energies as well as cosmological observations. 2) The dependence of the nucleon cross section on quarks other than u and d. 3) A proper nuclear model, which involves the nuclear form factor in the case of the the scalar interaction and the spin response function for the axial current. 4) Information about the density and the velocity distribution of the neutralino (halo model).

Using the present experimental limits on the event rate and suitable inputs in 3)-4) we have derived constraints in the nucleon cross sections. Since the obtained event rates are extremely low, we have examined some additional signatures inherent in the neutralino nucleus interaction, such as the periodic behavior of the rates due to the motion of Earth (modulation effect). Since, unfortunately, this is characterized by a small amplitude, we were lead to examine the possibility of directional experiments. Tese, in addition to the recoil energy, will also attempt to measure the direction of the recoiling nuclei. The event rate in a given direction is $\sim 6\pi$ smaller than that of the standard experiments, but one maybe able to exploit two novel characteristic signatures: a) large asymmetries and b) interesting modulation patterns.

Proceeding further we extended our study to include evaluation of the rates for other than recoil searches such as: i) Transitions to excited states and the observation of de-excitation γ rays, ii) detection of the recoiling electrons produced during the neutralino-nucleus collision and iii) observation of hard X-rays, following the de-excitation of the ionized atom.

With all the above signatures one hopes that, if the supersymmetric models do not conspire to lead to large suppression of the amplitudes, the direct direction of dark matter may soon follow.

Acknowledgements

This work was supported by European Union under the contract MRTN-CT-2004-503369 as well as the program PYTHAGORAS-1. The latter is part of the Operational Program for Education and Initial Vocational Training of the

Hellenic Ministry of Education under the 3rd Community Support Framework and the European Social Fund. The author is indebted to Professor Lefteris Papantonopoulos for support and hospitality during the Aegean Summer School.

References

1. S. Hanary et al., Astrophys. J. **545**, L5 (2000);
   J.H.P Wu et al., Phys. Rev. Lett. **87**, 251303 (2001);
   M.G. Santos et al., Phys. Rev. Lett. **88**, 241302 (2002).
2. P.D. Mauskopf et al., Astrophys. J. **536**, L59 (2002);
   S. Mosi et al., Prog. Nuc.Part. Phys. **48**, 243 (2002);
   S.B. Ruhl al, astro-ph/0212229 and references therein.
3. N.W. Halverson et al., Astrophys. J. **568**, 38 (2002);
   L.S. Sievers et al., astro-ph/0205287 and references therein.
4. G.F. Smoot et al. (COBE Collaboration), Astrophys. J. **396**, L1 (1992).
5. A.H. Jaffe et al., Phys. Rev. Lett., **86**, 3475 (2001).
6. D.N. Spergel et al., Astrophys. J. Suppl. **148**, 175 (2003).
7. M. Tegmark et al., Phys. Rev. D **69**, 103501 (2004).
8. G. Jungman, M. Kamionkowski and K. Griest, Phys. Rep. **267**, 195 (1996).
9. D.P. Bennett et al., Phys. Rev. Lett. **74**, 2967 (1995).
10. R. Bernabei et al., Phys. Lett. B **389**, 757 (1996).
11. R. Bernabei et al., Phys. Lett. B **424**, 195 (1998).
12. A. Benoit et al., [EDELWEISS collaboration], Phys. Lett. B 545, 43 (2002);
    V. Sanglar,[EDELWEISS collaboration] arXiv:astro-ph/0306233; D.S. Akerib
    et al., [CDMS Collaboration], Phys. Rev D **68**, 082002 (2003) [arXiv:astro-ph/0405033].
13. G.L. Kane et al., Phys. Rev. D **49**, 6173 (1994).
14. J. Ellis, K.A. Olive, Y. Santoso and V.C. Spanos, Phys. Rev. D **70**, 055005 (2004).
15. A. Bottino et al., Phys. Lett B **402**, 113 (1997). R. Arnowitt. and P. Nath, Phys. Rev. Lett. **74**, 4952 (1995); Phys. Rev. D **54**, 2394 (1996), hep-ph/9902237 V.A. Bednyakov, H.V. Klapdor-Kleingrothaus and S.G. Kovalenko, Phys. Lett. B **329**, 5 (1994).
16. M.E.Gómez and J.D. Vergados, Phys. Lett. B **512**, 252 (2001) [hep-ph/0012020] M.E. Gómez, G. Lazarides and C. Pallis, Phys. Rev. D **61**, 123512 (2000); Phys. Lett. B **487**, 313 (2000).
17. M.E. Gómez and J.D. Vergados, hep-ph/0105115.
18. M.W. Goodman and E. Witten, Phys. Rev. D **31**, 3059 (1985).
19. K. Griest, Phys. Rev. Lett **61**, 666 (1988).
20. J. Ellis and R.A. Flores, Phys. Lett. B **263**, 259 (1991); Phys. Lett. B **300**, 175 (1993); Nucl. Phys. B **400**, 25 (1993).
21. J. Ellis and L. Roszkowski, Phys. Lett. B **283**, 252 (1992).
22. J.D. Vergados, Part. Nucl. Lett. **106**, 74 (2001), [hep-ph/0010151].
23. U. Chattopadhyay, A. Corsetti and P. Nath, Phys. Rev. D **68**, 035005 (2003).
24. U. Chattopadhyay and D.P. Roy, Phys. Rev. D **68**, 033010 (2003), [hep-ph/0304108].

25. B. Murakami and J.D. Wells, Phys. Rev. D 015001, (2001) [hep-ph/0011082].
26. J.D. Vergados, J. Phys. G **30**, 1127 (2004).
27. Dark Matter Candidates in Supersymmetric Models K.A. Olive, Summary of talk at Dark 2004, proceedings of 5th International Heidelberg Conference on Dark Matter in Astro and Particle Physics, hep-ph/0412054.
28. J. Hisano, S. Matsumoto, M.M. Nojiri and O. Saito, Phys. Rev. D **71**, 015007 (2005).
29. Update on the Direct Detection of Supersymmetric Dark Matter, J. Ellis, K. A. Olive, Y. Santoso, V.C. Spanos, Phys. Rev. D **71**, 095007 (2005) [hep-ph/0502001].
30. J.D. Vergados, J. of Phys. G **22**, 253 (1996).
31. A. Arnowitt and B. Dutta, Supersymmetry and Dark Matter, hep-ph/0204187.
32. E. Accomando, A. Arnowitt and B. Dutta, Dark Matter, muon G-2 and other accelerator constraints, hep-ph/0211417.
33. A.K. Drukier, K. Freese and D.N. Spergel, Phys. Rev. D **33**, 3495 (1986).
34. K. Frese and J.A Friedman and A. Gould, Phys. Rev. D **37**, 3388 (1988).
35. J.D. Vergados, Phys. Rev. D **58**, 103001-1 (1998).
36. J.D. Vergados, Phys. Rev. Lett. **83**, 3597 (1999).
37. J.D. Vergados, Phys. Rev. D **62**, 023519 (2000).
38. J.D. Vergados, Phys. Rev. D **63**, 06351 (2001).
39. J.I. Collar et al., Phys. Lett. B **275**, 181 (1982).
40. P. Ullio and M. Kamioknowski, JHEP **0103**, 049 (2001).
41. M. Drees and N.N. Nojiri, Phys. Rev. D **48**, 3843 (1993); Phys. Rev. D **47**, 4226 (1993).
42. A. Djouadi and M.K. Drees, Phys. Lett. B **484**, 183 (2000); S. Dawson, Nucl. Phys. B **359**, 283 (1991); M. Spira it et al., Nucl. Phys. B **453**, 17 (1995).
43. T.P. Cheng, Phys. Rev. D **38**, 2869 (1988); H-Y. Cheng, Phys. Lett. B **219**, 347 (1989).
44. M.T. Ressell et al., Phys. Rev. D **48**, 5519 (1993); M.T. Ressell and D.J. Dean, Phys. Rev. C **56**, 535 (1997).
45. J.D. Vergados and T.S. Kosmas, Physics of Atomic nuclei, Vol. 61, No 7, 1066 (1998) from Yadernaya Fisika, Vol. 61, No 7, 1166 (1998).
46. P.C. Divari, T.S. Kosmas, J.D. Vergados and L.D. Skouras, Phys. Rev. C **61**, 044612-1 (2000).
47. K.N. Buckland, M.J. Lehner and G.E. Masek, in Proc. *3nd Int. Conf. on Dark Matter in Astro- and part. Phys.* (Dark2000), Ed. H.V. Klapdor-Kleingrothaus, Springer Verlag (2000).
48. The NAIAD experiment B. Ahmed et al., Astropart. Phys. 691, (2003) 19, [hep-ex/0301039]; B. Morgan, A.M. Green and N.J.C. Spooner, Phys. Rev. D **71** 103507, (2005) [astro-ph/0408047].
49. H. Ejiri, K. Fushimi and H. Ohsumi, Phys. Lett. B **317** 14 (1993).
50. J.D. Vergados, P. Quentin and D. Strottman, IJMPE **14**, 751 (2005).
51. E. Homlund, M. Kortelainen, T.S. Kosmas, J. Suhonen and J. Toivanen, Phys. Lett B **584** 31 (2004); Phys. Atom. Nucl. **67**, 1198 (2004).
52. T.S. Kosmas and J.D. Vergados, Phys. Rev. D **55**, 1752 (1997).
53. A.K. Drukier et al., Phys. Rev. D **33**, 3495 (1986), J.I. Collar et al., Phys. Lett B **275**, 181 (1992).
54. A. Green, Phys. Rev. D **66**, 083003 (2002).
55. P. Sikivie, I. Tkachev and Y. Wang, Phys. Rev. Let. **75**, 2911 (1995); Phys. Rev. D **56**, 1863 (1997) P. Sikivie, Phys. Let. B **432**, 139 (1998) [astro-ph/9810286].

56. D. Owen and J.D. Vergados, Astrophys. J. **589**, 17 (2003), [astro-ph/0203923].
57. J.D. Vergados and H. Ejiri, Phys. Lett. B **606**, 305 (2005), [hep-ph/0401151].
58. Ch.C. Moustakidis, J.D. Vergados and H. Ejiri, Nucl. Phys. B **727**, 406 (2005).
59. H. Ejiri, Ch.C. Moustakidis and J.D. Vergados, Dark matter search by exclusive studies of X-rays following WIMPs nuclear interactions, (to appear in Phys. Lett.), hep-ph/0507123.
60. J. Gasser, H. Leutwyler and M.E. Sainio, Phys. Lett. B **253**, 260 (1991); Lect. Notes phys. B **253**, 260 (1991).
61. J. Gasser and H. Leutwyler, Phys. Rep. **87**, 77 (1982).
62. W.B. Kaufmann and G.E. Hite, Phys. Rev. C **60**, 055294 (1999).
63. M.G. Olsson, Phys. Lett. B **482**, 50 (2000), [arXiv:hep-ph/0001203].
64. M.M. Pavan, I.I. Strakovsky, R.L. Workman and R.A. Arndt, PiN Newslett. 16, 110 (2002) [hep-ph/0111066].
65. M.G. Olsson and W.B. Kaufmann, PiN Newslett. **16**, 382 (2002), [hep-ph/0111066].
66. The Strange Spin of the Nucleon, J. Ellis and M. Karliner, hep-ph/9501280.
67. E. Moulin, F. Mayet and D. Santos, Phys. Lett. B **614**, 143 (2005).
68. D. Santos et al., The MIMAC-He3 Collaboration, A New ^{3}He Detector for non Baryonic Dark Matter Search, Invited talk in idm2004 (to appear in the proceedings).
69. J.D. Vergados, Direct SUSY Dark Matter Detection- Constraints on the Spin Cross Section, hep-ph/0512305.
70. P. Belli, R. Cerulli, N. Fornego and S. Scopel, Phys. Rev. D **66**, 043503 (2002), [hep-ph/0203242].
71. C. Savage, P. Gondolo and K. Freese, Phys. Rev. D **70**, 123513 (2004).
72. F. Giuliani and T.A. Girard, Phys. Lett. B **588**, 151 (2004).
73. D.S. Akerib et al. (CDMS Collaboration), Phys. Rev. Lett. **93**, 211301 (2004).
74. H. Pagels, Phys. Rep. **16**, 219 (1975).
75. E. Reya, Rev. Mod Phys. **40**, 545 (1974).
76. ULF-G. Meissner et al., hep-ph/0011277.
77. M.M. Pavan, R.A. Arndt, I.I. Strakovsky and R.L. Workman, Phys. PiN Newslett. **16**, 110 (2002), [hep-ph/0111066].

4

Galaxy Formation and Dark Matter

Joseph Silk

Department of Physics, Denys Wilkinson Building, Keble Road, Oxford, OX1
3RH, UK
silk@astro.ox.ac.uk

Abstract. The challenge of dark matter may be addressed in two ways; by studying
the confrontation of structure formation with observation and by direct and indirect searches. In this review, I will focus on those aspects of dark matter that are
relevant for understanding galaxy formation, and describe the outlook for detecting
the most elusive component, non-baryonic dark matter. Galaxy formation theory is
driven by phenomenology and by numerical simulations of dark matter clustering
under gravity. Once the complications of star formation are incorporated, the theory
becomes so complex that the brute force approach of numerical simulations needs
to be supplemented by incorporation of such astrophysical processes as feedback by
supernovae and by active galactic nuclei. I present a few semi-analytical perspectives
that may shed some insight into the nature of galaxy formation.

4.1 Introduction

Dark matter dominates over ordinary matter. The observations are compelling. Of course, by definition we do not observe matter if it is dark. Minimal
gravitational theory is needed to take us from the observational plane to conclude that dark matter is required. Gravity has been tested over scales that
range from millimetres to megaparsecs. Newton's description of gravity is perfectly adequate, apart from generally small deviations due to the curvature
of space near massive objects, such as stars, or more radically, black holes.
Einstein's theory of gravity tells us that gravity curves space and measuring
this effect was one of the great triumphs of 20th century physics. Nevertheless,
pending its direct detection, dark matter remains a hypothesis that depends,
inevitably, on our having the correct theory of gravitation. For the remainder
of this review, however, I will assume the reality of dark matter dominance
on scales from galactic to those spanning the entire universe.

The standard (or concordance) model of cosmology has a predominance
of dark energy. which amounts to 65% of the mass energy today whereas
non-baryonic matter is 30%. In contrast, luminous baryons (mostly in stars)
constitute 0.5% towards the total. An important component of the standard

J. Silk: *Galaxy Formation and Dark Matter*, Lect. Notes Phys. **720**, 101–121 (2007)
DOI 10.1007/978-3-540-71013-4_4 © Springer-Verlag Berlin Heidelberg 2007

model is the spectrum of primordial density fluctuations, measured in the linear regime via the temperature anisotropies of the CMB. This provides the initial conditions for large-scale structure and galaxy formation via gravitational instability once the universe is matter-dominated. Dark matter consequently provides the gravitational potential wells within which galaxies formed. The dark matter and galaxy formation paradigms are inextricably interdependent. Unfortunately we have not yet identified a dark matter candidate, nor do we yet understand the fundamental aspects of galaxy formation. Nevertheless, cosmologists have not been deterred, and have even been encouraged to develop novel probes and theories that seek to advance our understanding of these forefront issues.

Progress has been made on the baryonic dark matter front. Only about half of the baryons initially present in galaxies, or more precisely, on the comoving scales over which galaxies formed, are directly observed. We cannot predict with any certainty the mass fraction in dark baryons. Yet there are excellent candidates for the dark baryons, both compact and especially diffuse.

In contrast, we have at least one elegant and moderately compelling theory of particle physics, SUSY, that predicts the observed fraction of nonbaryonic dark matter. Unfortunately, we have no idea yet as to whether the required stable supersymmetric particles actually exist.

In this review, I will first describe the increasingly standard precision model of cosmology that enables us to provide an inventory of cosmic baryons. I summarise the current situation with regard to possible baryonic dark matter. I discuss how nonbaryonic matter has been successfully used to provide an infrastructure for galaxy formation, and review the astrophysical issues, primarily centering on star formation and feedback. I conclude with the outlook for future progress. for nonbaryonic dark matter detection and galaxy formation.

4.2 Precision Cosmology

Modern cosmology has emphatically laid down a challenge to theorists. A combination of new experiments has unambiguously measured the key parameters of our cosmological model that describes the universe. These include the temperature fluctuations in the cosmic microwave background, the large galaxy redshift surveys, gravitational shear distortions of distant galaxies by lensing, the studies of the intergalactic medium via the distribution of absorbing neutral clouds along different lines of sight and the use of distant Type Ia supernovae as standard candles. Cosmologists now debate the error bars of the standard model parameters. The ingredients of the standard model in effect define the model. These most crucially are the Friedmann-Robertson-Walker metric and the Friedmann-Lemaitre equations, and the contents of the universe: baryons, neutrinos, photons, baryons, dark matter and dark energy. On these constituents is superimposed a distribution of primordial adiabatic density (scalar) fluctuations characterised by a power spectrum of specified amplitude and spectral index. In addition, there may be a primordial gravity wave tensor mode of fluctuations. The number of free parameters in the standard model is 14, of which the most significant are: H_0, Ω_b, Ω_m, Ω_Λ, Ω_γ, Ω_ν, σ_8, n_s, r, n_T, and τ. One can also add an equation of state for dark energy parameter, $w = -p_\Lambda/\rho_\Lambda$, in effect really a function of redshift, and a rolling scalar (and possibly tensor) index, $dn_s/dlnk$.

No single observational set constrains all, or even most, of these parameters. There are well-known degeneracies, most notably between Ω_Λ and Ω_m, σ_8 and τ, and σ_8 and Ω_m. However use of multiple data sets helps to break these degeneracies. For example, CMB anisotropies fix the combination $\Omega_m + \Omega_\Lambda$ if a Hubble constant prior is adopted, as well as $\Omega_b h^2$ and $\Omega_m h^2$, and SNIa constrain the (approximate) combination $\Omega_m - \Omega_\Lambda$. Both weak lensing and peculiar velocity surveys specify the product $\Omega_m^{-0.6}\sigma_8$. Lyman alpha forest surveys extend the latter measurement to Mpc comoving scales, probing the currently nonlinear regime. Finally, baryon oscillations are providing a measure of Ω_m/Ω_b, independently of the CMB. Interpretation in terms of a standard model (Friedmann-Lemaitre plus adiabatic fluctuations) yields the concordance model with remarkably small error bars [1].

The flatness of space is measured to be $\Omega_{total} = 1.02 \pm 0.02$. Dark energy in the form of a cosmological constant dominates the universe, with $\Omega_\Lambda = 0.72 \pm 0.02$. The dark energy equation of state is indistinguishable from that of a cosmological constant, with $w \equiv p_\Lambda/\rho_\Lambda c^2 = -0.99 \pm 0.1$, this uncertainty holding to $z \sim 0.5$. Even at $z \sim 1$, the claimed uncertainty around $w = -1$ is only 20 percent. Non-baryonic dark matter dominates over baryons with $\Omega_m = 0.27 \pm 0.02$ and $\Omega_b = 0.044 \pm 0.004$. Most of the baryons are non-luminous, since $\Omega_* \approx 0.005$.

The spectrum of primordial density fluctuations is unambiguously measured both in the CMB and in the large-scale galaxy distribution from deep redshift surveys, and found to be approximately scale-invariant, with scalar index $n_s = 0.98 \pm 0.02$. One can also constrain a possible relic gravitational wave background, a key prediction of inflationary cosmology, by the tensor mode limit on relic gravitational waves: $T/S < 0.36$. It has been argued that a fundamental test of inflation requires sensitivity at a level $T/S \gtrsim 0.01$ [2]. Neutrinos are known to have mass as a consequence of atmospheric (ν_τ, ν_μ) and solar (ν_μ, ν_e) oscillations, with a deduced mass in excess of 0.001 eV for the lightest neutrino. From the power spectrum of the density fluctuations, the inferred mass limit (on the sum of the 3 neutrino masses) is $\Sigma m_\nu < 0.4$eV.

However one note of caution should be added. These tight error bars all depend on adoption of simple priors. If these are extended, to allow, for example, for an admixture of generic primordial isocurvature fluctuations, the error bars on many of these parameters increase dramatically, by up to an order of magnitude.

Clearly, the devil is in the observational details [3, 4, 5, 6]. Popular models of inflation predict that $n \approx 0.97$. Space is expected to be very close to flat, with $\Omega = 1 + \mathcal{O}(10^{-5})$. The numbers of rare massive objects at high redshift is specified by the theory of gaussian random fields applied to the primordial linear density fluctuations. The universe as viewed in the CMB should be isotropic [7, 8, 9, 10, 11, 12]. Any deviations from these predictions would be immensely exciting.

Suppose deviations were to be found. This would allow all sorts of possible extensions to the standard model of cosmology. One might consider the signatures of string relics of superstrings or transplanckian features in $\delta T/T|_k$ [14]. Large-scale cosmology might be affected by compact topology or global anisotropy with observable signatures in CMB temperature and polarisation maps [15]. The initial conditions might involve primordial nongaussianity. Anthropically constrained landscape scenarios of the metauniverse prefer a slightly open universe [16]. Some of these features, and others, could be a consequence of compactification from higher dimensions.

4.3 The Global Baryon Inventory

There are several independent approaches to obtaining the baryon abundance in the universe. At $z \sim 10^9$, primordial nucleosynthesis of the light elements yields $\Omega_b = 0.04 \pm 0.004$. At the epoch of matter-radiation decoupling, $z \sim 1000$, the ratios of odd and even CMB acoustic peak heights set $\Omega_b = 0.044 \pm 0.003$. At more recent epochs, Lyman alpha forest modelling of the intergalactic medium at $z \sim 3$ as viewed in absorption along different lines of sight towards high redshift quasars at $z \sim 3$ yields $\Omega_b \approx 0.04$. At the present epoch, on very large scales, of order 10 Mpc comoving linear regime equivalent, the intracluster baryon fraction measured via x-ray observations of massive galaxy clusters provides a baryon fraction of 15%. This translates into $\Omega_b \approx 0.04$. In summary, we infer that $\Omega_b = 0.04 \pm 0.005$ and $\Omega_b/\Omega_m = 0.15 \pm 0.02$.

One's immediate impression is that, at least until very recently, most of the baryons in the universe today are not accounted for. The reasoning is as follows. The luminous content in the form of stars sums to $\Omega_b \approx 0.004$ or 10% in spheroids, and $\Omega_b \approx 0.002$ or 5% in disks. There is also hot intracluster gas amounting to $\Omega_b \approx 0.002$ or 5%. Current epoch observations of the cold/warm photo-ionised IGM via the nearby Lyman alpha/beta forest at $10^4 - 10^5$K as well as CIII (at $z \sim 0$) yield a much larger baryonic reservoir of gas, $\Omega_b \approx 0.012$ or 30%. This gas is metal-poor, with an abundance of about 10% solar [17]. So far, we have only accounted for 50% of current epoch baryons.

The probable breakthrough, however, has come with recent detections of the warm-hot intergalactic medium at $T \lesssim 10^5 - 10^6$K at $z \sim 0$, observed in OVI absorption in the UV and especially via x-ray absorption via OVII and OVIII hydrogen-like transitions towards low redshift luminous AGN. Something like $\Omega_b \approx 0.012$ or 30% of the primordial baryon fraction appears to be in this form, enriched (in oxygen, at least) to about 10% of the solar value [18].

We now have $\gtrsim 80\%$ of the baryons accounted for today. The total baryon content sums to $\Omega_b = 0.032 \pm 0.005$. Given the measurement uncertainties, this would seem to remove any strong case for more exotic forms of dark baryons being present.

However, the situation is not so simple. The Andromeda Galaxy and our own galaxy are especially well-studied regions, where dark matter and baryons can be probed in detail. In the Milky Way Galaxy, the virial mass out to 100 kpc is $M_{virial} \approx 10^{12} M_\odot$, whereas the baryonic mass, mostly in stars, is $M_* \approx 6 - 8 \times 10^{10} M_\odot$. The inferred baryon fraction is at most 8% [19]. Similar statements may be made for massive elliptical galaxies [20]. These in fact are upper limits as the dark mass estimate is a lower bound.

I infer that globally, there is no problem. Nevertheless the outstanding question is: where are the galactic baryons? Most of the baryons are globally accounted for. But this is not the case for our own galaxy and most likely for all comparable galaxies. We cannot account for a mass in baryons comparable to that in stars. It is possible that up to 10% of all the baryons *may* be dark, and that the dark baryons are comparable in mass to the galactic stars.

4.4 The "Missing" Baryons

There are several possibilities for the "missing" baryons. Perhaps they never were present in the protogalaxy. Or they are in the outer galaxy. Or, finally, they may have been ejected.

The first of these options seems very unlikely (although we return below to a variant on this). Consider the second option. The most likely candidates for dark baryons are massive baryonic objects or MACHOs. These are constrained by several gravitational microlensing experiments. The allowed mass range is between 10^{-8} and $10 \, M_\odot$, and the best current limit on the MACHO abundance is $\lesssim 20\%$ of the dark halo mass. In fact, one experiment, that of the MACHO Collaboration, claims a detection from some 20 events seen towards the LMC, most of which cannot be accounted for by star-star microlensing. The observed range of amplification time-scales specifies the mass of the lensing objects. The preferred MACHO mass is around $\sim 0.5 \, M_\odot$.

This mass favours an interpretation in terms of old halo white dwarfs. Main sequence stars in this mass range can be excluded. Current searches for halo high velocity old white dwarfs utilise the predicted colours and proper motions as a discriminant from field dwarfs, and set a limit of $\lesssim 4\%$ of the dark halo mass on a possible old white dwarf component in the halo [21]. However even if this limit were to apply, an extreme star formation history and protogalactic IMF would be required. Observations at high redshift both of star-forming galaxies and of the diffuse extragalactic light background, combined with chemical evolution and SNIa constraints, make such an hypothesis extremely implausible.

If the empirical mass range constraint is relaxed, theory does not exclude either primordial brown dwarfs $(0.01 - 0.1\,M_\odot)$, primordial black holes (mass $\gtrsim 10^{-16}M_\odot$) or even cold dense H_2 clumps $\lesssim 1\,M_\odot$. The latter have been invoked in the Milky Way halo in order to account for extreme halo scattering events [22] or unidentified submillimetre sources [23]. However these possibilities seem to be truly acts of the last resort in the absence of any more physical explanation.

There is indeed another possibility that seems far less ad hoc. The nearby intergalactic medium is enriched to about 10% of the solar metallicity, and contains of order 50% of the baryons in photo-ionised and collisionally ionised phases. This strongly suggests that ejection from galaxies via early winds must have occurred, and moreover would inevitably have expelled a substantial fraction of the baryons along with the heavy elements. Supporting evidence comes from x-ray observations of nearby galaxy groups, which demonstrate that many of these are baryonically closed systems, containing their prescribed allotment of baryons.

There are candidates for young galaxies undergoing extensive mass loss via winds. These are the Lyman break galaxies at $z \sim 2 - 4$. Observations of spectral line displacements of the interstellar gas relative to the stellar component as well as of line widths are indicative of early winds from L_* galaxies [24]. Studies of nearby starburst galaxies, essentially lower luminosity counterparts of the distant LBGs, show that the gas outflow rate in winds is of order the star formation rate. The intracluster medium to $z \sim 1$ is enriched to about a third of the solar metallicity, again suggestive of massive early winds, in this case from early-type galaxies. Hence the "missing" baryons could be in the IGM, with about as much mass ejected in baryons as in stars remaining.

The ejection hypothesis however has to confront a theoretical difficulty. Winds from L_* galaxies cannot be reproduced by hydrodynamical simulations of forming galaxies [25]. The momentum source for gas expulsion appeals to supernovae. SN feedback works for dwarf galaxies and can explain the observed outflows in these systems. However an alternative feedback source is needed for massive galaxies. This most likely is associated with AGN, and the ubiquitous presence of central supermassive black holes in galaxy spheroids.

First, however, I address a more pressing and not unrelated problem, namely given that 90 percent of the matter in the universe is nonbaryonic and cold, how well does CDM fare in confronting galaxy formation models?

4.5 Large-scale Structure and Cold Dark Matter: The Issues

The cold dark matter hypothesis has had some remarkable successes in confronting observations of the large-scale structure of the universe. These have stemmed from predictions, now verified, of the amplitude of the temperature fluctuations in the cosmic microwave background that are directly associated with the seeds of structure formation. The initial conditions for gravitational instability to operate in the expanding universe were measured. The formation of galaxies and galaxy clusters was explained, as was the filamentary nature of the large-scale structure of the galaxy distribution. Nor was only the amplitude confirmed as a prerequisite for structure formation. The Harrison-Zeldovich-Peebles ansatz of an initially scale-invariant fluctuation spectrum, later motivated by inflationary cosmology, has now been confirmed over scales from 0.1 to 10000 Mpc, via a combination of CMB, large-scale galaxy distribution and IGM measurements.

Despite these stunning successes, difficulties remain in reconciling theory with observations. These centre on two aspects: the uncertainties in star formation physics that render any definitive predictions of observed galaxy properties unreliable, and the detailed nature of the dark matter distribution on small scales, where the simulations are also incomplete.

The former issues include such observables as the galaxy luminosity function, disk sizes and mass-to-light ratios, and the presence of old, red massive galaxies at high redshift. These difficulties in the confrontation of galaxy formation theory and observational data are plausibly resolved by improving the prescriptions for star formation and feedback, although there are as yet no definitive answers. The latter issues require high resolution dark matter simulations combined with hydrodynamic simulations of the baryons including star formation and feedback.

I will focus first on the dark matter conundrums, and in particular on the challenges posed by theoretical predictions of dark matter clumpiness, cuspiness and concentration. Implementation of numerical simulations of dark halos of galaxies in the context of hierarchical galaxy formation yields repeatable and reliable results at resolutions of up to $\sim 10^5 \, M_\odot$ in M_* halos. It is clear that the simulations predict an order of magnitude or more dwarf galaxy halos than are observed as dwarf galaxies. It is more controversial but probably true that the dark halos of dwarf galaxies and of barred galaxies do not have the $\sim r^{-1}$ central cusps predicted by high resolution simulations. The dark matter concentration parameter, defined by the ratio of r_{200}, approximately the virial scale, to the scale length, within which the cusp profile is found, measures the cosmological density at virialisation, and hence should be substantially lower for late-forming galaxy clusters than for galaxies. This may not be the case in the best-studied examples of massive gravitationally lensed clusters, cf. [26]. There are also examples of early-forming massive clusters [27]

4.6 Resurrection via Astrophysics

There are at least two viewpoints about resolving the dark matter issues, involving either fundamental physics or astrophysics. Tinkering with fundamental physics, in essence, opens up a Pandora's box of phenomenology. It seems to me that one should first take the more conservative approach of examining the impact of astrophysics on the dark matter distribution before advocating more fundamental changes. Of course if one could learn about fundamental physics, such as a new theory of gravity or higher dimensional dark matter relics from dark matter modelling, this would represent an unprecedented and unique breakthrough. But the prospect of such revelations may be premature.

Astrophysical resolution involves two complementary approaches. One incorporates star and AGN feedback in the dense baryonic core that forms by gas dissipation. Massive gas outflows can effectively weaken the dark matter gravity, at least in the central cusp. These may include stellar feedback driving massive winds via supernovae augmented by a top-heavy IMF and/or by hypernovae, or the impact of supermassive black hole-driven outflows. Another mechanism that shows some promise in terms of generating an isothermal dark matter core is dynamical feedback, via a central massive rotating gas bar. Such bars may form generically and dissolve rapidly, but their dynamical impact on the dark matter has not yet been fully evaluated [28, 29, 30].

All of these are radical procedures, but some are more radical than others. To proceed, one has to better understand when and how galaxies formed. Fundamental questions in galaxy formation theory still remain unresolved. Why do massive galaxies assemble early? And how can their stars form rapidly, as inferred from the α/Fe abundance ratios? Where are the baryons today? And if, as observations suggest, they are in the intergalactic medium, including both the photo-ionised Lyman α forest and the collisionally ionised warm-hot intergalactic medium (WHIM), how and when is the intergalactic medium (IGM) enriched to 0.1 of the solar value? Can the galaxy luminosity function be reconciled with the dark matter halo mass function? Does the predicted dark matter concentration allow a simultaneous explanation of both the Tully-Fisher relation, the fundamental plane and the galaxy luminosity function? And for that matter, is the dark matter distribution consistent with barred galaxy and low surface brightness dwarf galaxy rotation curves?

The observational data that motivates many of these questions can be traced back to the colour constraints on the interpretation of galaxy spectral energy distributions by population synthesis modelling [31, 32]. The galaxy distribution is bimodal in colour, and this can be seen very clearly in studying galaxy clusters. The presence of a red envelope in distant clusters of galaxies testifies to the early formation of massive ellipticals. A major recent breakthrough has been the realisation from UV observations with GALEX that many ellipticals, despite being red, have an ongoing trickle of star formation. Most field galaxies and those on the outskirts of clusters are blue, and are actively forming stars.

The general conclusion is that there must be two modes of global star formation: quiescent and starburst. The inefficient, long-lived, disk mode is motivated by cold gas accretion and global disk instability. The low efficiency is due to negative feedback. The disk mode is relatively quiescent and continues to form stars for a Hubble time. The violent starburst mode is necessarily efficient as inferred from the $[\alpha/Fe]$ clock. It is motivated by mergers, including observations and simulations, as well as by CDM theory. The high efficiency is presumably due to positive feedback, but it is not clear how the feedback is provided.

4.7 What Determines the Mass of a Galaxy?

The luminosity function of galaxies describes the stellar mass function of galaxies. It is biased by star formation in the B (blue) band but is a good tracer in the near-infrared (K) band. It is sensitive to the halo mass, at least for spiral galaxies, as demonstrated by rotation curves. There is a characteristic luminosity, and hence a characteristic stellar mass, associated with galaxies: $L_* \approx 3 \times 10^{10} L_\odot$ and $M_* \approx 10^{11} M_\odot$. The luminosity function declines exponentially at $L > L_*$. This is most likely a manifestation of strong feedback.

Consider first the mass-scale of a galaxy. There is no difference in dark matter properties between galaxy, group or cluster scales, but there is a very distinct difference in baryonic appearance. Specifically, the baryons are mostly in stars below a galaxy mass scale of M_* and mostly in hot gas for systems much more massive than M_*, such as galaxy groups [33] and clusters. A simple explanation comes from considerations of gas cooling and star formation efficiency. It does not matter whether the gas infall initially is cold or whether it virialises during infall. The gas generically will be clumpy, and cloud collisions will be at the virial velocity. In order for the gas to form stars efficiently, a necessary condition is that the cooling time of the shocked gas be less than a dynamical time, or $t_{cool} \lesssim t_{dyn}$.

The inferred upper limit on the stellar mass, for stars to form within a dynamical time in a halo of baryon fraction f_b and mean density ρ_h, can be written as

$$M_* = A^\beta m_p^{2\beta} G^{-(3+\beta)/2} (t_{cool}/t_{dyn})^\beta f_b^{1-\beta} \rho_h^{(\beta-1)/2} ,$$

where the cooling rate has been taken to be $\Lambda = A v_s^{2-3/\beta}$, with $\beta \approx 1$ being appropriate for metal-free cooling in the temperature range $10^5 - 10^6$K. This yields a characteristic mass $M_*/m_p \approx 0.1\alpha^3 \alpha_g^{-2}(m_p/m_e)(t_{cool}/t_{dyn}) \approx 10^{68}$, where $\alpha_g = Gm_p^2/e^2$. This is comparable to the stellar mass associated with the characteristic scale in the Schechter fit to the luminosity function, and also the scale at which galaxy scaling relations change slope. However there is no reason to believe that the dynamical time argument gives as sharp a feature as is observed in the decline of the galaxy luminosity function to high luminosities. Additional physics is needed.

4.8 Disk Galaxy Formation

In the quiescent mode, the clumpy nature of accretion suggests that ministarbursts might occur. In fact, what is more pertinent is the runaway nature of supernova feedback in a cold gas-rich disk. Initially, exploding stars compress cold gas and stimulate more star formation. Negative feedback is eventually guaranteed in part as the cold gas supply is exhausted and also as the cold gas is ejected in plumes and fountains from the disk, subsequently to cool and fall back.

Global simulations have inadequate dynamical range to follow the multiphase interstellar medium, supernova heating and star formation. The following toy model provides an analytical description of disk star formation. I assume that self-regulation applies to the hot gas filling factor $1 - e^{-Q}$, where Q is the porosity and is defined by

$$(SN\,bubble\,rate) \times (maximum\,bubble\,4\text{-}volume)$$
$$\propto (star\,formation\,rate) \times \left(turbulent\,pressure^{-1.4}\right) .$$

One can now write the star formation rate as [34]

$$\alpha_S \times rotation\,rate \times gas\,density$$

with $\alpha_S \equiv Q \times \epsilon$. Here $\epsilon = (\sigma_{gas}/\sigma_f)^{2.7}$, where the fiducial velocity dispersion $\sigma_f \approx 20\,\text{kms}^{-1}\left(E_{SN}/10^{51}\text{ergs}\right)^{0.6}(200M_\odot/m_{SN})^{0.4}$. Here m_{SN} is the mass in stars formed per supernova and E_{SN} is the initial kinetic energy in the supernova explosion. The star formation efficiency $\alpha_S \equiv Q\epsilon$ is

$$0.02\left(\frac{\sigma_{gas}}{10\,\text{kms}^{-1}}\right)\left(\frac{v_c}{400\,\text{kms}^{-1}}\right)\left(\frac{m_{SN}}{200M_\odot}\right)\left(\frac{10^{51}\text{ergs}}{E_{SN}}\right) .$$

The observed mean value is 0.017 [35]. Also, the analytic expression derived for the star formation rate agrees with that found in 3-D multiphase simulations [36]. In fact, the observed distribution of young stars in merging galaxies cannot be fit by modelling the star formation rate with a Schmidt-Kennicutt law, but requires the incorporation of a turbulence-like term [37], as incorporated in this simple model. It follows that porosity is small in merger-induced star formation, where the turbulence will be high, since $Q = \alpha_S/\epsilon \propto \sigma_{gas}^{-1.7}$.

4.8.1 Outflows from Disks

To extract the wind rate, we may assume that the hot gas phase vents out of the disk in a phenomenon resembling a fountain, if the energy input is insufficient to drive a galactic wind. One expects that the disk outflow rate equals the product of the gas flow rate, the hot gas mass, the hot gas surface filling factor and the mass loading factor (f_L). One may rewrite the outflow rate as

$$\dot{M}_{outflow} \approx f_L M_{hot} \Omega (1 - e^{-Q}) = f_L f_{hot} Q M_{gas} \Omega = f_L f_{hot} Q \epsilon^{-1} \dot{M}_* .$$

This reduces to

$$\dot{M}_{outflow} = f_L f_{hot}\left(\frac{\sigma_f}{\sigma_{gas}}\right)^{2.7}\dot{M}_* .$$

In dwarfs, $\dot{M}_{outflow} \sim \dot{M}_*$ if Q is of order 50% and both f_L and f_{hot} take on typical values: the outflow rate is of order the star formation rate. This evidently only is the case for dwarf galaxies. Once $\epsilon \gg 1$, or $\sigma_{gas} \gg \sigma_f$, the wind is suppressed, that is, outflows are suppressed in massive galaxies.

This begs the question of how massive disks such as our own and M31 have depleted their initial baryon content by of order 50 percent. One cannot appeal to protospheroid outflows initiated by AGN (see below) to resolve this issue. Presumably baryon depletion in late-type massive disks (with small spheroids) must have occurred during the disk assembly phase. A collection of gas-rich dwarfs most likely assembled into a current epoch massive disk, and outflows from the dwarfs could plausibly have expelled of order half of the baryons into the Local Group or even beyond. However weak lensing studies find that the typical late-type galaxy in a cluster environment appears to have utilised its full complement of baryons over a Hubble time [38], whereas an early-type galaxy may indeed have expelled about half of its baryons into the intracluster medium. In rich clusters, the baryon complement is complete, and in the field there is increasing evidence that the warm-hot intergalactic medium at $\sim 10^6$K, perhaps in combination with the clouds responsible for the Lyman alpha forest, accounts for up to 50% of the baryons.

4.9 Spheroidal Galaxy Formation

Galaxy spheroids formed early. The inferred high efficiency of star formation on a short time-scale, as inferred from the α/Fe enhancement as well as from the occurrence of old galaxies at high redshift, is suggestive of a feedback mechanism distinct from, and much more efficient than, supernovae. This may be negative, star formation being quenched, or it may be positive, star formation being enhanced. Either process speeds up the star formation phase.

The preferred context for such a mechanism is that of ultraluminous starbursts. Major mergers between galaxies produce extreme gas concentrations that provide an environment for the formation of supermassive black holes. The observed correlation between SMBH mass and the spheroid velocity dispersion suggests contemporaneous SMBH growth and coupled formation of the oldest galactic stars. The spheroid stars are old and formed when the galaxy formed. Hence the SMBHs, which account via the empirical correlation for approximately 0.001 of the spheroid mass, must have formed in the protogalaxy more or less contemporaneously with the spheroid. Supermassive black hole growth is certainly favoured in the gas-rich protogalactic environment.

4.9.1 Outflows from Protospheroids

One may actually be seeing the AGN-triggering phenomenon at work in ultraluminous infrared galaxies (ULIGs), which plausibly are the sites of spheroid formation and SMBH growth, as well as in powerful radio and in some sub-millimetre galaxies. High velocity neutral winds are found both in NaI [44] and in HI absorption [45] against the central bright nuclei. The rate of mass ejected in these superwinds is inferred to be a significant fraction of the star formation rate. Hence the baryon mass ejected is likely to be of order the stellar mass formed. This helps account for the baryon budget, with a complementary mechanism involving supernovae operative in dwarf galaxies and the precursor phase of massive disks.

A simple analytic model of this phenomenon may be constructed as follows. AGN momentum-driven outflow is inevitable once the mechanical momentum luminosity $\dot{M}_w v_w$ or the radiative momentum luminosity $f_{Edd} L_{Edd}/c$ exceeds GMM_g/r^2, i.e. σ^4/Gf_b. Here, $f_{Edd} = L/L_{Edd}$, where the AGN luminosity L includes both the nonthermal contribution from the AGN and the luminosity associated with the enhanced rate of associated galactic star formation. There is a chicken versus egg problem here: does star formation induced by a recent merger feed the AGN or does the impact of the AGN, itself fueled via a merger, on its environment induce star formation? The prevalent view favours the former hypothesis, with the ensuing AGN outflow helping to sweep out residual gas and terminate star formation [13].

Assume now that outflows lead to saturation of the star formation rate by exhausting the cold gas supply. I infer that $M_{bh} = f_b f_{Edd}^{-1} \frac{\kappa \sigma^4}{4\pi G^2}$. The cooling criterion for star formation efficiency guarantees that this relation must saturate for black hole masses of around $10^8 M_\odot$ if the relevant dynamical time-scale is gravitational (corresponding to a spheroid mass of $\sim 10^{11} M_\odot$), but the reduced time-scale of AGN feedback increases the saturation limit to $10^9 - 10^{10} M_\odot$.

Now whether the AGN wind is driven by radiation pressure or mechanical outflow, one still expects that $\dot{M}_w^{AGN} \propto L/c$ with $0.1 \lesssim f_{Edd} \lesssim 1$ and $L_{Edd} = 4\pi GcM_{bh}/\kappa$. In fact, $\dot{M}_w^{AGN} = \frac{L}{cv_w} = \frac{\sigma^3}{G}\frac{\sigma}{v_w}$. In contrast, for a supernova-driven wind:

$$\dot{M}_{outflow}^{SN} = f_L f_{hot} \left(\frac{\sigma_f}{\sigma_{gas}}\right)^{2.7} \dot{M}_* \propto \dot{M}_* \sigma_{gas}^{-2.7} .$$

Comparing the two outflows gives

$$\frac{\dot{M}_w^{SN}}{\dot{M}_w^{AGN}} = f_L f_{hot} f_b Q(v_w/\sigma_{gas}) \ll 1 .$$

This outflow suffices to limit the period of star formation by removing the gas supply.

4.10 Numerical Simulations

Numerical simulations of large-scale structure have made immense progress in the past decade. There are at least two major insights that have emerged from the simulations. One is the filamentary, web-like nature of the dark matter and galaxy distributions in a cold dark matter-dominated universe. A second is the clumpy substructure of galaxy halos. Galaxy redshift surveys reveal the bubbles, sheets and filaments that characterise large-scale structure. This is quantified by studying the two-point and higher order correlation functions, as well as other statistical measures. Semi-analytical techniques are introduced in order to specify prescriptions for star formation, along the lines of the discussion in the previous sections. Only the simplest prescriptions have so far been applied. Theory and data agree well, with the normalisation of the fluctuation power spectrum and its slope being empirically determined. However there is little doubt that until now, simulations have failed to cope adequately with the complex physics of galaxy formation.

The clumpiness of dark matter halos is a prediction that has yet to be verified, pending the detection of halo dark matter. It has important consequences, because the annihilation flux is proportional to the square of the dark matter density, and the clumpiness boost can amount to a factor of 100 or even more.The actual value depends on the complexity of clump survival and disruption due to tidal interactions with the galaxy, including the disk for massive clumps, and stars for the lowest mass clumps [46, 47]. Multiple lensing of quasars by intervening massive halos has provided hints of halo substructure, due to anomalous flux ratios [48].

The starting point for the semi-analytical simulations is a condition for star formation. This is the condition that the coolng time-scale be less than the local dynamical time for the baryons. Unfortunately this is only a necessary condition: it does not guarantee star formation. Assuming a local initial stellar mass function and efficiency of star formation, the first generation of semi-analytics found that the number of dwarf galaxies was greatly overproduced.In hindsight, this was obvious from the original Press-Schechter prescription, updated by Sheth and Tormen [49], which yields the galaxy halo mass function. An even more worrying result was that the more massive galaxies were found to be younger than the less massive galaxies. Again this is obvious in retrospect, since the dynamical time increases with epoch, as does the typical mass of a forming galaxy.

Progress was made in the next generation of simulations which incorporated more realistic feedback prescriptions. The gas evolution was followed in more detail, with differing prescriptions for isolated systems, undergoing cold gas infall, from merging systems, where the gas is gravitationally heated. The first supernovae to form in dwarf galaxies are assumed to eject a substantial fraction of the baryons. This assumption leads to a a strong suppression of the number of dwarf galaxies. This comes at a cost, however. The mass-to-light ratio of disk galaxies, as evidenced by the normalisation of the Tully-Fisher relation, is overestimated. The two problems are connected: solving one makes the other worse. Another problem that appeared with increasing dynamic range was that massive galaxies were overproduced. This is due to the so-called overcooling problem: baryons that can cool will cool, and, at least in the simulations, form stars.

Detailed simulations with cosmological initial conditions confirmed the worst fears of the sceptics. While the initial specific angular momentum matched that estimated analytically from 2nd order perturbation theory, some 90% was lost via baryonic interactions with the clumpy dark matter halo. the resulting galaxies have large spheroidal components. To date, there is no consensus on how a galaxy with low bulge-to-disk ratio, characteristic of a Milky-way type spiral galaxy, is formed. The angular momentum problem is further compounded by the fact that the observed angular momentum distribution of a disk galaxy, found to be universal, bears little resemblance to the the the predicted distribution [50]. which has a significant fraction of negative angular momentum gas relative to the total angular momentum of the dark halo.

Even the relatively simpler pure n-body simulations of individual halo formation pose a challenge. There is a consensus on the resulting dark matter profile. It is fit phenomenologically by the function

$$\rho = Ar^{-\alpha}(1 + r/r_s^2)^{1-\alpha} ,$$

where $\alpha \approx 1.2$ and $C = r_v/r_s$ is the measured concentration parameter with r_v the virial radius, roughly the radius at which the halo overdensity is 200. The simulations show that C is a function of halo formation redshift (and equivalently, a function of halo mass) and the hierarchical evolution is fit by $C \propto 1 + z$. [51] The difficulty here is that low surface brightness dwarf galaxies, inferred from the rotation curves to be dark matter-dominated, have low concentration parameters for their masses. Moreover the central density profiles are often flatter than predicted. The resolution here may be in part due to noncircular orbits and inadequate disk modelling. Another possibility is a complex dynamical history of baryon-dark matter interactions, given that the chemical abundances require severe mass loss to have occurred via supernova-driven outflows. For example, sustained gas flows can heat the dark matter cores [52]. Similar processes can even affect dark halo cusps in massive disk galaxies, via bar dissipation and dissolution [53].

The latest generation of semi-analytic simulations explore new astrophysics to help resolve some of the most persistent problems concerning massive galaxy formation. The new ingredient is that of massive gas outflows triggered by AGN, and in particular during the final phases of supermassive black hole growth by gas accretion. one observes immensely energetic outflows in the broad emission line regions of quasars, amounting to of order a solar mass per year at a velocity in excess of 0.2c, or 10^{46} ergs/s. The initial flow must be adiabatic, and the energy available for heating the interstellar medium is some 2 orders of magnitude per unit gas mass more than that available from supernovae. At the Eddington luminosity, a quasar is expected to have a mechanical outflow of order $0.1L_{Edd} \approx 10^{46}M_9$ ergs/s for a central black hole of $10^9 M_9 M_\odot$. For example, the AGN energy input in a luminous quasar with a $3 \times 10^8 M_\odot$ central black hole, appropriate to an L_* elliptical host galaxy, is at least 4×10^{54} ergs per century, whereas the SN energy input is 2×10^{51} ergs per century per $10^{10}L_\odot$. For an L_* elliptical, the luminosity in the formation phase, assumed to last $10^8 t_8$ years, is $10^{12} t_8^{-1} L_\odot$.

The fuelling rate of the black hole is of order the outflow rate, namely around a solar mass per year, and this is a small fraction, of order a percent of the mass in protogalactic gas flows, which in turn are of order the inferred protogalactic star formation rate. Star formation must occur with high efficiency during massive spheroid formation. Major mergers provide a more than adequate fuel supply [33] that is plausibly self-regulated by the quasar luminosity.

Incorporation of black hole growth, feeding and outflows into the semi-analytics has led to the resolution of several outstanding issues at the cost of greatly increasing the astrophysical complexity. One can easily avoid the overcooling problem in massive galaxy halos. A related problem is also resolved in clusters of galaxies, where the cooling flows are largely quenched by AGN activity. The negative feedback also helps account for the most massive galaxies in clusters being bulge-dominated [42, 43].

The models also predict that the most massive early-type galaxies in clusters are older than their lower mass counterparts. This anti-hierarchical behaviour is a consequence of the negative feedback being assumed only to be relevant for the most massive galaxy in a cluster or group. It is not yet clear if the inferred age difference is sufficient to account for the $[\alpha/Fe]$ excess in massive ellipticals. A clear prediction of the latest generation of semi-analytics is that cluster galaxies are significantly older than field galaxies of similar mass. This is one more example of how models are rapidly overtaken by observation. The most recent study of the fundamental plane in massive clusters finds only a very small present-day age difference ($\lesssim 0.4$Gyr) between massive cluster and field early-type galaxies [55]. At least one fundamental weakness is common to all semi-analytic simulations. There is no reason to assume that local prescriptions are valid in the more extreme conditions of the early universe. The latest feedback models utilise a local IMF. However adopting a top-heavy IMF in merger-induced starbursts is one explanation proffered [56] for the excess of powerful submillimeter and infrared galaxies at $z \sim 2$ compared to model predictions. This suggests one should be cautious about the degree of robustness of semi-analytical galaxy formation simulation predictions.

4.11 The Case for Positive Feedback

Another clue is that both SMBHs, as viewed in AGN and quasars, and massive galaxy spheroids formed anti-hierarchically at a similar epoch, peaking at $z \sim 2$. Massive systems form before less massive systems. This could be a consequence of the same feedback mechanism, which necessarily must be positive in order to favour the massive systems. Supernova feedback is negative and is most effective in low mass systems. SMBH outflows provide an intriguing possibility for positive feedback that merits further exploration. What is lacking for the moment is quantitative evidence for the frequency with which AGN activity is associated with ultraluminous infrared galaxies. Nevertheless, AGN feedback seems to provide the most promising direction for progress.

A specific mechanism for positive feedback appeals to SMBH-induced outflows interacting with the clumpy protogalactic medium. Twin jets are accelerated from the vicinity of the SMBH along the minor axis of the accretion disk. These jets are the fundamental power source for the high non-thermal luminosities and the huge turbulent velocities measured in the nuclear emission line regions in active galactic nuclei and quasars. The jets drive hot spots at a velocity of order 0.1c that impact the protogalactic gas. In a cloudy medium, the jets are frustrated and generate turbulence. The jets are surrounded by hot cocoons that engulf and overpressure ambient protogalactic clouds [39]. These clouds collapse and form stars. The speed of the cocoon as it overtakes the ambient gas clouds greatly exceeds the local gravitational velocity. In this way, a coherent and positive feedback is provided via triggering of massive star formation and supernovae on a time-scale shorter than the gravitational crossing time [40]. The short duty cycle for the AGN phase relative to the longer duty cycle for the induced starburst must be incorporated into inferences from surveys about the frequency of associated AGN activity, if any.

Let us formulate these ideas more precisely. The AGN-induced star formation rate is $\dot{M}_*^{AGN} = \epsilon M_{gas}/t_{jet}$. One can rewrite this as $f_g \epsilon \sigma^2 v_{jet}/G$. Hence for some fiducial spread of jet velocities, one deduces (since $\epsilon \propto \sigma$) that $\dot{M}_*^{AGN} \propto \sigma^3$. Moreover the outflow criterion $L^{AGN} \propto \sigma^4$ can be applied to further predict that $\dot{M}_*^{AGN} \propto (L^{AGN})^{3/4}$.

The star formation luminosity is predicted to be of order $L_{stellar} \approx \dot{M}_* \epsilon_{nuc} f_{core}$, where f_{core} is the mass in nuclear-burning stellar cores and $\epsilon_N \equiv \epsilon_{nuc}$ is the nuclear-burning efficiency. This is best deduced empirically: e.g. for submillimeter galaxies the median IR (8-1000μm) luminosity is 10^{12}L$_\odot$ and the median star formation rate is 1000M$_\odot$/yr, yielding the conversion factor $\epsilon_N = 3.8 \times 10^{-4}$. We can now estimate the ratio of AGN to star formation luminosity in a forming galaxy, namely

$$L^{AGN}/L_*^{AGN} \approx \frac{\sigma^2}{cv_{jet}} \frac{f_{Edd}f_b}{\epsilon \epsilon_N} \sim 1 \,.$$

These represent predictions for ultraluminous star-forming galaxies at high redshift that should eventually be verifiable: the star formation rate is proportional roughly to the cube of the virial (or roughly wind) velocity and also to the 3/4 power of the quasar luminosity.

4.11.1 ULIGs and Spheroid Formation

If the preceding ideas have some validity, the ULIG/ULIRG phenomenon involves both spheroid formation and SMBH growth associated with the gas-rich proto-spheroid phase. The superwinds are AGN momentum-driven and are self-limiting, with the rate of mass ejected inevitably being of order the star formation rate. The SMBH-triggered associated outflows generate the $M_{SMBH} \approx 10^6 \sigma_7^4 M_\odot$ relation, where σ_7 denotes the spheroid velocity dispersion in units of 100 km/s. This is in fact the observed correlation between M_{SMBH} and σ_g in both slope and normalisation, naturally cutting off above $10^9 - 10^{10} M_\odot$ [54].

Supernova-triggered galactic outflows are prevalent until $\sigma_{gas} \approx 100 \mathrm{km\,s^{-1}}$; at larger gas turbulence velocities, black hole outflow-initiated outflows must dominate. Eventually, the input of energy must be highly disruptive for the protogalaxy. When the SMBH is sufficiently massive, its Eddington-limited outflow drives out the remaining protogalactic gas in a wind. This curtailing of spheroid growth allows one to understand the quantitative correlation between SMBH mass and the spheroid gravitational potential [41]. Such negative feedback has been extensively applied in semi-analytic galaxy formation simulations to stop the gas cooling that otherwise results in excessive star formation in massive galaxies [42, 43]. However the possibility of positive feedback has not hitherto been implemented. In the positive feedback model advocated here, star formation triggered by the jet outflow occurs on a time-scale t_j that is much less than the dynamical time-scale or crossing time t_{dyn} associated with disk gravitational instability-induced star formation. This boost in star formation rate (and hence efficiency) can help account for observations of both ULIRGs and old stellar populations in massive galaxies at high redshifts.

4.12 Observing Cold Dark Matter: Where Next?

There is a motivated dark matter candidate, the lightest stable SUSY particle under R parity conservation, or WIMP. As yet, direct detection experiments have not found any unambiguous evidence for its existence. The Milky Way halo provides a laboratory par excellence for indirect WIMP searches via annihilations into high energy particles and photons.

The relic WIMP freezes out at $n_\chi < \sigma_{ann} v > t_H \lesssim 1$, corresponding to a temperature $T \lesssim m_\chi/20k$. The resulting CDM density is $\Omega_\chi \sim \sigma_{weak}/\sigma_{ann}$. Halo annihilations of the LSSP occur into γ and ν, as well as $\bar{p}, p$ and e^+, e^- pairs. In fact, halo detectability may require clumpiness $\langle n^2 \rangle / \langle n \rangle^2 \sim 100$. SUSY modelling of parameter space supplies the relation between σ_{ann} and m_χ. There is an uncertainty of some 2 orders of magnitude in the annihilation cross-section at specified WIMP mass. The WIMP mass most likely lies in the range 0.1–10 TeV, and annihilations provide possible high energy signatures via indirect detection for astronomy experiments. The only claimed evidence for direct detection relies on annual modulation in the DAMA NaI scintillation experiment, which is marginally viable for a spin-independent annihilation cross-section and a low WIMP mass ($\sim 1 - 10 \mathrm{GeV}$) [57]. The uncertainties are large however, and improved data is urgently needed to assess these issues.

One can envisage progress on a variety of fronts. In particle theory, one can readily imagine more than one DM candidate. Why not have 2 stable dark matter particles, one light, one heavy, as motivated by $N = 2$ SUSY? If one took the light dark matter and any of the possible heavy dark matter detections seriously, one could have a situation in which the light (a few MeV) spin-0 particle is subdominant but a $\sim 0.1 - 100\,\mathrm{TeV}$ neutralino is the dominant relic [58].

Because a neutralino of mass $\gtrsim 1\,\mathrm{TeV}$ is beyond the range of the LHC or even the ILC, astrophysical searches for DM merit serious consideration and modest funding. In direct detection, one might eventually hope to see a modulated signal, due to the effect of the Earth's motion through directed streams of CDM [59]. The streams are generic to tidal disruption of dark matter clumps. As for indirect detection, the prospects are exciting, because of the many complementary searches that are being launched. Evidence of neutralino annihilations may come from searches for γ, ν, e^+ and $\bar{p}$ signatures. Experiments under development include HESS2, MAGIC, VERITAS, GLAST (γ-rays), ICECUBE, ANTARES, KM3NET (ν), and PAMELA and AMS ($e^+, \bar{p}$). Targets include the Galactic Centre, the halo and even the sun, where neutralino annihilations in the solar core yield a potentially observable high energy neutrino flux [60].

Refined numerical simulations will soon explore the impact of supernova and SMBH-driven outflows and bar evolution on the distribution and especially the concentration of CDM. A better understanding of intermediate mass black holes as well as the SMBH in the Galactic Centre could eventually provide "smoking guns" where spikes of CDM were retained: the enhanced neutralino annihilations measure CDM where galaxy formation began, 12 Gyr ago. Fundamental physics could be probed: for example a higher dimensional signature, Kaluza-Klein dark matter, would have a spectral signature and branchings that are distinct from those of neutralinos. The prospect of multi-TeV dark matter is another tantalising probe. This provides a challenge for SUSY but is possibly a natural and fundamental scale for any stable relics surviving from n=3 extra dimensions.

4.13 Summary

Galaxy formation is still poorly understood despite its apparent successes. There is no fundamental theory of star formation. One can adopt various empirical parameters and functions, incorporate plausible assumptions and prescriptions and add new ingredients until satisfactory explanations are obtained of any specified observations. Beautiful images are often simulated at such vast cost in computer time that it is impossible to test the robustness of the favoured location in multidimensional parameter space.

Dark matter searches are not in a much healthier state. They rely on plausible assumptions about the dark matter candidates and on the theory of gravity. There is a vast parameter space that admits undetectable particles, such as the gravitino. One has to hope that the likely culprit has electromagnetic couplings.

This is the down side. Bayesians would abandon hope at this juncture, and argue that more science return per dollar will come, for example, by sending men to Mars. Yet to conclude on a more positive note, there is every prospect that potential advances in supercomputers, with virtually no limit to the size of future simulations, will allow us to reproduce our local universe in detail, thereby providing a firmer basis for extrapolation to the remote past. And this extrapolation could be largely phenomenological, driven by the data flow from ever larger and more powerful telescopes that peer further into the universe and hence into our past.

Likewise, the forthcoming LHC and the eventual construction of the ILC will pose tighter constraints on the underlying particle physics that provides the infrastructure for speculations about dark matter. With any luck, supersymmetry will be discovered, thereby setting dark matter candidates on a far firmer footing. And the complementary experiments in direct and in indirect detection should, within a decade, probe all of the allowable SUSY parameter space.

This is an exciting moment in cosmology. We are at the threshold of confirming a standard model, which seems boring and even ugly. Yet the the prospect beckons of finding new physics in the unexpected deviations from the model. A convergence of particle physics and astronomy, in experiment and in theory, will inevitably lead us onto uncharted territory. There can be no greater challenge than in deciphering what awaits us.

Acknowledgements

I thank my collaborators for discussions and exchanges on many of the issues covered here.

References

1. U. Seljak, Phys. Rev. D **71**, 103515 (2005).
2. L. Boyle, P. Steinhardt and N. Turok, Phys. Rev. Lett. **96**, 111301 (2006), astro-ph/0507455 (2005).
3. M. Blanton et al., ApJ **592**, 819 (2003).
4. J. Peacock et al., Nature **410,** 169 (2001).
5. H. Dahle, ApJ **653**, 954 (2006), astro-ph/0608480 (2006).
6. C. Heymans et al., MNRAS, **361**, 160 (2005).
7. I. McCarthy et al., MNRAS in press, astro-ph/0609314 (2006).
8. D. Spergel et al., ApJ in press, astro-ph/0603449 (2006).
9. W. Percival et al., ApJ **657**, 51 (2007), astro-ph/0608635 (2006).
10. H. Eriksen et al., ApJ **656**, 641 (2007), astro-ph/06088.

11. J. Kristiansen et al., preprint astro-ph/0608017.
12. G. Fogli et al., preprint astro-ph/0608060.
13. P. Hopkins et al., ApJS **163**, 1 (2006).
14. M. Gasperini and N. Nicotri, Phys. Lett. B **633**, 155 (2006), hep-th/0511039 (2005).
15. A. Riazuelo et al., preprint astro-ph/0601433 (2005).
16. B. Freivogel et al., JHEP **0603**, 039 (2006), hep-th/0505232 (2005).
17. C. Danforth, J. Shull, J. Rosenberg and J. Stocke, ApJ **640** 716 (2006) astro-ph/0508656 (2005).
18. F. Nicastro et al., ApJ **629**, 700 (2005).
19. A. Klypin, H. Zhao and R. Somerville, ApJ **573**, 597 (2002).
20. C. Lintott, I. Ferreras and O. Lahav, ApJ **648**, 826 (2006), astro-ph/0512175 (2005).
21. M. Creze et al., A&A **426**, 65 (2004).
22. M. Walker and M. Wardle, ApJ **498L**, 125 (1998).
23. A. Lawrence, MNRAS **323**, 147L (2001).
24. K. Adelberger, C. Steidel, A. Shapley and M. Pettini, ApJ **584**, 45 (2003).
25. V. Springel and L. Hernquist, MNRAS **339**, 289 (2003).
26. M. Oguri et al., ApJ **632**, 8410 (2005).
27. R. Overzier et al., ApJ in press, astro-ph/06011223 (2006).
28. M. Weinberg and N. Katz, ApJ **580**, 627 (2002).
29. O. Valenzuela and A. Klypin, MNRAS **345**, 406 (2003).
30. E. Athanassoula, IAU Symposium **220**, eds. S.D. Ryder, D.J. Pisano, M.A. Walker and K.C. Freeman. San Francisco: Astronomical Society of the Pacific (2004), p. 273.
31. A. Sandage, A&A **161**, 89 (1986).
32. R. Larson, MNRAS **218**, 409L (1986).
33. W. Mathews et al., ApJ **634**, L137 (2005).
34. J. Silk, MNRAS **343**, 249 (2003).
35. R. Kennicutt, ApJ **498**, 541 (1998).
36. A. Slyz, J. Devriendt, G. Bryan and J. Silk, MNRAS **356**, 735 (2005).
37. J. Barnes, MNRAS **350**, 798 (2004).
38. H. Hoekstra, B. Hsieh, H. Yee, H. Lin and M. Gladders, ApJ **235**, 73 (2005).
39. C. Saxton, G. Bicknell, R. Sutherland and S. Midgley, MNRAS **359**, 781 (2005).
40. J. Silk, MNRAS **364**, 1337 (2005).
41. J. Silk and M. Rees, A&A **331L**, 1 (1998).
42. D. Croton et al., MNRAS **365**, 11 (2006).
43. R. Bower et al., MNRAS **370**, 645 (2006).
44. C. Martin, ApJ, **621**, 227 (2005).
45. R. Morganti, C. Tadhunter and T. Oosterloo, A&A **444**, L9 (2005).
46. G. Angus and H. Zhao, preprint astro-ph/0608580 (2006).
47. T. Goerdt et al., Mon. Not. Roy. Astron. Soc. **375**, 191 (2007), astro-ph/0608495 (2006).
48. R. Metcalf, ApJ **622**, 72 (2005).
49. R. Sheth and G. Tormen, MNRAS **329**, 61 (2002).
50. F. van den Bosch, ApJ **576**, 21 (2002).
51. R. Wechsler et al., ApJ **568**, 52 (2002).
52. S. Mashchenko, H. Couchman and J. Wadsley, Nature **442**, 539 (2006).
53. K. Holley-Bockelmann, M. Weinberg and N. Katz, MNRAS **363**, 991 (2005).

54. T. Di Matteo, V. Springel and L. Hernquist, Nature **433**, 604 (2005).
55. P. van Dokkum and R. van der Marel, ApJ. **655**, 30 (2007), astro-ph/0609587 (2006).
56. C. Baugh et al., MNRAS **356**, 1191 (2005).
57. D. Akerib et al., PRL **96**, 011302 (2006).
58. C. Boehm, P. Fayet and J. Silk, PRD **69**, 101302 (2004).
59. K. Freese, P. Gondolo and H. Newburg, PRD **71**, 043516 (2005).
60. J. Silk, K. Olive and M. Srednicki, PRL **55**, 257 (1986).

Part II

Dark Energy: The Energy Balance
of the Universe
within the Standard Cosmological Model

5

Cosmological Parameters from Galaxy Clusters: An Introduction

Paolo Tozzi

INAF – Osservatorio Astronomico di Trieste, via G.B. Tiepolo 11, 34131
Trieste – Italy INFN – National Institute for Nuclear Physics, Trieste, Italy
tozzi@ts.astro.it

Abstract. This lecture is an introduction to cosmological tests with clusters of
galaxies. Here, I do not intend to provide a complete review of the subject, but
rather to describe the basic procedures to set up the *fitting machinery* to constrain
cosmological parameters from clusters, and to show how to handle data with a
critical insight. I will focus mainly on the properties of X–ray clusters of galaxies,
showing their success as cosmological tools, to end up discussing the complex thermodynamics of the diffuse intracluster medium and its impact on the cosmological
tests.

5.1 Introduction

This lecture concerns a classic topic of observational and theoretical astrophysics: investigating the global properties of the Universe by looking at its
large scale structure. In particular, we are interested in using our knowledge
on the physical properties of clusters of galaxies and their distribution with
mass and cosmic epoch to put constraints on the cosmological parameters,
namely the matter density parameter Ω_0 and the dark energy component
(parameter w or, in the simplest case $w = -1$, the cosmological constant Λ).
Our journey will be a round trip: starting from a simple theoretical approach,
we will build a powerful tool to interpret the data and measure the cosmological parameters, but then, we will be forced to go back to theory for a more
complex approach to the physics of clusters of galaxies.

The theoretical starting point (Sect. §5.1) provides a reasonable framework to understand the formation and evolution of clusters, which are the
most massive bound and quasi–relaxed objects in the Universe, in a cosmological context. The observational part (Sect. §5.2) will focus mostly on X–ray
observations, which offered the most important observational window for this
kind of test for the last 15 years. As often happens in astrophysics, we will
find that the increasing quality of the data sheds light on a situation much
more complex than previously thought. The most recent data, collected in

P. Tozzi: *Cosmological Parameters from Galaxy Clusters*, Lect. Notes Phys. **720**, 125–156
(2007)
DOI 10.1007/978-3-540-71013-4_5   © Springer-Verlag Berlin Heidelberg 2007

the last five years by the Chandra and XMM–Newton satellites, calls for a much deeper understanding of the physics of baryons in clusters of galaxies, forcing us to reconsider the basic physical ingredients to make a more robust connection between clusters and cosmology (Sect. §5.3). This effort is worth, since clusters are an invaluable tool for cosmology, and they can significantly constrain the cosmological parameters in a way which is complementary to the other classic cosmological tests (the Cosmic Microwave Background, hereafter CMB, and Type Ia Supernovae, SneIa).

5.2 Clusters of Galaxies in a Cosmological Context

5.2.1 What is a Cluster of Galaxies

We start with a simple definition of what is a cluster of galaxies. The simplest approach is to identify a cluster as an overdensity in the projected distribution of galaxies in an optical image. The first catalog was indeed a compilation of galaxy concentrations found by eye in optical images [1]. Today, the quality of optical images, especially that from the Hubble Space Telescope, are such that bright (or, in terms of galaxies, rich) clusters of galaxies are among the most spectacular objects of the extragalactic sky. In Fig. 5.1, first panel, we show an optical image of Abell 1689, a massive cluster at redshift $z = 0.18$. The bright, yellowish galaxies are the massive ellipticals which typically populate the inner part of rich clusters. In this image it is also possible to see background galaxies distorted by gravitational lensing.

However, the stars in the cluster galaxies, visible in the optical light, are not at all the dominant mass component. The X–ray image of Abell 1689 obtained with the Chandra satellite (second panel of Fig. 5.1) shows the distribution of hot gas, which is the dominant baryonic component. The total mass, anyway, is dominated by the non–baryonic component called dark matter (see the reconstructed distribution in the third panel). To review the properties and the hypothesis on the nature of the dark matter see [29]. Here we need to know only that dark matter is collisionless and that it dominates gravitationally large objects like groups and clusters of galaxies.

To be more quantitative, the composition of a cluster of galaxies is roughly as follows: 80% of the mass is in dark matter; 17% in hot diffuse baryons, the so–called IntraCluster Medium (ICM); 3% in the form of cooled baryons, meaning stars or cold gas. The total mass of clusters ranges from few $\times 10^{13} M_\odot$ (small groups) to more than $10^{15} M_\odot$. While the baryonic components can be directly observed (mainly in the optical, infrared and near infrared bands for the stars and in the X–ray band for the ICM), the dark matter can be measured only through the effect of gravitational lensing on the background galaxies or by other dynamical properties of clusters. Needless to say, the total mass of a cluster is the fundamental quantity we need to know. A useful definition of the dynamical mass of a cluster will be given after briefly discussing the physics of gravitational collapse.

  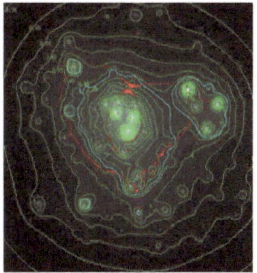

Fig. 5.1. The rich cluster of galaxies Abell 1689 ($z = 0.18$). The three images show the three main components in terms of mass. In order of increasing mass fraction, *from left to right:* an optical image (stars) taken with the Hubble Space Telescope (credits ACS Science Team, ESA NASA); an X–ray image taken with Chandra (showing the diffuse Intra Cluster Medium); the dark matter map reconstructed from lensing (after [9])

5.2.2 The Linear Theory of Gravitational Collapse

Clusters form through gravitational collapse, which is driven by dark matter. This is strongly simplifying our problem, since the dark matter, whatever it is, must behave as a collisionless fluid, and therefore it is not affected by dissipative processes, unlike the baryons, which are pressure supported, and experience radiative cooling. Since we are interested in the total mass, we can neglect, on a first instance, the physical processes affecting only the baryons.

To describe the evolution of a collisionless fluid under its own gravity, we can use the Eulerian equations of motion describing a perfect fluid assuming spherical symmetry (continuity, Euler and Poisson equations, see [21]:

$$\frac{\partial \rho}{\partial t} + \boldsymbol{\nabla} \bullet (\rho\, \boldsymbol{v}) = 0 \tag{5.1}$$

$$\frac{\partial \boldsymbol{v}}{\partial t} + (\boldsymbol{v} \bullet \boldsymbol{\nabla})\boldsymbol{v} + \frac{1}{\rho}\boldsymbol{\nabla}p + \boldsymbol{\nabla}\phi = 0 \tag{5.2}$$

$$\nabla^2 \phi = 4\pi G \rho\,, \tag{5.3}$$

where ρ is the density field, $\boldsymbol{v}$ is the velocity field, p is the pressure and ϕ is the gravitational potential generated by the density field itself. We are interested in how the density evolves with time. First, we consider small positive density perturbations with respect to a uniform and static background with density ρ_0, so that we can easily linearize the system of equations. We define our interesting variable as the overdensity $\delta \equiv (\rho - \rho_0)/\rho_0$, and assume that the unperturbed solution is a static background, $\rho = \rho_0 = const$[1]. We just need a little algebra to linearize the equations and derive the solution for the density

---

[1] This is not correct since the Poisson equation is not satisfied; however this assumption, called the *Jeans swindle*, leads to correct consequences.

contrast in terms if its linear components $\delta_k = A \exp[-i\mathbf{k}r + i\omega t]$. After defining the sound speed as $v_s^2 \equiv (\frac{\partial p}{\partial \rho})_{adiabatic}$, the solution can be written as follows:

$$\ddot{\delta}_k = (4\pi G\rho_0 - v_s^2 k^2)\delta_k \ , \tag{5.4}$$

or, in a very familiar way, $\ddot{\delta} = -\omega^2 \delta$. The solution is therefore an harmonic oscillator with dispersion relation $\omega^2 = v_s^2 k^2 - 4\pi G\rho_0$. Note that for dark matter, v_s is substituted by the velocity dispersion of the collisionless particles v_*. When ω^2 is negative, the solution behaves exponentially. This qualitative result is largely expected in this extremely simplified situation: in a static background, the gravitational force is proportional to the overdensity itself, and the gravitational instability evolves rapidly. The dispersion relation defines a length scale $\sim 1/k$ for which the perturbation is unstable.

However, we are interested in the realistic solution in an expanding background. This can be obtained by substituting a varying background density $\rho_0 = \rho_0(t_0)R^{-3}(t)$ in the equations, where $R(t)$ is the scale factor satisfying the usual Friedmann equation. The expansion of the Universe is conveniently expressed trough the fractional growth of $R(t)$ which is the Hubble function $H(t) = \dot{R}/R$. The solution of the linearized problem satisfies:

$$\ddot{\delta}_k + 2\frac{\dot{R}}{R}\dot{\delta}_k + \left(v_s^2 k^2 - 4\pi G\rho_0\right)\delta_k = 0 \ . \tag{5.5}$$

The additional term $2\frac{\dot{R}}{R}\dot{\delta}_k$ changes considerably the qualitative behaviour of the solution, depending on the behaviour of $R(t)$. To show a specific example, we adopt $R \propto t^{2/3}$, or $\dot{R}/R = (2/3)(t/t_0)^{-1}$, appropriate for an Einstein–de–Sitter Universe (EdS, $\Omega_0 = 1$), to find:

$$\ddot{\delta}_k + \frac{4}{3t}\dot{\delta}_k - \frac{2}{3t^2}\delta_k = 0 \tag{5.6}$$

(note that here we assumed a negligible v_s or v_* as appropriate for Cold Dark Matter). The growing mode solution is $\delta_+(t) = \delta_+(t_i)(t/t_i)^{2/3}$. Therefore, in an EdS Universe, we have the remarkably simple result that the linear growth of a density perturbation is proportional to the expansion factor $(1+z)$. One may wonder why we show the solution for $\Omega_0 = 1$, while we are here in this School to learn that dark energy is the dominant component in the Universe, while the matter component is $\Omega_0 \leq 0.3$. The fact is that the case for $\Omega_0 = 1$ gives simple analytical solutions, an occurrence that contributed substantially to the success of the EdS Universe until the early 90s, when observational evidences started to point towards a low matter density, making room for the debout of dark energy.

More in general, we find that the fastest is the expansion, the slowest is the linear growth of perturbations. The link between the expansion rate of the Universe and the rate of collapse of density perturbations is strongest at the largest scales. This is because large–scale perturbations are the last to leave the linear phase, while smaller scales (the one from which galaxies form, e.g.), collapsed earlier. This, in turn, is a consequence of the shape of the primordial spectrum of the density perturbations, and it is true in any cold dark matter (CDM) dominated Universe. We will discuss this aspect in greater detail later.

5.2.3 Non Linear Evolution of Density Perturbations and Virialization

Now we have a simple framework which allows us to compute the linear phase of collapse of a spherical density perturbations in an expanding universe. However, our final goal is to describe clusters of galaxies, which are definitely non linear (and non–spherical, but spherical symmetry is too convenient to be dropped!). In addition, we need to define accurately the total dynamical mass of a relaxed object. Should we abandon the simple linear treatment to look for a more complex and computationally heavier approach? Luckily for us, we can define a relaxed object in terms of the same parameters entering the linear theory, as shown in the following pages.

Thanks to the Birkhoff theorem, we can ignore what is outside a perturbation, and we can describe a uniform spherical (*top–hat*) perturbation like a sub–universe with a density larger than the critical one $\Omega \geq 1$. Such a universe would expand and recollapse in a finite time. If we consider a spherical shell encompassing the overdensity, we can use the Friedmann–Robertson Walker (FRW) model for the evolution of each shell in a parametric form:

$$R = \frac{GM}{2E_0}\left(1 - \cos(\eta)\right), \quad \mathbf{t} = \frac{GM}{(2E_0)^{3/2}}\left(\eta - \sin(\eta)\right). \quad (5.7)$$

The maximum of the expansion radius defines the turn–around time, which is the epoch when the shell starts to collapse, after decoupling from the cosmic expansion. Due to the symmetry of the solution, the time of collapse is twice the turn–around time. In our spherical approximation, the collapse ends into a singularity. What is actually happening to a real, non–spherical perturbation, is that the different shells cross each other and start oscillating across the center. However, we can bravely assume that, by that time, the perturbation (meaning all the mass included in the outermost spherical shell) is evolved into a spherical, self–gravitating *virialized* halo.

A virialized halo is a region of space where matter is gravitationally bound, and where a statistical equilibrium between the potential and the kinetic energy is established. Every mass component participates to the equilibrium: both the diffuse, ionized gas, the galaxies, and the dark matter particles, have random velocities described by a maxwellian distribution with the same temperature. The virialization condition in its simplest form reads $2T+U = 0$, where $T = M_{tot}\langle v^2\rangle/2$ is the average kinetic energy per particle, and $U = -GM_{tot}^2/R_c$ is the average potential energy. Energy conservation argument fix the relation between mass and the characteristic radius R_c of the halo, so that the virial theorem effectively establish a one–to–one correspondence between the total dynamical mass and the virial temperature.

Going back to the linear solution, how can we describe the formation of virialized halos in terms of the linear solutions? From Fig. 5.2, we learn that virialization is flagged by the recollapse of the outermost shell. The value of the linear δ at the time of collapse depends on the cosmic expansion rate, and therefore on the cosmological parameters. Thus, the linear value of the over-density can be used as a flag for collapse, providing a simple and convenient criterion to decide when a perturbation is virialized.

If we assume that the radius of the virialized halo is about half of the radius of maximum expansion, the reader should be able to derive the actual average density contrast within the virialized halo with respect to the ambient density, as well as the linear value of the density contrast at the time of collapse. This can be left as a useful exercise, in the simple case of an EdS universe (see [28] for the solution and much more on cosmic structure formation). It turns out that the linear threshold for collapse in an EdS universe is $\delta_{c0} = 1.686$, while the actual density contrast of a virialized halo is $\Delta_{vir} = 178$. These numbers, particularly the linear threshold, generalized for different choices of the cosmological parameters, will be relevant for the following analysis. One may wonder how few magic numbers can describe a plethora of complex physical processes. However, as we will see, these numbers allow us to make several predictions, whose reliability is supported by numerical experiments. We have many reasons to proceed confidently.

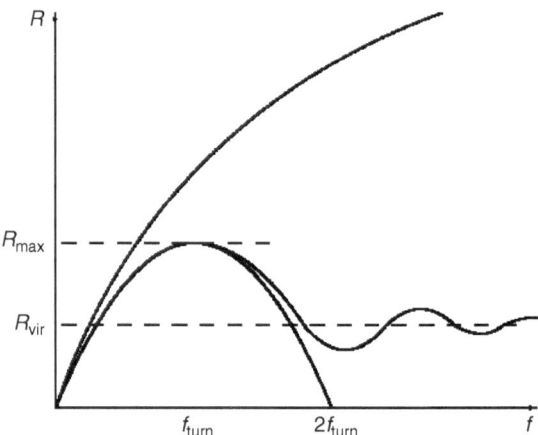

Fig. 5.2. The evolution with time of a *top–hat* perturbation. The upper curve is the expansion of the exterior mass shell, while the closed curve is the solution which behaves like a closed FRW model. The wavy curve is the radius of a realistic perturbation which bounces back and virializes after few oscillations (from [28])

5.2.4 Clusters of Galaxies Reflect the Expansion Rate of the Universe

As we saw, the expansion rate of the Universe, entering (5.5), affects the evolution of the linear perturbations. It follows that the growth of a perturbation is slower when the expansion is faster. The Hubble parameter in its general form writes $H(z) = H_0[\Omega_0(1+z)^3 + \Omega_k(1+z)^2 + \Omega_\Lambda(1+z)^{3+3w}]^{1/2}$, where w is the ratio between the pressure and the energy density in the equation of state of the dark energy component [10]. The special case $w = p/\rho = const = -1$ corresponds to the quantum vacuum energy, aka the cosmological constant. We find that in a low density Universe the expansion is faster than in the EdS case, so we expect that clusters form much later for the same initial conditions. We have a situation more similar to a low density Universe in the case of a flat Universe with cosmological constant. This last case is the favorite choice, since today many observational evidences tell us that the Universe is accelerating (as shown in several lectures at this School), and in the context of general relativity, this can be explained by the presence of dark energy.

Therefore, if we set the initial conditions and the cosmological parameters, we can predict the redshift when virialized halos of a given mass are expected to form. At this point we can reverse the problem: given a measure of the initial conditions (the fluctuations in the CMB are providing them at a redshift $z \sim 1500$) and after counting clusters of galaxies at each redshift, we can infer the expansion rate of the Universe and therefore the cosmological parameters. Clusters are much more useful for this kind of test than, e.g. galaxies, since they are the largest virialized structure in the universe, therefore the closest to the initial linear spectrum of density perturbations and most affected by the expansion rate.

To play this game, obviously we should not focus on single objects, rather we should measure the evolution of the number density of clusters with the cosmic epoch and their distribution with mass. Let's see this in detail.

5.2.5 Where Cosmological Parameters Enter the Game

We are interested in the statistical properties of the initial conditions, in other words, to the average value of δ on a given scale. Since the majority of inflationary models predict that the fluctuations in the density field ρ should be Gaussian, we need to know only its variance. To define operationally the variance on a given scale, we can imagine to smooth the linear field by measuring the overdensity around each point in space within a sphere of radius R (the top–hat filter function). Since the density field is linear, a spatial scale is related to a mass scale simply by $M = (4\pi/3)\rho_0 R^3$ where ρ_0 is the average density. If we express the fluctuations field in terms of its Fourier power spectrum $P(k)$, the variance reads:

$$\sigma^2(M) = \frac{1}{8\pi^3} \int W^2(kR)P(k)d^3k \,, \tag{5.8}$$

where $W(kR)$ is the filter function in the Fourier space. The filter function corresponding to a top–hat in real space is oscillating due to the sharp edges (see [28, 29]). Since every mode grows independently from each other in the linear regime, we expect that $\langle \delta \rangle$ is proportional to the linear growth factor $D(t)$. If we call δ_c the critical value corresponding to the collapse, the epoch of collapse of an overdensity of a mass M is implicitly defined by the relation:

$$\sigma(M)D(z_{coll}) = \delta_c \,. \tag{5.9}$$

The linear growth factor $D(z)$, which we know since it is the solution of (5.5), can be written in a generic cosmology as follows (see [36]):

$$D(z) = \frac{5}{2}\Omega_0 E(z) \int_z^\infty \frac{1+z'}{E(z')^3}dz' \,, \tag{5.10}$$

where $E(z) \equiv H(z)/H_0$. In the general case it is not possible to invert analytically (5.9). Again, since we are still in the theoretical mood, we can assume the EdS case and enjoy its analytical formulae (as you see, simplicity sometimes attracts theoreticians against any evidence!). Another useful step is to approximate the linear spectrum of the density perturbations with a power law, $P(k) \propto k^n$ with $n \simeq -1; -2$. In this case, from 2(40), $\sigma(M) \propto M^{-a}$ with $a = (n+3)/6 > 0$. Since the linear growth is $D(t) = (t/t_0)^{2/3} = 1/(1+z)$, we easily can invert the condition $D(t)\sigma(M) = \delta_c$ to obtain the typical mass which is collapsing at a given epoch:

$$M_c(t) = M_{c0}(t/t_0)^{\frac{4}{n+3}} \,. \tag{5.11}$$

Here we meet a fundamental property of any model based on CDM: the hierarchical clustering. For any $\sigma(M)$ which is decreasing with mass (which implies $a > 0$, or $n > -3$), more massive objects form at later times. The hierarchical clustering, i.e. the progressive assembling of larger and larger structures with cosmic time, is the direct consequence of this property. Actually, the preferred choice is the spectrum for adiabatic fluctuations in a CDM universe, and it is the result of a detailed computation involving fluid equations for relativistic and non–relativistic components in an expanding universe (see the software CMBFAST, http://cmbfast.org/, by U. Seljak and M. Zaldarriaga). Unsurprisingly, a realistic CDM spectrum is not as simple as a power law. The resulting $\sigma(M)$ shows a varying slope as shown in Fig. 5.3.

Before ending this section, we remark that few years ago, the hierarchical clustering hypothesis was not so radicated into cosmological models. Imagine that we have a kind of dark matter which has no power at all at small scales. From 2(40), it is easy to see that $\sigma(M) = const$ below some threshold $M < M_{th}$. As a consequence, all the scales with $M < M_{th}$ collapse at the same time. If M_{th} is large enough, let's say the scale of a cluster of galaxies, then

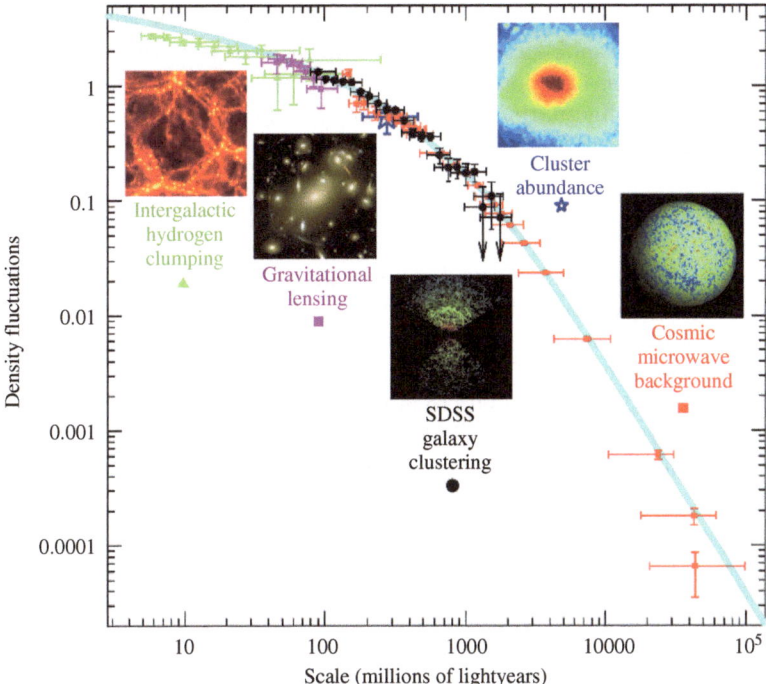

Fig. 5.3. The typical value of the linear fluctuations $\sigma(R)$ predicted for an $\Omega_0 = 0.3$, $\Omega_\Lambda = 0.7$ Universe compared with the values obtained from observations on different scales (from Tegmark 2002 [53]. See lectures by W. Percival for constraints from SDSS, and by R. Caldwell for constraints from the CMB)

clusters form at the same time or even before galaxies. This is the situation we have when we consider light particles like massive neutrinos as candidates for dark matter. Now we know that neutrinos give a negligible contribution to the density of the Universe (see [31]). Given the success of the CDM spectrum in reproducing observations on several scales (as shown in Fig. 5.3) the common wisdom is that cosmic structures follow hierarchical clustering, at least as far as dark matter is concerned (but beware of the baryons![2]).

5.2.6 The Mass Function

Now we can predict the typical mass scale which is virializing at a given redshift. Is this enough to efficiently constrain the cosmological parameters?

---

[2] As a further complication, there are now strong hints that massive galaxies form earlier than smaller ones, and bright quasars peaks earlier than weaker AGN. This anti–hierarchical behaviour of the stellar mass component and nuclear activity could, in principle, be reconciled with the hierarchical clustering of dark matter halos. But this is a debated issue, known as the *hierarchical versus monolithical controversy*. People use to get very aggressive on this topic.

Not yet: we can work out much better observables which will allow us to perform efficient cosmological tests. The important step we have to take now is to derive the mass function. To do that, first we must write the probability distribution of the fluctuations δ, which we already assumed to be a Gaussian with dispersion σ:

$$P(\delta) = \frac{1}{\sqrt{2\pi}\sigma} exp\{\frac{1}{2}\delta^2/\sigma^2\} . \tag{5.12}$$

Since we are dealing with a linear field, it seems safe to say that the fraction of mass which is in regions with overdensity larger than a given δ, is equal to the fraction of volume that, filtered with our top–hat filter of size R, is overdense above the same threshold. This fraction is simply the integral of the Gaussian from the overdensity threshold up to infinity. If we set $\delta_c(z) = \delta_{c0} \times D(0)/D(z)$, we obtain the fraction of mass which is in virialized halos at a given epoch z. This condition reads:

$$N(M)MdM = \int_{\delta_c(z)}^{\infty} P(\delta, \sigma(M))d\delta , \tag{5.13}$$

where $N(M)$ is the number density of virialized halos in the mass range M and $M + dM$. Then we can obtain an expression for $N(M)$ simply deriving the integral on the right hand side with respect to mass:

$$N(M) = \frac{\rho}{M}\frac{d}{dM}\int_{\delta_c(z)}^{\infty} P(\delta, \sigma(M)) . \tag{5.14}$$

Our tidy theoretical attitude is rewarded again: the solution is analytic. Leaving the mathematics to the reader, we write the final, famous result, the [42] (PS) mass function (1974):

$$N(M)\,dM = \sqrt{\frac{2}{\pi}}\frac{\rho}{M}\frac{\delta_c(z)}{\sigma^2}\frac{d\sigma}{dM} exp\left(-\frac{\delta_c(z)^2}{2\sigma^2}\right) dM . \tag{5.15}$$

Its typical shape is characterized by a power law at low masses, and an exponential cutoff at large masses. Given its simplicity, its success is often referred as *the Press & Schechter miracle*.

5.2.7 Is the Press & Schechter Approach Accurate Enough?

Unfortunately miracles are not allowed in science. You may think that this approach is just a didactical exercise to understand the basic concepts, while cosmologists actually use terribly complicated formulae or awfully long numerical computations for the mass function. Well, the truth is that this formula is still at the core of the majority of the works deriving cosmological parameters from clusters of galaxies. Indeed, many numerical experiments (N–body simulations) actually support the validity of the PS approach. Clearly, some differences with respect to the original PS approach were found. Discrepancies are mostly due to the many non linear effects which are not included in the PS formalism. A recent example of a comparison between N–body and the PS formula is shown in Fig. 5.4. We note that the PS formula tends to underestimate the number density of halos at very high redshift. However, if we consider that clusters are observed today up to redshifts slightly above 1, we have to admire the remarkable similarity with the results from the time–expensive, brute–force approach of N–body simulations.

To improve the PS model, some empirical fitting formulae were proposed on the basis on N–body simulations [26]. However, this approach is heavy, because in principle it requires a new simulation every time the cosmological parameters are varied. The PS mass function has the great advantage that the cosmological parameters space can be explored rapidly. Finally, I just mention here that the PS formalism can be extended to give complete merger histories of single halos [30], conditional probability function of progenitor halos [8], biased distribution of halos within halos [35], all topics we do not explore here, but that proved to be very useful in interpreting data. As a final comment, the PS approach after more of 30 years, is still extensively used in the large majority of the papers on precision cosmology with clusters of galaxies.

5.2.8 From the Mass Function to the Distribution of Observables

Let's take a closer look to the behaviour of the mass function. We identify two sets of ingredients: the initial conditions (normalization and shape of the power spectrum) entering $\sigma(M)$, and the cosmological parameters $(\Omega_0, \Omega_\Lambda, w)$ entering $\delta_c(z)$ and the overall normalization of the mass function. As you can

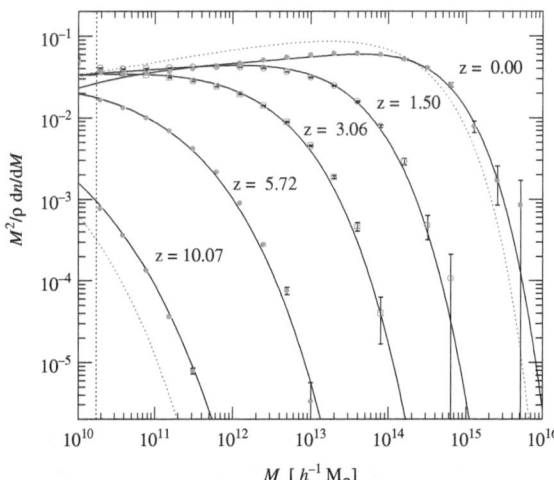

Fig. 5.4. The Press & Schechter mass function (*dotted lines*) tested against N–body simulations (*dots* and *solid lines*, from [50])

easily see, the exponential cut off at the massive end is where the function is most sensible at both sets of parameters, through the normalization of the power spectrum (expressed conventionally as σ_8, which is the amplitude of the spectrum at the scale of $8h^{-1}$ Mpc), and the linear growth factor. We can now see in much more detailed terms the behaviour we already appreciated qualitatively: the evolution of cosmic structures is slower in a universe with lower density with respect to and EdS universe; the same for a Λ–dominated, flat universe. In quintessence models, for higher values of the parameter w, the growth ceases earlier. If we normalize our mass function in order to have the same local density of clusters today, the evolution with z appears faster for $\Omega_0 = 1$ than for open or Λ–dominated universes. This is shown visually by the N–body simulation in Fig. 5.5 (upper panel). Quantitatively, the expected

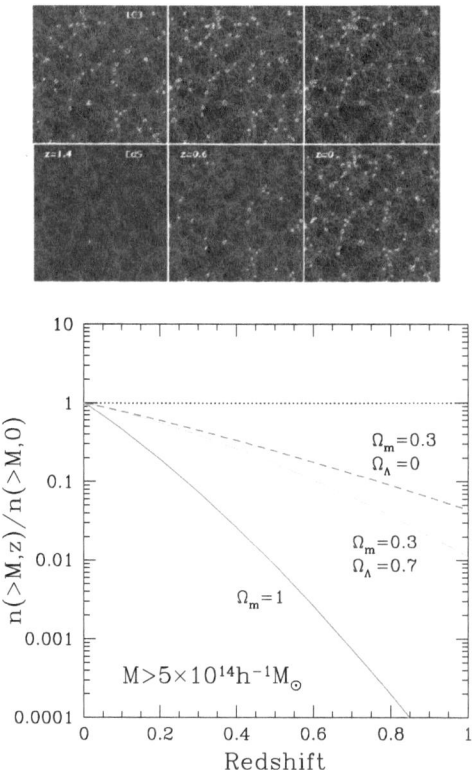

Fig. 5.5. Top: clusters of galaxies (*circles*) in an N–body simulation for an EdS universe (*bottom panels*) compared with clusters in an open FRW universe, with statistically equivalent conditions at $z = 0$. The evolution backward in time of the mass function is strikingly different [6]. **Bottom**: the evolution of the number density of clusters with virial mass $M > 5 \times 10^{14} h^{-1} M_\odot$ for different choices of the cosmological parameters [45]

evolution of the number density of massive clusters (with virial mass $M > 5 \times 10^{14} h^{-1} M_\odot$) is also shown in Fig. 5.5 (lower panel). Now our observational side can take over, and note that, after all, an EdS universe is not more appealing than a low density one since, after all, in the last case we expect much more clusters at high redshifts. And observers love to find high–redshift objects.

We are almost ready to handle real data, except for a final, small step, which consists in a simple change of variables. As you know, in most cases we do not measure directly the virial mass. What an astronomer typically measures is the emitted light in a given band. In our case, as we will see shortly, we will focus on measuring the total luminosity L in the X–ray band and the virial temperature T of the diffuse gas. Therefore, we prefer to have a prediction for the luminosity or the temperature function. This is straightforward if we have a relation M–L or M–T. We know, from the virial theorem, that these relations can be obtained from our spherical collapse model. Once we have the relationships between the observables and the mass, we can write the luminosity (XLF) and the temperature (XTF) functions as:

$$\Phi(L)dL = N(M)\frac{dM}{dL}dL , \quad \Phi(T)dT = N(M)\frac{dM}{dT}dT . \quad (5.16)$$

Enough theory.

5.3 From Observations to Cosmological Parameters

5.3.1 The Observer's Mood

We can start the second part of this lecture, where we will encounter different kind of problems. We are about to look at data, therefore we will face reality, which is always somewhat shocking when coming from the ideal, linear and spherical world of theory.

As we already know, we need a good measure of the actual number density of clusters of galaxies as a function of mass and redshift. We also know that we will get the luminosity or, in the best case, the temperature function of clusters. This implies that we need to be able to: find clusters, measure with high accuracy the quantity of interest, and define the *completeness* of our survey. Completeness is a key quantity in observational cosmology. A well defined completeness means that, for the solid angle of the sky covered by our survey, we are able to detect all the objects with luminosity (or temperature, or mass) above a given value and within a given redshift. This is mandatory to compute the volume we actually explore in the survey and, therefore, the comoving number density. Needless to say, a survey with few objects but a well defined completeness is way much better than a survey with hundreds of objects but a poorly defined completeness. Therefore, we need a strategy to find as many clusters as possible with a well defined completeness. Which is the best observational window to do that? Let's start examining some options.

5.3.2 Optical Band

Searching for clusters in optical images is basically counting galaxies and looking for overdensities with respect to the background value (see [20] for a review). In doing this, the optical colors of the member galaxies are a very useful information. Passive, red galaxies preferentially populate the central regions of clusters, and they form a well defined color–magnitude relation. A galaxy selection picking the reddest galaxies in the field, helps in reducing the contamination by the field galaxies. These techniques can give efficient clusters detection out to $z \geq 1$ (see [22]).

Optical surveys are very convenient to find many cluster candidates. We remind that clusters are rare objects (especially the massive ones) and therefore we need to survey large area to find many of them. The optical band offers the opportunity to cover wide area with large CCD frames, coupled to the availability of ground–based telescopes with large field of view. However optical observations have the drawback of a difficult calibration of the selection function, and therefore the completeness of an optical survey of clusters is very hard to define. This is because the detectability of a cluster depends on the luminosity, the number, and the concentration of its galaxies, three aspects that can vary from cluster to cluster. In addition, projection effects cause severe contamination from background and foreground galaxies: filamentary structures and small groups along the line of sight can mimic a rich cluster. For the same reason, in the presence of a positive fluctuations of the background galaxies, moderately rich cluster can be missed.

More troubles when we try to relate the optical light to the total mass. The total optical luminosity of a cluster is somehow proportional to the total mass. But we know that the stellar mass in the galaxies represents a tiny fraction of the total, and usually only the brightest galaxies are detected, so that a lot of stars in small, undetected galaxies must be accounted for, by assuming a model for the galaxy luminosity function. So, we should not be surprised to know that the relation between the total optical luminosity of a cluster and its total mass is very loose. In order to obtain an accurate measure of the mass, we may use optical spectroscopy to measure the velocity dispersion of the galaxies and then apply the virial theorem. However, this requires a lot of observing time, and obviously it is still affected by contamination from interlopers. All these problems become more severe at high redshift, as the field galaxy population overwhelms galaxy overdensities associated with clusters. A completely different technique is to measure the mass directly through strong and weak lensing (see, e.g. [13]). This is a very promising tool, but it has its own problems, like severe projections effects (the lensing depends on all the mass along the line of sight towards the clusters and on its position) and the difficulty to obtain clean lensing signal.

5.3.3 Millimetric Band (SZ effect)

Among the many virtues of clusters of galaxies, there is this peculiar feature: clusters can be seen as shadows on the cosmic background radiation. This is due to the Sunyaev–Zeldovich (SZ) effect [51]. We know that most of the baryons in clusters are in the form of very hot, ionized gas. Photons from the CMB passing through a cluster find many high–speed electrons and therefore experience Inverse–Compton scattering. In this process, the energy is transferred from the electrons to the much colder CMB photons. Since this process preserves the number of photons, the net result is that the black–body spectrum of the CMB is slightly distorted and shifted to larger frequencies by an amount that depends on the temperature, and on the column density of the ICM. The net effect on the CMB is the production of a cold spot at low and a hot spot at high frequencies, where the pivotal frequency is about 217 GHz (see [24]). This sounds very promising, since we have both a spatial and a spectral signature. Actually, several clusters have been imaged with the OVRO and BIMA arrays [11]. Indeed, the scientific community is making a strong effort to build instruments that can study both CMB and the SZ effect from the ground (like AMI, ACT, AMiBA, APEX, SPT), or from space (like the Planck satellite, whose full–sky survey is expected to detect thousands of clusters).

Among the positive aspects of SZ observations, we find the absence of the redshift dimming, which allows one to identify clusters virtually at any redshift. This means that the selection criteria are essentially equivalent to a completeness in mass, which is very desirable. However, severe contamination from foreground and background radio sources is expected. Multi–frequency observations can help a lot in disentangling the spectral signature of the SZ effect from the spectrum of radio sources. However, the difficulties in detecting clusters via the SZ effect are still significant (see [3]). An easy prediction is that in five years, the SZ effect will be one of the main observational window to find and study clusters of galaxies.

5.3.4 X–ray Band

At present, in my view, the X–ray band is the most convenient to find and investigate clusters. Anyway, it is the field in which I spent most of my activity, and therefore, for a mix of objective and private reasons, since now on, I will focus mostly on X–ray.

The first thing to say is that clusters appear as strong–contrast sources in the X–ray sky up to high redshifts, thanks to the dependence of the X–ray emission on the square of the gas density (see §5). Given the relatively small number of sources, X–ray images of clusters are virtually free from contamination from foreground and background structures. In other words, clusters are the second most prominent sources in the X–ray sky (after Active

Galactic Nuclei), at striking difference with the optical and millimetric bands where they have to struggle to emerge above other stronger signals. This can be clearly appreciated in Fig. 5.6 (left) where almost all the point sources in the image are AGN, while the bright, extended source in the center is a cluster at $z = 0.79$. The image has been taken with the ACIS–I detector, covering a square of 16 arcmin by side. X–ray emission from clusters can be detected up to redshift larger than one, as shown in Fig. 5.6 (right) where the X–ray emission (red) from the $z = 1.235$ cluster RXJ1252 is shown on top of the optical image.

A flux–limited X–ray survey can provide a sample of clusters with a well defined completeness, thanks to the fact that the X–ray emission from clusters is continuous (at variance with the optical emission associated to the single galaxies) and centrally peaked towards the center. Therefore we just need to establish a robust connection between the X–ray luminosity and the total mass. A potential problem with X–ray clusters is that the X–ray flux is sensitive to irregularities in the gas distribution. However, this problem does not seem dramatic, given that most of the clusters appear smooth and round, and the theory provides us with a robust connection between the total mass and the ICM properties. Thus, for the moment, we just need to fully appreciate the advantages in looking at clusters with X–ray satellites, which became possible since the 60s thanks to the first X–ray missions leaded by Riccardo Giacconi. In the spirit of constraining the cosmological parameters, X–ray surveys of clusters of galaxies had a large success in the 90s, thanks to ROSAT and other satellites, and provided consistent but sometimes debatable

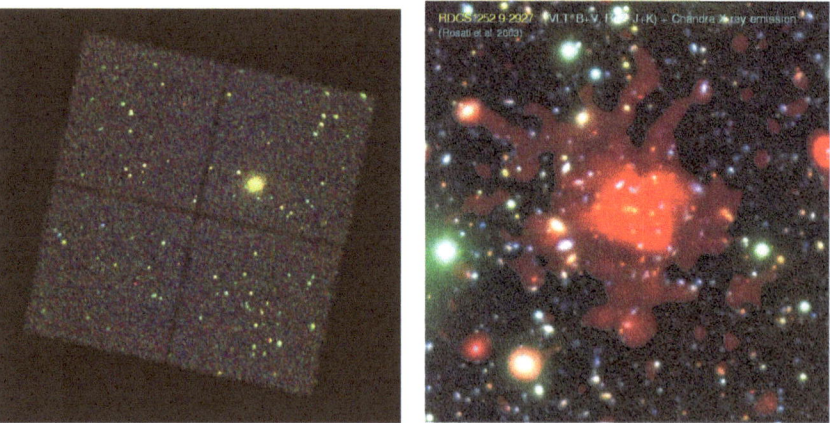

Fig. 5.6. Left: The cluster MS1137, $z = 0.79$, in a field observed for 116 ks with the X–ray telescope *Chandra*. The cluster is the bright extended source in the center, while most of the remaining sources are AGN. **Right**: the X–ray emission from the $z = 1.235$ cluster RXJ1252 is shown on top of the optical image taken with the VLT telescope [46]

results. For a review of the many surveys with cosmological impact see the review by [45].

We are now in the era of the XMM–Newton and Chandra satellites. These two telescopes are mostly performing pointed observations of clusters discovered in the previous surveys. No wide area surveys are currently planned, given the small field of view of these satellites (nonetheless, some serendipitous surveys are underway with both of them). These pointed observations are bringing to us many beautiful images, along with many uncomfortable news that we will discuss in §3. Before stepping further, let's remind the basics of X–ray emission from the ICM.

5.3.5 The X–ray Emission from Clusters of Galaxies

We know that most of the baryons in clusters are in the form of hot plasma. This plasma is optically thin and it radiates by free–free (bremsstrahlung) emission. It is in collisional equilibrium, therefore its typical temperature is set by the large dynamical masses of clusters ($10^{14} - 10^{15} M_\odot$) to be in the range of 10–100 millions K (corresponding to 1–10 keV). This implies that most of the emission is in the X–ray band. The total X–ray emissivity due to thermal bremsstrahlung is obtained by integrating over the distribution of speeds of the plasma electrons, and, after a further integration over frequencies, it can be written as (see [43]):

$$\frac{dL}{dV} = 1.4 \times 10^{-27} T^{1/2} n_e^2 Z^2 \bar{g}_B \text{ erg s}^{-1}\text{cm}^{-3} , \qquad (5.17)$$

where Z is the atomic number of the ions and $\bar{g}_B$ is the velocity–averaged Gaunt factor averaged over frequencies. First, we notice the dependence of the total emissivity on the square of the electron density. This is the main reason why clusters are high–contrast sources in the X–ray sky, and also why superposition or confusion effects due to smaller background or foreground halos, are less important than in the optical band, where the total luminosity scales linearly with the (stellar) mass. We also note the weaker dependence on the temperature ($T^{1/2}$).

Another contribution to the X–ray luminosity comes from the line emission due to heavy ions. This contribution is generally negligible in terms of total emission, since at temperatures larger than 5 keV, almost all the heavy nuclei are fully ionized. However, the line–emission contribution is increasing at low temperatures, and starts to be relevant below 2 keV. This aspect is important when studying the production of metals in cluster galaxies and their diffusion into the ICM. A typical X–ray spectrum of a cluster, with the typical Iron line at 6.7 keV rest–frame, is shown in Fig. 5.7 (right).

Equation (5.17) gives the luminosity per unit volume, therefore, the total luminosity must be obtained by integrating up to the virial radius. In the simplest assumption of isothermality ($kT = const$ at any radius in the cluster), the only relevant quantity is the square of the electron density $n_e^2(r)$, which

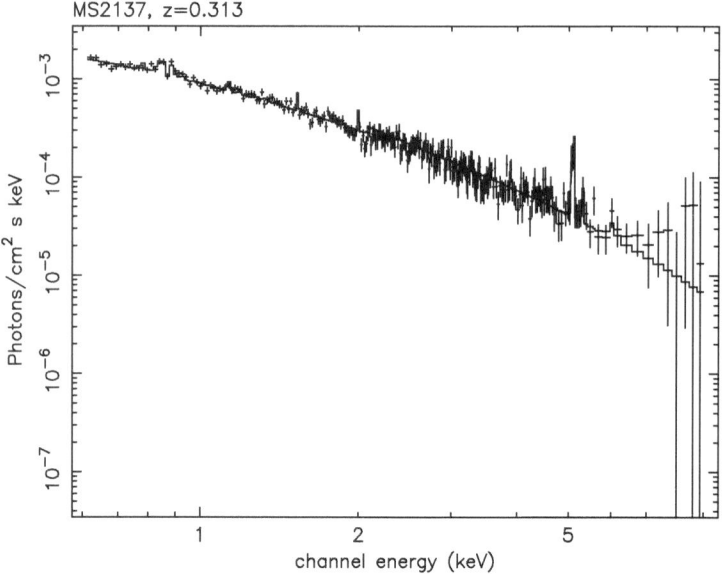

Fig. 5.7. The spectrum of MS2137 observed with ACIS-I onboard of the *Chandra* satellite. MS2137 is a bright X–ray cluster at z=0.313, with an average temperature of about 5 keV

is generally assumed proportional to the gas density n_g. In general the gas density profile is described with the so–called β–model [12], which consists in a flat central core and a steep decrease in the outer regions:

$$n_g \propto 1/(1 + (r/r_c)^2)^{3\beta/2} , \qquad (5.18)$$

where r_c is the core radius, and the parameter $\beta \sim 0.5$–1 can be interpreted as the ratio of the specific energy of the dark matter particles (often measured through the galaxies velocity dispersion) over the gas temperature. Given the steep slope outside the core, and the n_g^2 dependence of the luminosity, only the central regions (few core radii) are clearly detected in the X–ray images. The outer regions are hardly detected even with present–day satellites. Observers often prefer to quote all the quantities within the observed radius, which is typically half or less than the virial one.

As we know, X–ray detectors onboard of the Chandra and XMM satellites are CCD cameras, which read the collected photons every few seconds, recording both the position and the energy (with a reasonable error of few percent). Therefore X–ray astronomy has the big advantage of recording images and spectra at the same time. High resolution X–ray spectroscopy is still feasible through gratings, however the energy resolution of the CCD is good enough to our purposes of measuring the temperature of the baryons.

Once we obtain the baryon density from the X-ray surface brightness, and the temperature of the gas, we can measure the total mass simply by applying the condition of hydrostatic equilibrium:

$$M(< r) = -\frac{k_B T R}{G \mu m_p} \left(\frac{d\log(\rho_g)}{d\log(r)} + \frac{d\log(T)}{d\log(r)} \right) , \qquad (5.19)$$

where μ is the mean molecular weight ($\mu \sim 0.6$) and m_P is the proton mass (see [45]). Here we let the temperature free to change with the radius. Of course this equation is particularly simple in the isothermal case. In general, the masses obtained in this way are pretty close to that obtained simply through the virial theorem $T \propto M^{2/3}$. On the other hand, it is well known that clusters do have a temperature structure, which is often well described by a mild decrease outwards (see [57]), and, in more than half of the local clusters, a drop of about a factor of three in the very inner regions (the cold core, see [38]). The temperature profile is quite important, but its measure is increasingly difficult at increasingly high redshifts. Indeed, we need a lot of photons in order to measure the temperature in several concentric regions (at least one thousand for each independent spectrum), and to obtain the deprojected temperature profiles. For this reason, virial masses of distant clusters are often derived assuming isothermality.

Our framework allows us to relate the basic X-ray observables, luminosity and temperature, to the dynamical mass. We already know that luminosity is more affected by the details of the gas distribution, while the M-T relation appears more stable since it is directly based on the virial theorem. But we also know that luminosity is much easier to measure, since we need much less photons to measure a luminosity, and therefore we can observe many more clusters within a given amount of telescope time. A shortcut is to build phenomenologically the L-T relation, fitting the data with a formula of the kind:

$$L_{bol} = L_6 \left(\frac{T_X}{6keV} \right)^\alpha (1 + z)^A \left(\frac{d_L(z)}{d_{L,EdS}(z)} \right)^2 10^{44} h^{-2} ergs^{-1} , \qquad (5.20)$$

where α is measured to be about 3, while the evolutionary parameter A is more uncertain and varies between 1 and 0 (see [17, 55]). Once the relations between the X-ray observables and the mass are established, we can compare the observed XLF and XTF to our predictions. For a review of the X-ray properties of X-ray clusters, see the book by [46].

5.3.6 Measuring Ω_0 from the Observed X-ray Luminosity Function

The luminosity function seems easy to measure: first we count all the clusters in our survey, then we measure their flux just counting the photons from each cluster. We also have to know the redshift of each cluster with a good approximation, in order to compute luminosities. The redshift can be obtained with an optical spectroscopic follow–up on a limited number of member galaxies, or with photometric techniques. As noted before, shallow X–ray surveys allows us to measure the luminosity with good accuracy, and to scan a wide area of the sky. Once we have a flux limited sample with measured luminosities, we build the XLF by adding the contribution to the space density of each cluster in a given luminosity bin ΔL:

$$\phi(L_X) = \frac{1}{L_X} \Sigma_{i=1}^{n} \frac{1}{V_{max}(L_i, f_{lim})} \qquad (5.21)$$

where V_{max} is the total search volume defined as:

$$V_{max} = \int_0^{z_{max}} S[f(L,z)] \left(\frac{d_L(z)}{(1+z)}\right)^2 \frac{c\,dz}{H(z)} , \qquad (5.22)$$

where $S(f)$ is the sky coverage, which depends on the flux (since the sensitivity of a survey can vary across the surveyed region of the sky), and $d_L(z)$ is the luminosity distance.

Remember that we expect to get information on the cosmological parameters both from the shape of the XLF and from its evolution with redshift. To begin with, the shape of the local XLF is well understood thanks to several different surveys giving consistent values, and it is shown in Fig. 5.8 (upper panel). This allows already to get some information from the data at $z = 0$, by finding the parameters which minimize the χ^2 computed on the binned luminosity function from (5.21), or by a maximum–likelyhood approach using the unbinned data (see [5]).

However, when only local data are used, we find a lot of degeneracy among cosmological parameters. Lower Ω_0 can be compensated by higher spectrum normalization σ_8 (see Fig. 5.8, lower panels). To break this degeneracy we can use the evolution with redshift. The evolution of the XLF is still debated: there is a hint of evolution at the very bright end, but for the typical L_* clusters and less luminous ones, there is no evolution almost up to $z \sim 1$ (see discussion in the review by [45]. In other words, most of the clusters, if we exclude the brightest ones, are already in place at high redshift. We know what does it mean, at least qualitatively: the matter density parameter Ω_0 is significantly lower than 1.

Our group, few years ago, applied this cosmological test to the RDCS survey [44], which is the deepest sample of X–ray selected clusters. This choice provide a good leverage in terms of cosmic epoch, but necessarily, given the relatively small solid angle surveyed with respect to shallower surveys, does not probe well the high luminosity end. The results, published by [4, 5] are shown in Fig. 5.9, where we used also data from the EMSS survey [21]. In these Figures we notice that some degeneracy is still present also when fitting the XLF in the high redshift bins. We also notice that the constraints on the cosmological parameters Ω_0 and σ_8, are weakened when the parameters α and A, describing the slope and evolution of the L–T relation, are allowed to vary within the observational uncertainties.

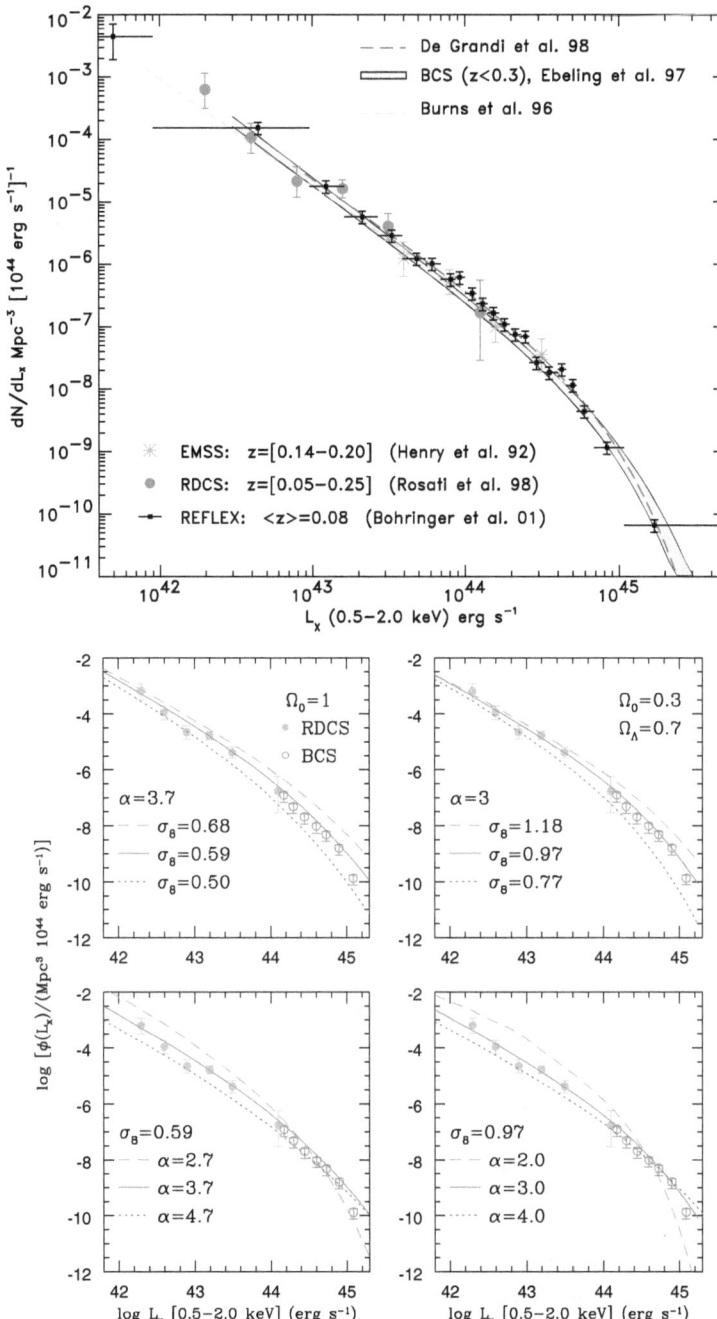

Fig. 5.8. Upper panel: the local X–ray luminosity function of clusters of galaxies from different samples computed for an EdS Universe with $H_0 = 50$ km s^{-1} Mpc^{-1} [45]. **Lower panels**: the local X–ray luminosity function of clusters of galaxies from RDCS (*filled circles*) and BCS (*open circles*) for different σ_8 and different parameter α for the slope of the L–T relation [4]

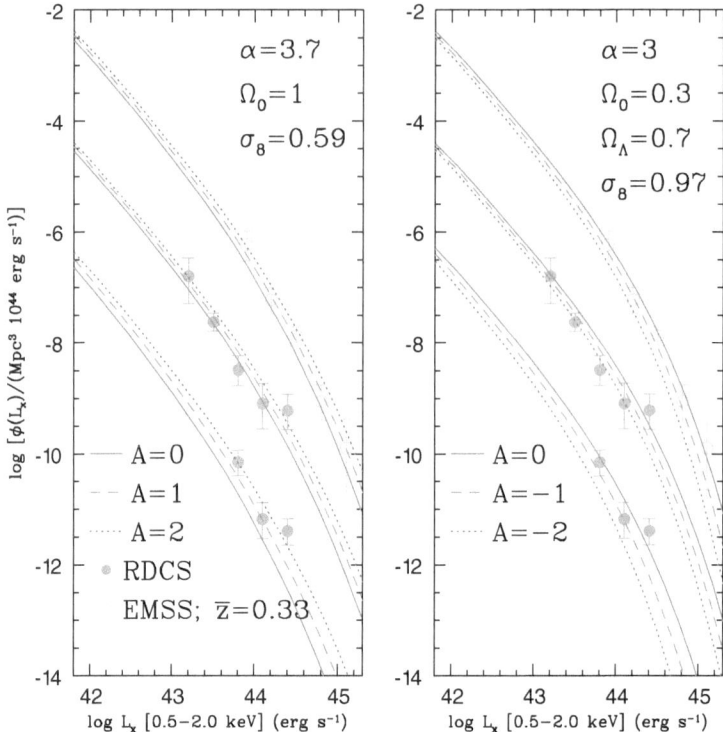

Fig. 5.9. The X–ray luminosity function of clusters of galaxies in three different redshift bins: z=0.3–0.6 (EMSS data); z=0.25–0.50 and z=0.50–0.85 (RDCS). For each model and at each redshift, different curves refer to different evolutions for the L–T relation [4]

The uncertainties on the cosmological parameters are better shown in terms of confidence contour levels, where we can also evaluate the effects of the uncertainties associated to the parameters describing the L–T relation. 5In Fig. 5.10 we show how the confidence contours in the Ω_0–σ_8 space dance around when the slope and evolution of the L–T relation (parametrized by α and A like in (5.20)), but also the normalization of the $M-T$ relation (parameter β), are allowed to vary. The displacements of the contours are at more than 3 σ, therefore we are learning uncomfortable news: the uncertainties on the properties of the ICM are affecting the cosmological tests at a significant level.

The situation is getting worse when we investigate the dark energy parameter w. While the density parameter Ω_0 is well constrained by clusters, w is hardly constrained at all. Recent works trying to constrain dark energy, combine constraints from both SneIa and clusters, to significantly improve the constraints on w due to the complementarity of the two tests in the Ω_0–w space (see Fig. 5.11).

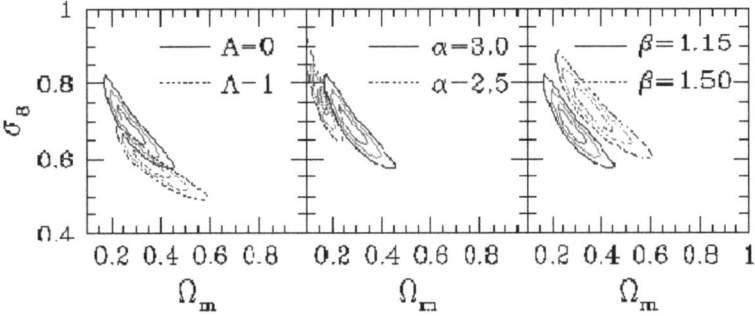

Fig. 5.10. Confidence contours in the density parameter Ω_0 and the normalization of the density fluctuations spectrum σ_8 from the fit of the high–z XLF for different choices of the parameters describing the physical relations L–T (α and A) and M–T (β; [5])

5.3.7 Measuring Ω_0 from the Observed X–ray Temperature Function

At this point you may ask: since our theoretical framework seems quite successful, why do we have such large uncertainties in the relations between L and T? Not only we showed that the relation between mass and luminosity is reasonably understood on the basis of the spherical collapse, but we also mentioned a possible shortcut through the direct measure of the L–T relation. Well, we knew that something wrong were lurking somewhere... However, before worrying too much, let's give a try to the XTF, which is based only on the more robust M–T relation. Indeed, the M–T relation relies directly on the virial theorem and it is observed to have smaller scatter with respect to that observed in the L–T relation.

When using the XTF, the price to pay, as we know, is that it is much more difficult to assemble a complete sample of clusters with temperatures

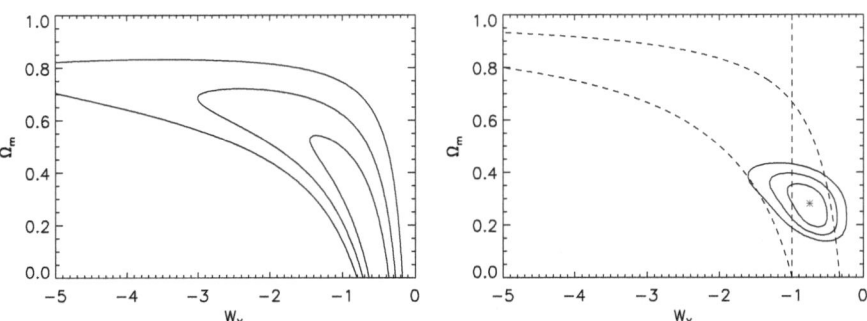

Fig. 5.11. Confidence contours (1–3 σ levels for two degrees of freedom) in the Ω_0–w plane obtained from SNeIa only [37] sample, *left panel*) and SNeIa plus REFLEX (*right panel*, from [48])

measured with reasonable errors. However the XTF is considered to be more effective in constraining cosmological parameters. The first good news is that the constraints from the XTF are similar to that from the XLF. The constraints obtained from the XTF point towards $\Omega_0 \sim 0.3$ for a flat universe (see [14]), providing at the same time significant constraints on the normalization of the power spectrum (see [39]). In Fig. 5.12 we show the results from [15]. We notice the tight constraints, but, again, also a significant degeneration in the $\sigma_8 - \Omega_0$ space.

An additional problem comes from a parameter which we considered, so far, pretty robust: the normalization of the M–T relation. It has been noticed that the value of β found in N–body simulations is higher than the observed one. This can be due to several effects (see [7]), but the net result is that the uncertainties on this parameter introduce uncertainties in the constraints from the XTF in the same way as the L–T parameters are weakening the constraints from the XLF (see, e.g. [25]).

It is clear at this point that the main uncertainties comes from the poor understanding of the scaling relations between the ICM observables and the mass, both from the theoretical and the observational points of view. A detailed investigation of the effects of such uncertainties is given in [40]. They conclude that the cosmological constraints from XLF and XTF, both the local and the evolved ones, are reliable and consistent with each other, but that the statistical errors on the cosmological parameters are larger than previously thought. The buzzword now is: we need to improve the quality of the data on single clusters to better understand the physics of the ICM. But why did clusters prove to be such a difficult topic, after being the best candidate for the most friendly objects in the Universe?

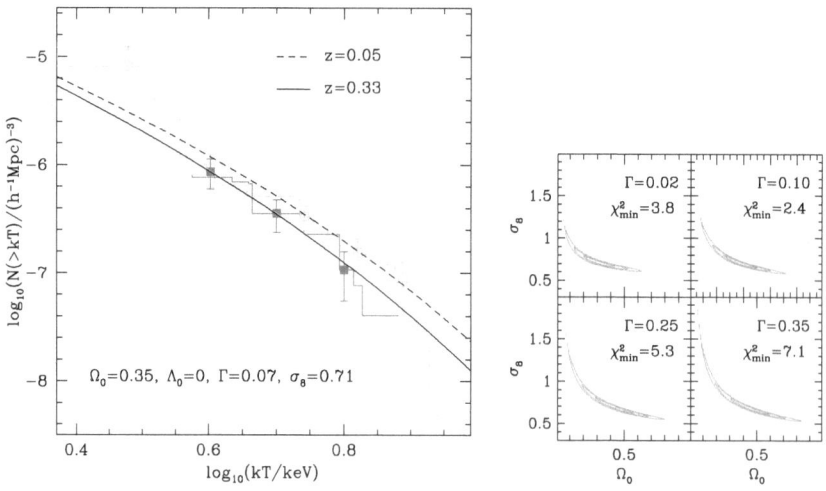

Fig. 5.12. Left: Fit to the evolved temperature function. **Right**: confidence contours in the Ω_0–σ_8 space (from [15])

5.4 New Physics and Future Prospects

5.4.1 Something is Missing: New Physics for the Clusters Baryons

Why do we have such a poor understanding of the L–T relation? From (5.17), assuming $n_e \propto \rho_{tot}$ (in other words, that the baryons follow the total matter distribution), and integrating over the volume, we obtain $L \propto T^2$ (without including line–emission). This is the L–T relation predicted in what is called the *self–similar* scaling [27]). As long as the baryons are distributed in the same way of the total mass, each X–ray observable scales like some power of the mass. Another way to say this, is that small clusters are the mass–rescaled version of massive clusters.

So far, we reasonably expected that the thermodynamics of the ICM, being dominated by dark matter, is driven by gravitational processes, like shocks and adiabatic compression occurring during the virialization phase and the subsequent growth in mass by accretion. This self–similar behaviour is also supported by N–body hydrodynamical simulations which do not include radiative cooling. But the observed slope of the L–T relation is much steeper then predicted ($\alpha \geq 3$ rather than 2 or lower when line emission is included) and it constitutes the first strong evidence of something wrong in the self–similar picture. That's why when performing the cosmological tests, we avoided this inconsistency by varying the parameters of the ICM scaling relations.

However, we learned that thawing the thermodynamic parameters introduces large uncertainties in the cosmological constraints. Obviously, we would appreciate a lot to have a physical basis for the observed scaling relations, in order to better control the uncertainties due to a poor description of the ICM thermodynamics. The first step is to invoke a physical process that leads naturally to an $L \propto T^3$ scaling, in other words, a process which implies a progressive decrease of the X–ray luminosity at low mass or temperatures, as shown in Fig. 5.13 (top). How can we obtain this? We know that we can efficiently decrease the predicted luminosity by imposing a lower density in the central regions of the clusters. To do that, we simply need to add an extra pressure, or some extra amount of energy in the center of clusters. *Extra* means in excess with respect to the energy acquired through shocks and adiabatic heating. This extra energy does not translate in an higher temperature; what happens, is that the pressure increases, and the gas distribution gets puffier, readjusting itself in the dark matter potential well. A useful quantity to describe such behaviour is $K \equiv T/n^{2/3}$. This is the normalization of the equation of state of the ICM, which is that of a perfect gas, $p = K\rho^{5/3}$. We remind that the entropy is $S = N ln(K)$. The entropy is also a very convenient thermodynamic variable, since it is constant during adiabatic compression, and it changes only in the presence of radiative cooling or shock heating. For this reason, another way of describing the break of the self–similarity in clusters, as shown by [41], is to plot the entropy as a function of the cluster temperature, as shown in Fig. 5.13 (bottom).

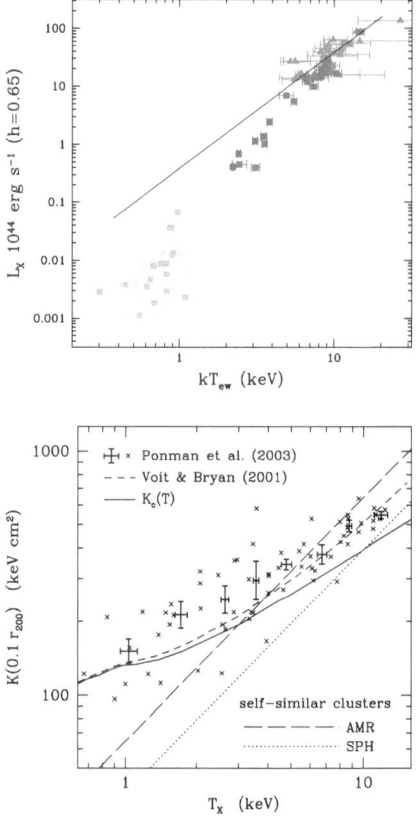

Fig. 5.13. Top: the L–T relation for groups and clusters showing the steeper slope with respect to the self–similar model $L \propto T^2$ (*continuous line*). **Bottom**: the entropy ramp, showing the higher entropy in low temperature systems with respect to the self–similar model [58]

The desired effect is obtained by giving about half or 1 keV to each gas particle. The effect is small in rich clusters, where the virial temperature is around 10 keV, and the energy budget is largely dominated by gravity, while it is increasingly large at lower temperatures, when the extra energy starts to be a significant fraction of the gravitational energy scale. In this way we solved the problem from the point of view of the thermodynamics (see, e.g. [54]). Of course, the real problem starts now: which is the physical mechanism responsible for the energy (or the entropy) excess?

We have two obvious candidates which can inject energy associated to non–gravitational processes: the prime candidate is feedback from star formation processes, whose effects are testified by the presence of heavy elements in the ICM. The second candidate is feedback from nuclear activity in the clusters galaxies. Actually, the interaction of AGN jets and the ICM has been directly observed. The most spectacular example is the Perseus cluster, where jets from the central AGN (visible in the radio emission) is pushing the ICM creating two large symmetric cavities towards the center ([18, 19]; see Fig. 5.14). Chandra and XMM added other surprises: the presence of *cold fronts* ([32, 54]) and of massive mergers strongly affecting the dynamical equilibrium. To this, we must add the puzzling discovery by XMM that the ICM in the central regions never cools down by more than a factor of 3 with respect to the virial temperature, despite the cooling time is much shorter than the age of the cluster. Again, another evidence that some homogeneous process heats the gas.

Today, we see clearly that the Chandra and XMM satellites changed our perspective of clusters of galaxies. If in the ROSAT era the main goal was to find as many clusters as possible with the aim of constraining cosmology, in the Chandra/XMM era the goal is to observe with much better spatial and spectral resolution the clusters previously discovered. The physics of the ICM is much more complex than expected and this forces us to reconsider all the relations between the X–ray observables and the dynamical mass. This aspect may cast some doubts on the use of X–ray clusters of galaxies as cosmological tools. One can also think to reverse the argument: the physics of the ICM is much more interesting, so let's investigate the evolutionary properties of clusters to understand the effects of feedback processes onto the ICM, and don't worry about cosmology.

In my view, the investigation of cosmology and of the ICM physics must proceed together. Actually, this is what is happening: if you go through the literature in the last six years, you discover indeed that there is still a strong interest in cosmological tests with clusters, which is supported by a growing amount of works on the ICM. It must be noticed in addition, that understanding the problem of the non–gravitational heating of the ICM by energetic feedback from star formation or nuclear activity, is a key issue in cosmic structure formation. Actually, feedback is the holy grail of structure formation today! If you go to a conference on galaxies, clusters, or anything on cosmic structure formation, you will hear everywhere the word "feedback". So, rather than saying that clusters became less interesting in a cosmological perspective,

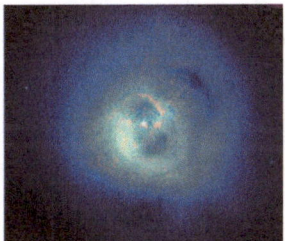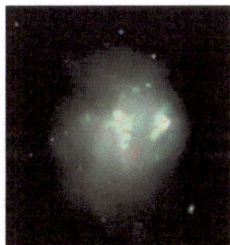

Fig. 5.14. *From left to right:* AGN activity creating cavities in the ICM of the Perseus cluster ([18];[19]); cold fronts in Abell 2142 [33]; an ongoing massive merger in 1E 0657–56, the *bullet cluster* [34]

I prefer to say that clusters became even more important to understand both structure formation and cosmology.

5.4.2 A Simpler Cosmological Test

There is not enough space here to describe the most recent progress in the understanding of the ICM thermodynamics. However, I want to mention another cosmological test that appears to be simpler than that discussed so far. Instead of relying on the knowledge of the dynamics of clusters, we can focus on a much simple quantity: the baryonic fraction f_B. We simply need to measure the total mass, and count all the baryons in the form of stars and ICM. From semianalytical models and numerical simulations, we expect that the physics of the ICM does not affect f_B if measured at a radius where the gravity dominates; therefore, it should be close to the cosmic value Ω_B/Ω_0. In other words, the baryons are allowed to behave wildly and decouple from the dark matter distribution in high density regions, but on large scales they are not displaced differently from dark matter. The virial radius is expected, then, to include a closed region where the average composition does not change during the evolution of the cluster. It is straightforward to see that the measure of f_B and the knowledge of Ω_B from nucleosynthesis or from the CMB [49] gives a straightforward measure of Ω_0 (see [59]).

But this is not all: for the same reasons, the baryonic fraction should not evolve with redshift. However, the actual measure of f_B does depend on the angular distance. The mass of baryons is recovered by measuring the flux and by knowing the physical size of the cluster. The relation between the measured flux S_X and the mass of gas reads as:

$$S_X = L_X(1+z)^{-4}/(4\pi d_{ang}^2) \propto M_{gas}^2 \theta_c^{-3} d_{ang}^{-3}/d_{ang}^2 . \tag{5.23}$$

On the other hand, the total mass depends on the angular distance as $M_{tot} \propto \theta_c d_{ang}$. It follows that $f_B = M_{gas}/M_{tot} \propto d_{ang}^{3/2}$. Thus, we have two advantages here: the value of the baryon density gives Ω_0, while its apparent evolution is depending on the cosmological parameters through d_{ang}. Therefore, the cosmological test consists in requiring no evolution in the observed f_B. Any apparent evolution in the baryonic fraction is the smoking gun of wrong cosmological parameters. It is important to perform this test on a redshift range as wide as possible (see Fig. 5.15, left).

This is not a dynamical test, but rather a geometrical test, and it is more sensitive to Ω_Λ (see [2]). However, we notice that the scatter in the baryonic fraction from cluster to cluster is somewhat larger than we would like, given the starting assumption of a universal value for f_B for all clusters at all epochs. This is probably due to the fact that the dynamical masses and the baryonic fraction measures are still affected by complexities in the ICM physics (see [23]). However, this kind of test is very promising, and it becomes very powerful when combined with CMB or SNeIa test, as shown in Fig. 5.15 (right).

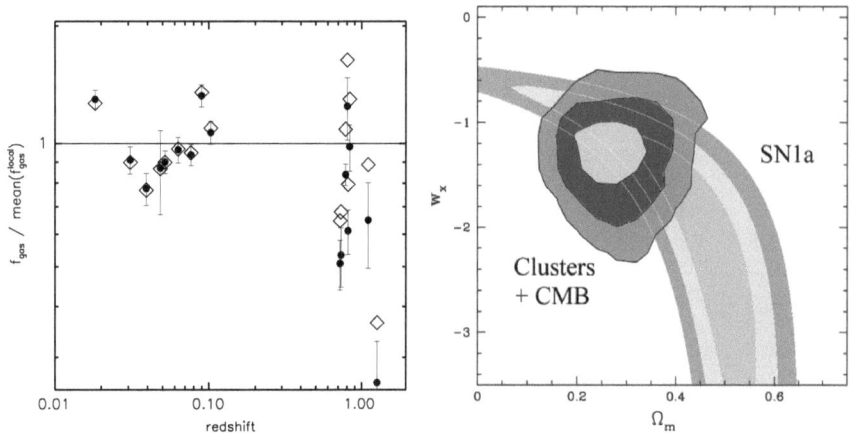

Fig. 5.15. Left: f_B measured for a sample of high–z clusters in an EdS cosmology (*dots*) and in a flat Λ universe (*empty diamonds*, from [16]). **Right**: constraints in the w–Ω_0 plane obtained by combining baryonic fraction in clusters and CMB [2]

5.4.3 Future Prospects for Precision Cosmology with Clusters

We are approaching the end of our brief introduction to cosmological tests with clusters of galaxies. A clear way to summarize it, is the cosmic triangle shown in Fig. 5.16. Each side represents one of the three main parameters: the mass density, the cosmological constant, and the curvature. Contours levels perpendicular to one side mean that a particular test is efficient in constraining that parameter. Cosmological tests based on clusters are mostly sensitive to Ω_0, while geometrical tests like CMB and SNeIa are more sensitive to the curvature and Ω_Λ. Roughly speaking, CMB can constrain $\Omega_0 + \Omega_\Lambda$, while SNeIa $\Omega_0 - \Omega_\Lambda$, mainly because of the different redshift range, 0.5–2 for SNeIa and 1000 for CMB. Obviously, the combination of the three tests is very powerful, but its application requires a good understanding of all the different systematics.

This picture is still valid after recent observations by Chandra and XMM showed that the physics of clusters is more complicated than expected. The key questions on the future of precision cosmology with clusters of galaxies are: do we need a new, large, all–sky survey of clusters? Or should we first understand better the physics of the ICM? Therefore, which is the best instrument we should build next? I think that the best answer is that a new, medium–depth all–sky survey of clusters is needed for both aspects. First, a large survey can help in obtaining strong constraints on the cosmological parameters, providing at the same time large samples to investigate the relationship between X–ray observables and the dynamical masses. A second crucial aspect, is that a large survey would discover new clusters, especially at high redshift. This is mandatory to provide targets for the future X–ray missions, which will provide sensitive, narrow–field instruments to investigate the physics of the

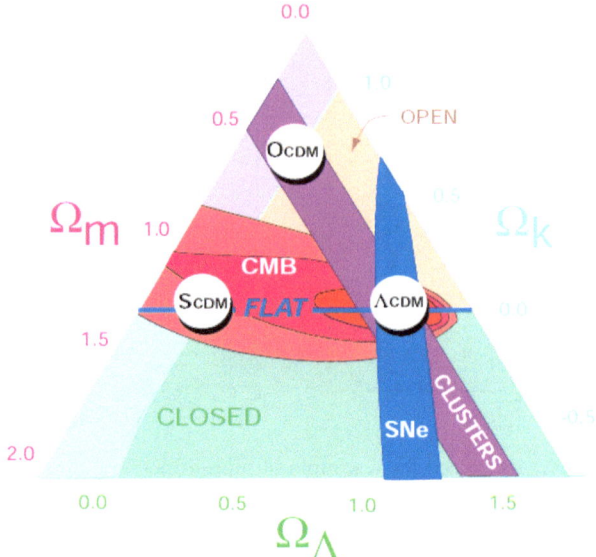

Fig. 5.16. The cosmic triangle [53], see http://wwwphy.princeton.edu/ steinh/). The complementarity of the three classic cosmological tests is clearly shown

ICM. Without a new wide survey, we will run out of clusters to observe! Several proposals of medium–size mission have been circulated so far, but at present there are no planned large–area surveys of the X–ray sky. The future of X–ray cluster astrophysics largely depends on this.

5.5 What to Bring Home

At the end of this introduction, we should be aware that clusters of galaxies constitute a cosmological tools to significantly constrain Ω_0 and the spectrum of primordial fluctuations, through tests based on dynamics or on geometry. X–ray observations offer the best tool to measure mass and collect complete sample of clusters. Main results points towards a flat Λ–dominated Universe ($\Omega \sim 0.3$ and $\Omega_\Lambda \sim 0.7$, or $w = -1$) and a normalization of the fluctuations power spectrum consistent with that measured from CMB for a CDM Universe ($\sigma_8 \simeq 0.8$).

If someone wants to start the business of cosmological tests with clusters, she/he just needs basic programming skills to put in a simple code all the formulae we discussed, and a good X–ray observer among the collaborators, in order to have access to a well defined, complete survey of clusters. However, one must know that this game was played a lot starting from the 90's, when it was realized that clusters constitute one of the most powerful cosmological tools. At present, in 2006, most of the best X–ray clusters surveys have been exploited in this sense. Therefore, if you want to start the business, you better have something smart in mind, mainly a way to deal with any possible systematics or with a better treatment of the effects of the poorly known thermodynamics of the ICM.

However, a noticeable contribution would be given by supporting the scientific case of future X–ray missions to obtain new data, in the form of a wide and complete sample of clusters. Larger samples indeed, will allow to study at the same time the thermodynamics of the ICM and the evolution of clusters as a population. Finally, one should not have the feeling that the physics of the ICM is now the hot topic at the expenses of precision cosmology, which should rely only on tests based on SNeIa and CMB. As a general comment, I would like to stress that clusters are probing a different cosmic epoch with respect to CMB, and a different physics with respect to SNeIa, therefore they will always be a complementary and useful test for cosmology. The complex ICM physics, instead of being an obstacle, must be seen as a further opportunity to learn about structure formation in the Universe.

References

1. G.O. Abell, (1958). ApJS **3**, 211.
2. Allen et al., (2004). MNRAS **353**, 457.
3. M. Birkinshaw, and K. Lancaster (2004) Proceedings of the International School of Physics "Enrico Fermi", Eds. Melchiorri, F. & Rephaeli, Y., astro-ph/0410336.
4. S. Borgani et al., (1999). ApJ **517**, 40.
5. S. Borgani et al., (2001). ApJ **561**, 13.
6. S. Borgani and L. Guzzo, (2001). Nature **409**, 39.
7. S. Borgani et al., (2004). MNRAS **348**, 1078.
8. R.G. Bower, (1991). MNRAS **248**, 332.
9. T. Broadhurst et al., (2005). ApJ **621**, 53.
10. R.R. Caldwell, Dave, R. and P.J. Steinhardt, (1998). PhRvL, **80**, 1582.
11. Carlstrom et al., (2002) ARA&A **40**, 643.
12. A. Cavaliere, and R. Fusco-Femiano (1976). A&A **49**, 137.
13. H. Dahle et al., (2003). ApJ **591**, 662.
14. M. Donahue, and G.M. Voit (1999). ApJL **523**, 137.
15. V.R. Eke, J.F. Navarro, C.S. Frenk (1998). ApJ, **503**, 569.
16. S. Ettori et al., (2003). A&A **398**, 879.
17. S. Ettori et al., (2004). A&A **417**, 13.
18. A.C. Fabian et al., (2002). MNRAS **331**, 369.
19. A.C. Fabian et al., (2003). MNRAS **344L**, 43.
20. R.R. Gal (2006). Guillermo Haro Summer School, astro-ph/0601195.
21. I.M. Gioia et al., (1990). ApJL **356** 35.
22. M.D. Gladders and H.K.C. Yee (2005). ApJS **157**, 1.
23. E.J. Hallman et al., (2005). ApJ submitted, astro-ph/0509460.
24. G.P. Holder, J.E. Carlstrom (2001). ApJ **558**, 515.
25. Y. Ikebe et al., (2002). A&A, **383**, 773.

26. A. Jenkins et al., (2001). MNRAS **321**, 372.
27. N. Kaiser (1986). MNRAS **222**, 323.
28. N. Kaiser (2002). *Elements of Astrophysics*, http://www.ifa.hawaii.edu/k̃aiser.
29. E.W. Kolb and M.T. Turner (1990). *The Early Universe*, Frontiers in Physics, v. 69, Addison–Wesley.
30. C. Lacey and S. Cole (1993). MNRAS **262**, 627.
31. O. Lahav and A.R. Lidlle (2006). in *The Review of Particle Physics 2006*, astro-ph/0601168.
32. P. Mazzotta et al., 2001, ApJ **555**, 205.
33. M. Markevitch et al., (2000). ApJ **541**, 542.
34. M. Markevitch et al., (2004). ApJ **606**, 819.
35. H.J. Mo and S.D.M. White (1996). MNRAS 282, 347.
36. P.J.E. Peebles (1993). *Physical Cosmology* (Princeton: Princeton Univ. Press).
37. S. Perlmutter et al., (1999). ApJ **517**, 565.
38. J.R. Peterson and A.C. Fabian (2005). Physics Reports astro-ph/0512549.
39. E. Pierpaoli, D. Scott and M. White (2001). MNRAS **325**, 77.
40. E. Pierpaoli et al., (2003). MNRAS **342**, 163.
41. T.J. Ponman, D.B. Cannon and J.F. Navarro, (1999). Nature **397**, 135.
42. W.H. Press and P. Schechter, (1974). ApJ **187**, 425.
43. G.B. Ribicky and A.P. Lighman (1979). *Radiative Processes in Astrophysics*, Wiley & Sons Eds.
44. P. Rosati et al., (1998). ApJL **492**, 21.
45. P. Rosati, S. Borgani and C. Norman (2002). ARA&A **40**, 539.
46. P. Rosati et al., (2004). AJ **127**, 230.
47. C.L. Sarazin (1988), *X–ray Emission from Clusters of Galaxies*, (Cambridge: Cambridge University Press).
48. P. Schuecker et al., (2003). A&A **402**, 53.
49. D.N. Spergel et al., (2003). ApJS **148**, 175.
50. D.N.Springel et al., (2005). Nature **435**, 629.
51. R. A. Sunyaev and Ya.B. Zeldovich (1972). CoASP **4**, 173.
52. Steinhardt (2003). http://wwwphy.princeton.edu/steinh.
53. M. Tegmark (2002). http://space.mit.edu/home/tegmark/sdss.html.
54. P. Tozzi and C. Norman (2001). ApJ **546**, 63.
55. A. Vikhlinin et al., (2001). ApJ **551**, 160.
56. A. Vikhlinin et al., (2002). ApJ **578**, 107.
57. A. Vikhlinin et al., (2005). ApJ **628**, 655.
58. G.M. Voit, S.T. Kay and G.L. Bryan (2005). MNRAS **364**, 909.
59. S.D.M. White et al., (1993). Nature **366**, 429.

6

Cosmological Constraints
from Galaxy Clustering

Will Percival

Institute of Cosmology and Gravitation, University of Portsmouth, Portsmouth
PO1 2EG, UK
will.percival@port.ac.uk

Abstract. In this manuscript I introduce the mathematics and physics that underpins recent work using the clustering of galaxies to derive cosmological model constraints. I start by describing the basic concepts, and gradually move on to some of the complexities involved in analysing galaxy redshift surveys, focusing on the 2dF Galaxy Redshift Survey (2dFGRS) and the Sloan Digital Sky survey (SDSS). Difficulties within such an analysis, particularly dealing with redshift space distortions and galaxy bias are highlighted. I then describe current observations of the CMB fluctuation power spectrum, and consider the importance of measurements of the clustering of galaxies in light of recent experiments. Finally, I provide an example joint analysis of the latest CMB and large-scale structure data, leading to a set of parameter constraints.

6.1 Introduction

The basic techniques required to analyse galaxy clustering were introduced in the 70s [48], and have been subsequently refined to match data sets of increasing quality and size. In this manuscript I have tried to summarise the current state of this field. Obviously, such an attempt can never be complete or unique in every detail, although it is still worthwhile as it is always useful to have more than one source of information. An excellent alternative viewpoint was recently provided by Hamilton [25, 26], which covers some of the same material, and provides a more detailed review of some of the statistical methods that are used. Additionally it is worth directing the interested reader to a number of good text books that cover this topic [11, 15, 37, 41]. In addition to a description of the basic mathematics and physics behind a clustering analysis I have attempted to provide a discussion of some of the fundamental and practical difficulties involved. The cosmological goal of such an analysis is consider in the final part of this manuscript, where the combination of cosmological constraints from galaxy clustering and the CMB is discussed, and an example multi-parameter fit to recent data is considered.

W. Percival: *Cosmological Constraints from Galaxy Clustering*, Lect. Notes Phys. **720**,
157–186 (2007)
DOI 10.1007/978-3-540-71013-4_6

6.2 Basics

Our first step is to define the dimensionless overdensity

$$\delta(\mathbf{x}) = \frac{\rho(\mathbf{x}) - \bar{\rho}}{\bar{\rho}} , \tag{6.1}$$

where $\bar{\rho}$ is the expected mean density, which is independent of position because of statistical homogeneity.

The autocorrelation function of the overdensity field (usually just referred to as the correlation function) is defined as

$$\xi(\mathbf{x_1}, \mathbf{x_2}) \equiv \langle \delta(\mathbf{x_1})\delta(\mathbf{x_2}) \rangle . \tag{6.2}$$

From statistical homogeneity and isotropy, we have that

$$\xi(\mathbf{x_1}, \mathbf{x_2}) = \xi(\mathbf{x_1} - \mathbf{x_2}) , \tag{6.3}$$
$$= \xi(|\mathbf{x_1} - \mathbf{x_2}|) . \tag{6.4}$$

To help to understand the correlation function, suppose that we have two small regions δV_1 and δV_2 separated by a distance r. Then the expected number of pairs of galaxies with one galaxy in δV_1 and the other in δV_2 is given by

$$\langle n_{\text{pair}} \rangle = \bar{n}^2 \left[1 + \xi(r) \right] \delta V_1 \delta V_2 , \tag{6.5}$$

where $\bar{n}$ is the mean number of galaxies per unit volume. We see that $\xi(r)$ measures the excess clustering of galaxies at a separation r. If $\xi(r) = 0$, the galaxies are unclustered (randomly distributed) on this scale – the number of pairs is just the expected number of galaxies in δV_1 times the expected number in δV_2. $\xi(r) > 0$ corresponds to strong clustering, and $\xi(r) < 0$ to anti-clustering. Estimation of $\xi(r)$ from a sample of galaxies will be discussed in Sect. 6.5.1.

It is often convenient to consider perturbations in Fourier space. In cosmology the following Fourier transform convention is most commonly used

$$\delta(\mathbf{k}) \equiv \int \delta(\mathbf{r}) e^{i\mathbf{k}\cdot\mathbf{r}} d^3 r \tag{6.6}$$

$$\delta(\mathbf{r}) = \int \delta(\mathbf{k}) e^{-i\mathbf{k}\cdot\mathbf{r}} \frac{d^3 k}{(2\pi)^3} . \tag{6.7}$$

The power spectrum is defined as

$$P(\mathbf{k_1}, \mathbf{k_2}) = \frac{1}{(2\pi)^3} \langle \delta(\mathbf{k_1})\delta(\mathbf{k_2}) \rangle . \tag{6.8}$$

Statistical homogeneity and isotropy gives that

$$P(\mathbf{k_1}, \mathbf{k_2}) = \delta_D(\mathbf{k_1} - \mathbf{k_2}) P(k_1) , \tag{6.9}$$

where δ_D is the Dirac delta function. The power spectrum is sometimes presented in dimensionless form

$$\Delta^2(k) = \frac{k^3}{2\pi^2} P(k) . \tag{6.10}$$

The correlation function and power spectrum form a Fourier pair

$$P(k) \equiv \int \xi(r) e^{i\mathbf{k}\cdot\mathbf{r}} d^3r \tag{6.11}$$

$$\xi(r) = \int P(k) e^{-i\mathbf{k}\cdot\mathbf{r}} \frac{d^3k}{(2\pi)^3} , \tag{6.12}$$

so they provide the same information. The choice of which to use is therefore somewhat arbitrary (see [25] for a further discussion of this).

The extension of the 2-pt statistics, the power spectrum and the correlation function, to higher orders is straightforward with (6.5) becoming

$$\langle n_{\text{tuple}} \rangle = \bar{n}^n \left[1 + \xi^{(n)} \right] \delta V_1 \cdots \delta V_n . \tag{6.13}$$

However, the central limit theorem implies that a density distribution is asymptotically Gaussian in the limit where the density results from the average of many independent processes. The overdensity field has zero mean by definition, so is completely characterised by either the correlation function or the power spectrum. Consequently, in this regime, measuring either the correlation function or the power spectrum provides a statistically complete description of the field.

6.3 Matter Perturbations

There are three physical stages in the creation and evolution of perturbations in the matter distribution. First, primordial perturbation are produced in an inflationary epoch. Second, the different forms of matter within the Universe affect these primordial perturbations. Third, gravitational collapse leads to the growth of these fluctuations. In this section we will discuss the form of the perturbations on scales where gravitational collapse can be described by a linear change in the overdensity. The gravitational collapse of perturbations will be considered in Sect. 6.4.

6.3.1 Why Are There Matter Perturbations?

A period of "faster than light" expansion in the very early Universe solves a number of problems with standard cosmology. In particular, it allows distant regions that appear causally disconnected to have been connected in the past and therefore explains the flatness of the CMB. Additionally it drives the energy density of the Universe close to the critical value and, most importantly for our discussion of perturbations, it provides a mechanism for producing seed perturbations as quantum fluctuations in the matter density are increased to significant levels. For a detailed examination of the creation of fluctuations see [36]. For now, we will just comment that the most basic inflationary models give a spectrum of fluctuations $P(k) \propto k^n$ with $n \sim 1$.

6.3.2 The Effect of Dark Matter

The growth of dark matter fluctuations is intimately linked to the Jeans scale. Perturbations smaller than the Jeans scale do not collapse due to pressure support – for collision-less dark matter this is support from internal random velocities. Perturbations larger than the Jeans scale grow through gravity at the same rate, independent of scale. In a Universe with just dark matter and radiation, the Jeans scale grows to the size of the horizon at matter-radiation equality, and then reduces to zero when the matter dominates. We therefore see that the horizon scale at matter-radiation equality will be imprinted in the distribution of fluctuations – this scale marks a turn-over in the growth rate of fluctuations. What this means in practice is that there is a cut-off in the power spectrum on small scales, dependent on $\Omega_M h$, with a stronger cut-off predicted for lower $\Omega_M h$ values. This is demonstrated in Fig. 6.1.

6.3.3 The Effect of Baryons

At early epochs baryons are coupled to the photons and, if we consider a single fluctuation, a spherical shell of gas and photons is driven away from the perturbation by a sound wave. When the photons and gas decouple, a spherical shell of baryons is left around a central concentration of dark matter. As the perturbation evolves through gravity, the density profiles of the baryons and dark matter grow together, and the perturbation is left with a small increase in density at a location corresponding to the sound horizon at the end of the Compton drag epoch [2, 3]. This real-space "shell" is equivalent to oscillations in the power spectrum. In addition to these acoustic oscillations, fluctuations smaller than the Jeans scale, which tracks the sound horizon until decoupling, do not grow, while large fluctuations are unaffected and continue to grow. The presence of baryons therefore also leads to a reduction in the amplitude of small scale fluctuations. For more information and fitting formulae for the different processes a good starting point is [17].

6.3.4 The Effect of Neutrinos

The same principal of gravitational collapse versus pressure support can be applied in the case of massive neutrinos. Initially the neutrinos are relativistic and their Jeans scale grows with the horizon. As their temperature decreases

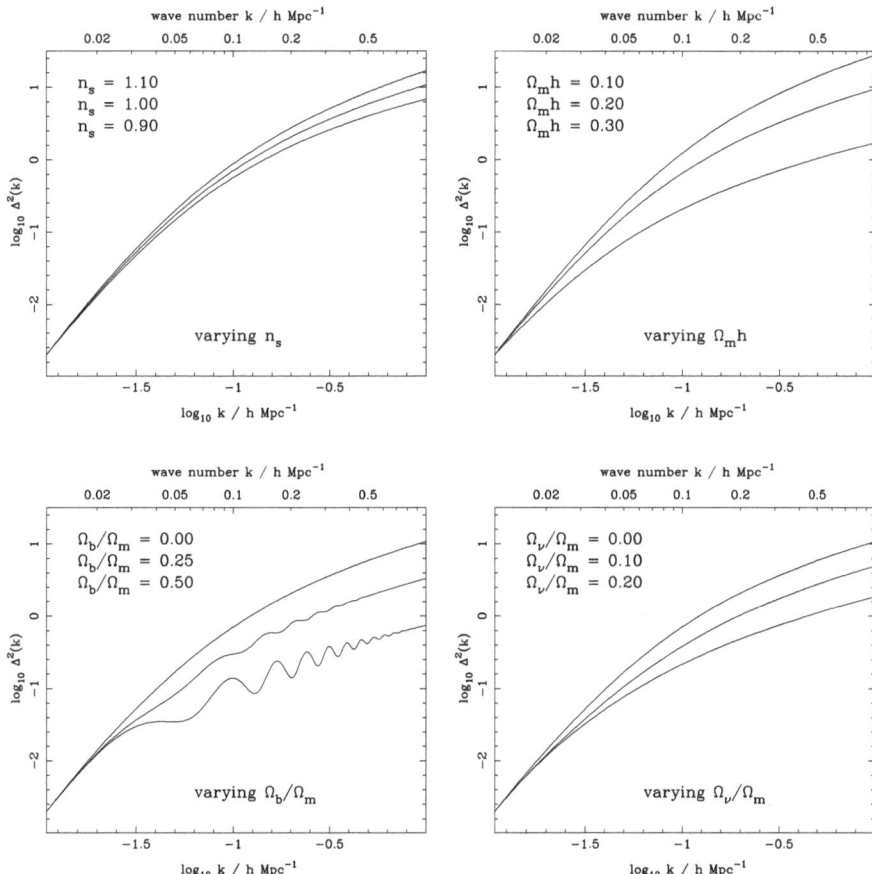

Fig. 6.1. Plots showing the linear power spectrum (*solid lines*) for a variety of different cosmological parameters. Only the shapes of the power spectra are compared, and the amplitudes are matched to the same large scale value. Our base model has $\Omega_M h = 0.2$, $n_s = 1$, $\Omega_b/\Omega_M = 0$ and $\Omega_\nu/\Omega_M = 0$. Deviations from this base model are given in each panel. As can be seen many of the shape distortions from changing different parameters are similar, which can cause degeneracies between these parameters when fitting models to observations

their momenta drop, they become non-relativistic, and the Jeans scale decreases – they can subsequently fall into perturbations. Massive neutrinos are interesting because even at low redshifts the Jeans scale is cosmologically relevant. Consequently the linear power spectrum (the fluctuation distribution excluding the non-linear collapse of perturbations) is not frozen shortly after matter-radiation equality. Instead its form is still changing at low redshifts. Additionally, the growth rate depends on the scale – it is suppressed until neutrinos collapse into perturbations, simply because the perturbations have lower amplitude. The effect of neutrino mass on the present day linear power spectrum is shown in Fig. 6.1. Note that in this plot the relative amplitudes of the power spectra have been removed – it is just the shape that is compared. The amplitude would also depend on the combined neutrino mass.

6.4 The Evolution of Perturbations

Having discussed the form of the linear perturbations, we will now consider how perturbations evolve through gravity in the matter and dark energy dominated regimes. To do this, we will use the spherical top-hat collapse model, where we compare a sphere of background material with radius a, with one of radius a_p which contains the same mass, but has a homogeneous change in overdensity. The ease with which the behaviour can be modelled follows from Birkhoff's theorem, which states that a spherically symmetric gravitational field in empty space is static and is always described by the Schwarzchild metric [8]. This gives that the behaviour of the homogeneous sphere of uniform density and the background can be modelled using the same equations. For simplicity we initially only consider the sphere of background material.

The sphere of background material behaves according to the standard Friedmann and cosmology equations

$$E^2(a) = \frac{1}{a^2} \left(\frac{da}{dH_0t} \right)^2 = \Omega_M a^{-3} + \Omega_K a^{-2} + \Omega_X a^{f(a)} , \tag{6.14}$$

$$\frac{1}{a} \frac{d^2 a}{dt^2} = -\frac{H_0^2}{2} \left[\Omega_M a^{-3} + [1 + 3w(a)] \Omega_X a^{f(a)} \right] . \tag{6.15}$$

These equations have been written in a form allowing for a general time-dependent equation of state for the dark energy $p = w(a)\rho$. Conservation of energy for the dark energy component provides the form of $f(a)$

$$f(a) = \frac{-3}{\ln a} \int_0^{\ln a} [1 + w(a')] \, d\ln a' . \tag{6.16}$$

The dark matter and dark energy densities evolve according to

$$\Omega_M(a) = \frac{\Omega_M a^{-3}}{E^2(a)} , \quad \Omega_X(a) = \frac{\Omega_X a^{f(a)}}{E^2(a)} . \tag{6.17}$$

Tracks showing the evolution of $\Omega_M(a)$ and $\Omega_X(a)$ are presented in Fig. 6.2 for $h = 0.7$ and constant dark energy equation of state $w = -1$. Of particular interest are solutions which predict recollapse, but that have $\Omega_X > 0$. Provided that $\Omega_M >> \Omega_X$, the perturbation will collapse before the dark energy dominates. For a cosmology with $\Omega_M \sim 0.3$ and $\Omega_X \sim 0.7$, these solutions correspond to overdense spheres that will collapse and form structure.

For the perturbation, the cosmology equation can be written

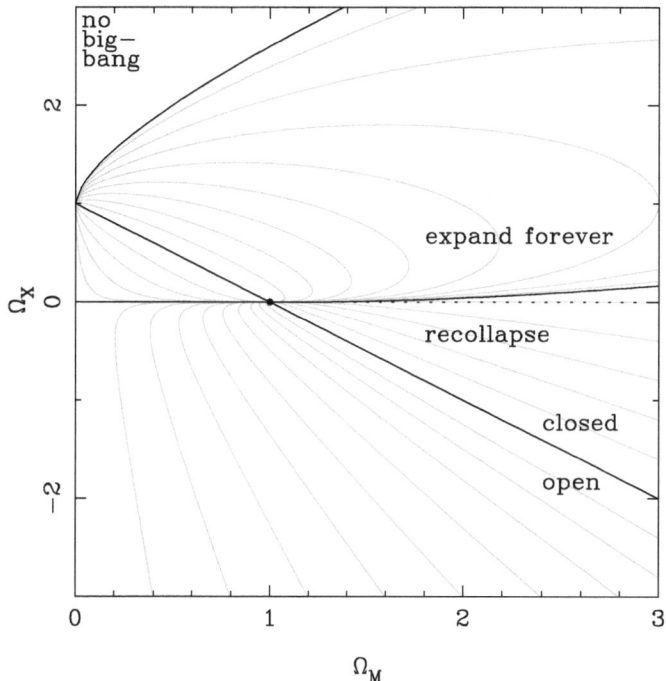

Fig. 6.2. Plot showing the evolution of the matter and vacuum energy densities for a selection of cosmologies (*grey lines*) with constant dark energy equation of state parameter $w = -1$. The critical models that border the different types of evolution are shown by the black lines. The dotted line highlights $\Omega_X = 0$

$$\frac{1}{a_p}\frac{d^2 a_p}{dt^2} = -\frac{H_0^2}{2}\left[\Omega_M a_p^{-3} + [1 + 3w(a)]\Omega_X a^{f(a)}\right] , \qquad (6.18)$$

where it is worth noting that the dark energy component is dependent on a rather than a_p. This does not matter for Λ-cosmologies as $f(a) = 0$, and the a dependence in this term is removed. For other dark energy models, this dependence follows if the dark energy does not cluster on the scales of interest. For such cosmological models, we cannot write down a Friedmann equation for the perturbation because energy is not conserved [63]. We also have to be more careful using virialisation arguments to analyse the behaviour of perturbations [47].

To first order, the overdensity of the perturbation $\delta = a^3/a_p^3 - 1$ evolves according to

$$\frac{d^2\delta}{d(H_0 t)^2} + \frac{2}{a}\frac{da}{d(H_0 t)}\frac{d\delta}{d(H_0 t)} - \frac{3}{2}\Omega_M a^{-3}\delta = 0 , \qquad (6.19)$$

which is known as the linear growth equation.

The evolution of the scale factor of the perturbations is given by the solid lines in Fig. 6.3, compared with the background evolution for a cosmology

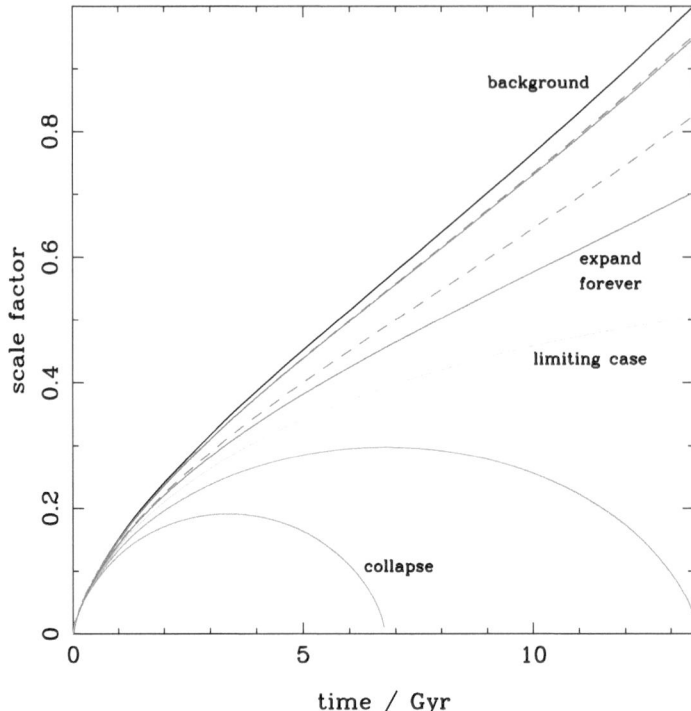

Fig. 6.3. Plot showing the evolution of the scale factor of perturbations with different initial overdensities. A standard cosmology with $\Omega_M = 0.3$, $\Omega_X = 0.7$, $h = 0.7$, $w = -1$ is assumed. The dashed lines show the linear extrapolation of the perturbation scales for the two least overdense perturbations

with $\Omega_M = 0.3$, $\Omega_X = 0.7$, $h = 0.6$, $w = -1$. These data were calculated by numerically solving (6.18). For comparison, the dashed lines were calculated by extrapolating the initial perturbation scales using the linear growth factor, calculated from (6.19). Dashed lines are only plotted for the two least overdense perturbations. In comparison, the most overdense perturbations are predicted to collapse to singularities. However, in practice inhomogeneities, and the non-circular shape of actual perturbations will mean that the object virialises with finite extent.

The evolution of perturbations has a profound affect on the present day power spectrum of the matter fluctuations on small scales. On the largest scales, the overdensities are small and linear theory (6.19) holds. This increases the amplitude of the fluctuations, but does not change the shape of the power spectrum, as the perturbation all grow at the same rate (except if neutrinos are cosmologically relevant – see Sect. 6.3.4). However, on the smallest scales, overdensities are large and collapse to virialised structures (e.g. cluster of galaxies). The effect on the power spectrum is most easily quantified using

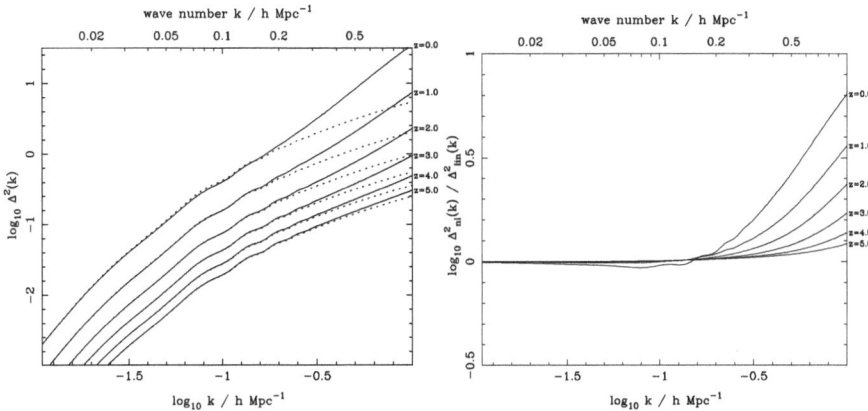

Fig. 6.4. Plots comparing non-linear (*solid lines*) and linear power spectra (*dotted lines*) at a series of redshifts from $z = 0$ to $z = 5$. In the left panel the raw dimensionless power spectra are plotted while in the right panel the ratio between non-linear and linear predictions is shown. As can be seen, on large scales linear growth simply increases the amplitude of the power spectrum, while on small scales we also see an increase in power as structures collapse at low redshifts. There is also a slight decrease in power on intermediate scales – it is this power that is transferred to small scales. Non-linear power spectra were calculated from the fitting formulae of [56] with $\Omega_M = 0.3$, $h = 0.7$, $n_s = 1$, and $\Omega_b/\Omega_m = 0.15$

numerical simulations, and power spectra calculated from fitting formulae derived from such simulations [56] are plotted in Fig. 6.4.

6.5 Galaxy Survey Analysis

6.5.1 Estimating the Correlation Function

First suppose that we have a single population of objects forming a Poisson sampling of the field that we wish to constrain. This is too simple an assumption for the analysis of modern galaxy redshift surveys, but it will form a starting point for the development of the analysis tools required.

First we define the (unweighted) galaxy density field

$$n_g(\mathbf{r}) \equiv \sum_i \delta_D(\mathbf{r} - \mathbf{r_i}) . \tag{6.20}$$

The definition of the correlation function then gives

$$\langle n_g(\mathbf{r})n_g(\mathbf{r}') \rangle = \bar{n}(\mathbf{r})\bar{n}(\mathbf{r}')[1 + \xi(\mathbf{r} - \mathbf{r}')] + \bar{n}(\mathbf{r})\delta_D(\mathbf{r} - \mathbf{r}') . \tag{6.21}$$

The final term in this equation relates to the shot noise, and only occurs for zero separation so can be easily dealt with.

In order to estimate the correlation function, we can consider a series of bins in galaxy separation and make use of (6.21). Suppose that we have created a (much larger) random distribution of points that form a Poisson sampling of the volume occupied by the galaxies, then

$$1 + \xi = \frac{\langle DD \rangle}{\langle RR \rangle} (1 + \xi_\Omega) ,$$

(6.22)

where DD is the number of galaxy-galaxy pairs within our bin in galaxy separation divided by the maximum possible number of galaxy-galaxy pairs (i.e. for n galaxies the maximum number of distinct pairs is $n(n-1)/2$). Similarly RR is the normalised number of random-random pairs, and we can also define DR as the normalised number of galaxy-random pairs.

If the true mean density of galaxies $\bar{n}(\mathbf{r})$ is estimated from the sample itself (as is almost always the case), we must include a factor $(1+\xi_\Omega)$ that corrects for the systematic offset induced. ξ_Ω is the mean of the two-point correlation function over the sampling geometry [34]. Given only a single clustered sample it is obviously difficult to determine ξ_Ω, and the integral constraint (as it is known) remains a serious drawback to the determination of the correlation function from small samples of galaxies.

Because the galaxy and random catalogues are uncorrelated, $\langle DR \rangle = \langle RR \rangle$, and we can consider a number of alternatives to (6.22). In particular

$$1 + \xi = \left(1 + \frac{\langle (D - R)^2 \rangle}{\langle RR \rangle} \right) (1 + \xi_\Omega) ,$$

(6.23)

has been shown to have good statistical properties [34].

6.5.2 Estimating the Power Spectrum

In this section we consider estimating the power spectrum by simply taking a Fourier transform of the overdensity field [5, 21, 45]. As for our estimation of the correlation function, suppose that we have quantified the volume occupied by the galaxies by creating a large random catalogue matching the spatial distribution of the galaxies, but with no clustering (containing α times as many objects). The (unnormalised) overdensity field is

$$F(\mathbf{r}) = n_g(\mathbf{r}) - n_r(\mathbf{r})/\alpha ,$$

(6.24)

where n_g is given by (6.20), and n_r is similarly defined for the random catalogue.

Taking the Fourier transform of this field, and calculating the power gives

$$\langle |F(\mathbf{k})|^2 \rangle = \int \frac{d^3 k'}{(2\pi)^3} [P(\mathbf{k}') - P(0)\delta_D(\mathbf{k})]|G(\mathbf{k} - \mathbf{k}')|^2 + \left(1 + \frac{1}{\alpha} \right) \int d^3 r \bar{n}(\mathbf{r}) ,$$

(6.25)

where $G(\mathbf{k})$ if the Fourier transform of the window function, defined by

$$G(\mathbf{k}) \equiv \int \bar{n}(\mathbf{r})e^{i\mathbf{k}\cdot\mathbf{r}}d^3r \, , \tag{6.26}$$

and the final term in (6.25) gives the shot noise. In contrast to the correlation function, there is a shot noise contribution at every scale. The integral constraint has reduced to subtracting a single Dirac delta function from the center of the unconvolved power – as before this allows for the fact that we do not know the mean density of galaxies.

6.5.3 Complications

There are two complications which constitute the main hindrance to using clustering in galaxy surveys to constrain cosmology. They are redshift space distortions – systematic deviations in measured redshift in addition to the Hubble flow, and galaxy bias – the fact that galaxies do not form a Poisson sampling of the underlying matter distribution. Denoting the measurement of a quantity in redshift space (galaxy distances calculated from redshifts) by a superscript ^{s} and in real space (true galaxy distances) by ^{r}, we can write the measured power spectrum P^s_{gal} as

$$\frac{P^s_{\text{gal}}}{P_{\text{mass}}} = \frac{P^s_{\text{gal}}}{P^r_{\text{gal}}} \times \frac{P^r_{\text{gal}}}{P_{\text{mass}}} \, . \tag{6.27}$$

The first of these terms corresponds to redshift space distortions, while the second corresponds to galaxy bias.

Redshift Space Distortions

There are two key mechanisms that systematically distort galaxy redshifts from their Hubble flow values. First, structures are continually growing through gravity, and galaxies fall into larger structures. The infall velocity adds to the redshift, making the distance estimates using the Hubble flow wrong. This means that clusters of galaxies appear thinner along the line-of-sight, causing an increase in the measured power. In the distant observer approximation, the apparent amplitude of the linear density disturbance can be readily calculated [31], leading to a change in the power corresponding to

$$P^s_{\text{gal}} = P^r_{\text{gal}}(1 + \beta\mu^2)^2 \, , \tag{6.28}$$

where $\beta = \Omega_M^{0.6}/b$, b is an assumed linear bias for the galaxies, and μ is the cosine between the velocity vector and the line-of-sight. In the small angle approximation, we average over a uniform distribution for μ giving

$$P^s_{\text{gal}} = P^r_{\text{gal}}\left[1 + \frac{2}{3}\beta + \frac{1}{5}\beta^2\right] \, . \tag{6.29}$$

For large redshift surveys of the nearby Universe, the small angle approximation breaks down, although a linear result can be obtained using a spherical
expansion of the survey (see Sect. 6.5.5).

When objects collapse and virialise they attain a distribution with some
velocity dispersion. These random velocities smear out the collapsed object
along the line of sight in redshift space, leading to the existence of linear structures pointing towards the observer. These structures, known as "fingers-of-
god" can be corrected by matching with a group catalogue and applying a
correction to the galaxy field before analysis [60]. Alternatively, if the pairwise distribution of velocity differences is approximated by an exponential
distribution, then

$$P^s_{gal} = P^r_{gal}(1 + k^2\mu^2\sigma_p^2/2)^{-1} , \qquad (6.30)$$

where $\sigma_p \sim 400\,\mathrm{km\,s^{-1}}$ is the pairwise velocity dispersion [28].

Galaxy Bias

By the simple phrase "galaxy bias" astronomers quantify the "messy" astrophysics of galaxy formation. It is common to assume a local linear bias
with $\delta_{gal} = b\delta_{mass}$, which leads to a simple relation between power spectra
$P^r_{gal} = b^2 P_{mass}$. If this bias is independent of the scale probed, then there is
nothing to worry about – the galaxy and matter power spectra have the same
shape. However, it is well known that galaxies of different types have different
clustering strengths – two recent analyses are [53, 64].

One simple way of understanding galaxy bias is to use the "halo model",
which has become popular over the last 5 years [13, 42, 54]. First, consider
the distribution of the underlying matter – the power spectrum was shown in
Fig. 6.4. There are two distinct regimes: on large scales, linear growth holds,
while on small scales the dark matter has formed into halos: it has either
undergone collapse and has virialised, or is on the way to virialisation. Galaxies pinpoint certain locations within the dark matter halos, according to an
occupation distribution for each galaxy type. This forms a natural environment in which to model galaxy bias, with galaxies of different luminosities
and types have different occupation distributions depending on the physics of
their formation.

For 2-pt statistics, then there are two possibilities for pairs of galaxies. We
could have chosen a pair where both galaxies lie in the same halo – this is
most likely on small scales. Alternatively, the galaxies might be in different
halos – this is most likely on large scales. On large scales, the halos themselves
are biased compared with the matter and we can use the peak-background
split model [9, 40, 55] to estimate the increase in clustering strength. This
limiting large scale value offers a route to determine the masses of the virialised
structures in which particular galaxies live.

Given a linear bias model for each type of galaxy in the sample to be
analysed, it is possible to multiply the contribution of each galaxy to the estimate of the overdensity field by the inverse of an expected bias [45]. Provided

the bias model is correct (and possibly altered for each scale observed), then this removes any systematic offset in the recovered power spectrum caused by galaxy bias. The problem is that we need to have an accurate model of the galaxy bias in order to remove it.

6.5.4 Weights

The procedure described in Sect. 6.5.2 can be extended to include weights for each galaxy in order to optimise the analysis [21]. Under the assumptions that the wavelength of interest $2\pi/k$ is small compared with the survey scale (i.e. the window is negligible), and that the fluctuations are Gaussian, then the optimal weight applied to galaxy i is

$$w_i = \frac{1}{1 + \bar{n}(\mathbf{r}_i)\hat{P}(k)} \,, \qquad (6.31)$$

where $\bar{n}(\mathbf{r}_i)$ is the mean galaxy density at the location of galaxy i. At locations where the mean galaxy density is low, galaxies are weighted equally. Where the galaxy density is high, we weight by volume. It is worth noting that the optimal weights also depend on an estimate of the power spectrum to be measured, and therefore depend on the scale of interest. However, in practice this dependence is sufficiently weak that very little information is lost by assuming a constant $\hat{P}(k)$.

It is possible to include galaxy bias when determining weights and optimising the analysis in order to recover the most signal. Given a bias for each galaxy b_i (which can be dependent on any galaxy properties and the scale of interest), then the optimal weighting is [45].

$$w_i = \frac{b_i^2}{1 + \sum_j \bar{n}(\mathbf{r_i}, b_j)b_j^2\hat{P}(k)} \,, \qquad (6.32)$$

which up-weights the most biased galaxies that contain the strongest cosmological signal.

6.5.5 Spherical Bases

In Sect. 6.5.2 we described the most simple analysis method for a 3-dimensional galaxy survey – decomposing into a 3D Fourier basis. However, as we discussed in Sect. 6.5.3 redshift-space distortions complicate the situation, and cannot easily be dealt with using a Fourier basis. By decomposing into a basis that is separable in radial and angular directions, we can more easily correct such distortions. A pictorial comparison of the Fourier basis with a radial-angular separable basis is presented in Fig. 6.5.

In this section we provide an overview of a formalism to do this based on work by [29, 46, 58]. For alternative formalisms see [20, 26, 60]. In comparison with the Fourier decomposition (6.6), we decompose into a 3D basis of Spherical Harmonics Y_{lm} and spherical Bessel functions j_l

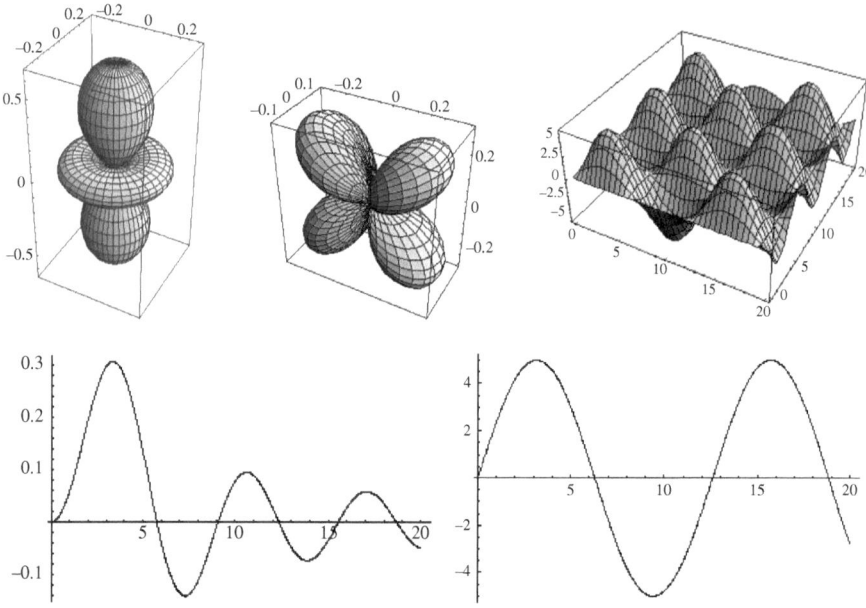

Fig. 6.5. Comparison of 3D Fourier basis split into 2D and 1D components (*right*) with basis of Spherical Harmonics (with $l = 2$ and $m = 0, 1$ – *top left*) and Spherical Bessel functions (*bottom left*)

$$\delta(\boldsymbol{x}) = \sqrt{\frac{2}{\pi}} \int_0^\infty \sum_{l,m} \delta_{lm}(k) j_l(kx) Y_{lm}(\theta, \phi) k dk \ . \tag{6.33}$$

Because of the choice of bases, the transformation $\delta_{lm}(k) \leftrightarrow k\delta(\boldsymbol{k})$ is unitary so we retain the benefit of working with the Fourier power spectrum

$$\langle \delta_{lm}(k) \delta_{l'm'}(k') \rangle = P(k) \delta_D(k - k') \delta_D(l - l') \delta_D(m - m') \ . \tag{6.34}$$

As in Sect. 6.5.2, we have simplified the analysis by not including any galaxy weights, although these can be introduced into the formalism. Additionally, it is easier to work with a fixed boundary condition - usually that fluctuations vanish at some large radius so that we are only concerned with radial modes that have

$$\frac{d}{dx} j_l(kx) \Big|_{x_{\max}} = 0 \ , \tag{6.35}$$

so that the decomposition becomes

$$\delta(\boldsymbol{x}) = \sum_{l,m,n} c_{ln} \delta_{lmn} j_l(k_{ln}x) Y_{lm}(\theta, \phi) \ , \tag{6.36}$$

where c_{ln} is a normalising constant.

In order to analyse the transformed modes, we need a model for $\langle \delta_{lmn} \delta_{l'm'n'} \rangle$. First we deal with the survey volume by introducing a convolution

$$\hat{\delta}_{lmn} = \sum_{l'm'n'} M^{l'm'n'}_{lmn} \delta_{l'm'n'} \,, \tag{6.37}$$

where

$$M^{l'm'n'}_{lmn} = c_{ln} c_{l'n'} \int d^3 x \bar{\rho}(x) j_l(k_{ln} x) j_{l'}(k_{l'n'} x) Y^*_{lm}(\theta, \phi) Y_{l'm'}(\theta, \phi) \,. \tag{6.38}$$

We can include the effect of linear redshift space distortions by a transform

$$j_l(k_{ln} x^s) \simeq j_l(k_{ln} x^r) + \Delta x_{\text{lin}} \frac{d}{dx^r} j_l(k_{ln} x^r) \,, \tag{6.39}$$

where

$$\Delta x_{\text{lin}} = \beta \sum_{lmn} \frac{1}{k^2_{ln}} c_{ln} \delta_{lmn} \frac{d j_l(k_{ln} x^r)}{dx^r} Y_{lm}(\theta, \phi) \,. \tag{6.40}$$

Here $\beta = \Omega_M^{0.6}/b$. The bias b corrects for the fact that while we measure the galaxy power spectrum, the redshift space distortions depend on the mass. We can also introduce a further convolution to correct for the small-scale fingers-of-god effect

$$\hat{\delta}_{l'm'n'} = \sum_{l''m''n''} S^{l''m''n''}_{l'm'n'} \delta_{l''m''n''} \,, \tag{6.41}$$

where

$$S^{l''m''n''}_{l'm'n'} = c_{l'n'} c_{l''n''} \delta^D_{l'l''} \delta^D_{m'm''} \int \int p(r - y) j_{l'}(k_{l'n'} r) j_{l''}(k_{l''n''} y) \, r \, dr \, y \, dy \,, \tag{6.42}$$

and $p(r - y)$ is the 1-dimensional scattering probability for the velocity dispersion. It is also possible to include bias and evolution corrections in the analysis method [46].

For a given cosmological model, we can use the above formalism to calculate the covariance matrix $\langle \delta_{lmn} \delta_{l'm'n'} \rangle$ for N modes, and then calculate the Likelihood of a given cosmological model assuming that $\hat{\delta}_{lmn}$ has a Gaussian distribution

$$\mathcal{L}[\hat{\delta}_{lmn}|\text{model}] = \frac{1}{(2\pi)^{N/2} |C|^{1/2}} \exp\left[-\frac{1}{2} \hat{\delta}^T_{lmn} C^{-1} \hat{\delta}_{lmn} \right] \,, \tag{6.43}$$

where C is the matrix of $\langle \delta_{lmn} \delta_{l'm'n'} \rangle$.

6.6 Practicalities

6.6.1 Brief Description of Redshift Surveys

The 2dF Galaxy Redshift Survey (2dFGRS), which is now complete, covers approximately 1800 square degrees distributed between two broad strips, one across the South Galactic pole and the other close to the North Galactic Pole, plus a set of 99 random 2 degree fields spread over the full southern galactic cap. The final catalogue contains reliable redshifts for 221 414 galaxies selected to an extinction-corrected magnitude limit of approximately $b_J = 19.45$ [12].

In contrast, the Sloan Digital Sky Survey (SDSS) is an ongoing photometric and spectroscopic survey. The SDSS includes two spectroscopic galaxy surveys: the main galaxy sample which is complete to a reddening-corrected Petrosian r magnitude brighter than 17.77, and a deeper sample of luminous red galaxy sample selected based on both colour and magnitude [18]. The SDSS has regular public data releases: the 4th data release in 2005 included 480000 independent galaxy spectra [1]. When completed, the SDSS will have obtained spectra for $\sim 10^6$ galaxies.

6.6.2 Angular Mask

Both the recent 2dF galaxy redshift (2dFGRS) and the ongoing Sloan Digital Sky Survey (SDSS) adopted an adaptive tiling system in order to target pho-

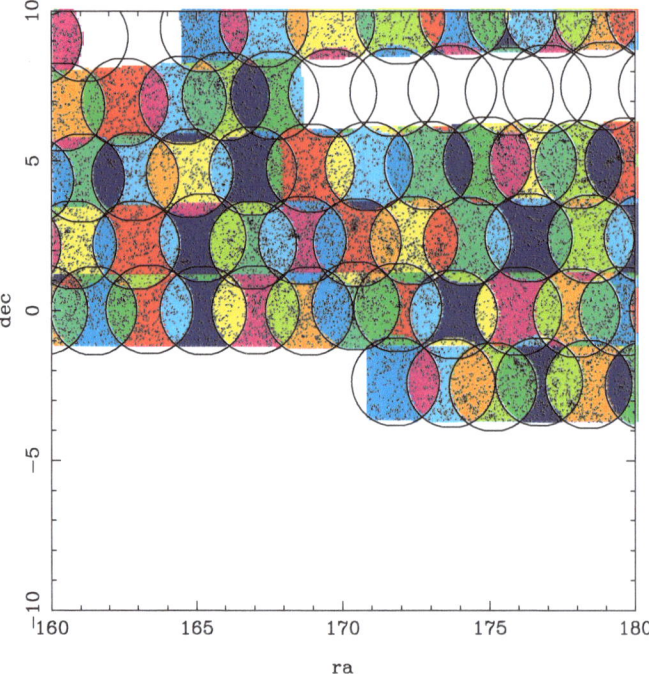

Fig. 6.6. Section in the SDSS DR4 angular mask showing the positions of galaxies with measured redshifts (*black dots*), the positions of the plates from which the spectra were obtained (*large black circles*) and the segments within the mask that have different completenesses (*coloured regions*)

tometrically selected galaxies for spectroscopic follow-up. The circular tiles within which spectra could be taken in a single pointing of the telescope were adaptively fitted over the survey region, with regions of high galaxy density being covered by two or more tiles. A region of such tiling is shown in Fig. 6.6. This procedure divides the survey into segments, each with a different completeness – the ratio of good quality spectra to galaxies targeted. It is usually assumed that this completeness is uniform across each of the segments formed by overlapping tiles. Understanding this completeness is a major consideration when performing a large-scale structure analysis of either of these surveys.

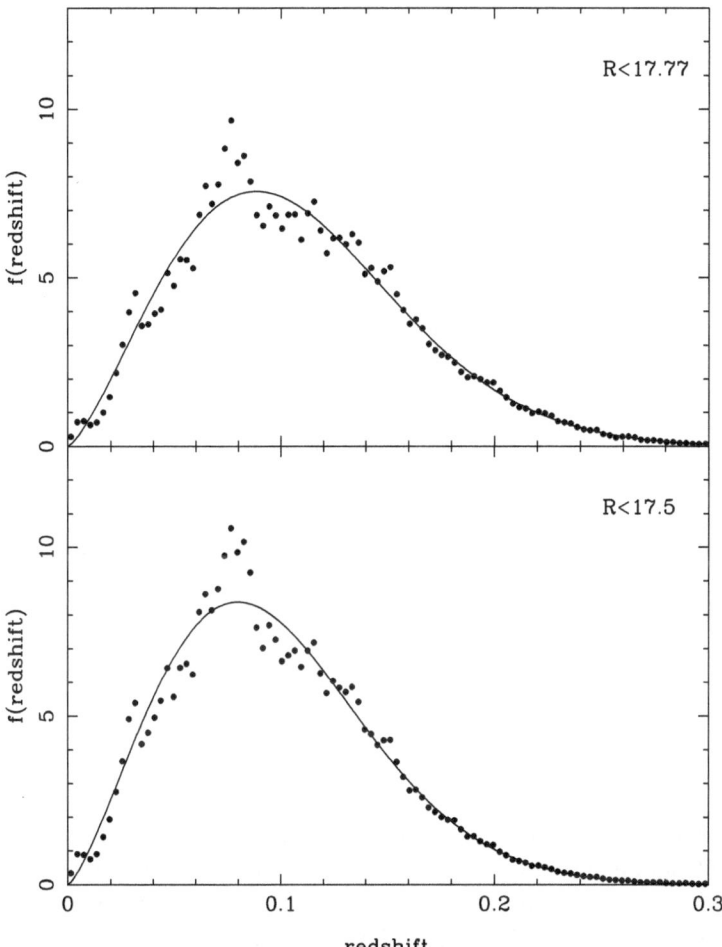

Fig. 6.7. Redshift distribution of spectroscopically observed galaxies within the SDSS DR4 with apparent R magnitude less than 17.5 and 17.77 (*solid circles*). For comparison we show the best fit model given by (6.44) for each distribution (*solid lines*)

Note that the distribution of segments depends on all adjoining targeted tiles, not just those that have been observed.

As well as understanding the completeness, we also need to consider the effect of the weather – spectra taken under bad observing conditions will tend to preferentially give redshifts for nearby rather than distant galaxies. We also need to worry about bad fields – regions near bright stars where photometric data is of poor quality. For the SDSS, there are hard limits for the spectroscopic region depending on how much photometric data was available when the targeting algorithm was run. All of these effects are well known and can be included in an analysis.

6.6.3 Radial Distribution

In addition to the angular distribution of galaxies, we also need to be able to model the radial distribution – in the formalism introduced in Sect. 6.5.2, we need this information in order to create the random catalogue. Perhaps the best way of doing this is to model the true luminosity function of the distribution of observed galaxies, and then apply a magnitude cut-off. This was the procedure adopted in [10]. However, the reduction in the amplitude of the recovered power spectrum caused by fitting to the redshift distribution is small and it is common to simply fit a functional form to the distribution. In Fig. 6.7 we present the distribution of galaxy redshifts in the SDSS DR4 sample compared with a fit of the form [4]

$$f(z) = z^g \exp\left[-\left(\frac{z}{z_c}\right)^b\right] , \tag{6.44}$$

where g, b and z_c are free parameters that have been fitted to the data.

6.7 Results from Recent Surveys

6.7.1 Results

In Table 6.1 we summarise recent cosmological constraints derived from the 2dFGRS and SDSS. In order to provide a fair test of different analyses, we have only presented best-fit parameters and errors for $\Omega_M h$, fixing the other important parameters. Degeneracies between parameters, caused by the similarity between power spectrum shapes shown in Fig. 6.1 mean that, it is only the most recent analyses of the largest samples that can simultaneously constrain 2 or more of these parameters. In Table 6.1 we also presented the number of galaxy redshifts used in each analysis.

The power spectra recovered from these analyses are compared in Fig. 6.8. We have corrected each for survey window function effects using the best-fit model power spectrum. The amplitudes have also been matched, so this plot

Table 6.1. Summary of recent cosmological constraints from 2dFGRS and SDSS galaxy redshift surveys. To try to provide a fair comparison, we only present the best-fit value and quoted error for $\Omega_M h$ assuming that all other cosmological parameters are fixed ($n_s = 1$, $h = 0.72$, $\Omega_b/\Omega_M = 0.17$, $\Omega_\nu/\Omega_M = 0.0$), and marginalise over the normalisation

survey	reference	galaxy redshifts	method	$\Omega_M h$
2dFGRS	[43]	166490	Fourier	0.206 ± 0.023
2dFGRS	[46]	142756	Spherical Harmonics	0.215 ± 0.035
2dFGRS	[10]	221414	Fourier	0.172 ± 0.014
SDSS	[49]	205484	KL analysis	0.207 ± 0.030
SDSS	[60]	205443	Spherical Harmonics	0.225 ± 0.040
SDSS LRG	[19]	46748	correlation function	0.185 ± 0.015

merely shows the shapes of the spectra. It is clear that the general shape of the galaxy power spectrum is now well known, and the turn-over is detected at high significance. The exact position of the turn-over is however, more poorly known and by examining the final column of Table 6.1, we see that there are discrepancies between recent analyses at the $\sim 2\sigma$ level.

6.8 Combination with CMB Data

In this section we consider recent CMB observations and see how the complementarity between CMB and large scale structure constraints can break degeneracies inherent in these data. The major steps required in a joint analysis are described, leading up to Sect. 6.8.5, in which we present the constraints from an example fit to recent data.

6.8.1 Cosmological Models

Before we start looking at constraining cosmological models using CMB and galaxy $P(k)$ data, it is worth briefly introducing the set of commonly used cosmological parameters (for further discussion see the recent review by [33]). It is standard to assume Gaussian, adiabatic fluctuations, and we will not discuss alternatives here. It is possible to parameterise the cosmological model using a number of related sets of parameters. It is vital in any analysis that the model that is being fitted to the data is fully specified – including parameters and assumed priors. Many parameters have values that simplify the theory from which the models are calculated (e.g. the assumption that the total density in the Universe is equal to the critical density). Whether the data justify dropping one of these assumptions is an interesting Bayesian question [38], which is outside the remit of the overview presented here, and we will simply introduce the parameters commonly used and possible assumptions about their values.

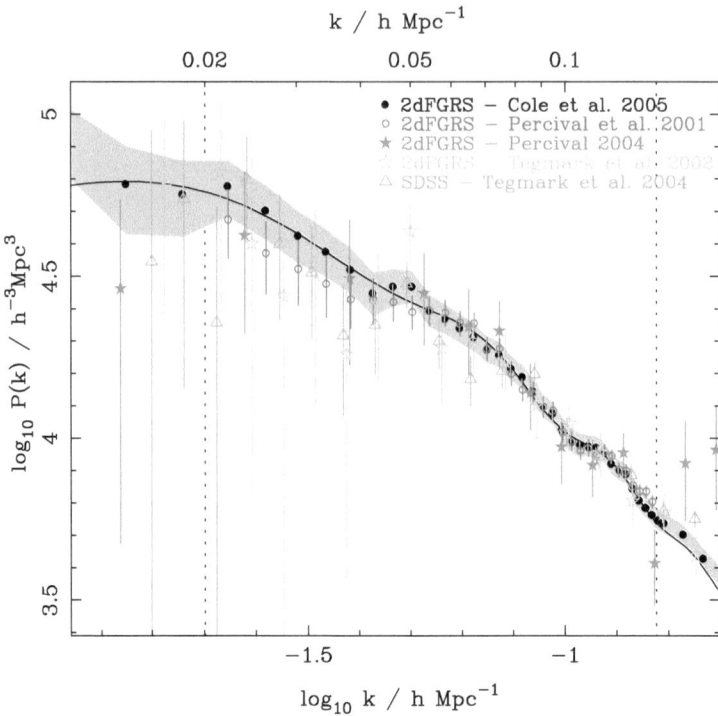

Fig. 6.8. Plot comparing galaxy power spectra calculated by different analysis techniques for different surveys. The redshift-space power spectrum calculated by [10] (solid circles with 1-σ errors shown by the shaded region) are compared with other measurements of the 2dFGRS power spectrum shape by [43] – open circles, [46] – solid stars, [59] – open stars. Where appropriate the data have been corrected to remove effects of the survey volume, by calculating the effect on a model power spectrum with $\Omega_M h = 0.168$, $\Omega_b/\Omega_M = 0.0$, $h = 0.72$ & $n_s = 1$. A zero-baryon model was chosen in order to avoid adding features into the power spectra. All of the data are renormalized to match the power spectrum of [10]. The open triangles show the uncorrelated SDSS real space $P(k)$ estimate of [60], calculated using their 'modeling method' with no FOG compression (their Table 3). These data have been corrected for the SDSS window as described above for the 2dFGRS data. The solid line shows a model linear power spectrum with $\Omega_M h = 0.168$, $\Omega_b/\Omega_M = 0.17$, $h = 0.72$, $n_s = 1$ and normalization matched to the 2dFGRS power spectrum

First, we need to know the geometry of the Universe, parameterised by total energy density $\Omega_{\rm tot}$, or the curvature Ω_K, with the "simplified" value being that the energy density is equal to the critical value ($\Omega_{\rm tot} = 1$, $\Omega_K = 0$). We also need to know the constituents of the energy density, which we parameterise by the dark matter density Ω_c, baryon density Ω_b, and neutrino density Ω_ν. Although it is commonly assumed that the combined neutrinos mass has negligible cosmological effect. The combined matter

density $\Omega_M = \Omega_c + \Omega_b + \Omega_\nu$ could also be defined as a parameter, re-
placing one of the other density measurements. We also need to specify
the dark energy properties, particularly the equation of state $w(a)$, which
is commonly assumed to be constant $w(a) = -1$, so this field is equivalent
to Λ. The perturbations after inflation are specified by the scalar spectral
index n_s, with $n_s = 1$ being the most simple assumption. Possible running of this spectral index is parameterised by $\alpha = dn_s/dk$ if included.
A possible tensor contribution parameterised by the tensor spectral index
n_t, and tensor-to-scalar ratio r is sometimes explicitly included. The evolution to present day is parameterised by the Hubble constant h, and for the
CMB the optical depth to last-scattering surface τ. Finally, three parameters that are often ignored and marginalised over are the galaxy bias $b(k)$
(often assumed to be constant) and the CMB beam B and calibration C
errors.

6.8.2 The MCMC Technique

Large multi-parameter likelihood calculations are computationally expensive
using grid-based techniques. Consequently, the Markov-Chain Monte-Carlo
(MCMC) technique is commonly used for such analyses. While there is publically available code to calculate cosmological model constraints [35], the basic
method is extremely simple and relatively straightforward to code.

The MCMC method provides a mechanism to generate a random sequence
of parameter values whose distribution matches the posterior probability distribution of a Bayesian analysis. Chains are sequentially calculated using the
Metropolis algorithm [39]: given a chain at position $\boldsymbol{x}$, a candidate point $\boldsymbol{x}'$ is
chosen at random from a proposal distribution $f(\boldsymbol{x}'|\boldsymbol{x})$. This point is always
accepted, and the chain moves to point $\boldsymbol{x}'$, if the new position has a higher
likelihood. If the new position $\boldsymbol{x}'$ is less likely than $\boldsymbol{x}$, then $\boldsymbol{x}'$ is accepted, and
the chain moves to point $\boldsymbol{x}'$ with probability given by the ratio of the likeli-
hood of $\boldsymbol{x}'$ and the likelihood of $\boldsymbol{x}$. In the limit of an infinite number of steps,
the chains will reach a converged distribution where the distribution of chain
links are representative of the likelihood hyper-surface, given any symmetric
proposal distribution $f(\boldsymbol{x}'|\boldsymbol{x}) = f(\boldsymbol{x}|\boldsymbol{x}')$ (the Ergodic theorem: see, e.g. [51]).

It is common to implement dynamic optimisation of the sampling of the
likelihood surface (see [24] e.g.). Again, it is simple to assume a multi-variate
Gaussian proposal function, centered on the current chain position. Given
such a proposal distribution, and an estimate of the covariance matrix for the
likelihood surface at each step, the optimal approach for a Gaussian likelihood
would proceed as follows.

Along each principal direction corresponding to an eigenvector of the covariance matrix, the variance σ^2 of the multi-variate Gaussian proposal function should be set to be a fixed multiple of the corresponding eigenvalue of
the covariance matrix. To see the reasoning behind this, consider translat-
ing from the original 17 parameters to the set of parameters given by the

decomposition along the principal directions of the covariance matrix each divided by the standard deviation in that direction. In this basis, the likelihood function is isotropic and the parameters are uncorrelated. Clearly an optimized proposal function will be the same in each direction, and we have adjusted the proposal function to have precisely this property. There is just a single parameter left to optimize – we are free to multiply the width of the proposal function by a constant in all directions. But we know that the optimal fraction of candidate positions that are accepted should be ~ 0.25 [23], so we can adjust the normalization of the proposal width to give this acceptance fraction. Note that the dynamic changing of the proposal function width violates the symmetry of the proposal distribution $f(\boldsymbol{x}'|\boldsymbol{x})$ assumed in the Metropolis algorithm. However, this is not a problem if we only use sections of the chains where variations between estimates of the covariance matrix are small.

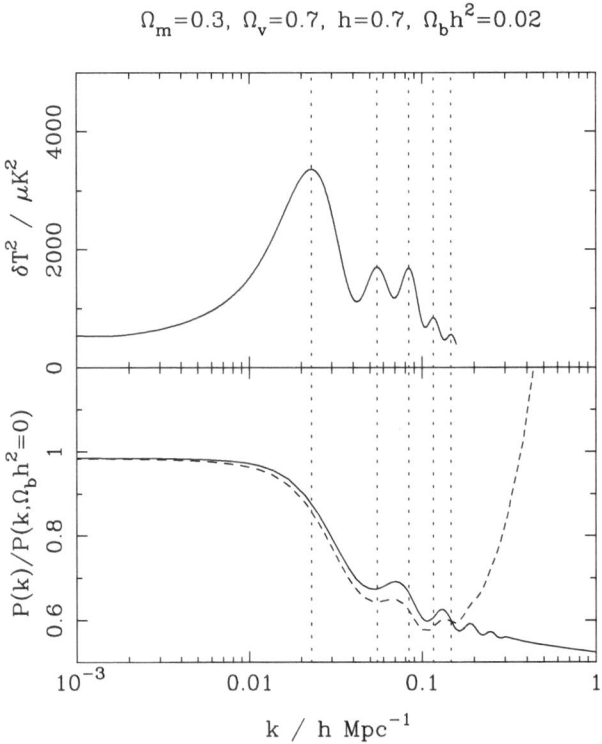

Fig. 6.9. Plot comparing large scale structure (*lower panel*) and CMB (*upper panel*) power spectra. The angular CMB power spectrum was converted to comoving scales using the comoving distance to the last scattering surface. The matter power spectrum (*solid* – linear, *dashed* – non-linear, present day), has been ratioed to a smooth model with zero baryons in order to highlight the baryonic features. Dotted lines show the positions of the peaks in the CMB spectrum

The remaining issue is convergence – how do we know when we have sufficiently long chains that we have adequately sampled the posterior probability.
A number of tests are available [22, 62], although it's always a good idea to
perform a number of sanity checks as well – e.g. do we get the same result
from different chains started a widely separated locations in parameter space?

6.8.3 Introduction to the CMB

Over the past few years there has been a dramatic improvement in the resolution and accuracy of measurements of fluctuations in the temperature of
the CMB radiation. The discovery of features, in particular, the first acoustic peak, in the power spectrum of the CMB temperature has led to a new
data-rich era in cosmology [7, 27]. More recently a significant leap forward
was made with the release of the first year data from the WMAP satellite

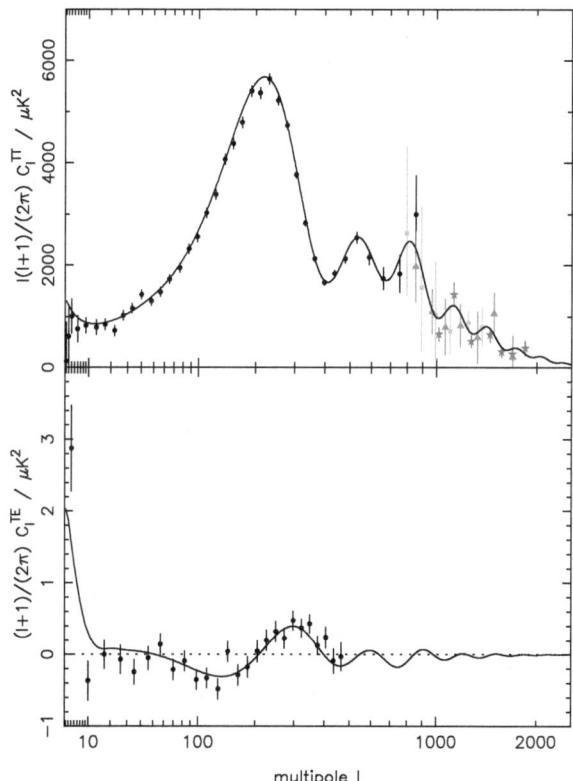

Fig. 6.10. Upper panel: The 1-yr WMAP TT power spectrum (*circles*) is plotted
with the CBI (*triangles*), VSA (*squares*) and ACBAR (*stars*) data at higher l.
Lower panel: The 1-yr WMAP TE power spectrum (*circles*). In both panels the
solid black line shows the best fit model calculated from fitting the CMB data

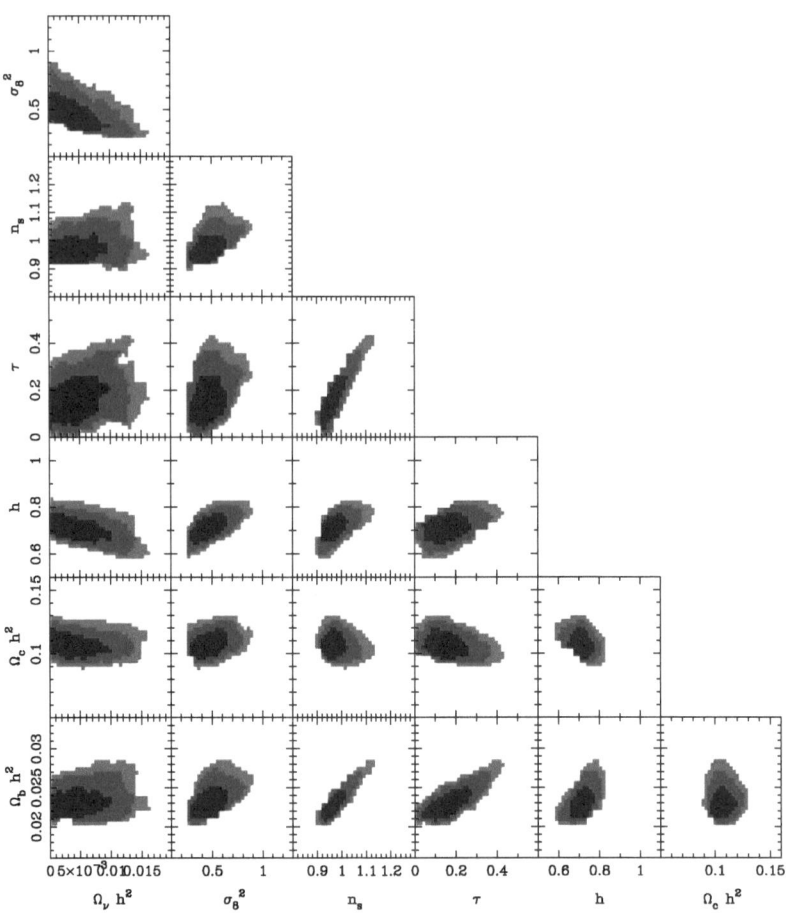

Fig. 6.11. 2D projections of the 7D likelihood surface resulting from a fit to the CMB data plotted in Fig. 6.10. The shading represents areas with $-2\Delta\mathcal{L} = 2.3, 6.0, 9.2$ corresponding to 1σ, 2σ and 3σ confidence intervals for multi-parameter Gaussian random variables. There are two primary degeneracies - between $\Omega_c h^2$ and h and between n_s, τ and $\Omega_b h^2$, which are discussed further in Sect. 6.8.4

[6, 30]. The relative positions and heights of the acoustic peaks encode information about the values of the fundamental cosmological parameters, as discussed for the matter power spectrum in Sect. 6.3. For a flat cosmological model with $n_s = 1$, $\Omega_M = 0.3$, $h = 0.7$ and $\Omega_b h^2 = 0.02$ the CMB and matter power spectra are compared in Fig. 6.9. In order to create Fig. 6.9, the angular CMB power spectrum was converted to comoving scales by considering the comoving scale of the fluctuations at the last scattering surface. In Fig. 6.9, the matter power spectrum has been rationed to a smooth zero baryon model in in order to highlight features – even so, the baryon oscillations are significantly more visible in the CMB fluctuation spectrum. The vertical dotted lines

in this plot are located at the peaks in the CMB spectrum and highlight the phase offset between the two spectra. The CMB peaks are $\pi/2$ out of phase with the matter peaks because they occur where the velocity is maximum, rather than the density at the last scattering surface – this is known as the velocity overshoot. Additionally there is a projection effect – the observed CMB spectrum is the 2D projection of 3D fluctuations, and so is convolved with an asymmetric function: the projection can increase, but not decrease the wavelength of a given fluctuation.

A compilation of recent CMB data is presented in Fig. 6.10. Here we have plotted both the temperature-temperature (TT) auto-power spectrum and the temperature-E-mode polarisation (TE) cross-power spectrum. The most significant current data set is, of course, the WMAP data shown by the solid circles in this figure. However, additional information is provided on small scales by a number of other experiments. In Fig. 6.10, we plot data from the CBI [50], VSA [14], and ACBAR [32] experiments.

Likelihood surfaces from a multi-parameter fit to these CMB data are shown in Fig. 6.11. For this fit, 7 parameters were allowed to vary: $\Omega_c h^2$, $\Omega_b h^2$, h, τ, n_s, σ_8, and $\Omega_\nu h^2$. Other cosmological parameters were set at their "model simplification" values as discussed in Sect. 6.8.1. In particular, we have assumed a flat cosmological model with $\Omega_{\text{tot}} = 1$ and that the tensor contribution to the CMB is negligible. In choosing this set of 7 parameters, and using the standard MCMC technique we have implicitly assumed uniform priors for each. The constraints on the 7 fitted parameters are given in Table 6.2.

6.8.4 Parameter Degeneracies in the CMB Data

By examining Fig. 6.11 we see that the CMB data alone do not constrain all of the fundamental cosmological parameters considered to high precision.

Table 6.2. Summary of cosmological parameter constraints calculated by fitting a 7-parameter cosmological model to the CMB data plotted in Fig. 6.10 and to the combination of these data with the measurement of the 2dFGRS power spectrum [10] – see text for details. Data are given with 1σ error, except for $\Omega_\nu h^2$ which is presented as a 1σ upper limit

parameter	CMB constraint	CMB+2dFGRS constraint
$\Omega_c h^2$	0.107 ± 0.015	0.106 ± 0.006
$\Omega_b h^2$	0.0238 ± 0.0021	0.0235 ± 0.00166
h	0.725 ± 0.096	0.718 ± 0.036
τ	$< 0.204 \pm 0.117$	$< 0.195 \pm 0.085$
n_s	1.00 ± 0.064	0.987 ± 0.046
σ_8	0.703 ± 0.125	0.696 ± 0.085
$\Omega_\nu h^2$	< 0.00700	< 0.006

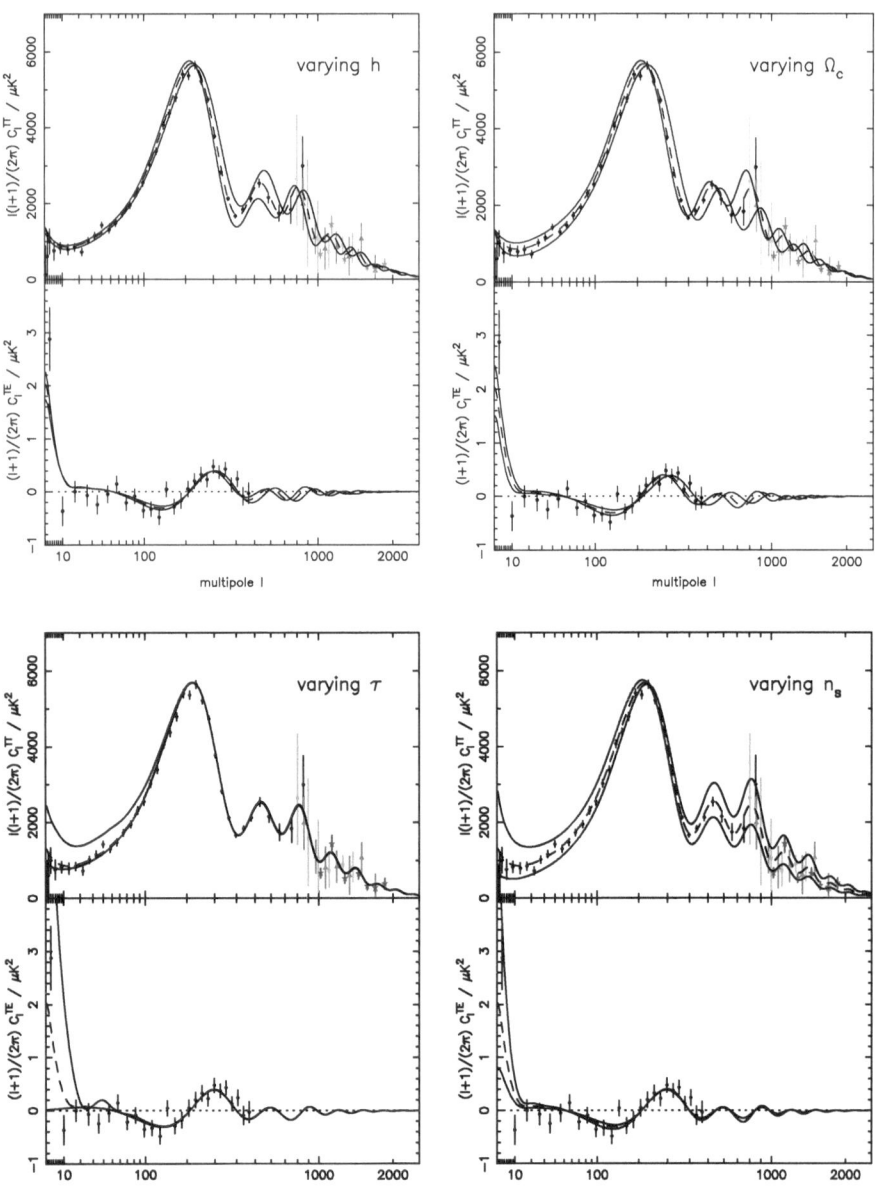

Fig. 6.12. As Fig. 6.10, but now showing 3 different models: the dashed line shows the best fit model in all panels – the model plotted in Fig. 6.10. The solid lines in the top-left panel were calculated with $h = \pm 0.1$, top-right $\Omega_c \pm 0.1$, bottom-left $\tau + 0.3$ and $\tau = 0$, and bottom-right $n_s \pm 0.2$

Degeneracies exist between certain combinations of parameters which lead to CMB fluctuation spectra that cannot be distinguished by current data [16]. To help to explain how these degeneracies arise, CMB models with different cosmological parameters are plotted in Fig. 6.12.

Constraining models to be flat does not fully break the geometrical degeneracy present when considering models with varying Ω_{tot}, and a degeneracy between the dark matter density Ω_c and the Hubble parameter h remains. Figure 6.12 shows that both Ω_c and h affect the location of the first acoustic peak. A simple argument can be used to show that models with the same value of $\Omega_m h^{3.4}$ predict the same apparent angle subtended by the light horizon and therefore the same location for the first acoustic peak in the TT power spectrum [44]. The degeneracy in Fig. 6.11 roughly follows this prediction.

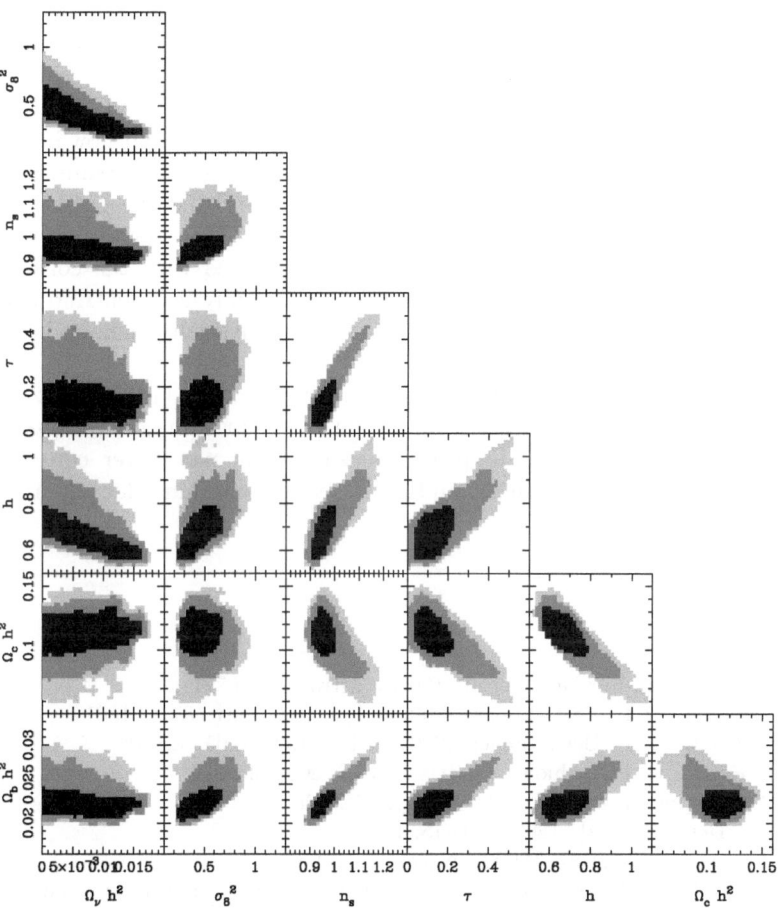

Fig. 6.13. As Fig. 6.11, but now including extra constraints from the 2dFGRS analysis of [10]. These constraints helps to break the primary degeneracies discussed in Sect. 6.8.4

There is another degeneracy that that can be seen in Fig. 6.11 between n_s, τ and $\Omega_b h^2$. From Fig. 6.12, we see that the effect of the optical depth τ on the shape of the TT power spectrum occurs predominantly at low multipoles. By adjusting the tilt of the primordial spectrum (n_s), the low-ℓ power spectrum can be approximately corrected for the change in τ, and the high-ℓ end can be adjusted by changing the baryon density. This degeneracy is weakly broken by the TE data which provide an additional constraint on τ.

6.8.5 Results from the Combination of LSS and CMB Data

The CMB degeneracy between Ω_c and h can be broken by including additional constraints from the power spectrum of galaxy clustering. There have been a number of studies using both CMB and large-scale structure data to set cosmological constraints, with a seminal paper coming from the WMAP collaboration [57]. Recently new small-scale CMB data and large-scale structure analyses have increased the accuracy to which the cosmological parameters are known. [52, 61].

In Fig. 6.13, we provide a likelihood plot as in Fig. 6.11, but now including the cosmological constraints from the final 2dFGRS power spectrum [10]. For this analysis, a constant bias was assumed and we fitted the galaxy power spectrum over the range $0.02 < k < 0.15 \, h \, \mathrm{Mpc}^{-1}$. The derived parameter constraints for the 7 parameters varied are compared with the constraints from fitting the CMB data only in Table 6.2. The physical neutrino density $\Omega_\nu h^2$ is unconstrained within the prior interval (physically, it must be > 0), so we only provide an upper limit.

A Table of parameter constraints, such as that presented in Table 6.2 represents the end point of our story. We have introduced the major steps required to utilise a galaxy survey to provide cosmological parameter constraints, and have ended up with an example of a set of constraints for a particular model.

References

1. J.K. Adelman-McCarthy et al., (2005) [astro-ph/0507711].
2. S. Bashinsky, E. Bertschinger, (2001) Phys. Rev. Lett., 87, 081301.
3. S. Bashinsky, E. Bertschinger, (2002). Phys. Rev.D, 65, 123008.
4. C. Baugh, G. Efstathiou, (1993). MNRAS, 265, 145.
5. D.J. Baumgart, J.N. Fry, (1991). ApJ, 375, 25.
6. C.L. Bennett, et al., (2003). ApJS, 148, 1.
7. P. de Bernardis et al., (2000). Nature, 404, 955.
8. G. Birkhoff, (1923), *Relativity and Modern Physics*, Harvard University Press, Cambridge, Mass.
9. S. Cole, N. Kaiser, (1989). MNRAS, 237, 1127.
10. S. Cole, et al., (2005). MNRAS, 362, 505.

11. P. Coles, F. Lucchin, (1995). *Cosmology, The Origin and Evolution of Cosmic Structure*, Wiley.
12. M. Colless et al., (2003), astro-ph/0306581.
13. A. Cooray and R. Sheth, (2002). Physics Reports, 372, 1.
14. C. Dickinson et al., (2004), MNRAS **353**, 732.
15. S. Dodelson, (2003)., *Modern Cosmology*, Academic Press.
16. G. Efstathiou and J.R Bond (1999). MNRAS **304**, 75.
17. D.J. Eisenstein, and W. Hu, (1998). ApJ **496**, 605.
18. D.J. Eisenstein et al., (2001). AJ **122**, 2267.
19. D.J. Eisenstein et al., (2005). ApJ **633**, 560.
20. K.B. Fisher, C.A Scharf and O. Lahav, (1994). MNRAS **266**, 219.
21. H.A. Feldman, N. Kaiser, and J.A. Peacock, (1994). MNRAS **426**, 23.
22. A. Gelman, and D. Rubin, (1992)., Statistical Science **7** 457.
23. A. Gelman, G.O Roberts and W.R. Gilks, (1996). in eds Bernardo J.M., Berger J.O., Dawid A., Smith A., Bayesian Statistics 5, 599, OUP.
24. W.R. Gilks, S. Richardson and D.J. Spiegelhalter, *Markov chain monte carlo in practice*, (1996). Chapman and Hall.
25. A.J.S. Hamilton, (2005). "Data analysis in Cosmology", ed. V.Matrinez, Springer-Verlag lecture notes in Physics, astro-ph/0503603.
26. A.J.S. Hamilton, (2005). "Data analysis in Cosmology", ed. V.Matrinez, Springer-Verlag lecture notes in Physics, astro-ph/0503604.
27. S. Hanany et al., (2000). ApJ **545**, L5.
28. E. Hawkins et al., (2003). MNRAS **346**, 78.
29. A.F. Heavens and A.N. Taylor, 1995. MNRAS **275**, 483.
30. G. Hinshaw et al., (2003). ApJS **148**, 135.
31. N. Kaiser, (1987). MNRAS **227**, 1.
32. C.L. Kuo et al., (2004). ApJ **600**, 32.
33. O. Lahav and A.R. Liddle, (2006). Phys. Lett. B 592, 1 (2004) and 2005 partial update for the 2006 edition available at the PDG WWW pages at http://pdg.lbl.gov/, astro-ph/0601168.
34. S.D. Landy and A.S. Szalay, (1993). ApJ **412**, 64.
35. A. Lewis and S. Bridle, (2002). Phys. Rev. D **66**, 103511.
36. A.R. Liddle and D.H. Lyth, Physics Reports **231**, 1, astro-ph/9303019.
37. A.R. Liddle and D.H. Lyth (2000). *Cosmological Inflation and Large-Scale Structure*, Cambridge University Press.
38. A.R. Liddle (2004). MNRAS **351**, L49.
39. N. Metropolis, A.W. Rosenbluth, M.N. Rosenbluth, A.H. Teller and E. Teller (1953). Journal of Chemical Physics **21**, 1087.
40. H.J. Mo and S.D.M. White (1996). MNRAS **282**, 347.
41. J.A. Peacock (1999). *Cosmological Physics*, Cambridge University Press.
42. J.A. Peacock and R.E. Smith (2000). MNRAS **318**, 1144.
43. W.J. Percival et al., (2001). MNRAS **327**, 1297.
44. W.J. Percival et al., (2002). MNRAS **337**, 1068.
45. W.J. Percival L. Verde, J.A. Peacock (2004). MNRAS **347**, 645.
46. W.J. Percival et al., (2004). MNRAS **353**, 1201.
47. W.J. Percival (2005). A&A **443**, 819.
48. P.J.E. Peebles (1973). ApJS **185**, 413.
49. A.C. Pope et al., (2004). ApJ **607**, 655.
50. A.C.S. Readhead et al., (2004). ApJ **609**, 498.

51. G.O. Roberts (1996). in eds W.R. Gilks, S. Richardson, D.J. Spiegelhalter, *Markov chain monte carlo in practice*, Chapman & Hall.
52. A.G. Sanchez et al., MNRAS accepted, (2006). astro-ph/0507583.
53. M.D. Seaborne et al., (1999). MNRAS **309**, 89.
54. U. Seljak (2000). MNRAS **318**, 203.
55. R.K. Sheth and G.Tormen (1999)., MNRAS **308**, 119.
56. R.E. Smith et al., (2003). MNRAS **341**, 1311.
57. D.N. Spergel et al., (2003). ApJS **148**, 175.
58. H. Tadros et al., (1999). MNRAS **305**, 527.
59. M. Tegmark, A.J.S. Hamilton and Y. Xu, (2002). MNRAS **335**, 887.
60. M. Tegmark et al., (2004). ApJ **606**, 702.
61. M. Tegmark et al., (2004b) Phys. Rev. D **69**, 103501.
62. L. Verde et al., (2003). ApJS **148**, 195.
63. L. Wang and P.J. Steidhardt, (1998), ApJ **508**, 483.
64. V. Wild et al., (2005). MNRAS **356**, 247.

7

Dark Energy and the Microwave Background

Robert Crittenden

Institute of Cosmology and Gravitation, University of Portsmouth, Portsmouth
PO1 2EG, UK
Robert.Crittenden@port.ac.uk

Abstract. Evidence for dark energy comes from a wide variety of data. Here I
discuss the role the CMB anisotropies have had in framing the dark energy problem.
After reviewing the physics of the CMB, I discuss the different methods that are
used in determining the dark energy's density, evolution and clustering properties
and the crucial role the microwave background plays in all of these methods.

7.1 Introduction

In the last few years, the accepted cosmological model has undergone a
paradigm shift. Despite much resistance, cosmologists have been forced to admit that the Universe has begun accelerating, perhaps as the result of a new
type of repulsive matter called 'dark energy.' From a theoretical standpoint,
it is hard to imagine a less attractive cosmological ingredient. This invisible
dark energy has become important only very recently, and seems require a
incredibly small mass scale, of order $10^{-3}eV$. In addition, we must cope with
the coincidence that we happen to be living in a rare epoch where the dark
energy and dark matter happen to have comparable densities.

Clearly, considerable evidence is required before we accept such a radical
element in our cosmological model. For at least a decade, there have been hints
that the dark matter we saw was not sufficient to account for the observed
expansion rate of the Universe. However, it is only recently that such hints
have been more definite and other possibilities, like an open universe, have
been excluded by the data. Arguably the 'tipping point' came when observations of supernovae at high redshifts showed them to be dimmer and thus
further away than expected. However, this might have been argued away as a
systematic effect, had it not been for the many other observations pointing in
the same direction. The goal of this review is to show the key role observations
of the microwave background have had in making the case for dark energy.

R. Crittenden: *Dark Energy and the Microwave Background*, Lect. Notes Phys. **720**, 187–217
(2007)
DOI 10.1007/978-3-540-71013-4_7

7.1.1 CMB and Dark Energy

Naively, the CMB seems like a terrible place to look for evidence for dark energy. Most of the CMB anisotropies we see were created at a redshift of $z \simeq 1000$, when the Universe was very much smaller. Given that the dark matter and dark energy densities are comparable today, the density fraction in dark energy at that time would have been of order 10^{-9}, far too small to make an important dynamical impact (see Fig. 7.1).

Why then are CMB observations so relevant to the study of dark energy? There are a number of reasons:

1. While we learn little about dark energy at $z \simeq 1000$, the CMB anisotropies can give us an inventory for virtually everything else in the Universe at that time: baryons, dark matter, photons, neutrinos. We can extrapolate these forward to the present to see if there is sufficient matter to explain the observed expansion rate.
2. The CMB also imprints a standard ruler in the sky of a known size. We use this information to constrain the Universe's curvature and the physical distance to the last scattering surface.

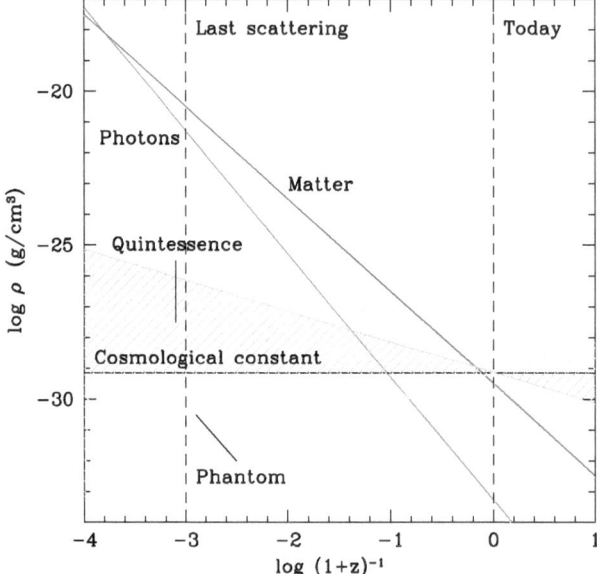

Fig. 7.1. Dark energy density versus the matter and photon densities. The quintessence and phantom regions assume a constant equation of state. In most models, the dark energy density was many orders of magnitude smaller than the matter and photon densities when the CMB photons were last scattered and play no role in the origin of the anisotropies. Only for a brief epoch are the matter and dark energy densities comparable

3. Not all the CMB anisotropies we see are ancient in origin. Some of them are actually created very recently through gravitational effects or rescattering. These are coeval with dark energy dominance and can be strongly affected by it.
4. Finally, some models for dark energy can actually have appreciable matter density at early times. For example, in 'tracker' models the dark energy density tracks the dominant energy density, staying at a fixed fraction of the total until very recently. Thus, they could potentially have an important dynamic contribution at last scattering.

Here I will focus on the first three of these effects, where the early CMB physics is largely unaffected by the presence of dark energy. I begin by giving a brief overview of some of the models for dark energy, because these will help identify its most important properties. Then I examine the physics of the microwave background, aiming to give a physical intuition for how the matter content of the Universe makes itself known in the CMB anisotropy spectrum. I then look at the various ways dark energy is probed and the role CMB observations have played, particularly with regard to the detection of the integrated Sachs-Wolfe effect.

7.2 Models for Dark Energy

It is worth investigating briefly some of the various alternatives for explaining the recent accelerated expansion, as it will highlight precisely which aspects of dark energy are the most interesting to try to observe. (A more complete overview of dark energy models can be found in the recent review by Sahni [1].)

7.2.1 Cosmological Constant

The simplest example of dark energy is the cosmological constant, which is constant term which can be consistently added to Einstein's equations. Unfortunately, the cosmological constant has a history of being introduced to help fit observations which later turned out to be wrong. (For reviews of the cosmological constant, see [2, 3, 4]). Einstein originally introduced it in 1917 to try to produce a static universe out of general relativity. However, when Hubble discovered the Universe to be expanding, Einstein discarded the term, reportedly describing it as his biggest blunder.

Unfortunately, the expansion rate measured by Hubble was much too large, resulting in a universe too young compared to the known age of the Earth. Some cosmologists resolved this age crisis by reintroducing the cosmological constant which, correctly tuned, can induce a nearly static 'loitering' phase. Eventually, the expansion rate was revised down, alleviating the age crisis and allowing the cosmological constant to be discarded. However, in 1967 the 'loitering universe' idea was resurrected to explain an apparent excess of quasars at redshift of $z \simeq 2$. Again, better observations ended up making the loitering phase unnecessary, which was where things stood until recently. It is understandable then that when the data began mounting again for a cosmological constant, there was some resistance to the idea.

More resistance came from particle physics, which actually provides a natural candidate for a physical origin of the cosmological constant: the zero-point vacuum fluctuations of bosonic or fermionic fields. However, the typical scale of the cosmological constant is of order M_{cutoff}^4, where M_{cutoff} is the ultraviolet cutoff of the theory describing the fields. Using the Planck mass as the ultraviolet cutoff gives $\Lambda_{planck} \sim (10^{19}GeV)^4$. The actual observations show a cosmological constant many orders of magnitude smaller,

$$\Lambda_{obs} \sim (10^{-3}eV)^4 = 10^{-120}\Lambda_{planck} \ . \tag{7.1}$$

Supersymmetry provides one way of removing these contributions because the fermionic and bosonic contributions exactly cancel out; however the fact that we have not yet seen direct evidence for supersymmetry means the breaking scale, which effectively becomes the new cutoff scale, must be very large $> 100GeV$. Even the lower energy QCD physics would produce a cosmological constant very much larger (10^{41} times) than that which is observed.

Before the recent evidence for dark energy, it was hoped that another symmetry might be found which could explain why such a large value for the cosmological constant was not seen. The expectation was that such a symmetry would cause the constant to be precisely zero. However, the recent observations have left us instead with the worse problem of having to explain an incredibly small number, and thus a large hierarchy in mass.

7.2.2 Dynamical Dark Energy

Until the evidence for dark energy became compelling, virtually the only candidate for accelerated expansion was a cosmological constant, a simple, one parameter model. However, the void was soon filled with a myriad of different dynamical dark energy models, where the energy density was no longer constant in space or time.

A useful way of parameterising dark energy models is through the effective equation of state, the ratio of the pressure to the energy density, $w \equiv p/\rho$. It can be easily seen from the Raychaudhuri equation,

$$\frac{\ddot{a}}{a} = -4\pi G(\rho + 3p) = -4\pi G\rho(1 + 3w) \ , \tag{7.2}$$

that to get accelerated expansion, one requires $w_{tot} < -1/3$. If one assumes that a third of the present density is in the form of dark matter ($p = 0$), then the equation of state of the dark energy itself needs to be $w_{DE} < -1/2$ to cause acceleration. For example, the cosmological constant has an equation of state $w = -1$.

Quintessence Models

Cosmologists were very familiar with accelerating dynamics from theories of the early universe. Inflationary theories, typically based on scalar fields, used a period of accelerated expansion to solve initial conditions problems of the big bang. These were quickly adapted to the late universe observations and collectively have become known as quintessence models [6, 11].

To produce acceleration, the energy density in the scalar fields is typically dominated by a very weak potential. This requires a very small effective mass, $m_\phi < H \sim 10^{-33} eV$ [7]. For some potentials, the field is frozen at early times, while for others (so-called 'tracker' models) it can slowly roll down the potential, its energy density tracking the energy density of the dominant fluid until very recently.

The equation of state in quintessence models generally evolves, though with a special choice of the potential it can be constant. Indeed, it has been shown that any $w(z) > -1$ history can be produced with a suitably tuned potential [8]. Rather than focusing on the potential, often phenomenological models for $w(z)$ have been used, attempting to capture the essence of the evolution by a one or two parameter fit, such as $w(z) = w_0 + w_1(1 - a)$, where a is the scale factor.

Tangled Defects

One well motivated class of dark energy models arises naturally from theories of cosmological defects. Generically a phase transition occurs in the early universe, producing either cosmic domain walls or cosmic strings [9] (or something even stranger [10].) Depending on the symmetries, these can become tangled, resulting in a static configuration.

As the universe expands, the defect network also expands. For strings, the total length will grow proportional to a, leading to an energy density dropping as a^{-2}. This leads to an equation of state of $w = -1/3$, just failing to produce acceleration.

In the domain wall picture, the defect area (and thus total energy) in a comoving volume grows as a^2, leading to an energy density which drops as a^{-1}. This is equivalent to an equation of state of $w = -2/3$ (though see [11]...) which can produce acceleration. This model however seems to be at odds with the most recent data which seems to prefer a more negative equation of state.

Chaplygin Gas

Another proposed model for dark energy actually from a kind of behavior seen in early studies of aerodynamics and the Chaplygin gas model [17] is named for the Russian who introduced the effective equation of state it in that context. This model effectively postulates an equation of state of $p = -A/\rho$, though often a more generalized form is used

$$p = -A/\rho^\alpha \ . \tag{7.3}$$

This leads to a density which scales as

$$\rho = \left(A + Ba^{-3(1+\alpha)} \right)^{1/(1+\alpha)} \ . \tag{7.4}$$

The attraction of the Chaplygin gas model is that the dark energy scales effectively like dark matter ($p = 0$) at high densities, but then can act like a cosmological constant at low densities. Thus, it was hoped that a single fluid could play the roles of both dark energy and dark matter. While this hope has been excluded now by observations, Chaplygin gas is still a possibility for explaining dark energy alone.

Phantoms and Ghosts

Motivated by early SN data which seemed to prefer $w < -1$, much work has been done looking at such models, known collectively as phantom dark energy models [13]. Such dark energy has the property that its density actually increases as the Universe gets bigger. It can in fact diverge at a finite time, it what has come to be called a 'big rip' [14].

Theoretically, such phantom models face many challenges. They violate both the weak energy condition, and lead to negative norm states. They are also classically and quantum mechanically unstable [15, 16].

7.2.3 Modified Gravity

Perhaps the most interesting explanation of the acceleration is not to modify the matter content of the Universe at all, but to change instead the gravitational attraction on large scales. This requires something besides general relativity. Many proposals for doing this are being investigated, including Brans-Dicke theory and extra-dimensional braneworld theories. (For reviews, see the notes by Maartens in this volume [17].)

7.3 The Physics of the Microwave Background

Before going into detail the various ways the CMB can probe the dark energy, it is worth reviewing the basics of microwave background: what it is, how it is described and how we think the observed fluctuations arose.

7.3.1 Basics of the Microwave Background

Discovered just forty years ago, the microwave background is the relic radiation left over from the big bang [18, 19]. When the Universe was much smaller, the radiation was much hotter; at early times it was in thermal equilibrium with the other matter in the Universe through interactions like Thomson scattering with free electrons. These interactions allowed the photon distribution to thermalize and this thermal spectrum has been preserved through the later expansion. Its present spectrum has been very accurately measured by the COBE FIRAS instrument to be a blackbody spectrum to within 50 parts per million, with a current temperature of $2.725 \pm 0.002K$ [20, 21].

While the free electron density was high, the photons were tightly coupled to them and could not travel far between interactions. However, once the Universe cooled enough to allow the electrons and protons to become bound in neutral hydrogen ($z \sim 1100$), the free electron density dropped dramatically and the photons were allowed to travel freely. Most of the CMB light we see was last scattered at these redshifts. Subsequent reionization of the hydrogen at low redshifts means that some photons can rescatter, but the low density that this occurs means that these events are relatively rare.

Anisotropy Maps

The microwave background is to a large degree very homogeneous over the sky. However, anisotropies have been observed at low levels, starting with a dipole anisotropy at a level of $\Delta T = 3.372 \pm 0.007 mK$ discovered in the 1970's. This is thought to be dominated by the Doppler effect from the motion of the solar system with respect to the rest frame of the microwave background, corresponding with a velocity of $v = \frac{\Delta T}{T_0} c = 360$ km/s.

After many years of searching, further anisotropies were found by the COBE satellite in 1992, at a level of one hundredth of the dipole [22]. These observations were at the level predicted for the cold dark matter model. Subsequent observations by balloon, ground based and the satellite-based Wilkinson Microwave Anisotropy Probe (WMAP) have refined these observations dramatically, from about thousand independent pixels in COBE to hundreds of thousands in WMAP [23, 24].

WMAP, benefitting from observing the full sky in many frequencies, has the best measurements of the CMB on large scales. But on scales smaller than the WMAP beam ($\theta < 0.3°$), the best observations come from an array of ground and balloon based experiments, including the VSA [25, 26], ACBAR [27], BOOMERANG [28], MAXIMA [29] and CBI [30].

While the observations typically produce two dimensional maps on the sky, in order to compare to theoretical predictions it is useful to expand these maps in terms of orthogonal spherical harmonic functions:

$$\frac{\delta T(\mathbf{n})}{T} = \sum_{\ell, m} a_{\ell m} Y_{\ell m}(\mathbf{n}) . \tag{7.5}$$

Such a transformation is analogous to a Fourier transform, approaching one on suitably small patches of the sky. The multipole ℓ value is an effective wavenumber, relating to the angular scale of the variations, $\Delta\theta \sim \pi/\ell$. For each ℓ, there are $2\ell + 1$ values of $m = -\ell, -\ell + 1, \ldots \ell$. The orthogonality of the spherical harmonic functions leads to the inverse transform,

$$a_{\ell m} = \int \frac{\delta T(\mathbf{n})}{T} Y_{\ell m}^*(\mathbf{n}) d\Omega_{\mathbf{n}} . \tag{7.6}$$

Two Point Statistics

Theoretical models do not predict the precise temperature fluctuation in any particular direction, but instead focus on the statistical properties of the maps. If the maps are Gaussian, then the statistical properties are entirely determined by the two point statistics of the map. These are described by either the two-point correlation function, or its Fourier analog, the temperature power spectrum. The power spectrum of the fluctuations is defined as the expectation of the multipole moments:

$$\langle a_{\ell m} a_{\ell' m'}^* \rangle = C_\ell \delta_{\ell\ell'} \delta_{mm'} , \tag{7.7}$$

which follows simply from the assumption of invariance under rotations.

The two point correlation function in real space is defined as

$$C(\theta) \equiv \left\langle \frac{\delta T(\mathbf{n})}{T} \frac{\delta T(\mathbf{n}')}{T} \right\rangle = \sum_{\ell,m} C_\ell Y_{\ell m}(\mathbf{n}) Y_{\ell m}^*(\mathbf{n}') = \frac{1}{4\pi} \sum_\ell (2\ell+1) C_\ell P_\ell(\cos\theta) ,$$
$$\tag{7.8}$$

where θ is the angle between $\mathbf{n}$ and $\mathbf{n}'$. The variance of the temperature field is given by the correlation at zero separation, $C(0) = \frac{1}{4\pi} \sum_\ell (2\ell + 1) C_\ell$.

Cosmic Variance

If the fluctuations are Gaussian and isotropic, the multipole moments are all independent. Measurement of the correlation function is fundamentally limited by the fact that we observe only one sky. The power spectrum C_ℓ is discrete, and each ℓ value is sampled only $2\ell + 1$ times. This results in an unresolvable uncertainty in the spectrum measurement of $\delta C_\ell \simeq \sqrt{2} C_\ell / (2\ell + 1)^{\frac{1}{2}}$. This is known as *cosmic variance*.

Cosmic variance is the dominant source of uncertainty for large modes; for example, WMAP is cosmic variance limited out to $\ell = 400$ [24]. Thus, further observations will not improve the measurement of these modes significantly, though they could aid by improving the removal galactic foregrounds.

Polarization

In addition to temperature fluctuations, the microwave radiation can also be linearly polarized. This polarization arises naturally in Thomson scattering if there is a quadrupole in the radiation which is incident on the scattering electrons. While the photons and electrons are tightly coupled, there is no polarization. However, once the photons are able to travel some distance between scattering events, polarization is generated at the level of the local quadrupole. Typically, the CMB radiation is polarized at 10–15% of the level of the anisotropy.

The two polarization degrees of freedom (amplitude and inclination angle) can be decomposed into two kinds of fluctuations, E-type (gradient) and B-type (curl) [31, 32]. Polarization created as the result of scalar perturbations are purely E-type, though later non-linear processes like gravitational lensing or Faraday rotation of the polarization planes can cause it to be mixed into B-type. Tensor perturbations (gravity waves) and vector perturbations create both E-modes and B-modes; thus B-modes, particularly on large scales where they cannot be produced by gravitational lensing, may be used to determine whether a significant portion of the observed CMB anisotropies were caused by primordial gravity waves.

The E-modes are also correlated with the temperature fluctuations [33]. The polarization data supplements the temperature data, allowing us to confirm the general picture of the generation of anisotropies. There are also some parameters for which the polarization is especially sensitive, most notably the optical depth for rescattering after reionization, which induces a large scale signal in the polarization auto and cross correlation functions.

The first detection of polarization has come very recently. In 2002, DASI detected the polarization at small angular scales [34], and these observations have been confirmed with CBI [35], CAPMAP [36] and Boomerang [37]; all these seem to confirm the cosmological model. On large scales, WMAP has measured the cross correlation spectrum [38] and recently produced the first large scale autocorrelation measurements [39]. Perhaps the biggest surprise in the first year data was the large amplitude of the cross correlation on the largest scales, which seemed to indicate a very early reionization of the Universe. The more recent observations however show that these were subject to foreground contamination and the reionization epoch has been revised down.

7.3.2 Origin of the Fluctuations

In order to understand the CMB constraints on the cosmological model, it is important to see just how the anisotropies arose. To calculate these in detail requires a Boltzmann code, following the full evolution of the photon distribution function [40]. A number of publically available codes are now available to do these calculations, including CMBfast [41] and CAMB [42]. Here, rather than reviewing the full formalism (see e.g. the recent review by Challinor [43]), I will simply try to provide some intuition for how the predictioned spectra come about.

Gauge Issues

There are two gauges that are typically used to describe the CMB anisotropies and it is useful to understand both. Most numerical work (including CMBfast and CAMB) is performed in the synchronous gauge,

$$ds^2 = a^2(\tau)(-d\tau^2 + (\delta_{ij} + h_{ij})dx^i dx^j) . \tag{7.9}$$

The conformal Newtonian gauge,

$$ds^2 = a^2(\tau)(-(1 - 2\Psi)d\tau^2 + (1 + 2\Phi)dx^i dx_i) , \tag{7.10}$$

is useful to get physical intuition, as Φ approaches the Newtonian gravitational potential within the horizon. In addition, unlike the synchronous gauge, it contains no residual gauge freedom, so it is useful for analytic calculations. A nice review of these two gauges and the Boltzmann equations relevant for the microwave background has been written by Ma and Bertschinger [44]. (Beware that there are a number of different sign conventions in the literature!)

Density and Potential Fluctuations

On scales large enough to be superhorizon at last scattering, two effects dominate the CMB anisotropies and are jointly referred to as the *intrinsic fluctuations*. The first arises due to the fluctuations in the density of the photons on the last scattering surface. Since the photon density is proportional to $\rho_\gamma \propto T^4$, fluctuations in the temperature are related to the density fluctuations by $\frac{\delta T}{T} = \frac{1}{4}\frac{\delta\rho}{\rho}$.

In addition, there will be fluctuations in the gravitational potential on the last scattering surface. Photons will have to climb out of (or roll down) the gravitational potential, and will lose or gain energy. Thus, there will be an additional fluctuation of $\frac{\delta T}{T} = \Phi$. Thus the total intrinsic term is,

$$\frac{\delta T}{T} = \frac{1}{4}\frac{\delta\rho}{\rho} + \Phi . \tag{7.11}$$

Note that in the absence of anisotropic stress, the Einstein equations imply $\Psi = \Phi$.

These two terms will be comparable for wavenumbers as they cross the horizon, with the density contribution dominating within the horizon. If the initial conditions are adiabatic, the photon density will be highest where the potential is deepest and these two terms will tend to cancel. (This is not necessarily true for isocurvature initial conditions.) In this case, it can be shown that on horizon scales, this leads to the Sachs-Wolfe contribution,

$$\frac{\delta T}{T} = \frac{1}{3}\Phi . \tag{7.12}$$

Note that in the synchronous gauge, only the density term is relevant; the potential is effectively mixed between this and the line of sight term discussed below.

Doppler Effect

Motion of the photon-baryon fluid induces an addition fluctuation via the Doppler effect. This leads to $\frac{\delta T}{T} = \mathbf{v}_\gamma \cdot \mathbf{n}$, where $\mathbf{n}$ is a unit vector in the direction of the last scattering surface.

As can be seen in Fig. 7.3, the Doppler term contributes most on scales inside the horizon. However, it is usually smaller than the density contribution. The so-called 'Doppler peaks' (see Fig. 7.2) are a misnomer; the peaks are actually dominated by the density fluctuations, not the Doppler term which actually tends to fill in the gaps between the peaks.

Line of Sight Effects

The contribution from redshifting along the photon path can be understood in a fairly simple way in synchronous gauge. Consider a small segment of the path that a photon transverses in time $\delta\tau$. This segment has a length given by

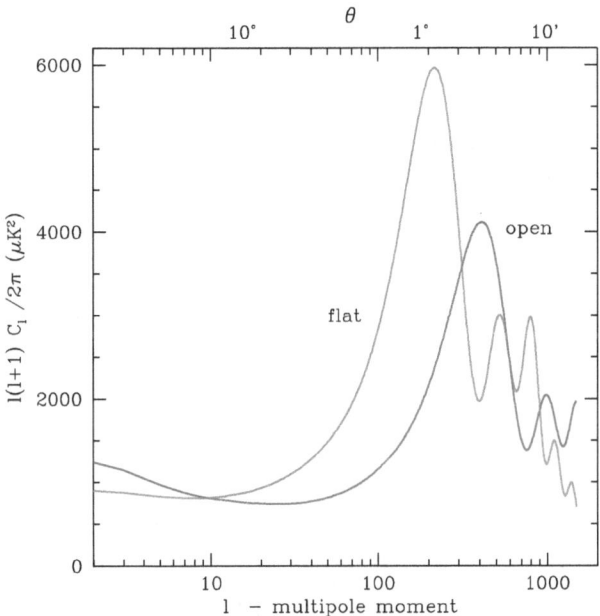

Fig. 7.2. Typical CMB spectra for flat and open cosmologies. The axes are chosen such that the spectrum on large scales ($\ell < 50$) is approximately flat for a scale invariant primordial spectrum. These large modes are outside the horizon at last scattering and microphysics cannot act. On smaller scales, modes have time to oscillate acoustically before last scattering, producing the so-called Doppler peaks. Geometrical effects cause these features to appear smaller in an open universe, and larger in a closed universe. On very large scales, CMB anisotropies are created at late times via the ISW effect, providing additional power at very low ℓ

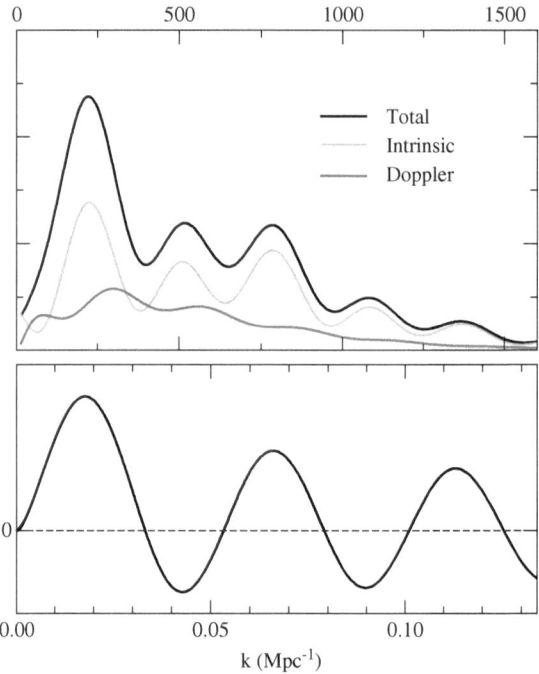

Fig. 7.3. The top figure shows the intrinsic and Doppler contributions to the CMB anisotropy spectrum in the synchronous gauge. The lower figure shows the amplitude of the photon density fluctuations at the last scattering epoch for a range of wavenumbers. Peaks of the density, where the fluid is most compressed, align with the odd Doppler peaks of the CMB spectrum. Troughs of the density (rarefactions) align with the even Doppler peaks. The true Doppler contributions are greatest when the rate of change of the density is greatest. However, structures in the CMB spectrum are generally smoothed out by the projection onto the last scattering surface

$$\delta L = (g_{ij}dx^i dx^j)^{1/2} = (a^2(\delta_{ij} + h_{ij})n_i n_j)^{1/2}\delta\tau \simeq a(1 + \frac{1}{2}h_{ij}n_i n_j)\delta\tau . \quad (7.13)$$

The shift in the photon temperature over this distance is given by the fractional change of the segment length in the period $\delta\tau$,

$$\frac{\delta T}{T}|_{\delta L} = -\frac{1}{\delta L}\frac{\partial(\delta L)}{\partial\tau}\delta\tau = -\left[\frac{\dot{a}}{a} + \frac{1}{2}\dot{h}_{ij}n_i n_j\right]\delta\tau . \quad (7.14)$$

The first term corresponds to the usual homogeneous redshifting of photons and the second, integrated over the whole photon path, gives the term in Sachs-Wolfe expression (see below). Here and below, the dots represent derivatives with respect to conformal time τ.

Sachs-Wolfe Equation

In synchronous gauge, these various effects are combined to give

$$\frac{\delta T}{T} = \frac{1}{4}\delta_\gamma + \mathbf{v}_r \cdot \mathbf{n} - \frac{1}{2}\int_{\tau_{rec}}^{\tau_f} \dot{h}_{ij}n_in_jd\tau \ , \tag{7.15}$$

where the dots are derivatives with respect to co-moving time and $\mathbf{n}$ is a unit vector along the line of sight. This is known as the Sachs-Wolfe equation [45].

The Sachs-Wolfe equation in Newtonian gauge can be shown to be

$$\frac{\delta T}{T} = \frac{1}{4}\delta_\gamma + \Psi + \mathbf{v}_r \cdot \mathbf{n} + \int_{\tau_{rec}}^{\tau_f} (\dot{\Phi} + \dot{\Psi})d\tau \ . \tag{7.16}$$

This is similar to the synchronous gauge expression, except the intrinsic piece is modified to include the initial gravitational potential. The final term, which is the contribution along the path of the photon, is known as the *integrated Sachs-Wolfe* contribution.

7.3.3 Acoustic Oscillations

On smaller scales, the modes have had time to evolve within the horizon prior to last scattering. On these scales, the anisotropies are dominated by the intrinsic anisotropy and to a lesser extent by the Doppler effect. As a result, the oscillations of the photon-baryon fluid take on a key roll.

Prior to recombination, the photons and baryons are strongly coupled and thus act as a single fluid. They then share the same velocity and their densities are related by the adiabatic condition that the number of photons per baryon is fixed. This fluid acts as one with pressure $\frac{1}{3}\bar{\rho}_\gamma$, but which has density $\bar{\rho}_\gamma + \bar{\rho}_b$, The speed of sound in this fluid is

$$c_s^2 = \frac{\delta P}{\delta \rho} = \frac{\frac{1}{3}\bar{\rho}_\gamma}{\frac{3}{4}\bar{\rho}_b + \bar{\rho}_\gamma} = \frac{1}{3(1+R)} \ ,$$

where $R \equiv 3\bar{\rho}_b/4\bar{\rho}_\gamma$.

The density and velocity equations can be combined into a single second order equation to show that the density acts as a forced, damped harmonic oscillator:

$$\ddot{\delta}_\gamma + c_s^2 k^2 \delta_\gamma = -\frac{2}{3}\ddot{h} \ . \tag{7.17}$$

Here for simplicity we have dropped the damping terms, which is appropriate in the radiation dominated regime ($c_s^2 \simeq 1/3$.) Following decoupling, the photons freestream and the baryons fall into the potential wells generated by the dark matter.

The largest scale on which microphysics can act is known as the sound horizon, and the sound horizon at the last scattering surface becomes imprinted into the CMB and large scale structure power spectra. This scale is given by

$$r_S = \int_0^{a_{rec}} \frac{dt}{a}c_s(a) = \int_0^{a_{rec}} c_s(a)\frac{da}{a^2 H(a)} = \int_0^{a_{rec}} c_s(a)\frac{da}{H_0\Omega_m^{\frac{1}{2}}(a + a_{eq})^{\frac{1}{2}}} \ , \tag{7.18}$$

where $a_{rec} = 1/1100$ is the scale factor at last scattering and $a_{eq} \simeq 1/3000$ is the scale factor at matter radiation equality. While the photon density dominates over the baryon density, the sound speed is very nearly constant $c_s = 1/\sqrt{3}$ (it is only slightly (20%) smaller than this at last scattering), so it is a good approximation to take it out of the integral. This gives,

$$r_S \simeq 2c_s H_0^{-1} \Omega_m^{-\frac{1}{2}} \sqrt{a_{rec}} \left[\sqrt{1 + a_{eq}/a_{rec}} - \sqrt{a_{eq}/a_{rec}} \right] . \tag{7.19}$$

It is straight forward to show that with the best determinations of H_0 and Ω_m the present size of this scale is approximately $150 Mpc$ (e.g. [46]).

7.3.4 Features of the Power Spectrum

Many important aspects of the cosmological model can be read off of the CMB power spectrum by looking at a few characteristic features, such as the positions and relative amplitudes of the Doppler peaks.

Peak Positions and the Angle-Distance Relation

The photon-baryon oscillations provide a useful physical yardstick for the early universe. By knowing the size of the sound horizon at last scattering and the observed angular size of these features, we can infer the total curvature of the universe from the angle-distance relation.

To get an intuition for the expected scale, let us assume that radiation-matter equality occurred much before recombination. Then the sound speed is constant and the sound horizon is given by $r_S = c_s \tau_{rec} = c_s 2 H_0^{-1} \sqrt{\Omega_m^{-1} a_{rec}}$. In a flat, matter dominated, Einstein-de Sitter universe, the comoving distance to the last scattering surface is given by $r_{rec} \simeq 2 H_0^{-1} \sqrt{\Omega_m^{-1}}$, so the angular scale of the sound horizon at last scattering is

$$\theta = r_S / r_{rec} = c_s \sqrt{a_{rec}} \simeq 1° . \tag{7.20}$$

Note that the matter density dependence of r_S and r_{rec} are the same, and cancel out in the expression for the angular scale. With a low matter density, the radiation epoch cannot be ignored; using the exact expression for r_S above yields a slightly smaller angle, $\theta \simeq 0.5°$.

If the universe is open, the comoving distance to the last scattering surface is no longer equal to the comoving time since last scattering, but is given instead by

$$r_{rec} = \frac{1}{\sqrt{k}} \sin \left(2 \sin^{-1} (\sqrt{k} H_0^{-1} \Omega_m^{-1/2}) \right) = 2 H_0^{-1} \Omega_m^{-1} , \tag{7.21}$$

where $k \equiv H_0^2 (\Omega_m - 1)$. (This very remarkable simplification occurs only when the dominant matter is dust-like.) Here, the resulting angular scale of the sound horizon is $\theta = c_s \sqrt{\Omega_m a_{rec}}$. The matter dependence fails to cancel, making this a very powerful probe of the curvature of the universe.

Finally, for universes with a cosmological constant,

$$r_{rec} \simeq \tau_0 = 2H_0^{-1}\Omega_m^{-1/2}F(\Omega_m) \,, \tag{7.22}$$

where the hyper-geometric function $F(\Omega_m)$ is a slowly (logarithmically) varying function of the matter density, ranging from 1 ($\Omega_m = 1$) to 0.8 ($\Omega_m = 0.1$); it is sometimes approximated by $\Omega_m^{0.1}$. As in the Einstein-de Sitter case, the dominant Ω_m dependence of r_S and r_{rec} cancel, leaving a weak dependence in the angular scale, $\theta \simeq 0.5°/F(\Omega_m)$; however, if the curvature is assumed to be zero, this weak dependence can be used constrain the matter density or dark energy equation of state.

Peak Heights: Probing the Matter and Baryon Densities

The heights of the Doppler peaks relative to each other and to the large scale (low ℓ) anisotropies offers other important information about the cosmological parameters.

The total matter density (baryons and dark matter) is particularly important in determining the overall amplitude of the peaks. Lowering the matter density makes the redshift of radiation-matter equality later, $z_{eq} = 2.3 \times 10^4 \Omega_m h^2$. This affects both the forcing term for the photon-baryon oscillations and the contributions from the early integrated Sachs-Wolfe effect. The forcing term is stronger near radiation-matter equality, so the heights of the Doppler peaks are higher when the matter density is lower.

The physical baryon density, $\Omega_b h^2$, is critical to the evolution of the microwave anisotropies. Recall that the photon-baryon fluid acts as a fluid with pressure $p = \rho_\gamma/3$ and density $\rho = \rho_b + \rho_\gamma$, so that increasing the baryon fraction at recombination makes the fluid heavier. This causes the fluid to compress more and bounce back less, shifting the effective zero point of the oscillations so that $|\delta_{compression}| > |\delta_{expansion}|$. (See Fig. 7.3.) Thus the odd numbered Doppler peaks associated with the compression will be higher than the even peaks associated with the rarefaction.

The shape of the primordial spectrum will also affect the relative heights of the Doppler peaks, raising the higher peaks relative to the first Doppler peak if the underlying spectrum is blue $n > 1$ and lowering them if the spectrum is red $n < 1$. A large optical depth for rescattering of the photons (see below) has a similar effect, as it will damp out more the higher peaks. These similar effects lead to a degeneracy between n and the optical depth if only the temperature anisotropies are observed.

Damping Effects

As the photons become less tightly coupled to the electrons near recombination, they are able to diffuse out of the matter fluctuations, effectively damping out the smallest scale modes [47]. In addition, the observed temperature anisotropies are blurred by the finite thickness of the last scattering surface. These effects lead to an exponential suppression of the high ℓ modes for which $k\Delta\tau_{rec} \geq 1$, where $\Delta\tau_{rec}$ is the thickness (in comoving time) of the surface of last scattering.

When the Universe is reionized at late times, the CMB photons can be rescattered before reaching us, which can potentially erase the original CMB fluctuations and create new ones. To evaluate whether reionization is important, we must look at the optical depth

$$\kappa = \int_{t_{rei}}^{t_0} \sigma_T n_e c dt , \qquad (7.23)$$

which is related to the probability that a photon is rescattered. (Here, σ_T is the Thomson cross section and n_e is the free electron density, integrated from the onset of reionization to the present. A significant optical depth will suppress structure on smaller scales and will also create large scale polarization. If $\kappa > 1$, then it is quite likely that a photon will scatter off of the ionized medium. Recent third year WMAP measurements of the large scale polarization indicate the optical depth is closer to $\kappa \sim 0.1$, corresponding with the beginning of reionization of about $z \simeq 10 - 12$, assuming that the reionization happened all at once [39].

7.3.5 Non-linear Effects and Foregrounds

At low redshifts, non-linear effects and foregrounds can be important sources of CMB anisotropies. These generate anisotropies in three ways: gravitationally, through the rescattering of the CMB photons from reionized electrons, and through the foreground processes which create microwave photons by other mechanisms.

The most interesting gravitational effects are gravitational lensing and the Rees-Sciama effect [48], which are closely related [49]. The Rees-Sciama effect is the non-linear version of the ISW effect, where the non-linear evolution of the gravitational potential in clusters leads to a line of sight CMB anisotropy. Gravitational lensing of the CMB by foreground structures cannot create anisotropies itself, but will distort any anisotropies that exist on small scales. Lensing of polarization is also important, as it will induce B-modes in a map of pure E-mode polarization; thus it is an important foreground for the search for primordial B-modes from gravity waves.

Once the Universe is reionized, anisotropies can be generated through the rescattering of the photons. The most important of these is the thermal Sunyaev-Zeldovich effect [50], where the photons are up-scattered from the hot gas in clusters. (This is discussed more below.) Anisotropies are also generated by scattering off of moving electrons, either in clusters (the kinetic Sunyaev-Zeldovich effect [51]) or by larger bulk flows (the Ostriker-Vishniac effect [52, 53].) Like the gravitational effects, the latter two effects produce anisotropies with a thermal spectrum. However, the thermal Sunyaev-Zeldovich effect produces a characteristic distortion of the spectrum, clusters appearing hotter at high frequencies and colder at low frequencies.

Many foreground processes can also emit in the microwave frequencies, and any anisotropies produced by such mechanisms must be subtracted before the cosmological anisotropies can be understood. Luckily, the significantly different frequency spectra of the foregrounds makes such a subtraction tractable. At high frequencies the most important foreground is from thermal emission from dust. The primary source of this dust is in our own galaxy, where it is typically of order 20K, heated by optical or UV radiation.

At lower frequencies, synchrotron and free-free radiation are important. Both magnetic fields and energetic electrons are required in order to generate synchrotron emission. Synchrotron contamination thus tends to come from the disk of our galaxy where the magnetic fields are highest; however, it can also come from other galaxies. Bremsstrahlung radiation, also known as free-free radiation, results from electrons scattering off of ionized hydrogen or helium atoms. Like synchrotron, it primarily originates from the disk of our galaxy, where the densities of ionized particles are highest.

A nice review of CMB observations and foregrounds can be found in the book by Partridge [54].

Sunyaev-Zeldovich Effect

For dark energy studies, the most important non-linear source is the Sunyaev-Zeldovich (SZ) effect, inverse Compton scattering of photons off of hot electrons in clusters. This effect gives a way to probe collapsed structures at high redshifts and can measure directly the baryonic mass in clusters. Scattering off of hotter electrons causes a simple energy shift parameterized by the Compton $y-$parameter,

$$y = \frac{\delta\nu}{\nu} = \int \sigma_T n_e \frac{kT_e}{m_e c^2} dl \ , \tag{7.24}$$

where σ_T is the Thomson scattering cross section, n_e is the free electron density and the integral is along the photon path through the cluster. Typical clusters have temperatures of order $10^7 - 10^8 K$, and yield Compton-y parameters of order $y \sim 10^{-4}$.

The photons are upscattered, causing the low energy tail of the spectrum to be shifted to higher frequencies,

$$\frac{\delta n_\nu}{n_\nu} = -y \frac{x e^x}{e^x - 1} \left[4 - x \coth(x/2) \right] \ , \tag{7.25}$$

where $x \equiv h\nu/kT_\gamma$. This results in a temperature deficit at frequencies below the blackbody peak $\Delta T/T = -2y$, and a temperature increase at higher frequencies, with the sign changing at $\nu \sim 220$ GHz.

A nice feature of the SZ effect is that the temperature decrement or increment does not depend on the distance to the cluster, so it provides a way of seeing clusters at high redshifts.

7.4 Ways of Probing Dark Energy

Like dark matter, the distinguishing feature of dark energy is that we have not seen it directly. In most models of dark energy, its principal interactions are gravitational and most of our probes are based on testing its gravitational influence indirectly. Unlike dark matter, however, dark energy is believed to be smooth on small scales. Thus, we must look for its effects on cosmological scales, and these primarily come through altering the expansion rate and its evolution through the Friedmann equation.

The observational effects of dark energy can be broken into four classes:

1. First, we can ask how much dark energy contributes to the present expansion rate. This requires an inventory of the matter in the Universe to discover the fraction Ω_{DE} which is unaccounted for.
2. Next, we can look directly at how this fraction has evolved in time, by examining $H(z)$. Measurements of the time evolution of global properties of the Universe (ages, distances, geometry) give us quantities which depend on different integrated functions of the Hubble parameter.
3. Changing the Hubble parameter has knock on effects onto other observable quantities, most notably the growth of density fluctuations. We can look for such effects in how large scale structures evolve with redshift.
4. Finally, on sufficiently large scales, the dark energy itself can cluster, and this can also affect the very large scale clustering of the ordinary matter and so can potentially be seen.

These tests effectively tell us what aspects of dark energy we might be able to measure. The first will tell us what the present density of dark energy is (Ω_{DE}), the second and third will tell us how it has evolved ($w(z)$) and the last how it clusters (the dark energy sound speed, c_{DE}.)

A wide array of possible probes are available to us, including supernovae, baryon oscillations, etc. But in all of these classes of probes, the microwave background plays an important role directly, or indirectly by helping to break parameter degeneracies.

7.4.1 Matter Inventory

The most direct way to search for dark energy is to compare the amount of matter in various forms and see if they are sufficient to account for the observed expansion. The Friedmann equation for the present epoch can be written as,

$$\rho_{DE} + \rho_{CDM} + \rho_b + \rho_\gamma + \rho_\nu + k = \rho_{crit} \equiv \frac{3H_0^2}{8\pi G} . \tag{7.26}$$

The first term, $\rho_{DE} \propto \Omega_{DE}h^2$, is the present dark energy density we wish to determine. The CMB anisotropies play a key role in determining virtually all the other terms, with the exception of the critical density [55]. Uncertainties in the determination of the Hubble constant [58] (and thus the critical density) are one of the biggst obstacles in determining the precise dark energy density.

By far the best measured term above is the photon density, ρ_γ, which is given to great accuracy by measurements of the present CMB temperature. While the photon density is small ($\Omega_\gamma h^2 = 2.56 \times 10^{-5}$), it is crucial, as the CMB anisotropies determine the densities of the other species relative to the photons, at the time of last scattering.

The total matter density determines how close to last scattering the matter domination epoch was, which is reflected in the overall amplitudes of the Doppler peaks. The CMB results confirm a host of other kinds of observations, including the large scale structure power spectrum and measurements of the baryon-to dark matter ratio in clusters. The measurements usually constrain the sum of the dark matter and baryons, $\Omega_{CDM}h^2 + \Omega_b h^2 = 0.127^{+0.007}_{-0.013}$ [55].

The baryon density can effect the relative heights of the Doppler peaks, and increasing it can make the odd peaks larger and the even peaks smaller. These constraints, $\Omega_b h^2 = 0.0223^{+0.0007}_{-0.0009}$ are now beginning to improve on those limits from nucleosynthesis.

While the cosmological limits on the neutrino density come primarily from the shape of the large scale power spectrum, CMB observations alone are capable of constraining the neutrino masses as well, if the neutrinos are heavy enough to to begin acting non-relativistically before last scattering [56]. The present constraints limit it to be less than a fraction of the dark matter density, about 1% of the total density [57].

Perhaps most importantly, the CMB observations are able to rule out large curvature as the explanation of the missing matter. The measurements effectively constrain $|k/\rho_{crit}|$ to be less than a few percent (assuming the Hubble constant is not extremely low.)

Assuming the curvature can be neglected, the most important terms in constraining the dark energy density are the matter density and the Hubble constant. This dark energy density constraint is approximately given by,

$$\Omega_{DE} = 1 - 0.13h^{-2} . \tag{7.27}$$

This could be consistent with no dark energy if the Hubble constant is sufficiently small, but such a value is strongly disfavored by direct measurements of the Hubble constant [58]. Using more standard values of the Hubble constant ($h = 0.72$) indicates a dark energy density of $\Omega_{DE} \simeq 0.75$.

7.4.2 Expansion History

There are many ways of trying to probe the evolution of the dark energy [59]. These include measurements of the cosmic age, the comoving volume, the luminosity distance, the angular diameter distance and the Alcock-Paczynski tests. Each of these provides constraints on a different integral of the Hubble parameter. (Given that the CMB indicates the Universe is nearly flat, I present only the flat space expressions below.)

For example, age constraints, such as from globular clusters [60], can be used to constrain

$$t(z_{form}) = \int_0^{z_{form}} \frac{dz}{H(z)(1+z)} \, . \tag{7.28}$$

Here, z_{form} is the redshift of the object's formation, which is often somewhat uncertain. Thus far the constraints are fairly weak, but there are prospects of using other age measurements, such as of elliptical galaxies assuming they evolve passively [61], which could improve these constraints.

Better constraints have arisen from the determination of luminosity distances to cosmologically distant objects, such as supernovae,

$$d_L(z) = (1+z) \int_0^z \frac{dz}{H(z)} \, . \tag{7.29}$$

Observations indicate that some supernovae can be used as standard candles by using correlations between their intrinsic brightness and the evolution of their light curves. Surveys using of order a hundred distant supernovae have been used to put constraints on the dark energy density and evolution [5, 6]. The advantage of SN observations is that they cover a range of redshifts, and so can provide more detailed information about the evolution of $H(z)$.

The number density of objects as a function of redshift can also provide a useful cosmology tool, using

$$\frac{d^2 V}{d\Omega dz} = \frac{1}{H(z)} \left(\int_0^z \frac{dz'}{H(z')} \right)^2 \, . \tag{7.30}$$

This has been used to constrain dark energy from strong gravitational lensing on the assumption that the comoving number density of the lens galaxies is constant [64]. In principle, one can also constrain models using Alcock-Paczynski tests, where one looks at things expected to be isotropic on average, such as the shapes of structures or correlations functions, and compare measurements parallel and perpendicular to the line of sight which will have different dependencies on the geometry [65].

The quantity which is most relevant to CMB studies is the angular diameter distance of the sound horizon at recombination, used earlier to show that the Universe is very close to flat. While this most sensitively constrains the curvature, if we assume the curvature is zero, then it can be used to constrain the dark energy model. Specifically what is constrained is

$$d_A(z) = \frac{1}{(1 + z_{rec})} \int_0^{z_{rec}} \frac{dz}{H(z)} .$$

(7.31)

This provides only a single integrated constraint on the equation of state, but because the CMB anisotropies are measured so well, it is an important one.

The sound horizon is also imprinted in the matter power spectrum, through gravitational interactions with the baryons. This feature has come to be called 'baryon oscillations.' Unfortunately the amplitude of this effect is suppressed because the baryons are only a fraction of the dark matter density, making the it difficult to observe. However, it has recently been discovered by the SDSS and 2dFRS surveys [66, 67]. A similar angular diameter constraint can be found on the dark energy. While poorly determined at present, this has the potential advantage of being observable in a range of redshifts, and so could provide a better constraint. (See the review by Percival in this volume [68].)

Thus far, these kinds of 'geometrical' tests (dominated by the SN, CMB and baryon oscillations) have shown that the dark energy is consistent with a cosmological constant ($w = -1$), with an error on the equation of state of order 10%.

7.4.3 Growth of Perturbations

The previous dark energy constraints directly use information about the expansion history of the Universe. However, an important class of tests constrain the expansion history indirectly, through its impact on the growth of structure. If the Universe starts to accelerate due to the presence of dark energy, this can make it harder for objects to collapse, changing the growth rate which can then be observed.

If we ignore possible clustering of the dark energy, the evolution of the growth rate of dark matter perturbations can be described by

$$\ddot{\delta}_m + 2H\dot{\delta}_m = 4\pi G \rho_m \delta_m .$$

(7.32)

In a matter dominated universe, this has the solution $\delta_m \propto a$. However, if the Hubble constant has additional contributions from dark energy, this increases the damping and slows down the growth of δ_m.

This change in the growth rate can be measured by looking at the amplitude of fluctuations as a function of time. The CMB strongly constrains the amplitude at recombination, and this can be compared to its present value. The amplitude today is usually parameterized by σ_8, the standard deviation of the matter fluctuations smoothed on $8h^{-1}Mpc$, the scale where the fluctuations in the light are approximately unity. This quantity is surprisingly difficult to measure accurately, and estimates are typically in the range $0.7 - 0.9$.

Much of the difficulty in determining the present level of fluctuations arises because we do not usually measure the matter fluctuations directly, but rather the fluctuations in light. Weak lensing offers a direct way out of this problem since it is sensitive to the matter distribution itself. These observations are still very new and still are prone to systematics, but in the future weak lensing should be very effective in constraining dark energy [69, 70]. (Weak lensing can also provide a geometrical test of dark energy [71].)

Another way of measuring the amplitude of the matter fluctuations is by looking at the number of objects of a given mass which have collapsed. If the fluctuations are Gaussian, one can calculate the fraction of volume which exceeds a given threshold to collapse [72],

$$F(\sigma) = \frac{2}{\sqrt{\pi}} \int_{\delta_c/\sqrt{2\sigma^2}}^{\infty} e^{-u^2} du \qquad (7.33)$$

(This is approximate, and corrections have been estimated taking into account non-spherical collapse [73] and using n-body simulations [74].) This can be used to estimate the number density of collapsed objects of a particular mass, and in particular the cluster mass function.

The cosmology dependence is in σ, the measure of the standard deviation of the density field which grows proportionately to the matter perturbation, δ_m. Thus, the number density of haloes of a given mass as a function of redshift is exponentially sensitive to the growth of perturbations. Surveys of clusters, particularly those found at high redshifts using the SZ effect, could potentially provide very strong constraints on dark energy. However, the data can be difficult to interpret, as some problems can arise in relating the cluster masses to the observed temperature fluctuations [75]. (See the review by P. Tozzi [76] in this volume.)

7.4.4 Dark Energy Clustering

The previous discussion focused on how the accelerated expansion triggered by the dark energy impacts the growth of structure. However, this is only part of the story. With the exception of a cosmological constant, dark energy is expected to cluster on sufficiently large scales. The scale on which this happens is determined by the sound speed of the dark energy fluid, c_{DE}. The dark energy sound speed is an important way to potentially discriminate between dark energy models (see Fig. 7.4).

For typical dark energy models, such as quintessence, the sound speed is very near the speed of light, meaning the scales on which the dark energy clusters are of order the horizon. Most measurements of the growth of perturbations, such as weak lensing or the abundance of clusters, are on scales much too small to be sensitive to the dark energy perturbations. One way to probe the structure on large scales is with the integrated Sachs-Wolfe effect.

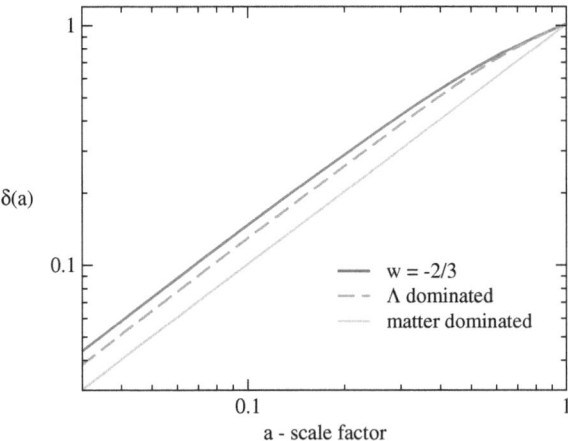

Fig. 7.4. The linear growth factor as a function of scale factor for different cosmologies. In the matter dominated cosmology, the growth rate is linear in the scale factor. With dark energy, the late acceleration makes it harder for structures to collapse, slowing the rate of growth. Normalized to the level of structure presently observed, dark energy models predict more structure at high redshifts. Thus, one expects to observe more high redshift clusters in such models

7.5 The Integrated Sachs-Wolfe Effect

The integrated Sachs-Wolfe (ISW) effect results from the line of sight integral in the Sachs-Wolfe equation. CMB photons pass through peaks and wells of the gravitational potential along their way to us. As they fall into a potential well, photons gain energy; if the well is not evolving, the photons lose the same energy when they climb out, leaving no net change. However, if the gravitational potentials decay while the photons pass through, then the energy that they lose climbing out is less that what they gained falling in, leaving a net shift in the photon temperature.

Both the ISW and Rees-Sciama effect arise in this way; the ISW effect is generally taken to be the contribution from the linear evolution of the gravitational potential, while the Rees-Sciama effect arises from the non-linear evolution of the potential in clusters. While the non-linear effect is inevitable, the linear effect depends on the cosmological model and requires that the background equation of state changes. This happens at early times as the universe goes from being radiation dominated to matter dominated, and can also occur at late times as the dark energy (or curvature) takes over from the matter.

The evolution of the gravitational potential can be related to the linear density perturbation via Poisson's equation in Fourier space,

$$\Phi = -\frac{4\pi G a^2}{k^2} \bar{\rho}_m \delta_m \,, \tag{7.34}$$

where k is the comoving wave number. Since the matter density is proportional to $\rho_m \propto a^{-3}$, we see that the gravitational potential evolves as $\Phi \propto \delta_m/a$. As we saw above, in the matter dominated regime $\delta_m \propto a$, meaning the gravitational potential is constant in time: the collapse of the perturbations is exactly balanced by the dilution of the matter. However, when dark energy or curvature begins to dominate, the growth of perturbations is slowed, and the gravitational potentials begin to decay, giving rise to the late time ISW effect.

Unlike the ISW perturbations generated at the earlier radiation-matter transition, the ISW anisotropies generated at late times are virtually uncorrelated with the CMB fluctuations generated at the last scattering surface. Thus, the CMB sky we see is effectively composed of two independent maps, those fluctuations created at last scattering or soon afterwards, and those created at low redshifts when dark energy or curvature has become dynamically important (see Fig. 7.5.)

The spectrum of the temperature anisotropies generated by the late-time ISW effect is shown in Fig. 7.6. It is predominantly on very large scales, and for typical models, it is not as large as the anisotropies from the last scattering surface. It is dominated by modes which are of the horizon size, because it is these modes which will have the most time for the potential to change as the photons pass through. For smaller scale perturbations, photons can get many positive and negative smaller amplitude contributions which will tend to cancel out.

7.5.1 Detecting the ISW Effect with Cross Correlations

How can we determine whether the CMB fluctuations we see are generated at early or late times? Unlike many foregrounds, the ISW fluctuations have the same frequency spectrum as the primordial anisotropies, so we cannot use different frequency observations to isolate them. We can attempt to look for the additional power in the CMB auto-correlation spectrum, but this is difficult because the ISW amplitude is small and where it is largest, cosmic variance is also large (Fig. 7.6.) This makes direct detection difficult.

However, we do know that the ISW anisotropies will be produced by local ($z < 2$) fluctuations in the gravitational potential, and this we can determine if we know how the matter is distributed on large scales. While the gravitational potential is difficult to reconstruct, we can use the observed galaxy density as a way of tracing it. If the gravitational potential is decaying, statistically we expect overdensities of galaxies to align with temperature hot spots and under densities with temperature coldspots. Thus we can constrain our cosmological model by looking for cross correlations between the CMB maps and large scale distribution of matter [77, 78].

Detecting the cross correlations is difficult, as it requires both a good map of the CMB on large scales and a map of the galaxy distribution which is both deep and covers a large fraction of the sky [77, 79, 80]. Large sky coverage is essential because the primordial fluctuations act effectively like noise when

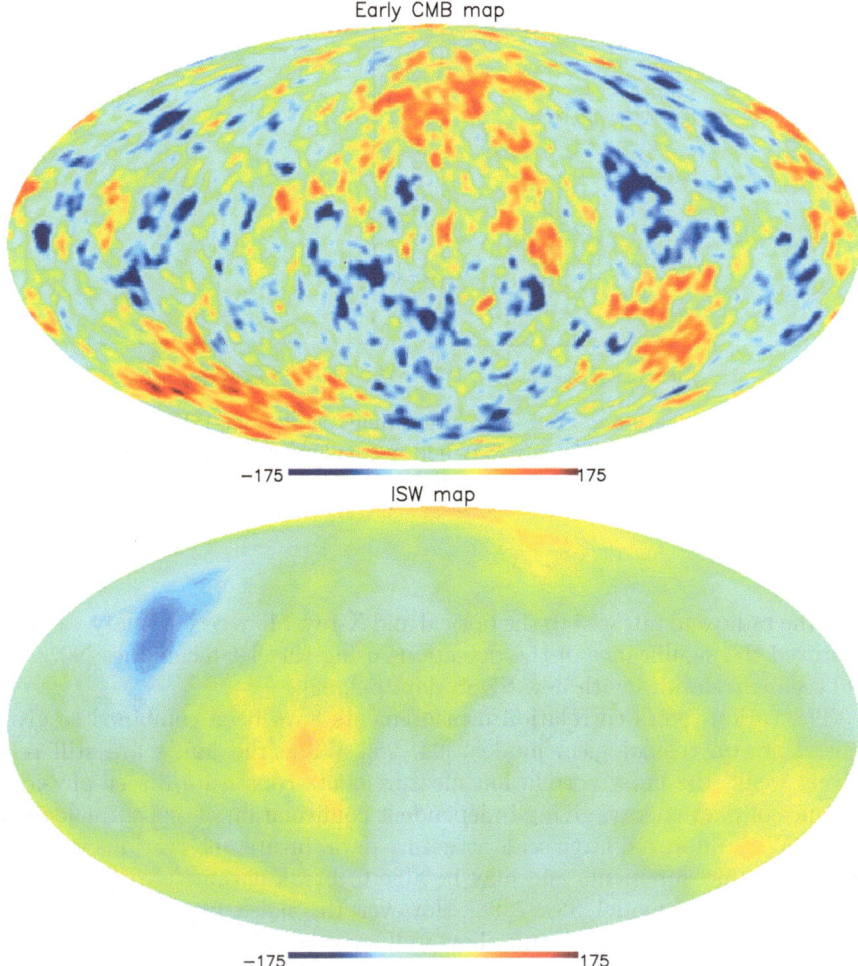

Fig. 7.5. Large scale CMB anisotropies arising in the early and late universe. Only the late contributions from the ISW effect on large scales are shown; non-linear effects like the Sunyaev-Zeldovich effect and the Rees-Sciama effect will arise on smaller scales

searching for anisotropies generated recently, and so the measurements are always 'noise' dominated. The first attempts of detecting the correlation using the COBE data and maps of the X-ray background (believed to trace AGN) or radio galaxy distribution produced no detections [81, 82].

However, the picture improved greatly with the WMAP observations. Correlations were quickly seen with the hard X-ray background [83], the NVSS radio galaxy survey [83, 84], the APM galaxy survey [85], the SDSS [86, 87, 88] and the 2MASS survey [89]. While all the detections are at a low significance $(2-3\sigma)$, it is encouraging that they are seen is such a broad range of surveys,

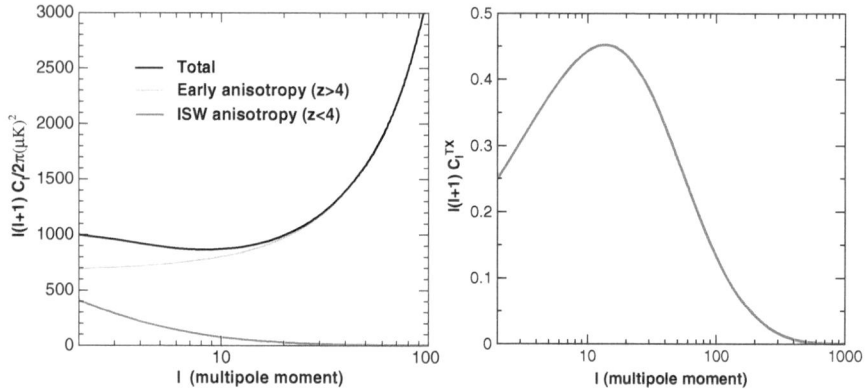

Fig. 7.6. Typical auto and cross correlation functions for the ISW effect in a cosmological constant model. The late ISW adds a small amount of large scale power to the temperature maps, largely uncorrelated with the anisotropy arising from early times. The cross correlation itself (shown in arbitrary units) peaks on scales of a few degrees

from the radio and infrared to the optical and X-ray. More recent analyses have improved the significance of the detections using wavelet techniques [90, 91], and seen correlations with new SDSS data [92, 93].

The various cross correlation measurements have been combined to give constraints on cosmological models [94, 95]. While the limits are still relatively weak, the cross correlation measurements constitute direct physical evidence of dark energy, giving independent confirmation of our cosmological model. With improved large scale structure data, future observations will improve these measurements and may be able to break parameter degeneracies of the dark energy model [38, 97, 98]. However, the 'noise' from the primordial anisotropies fundamentally limits how well cross correlations can ever be detected, with an ideal signal to noise of approximately 10 in the most optimistic models [77, 80].

Despite this limitation, cross correlation studies are crucial as they offer an important window on the clustering of the dark energy. Few other probes are on sufficiently large scales to see this clustering, and including the dark energy perturbations makes a significant impact on the ISW anisotropies [99, 100]. Given the uncertainties, the present constraints on the sound speed are fairly weak; however, ultimately cross correlation measurements should be able to discriminate between different sound speeds [101].

7.6 Conclusions and Future Prospects

Dark energy and the accelerated expansion of our Universe are clearly surprising results, dramatically changing our cosmological picture. Given our theoretical prejudices against most dark energy explanations, it is very important to investigate whether an alternative model might also be able to explain the same observations.

One possibility, is that the Universe is actually flat and matter dominated, but with a very low Hubble constant [102]. As discussed above, this can alleviate the constraints coming from the matter inventory. However, in addition to the tension with direct measurements of the Hubble constant, the model requires a well placed feature in the primordial power spectrum and massive neutrinos in order to reproduce the observed large scale clustering. Even so, other evidence supporting dark energy, such as the SN luminosity distance measurements and the cross correlation observations, must be accounted for by non-cosmological means. But perhaps the biggest challenge for this model is the angular scale of the observed baryon oscillations [103]. Though this model cannot be conclusively ruled out, the preponderance of the data seem to be against it.

The cosmological constant model is significantly simpler and seems to be favored if we accept all of the data at face value. However, we have no theoretical explanation for such a small cosmological constant. If we take the large theoretical priors, we seem to find ourselves in a very unlikely situation. While anthropic arguments may alleviate this concern for some, the data really seem to demand a better theory.

It is worth considering the implications of taking the prior on the cosmological constant seriously. Models are usually compared by looking at their Bayesian evidence, which is roughly the likelihood of the best fit model times the prior probability of the model being near the best fit. If we assume a uniform distribution on the cosmological constant value, this prior probability is incredibly tiny. Compare this to any other model, no matter how unlikely it might seem or how poorly it explains the observations; as long as it did not have this small prior factor, it would win out in comparison.

For example, suppose we believed that there could reasonably exist a symmetry which would cause the cosmological constant to be precisely zero. We would then need to choose between two hypotheses: either the cosmological constant is zero, or it is non-zero with a value anywhere between $-\Lambda_{planck}$ and Λ_{planck}. If we assume a uniform distribution of possible values for Λ_{obs}, then an extremely small fraction of the parameter space ($\sim 10^{-120}$) is consistent with the observations. Thus, the zero cosmological constant model would be preferred unless the data rule it out to a large significance; naively setting

$$e^{-(S/N)^2/2} \leq 10^{-120} \qquad (7.35)$$

we would require that a detection at the 24σ level to conclude that the cosmological constant was not zero. (Starkman and Trotta have made a similar argument [104].) If one believes in supersymmetry, the required significance drops to 16σ. At present, the data are not yet near this level, especially given the possible systematic errors involved; thus anyone who thought there was even a slight chance such a $\Lambda = 0$ symmetry existed (even at the one in a million level) should safely conclude there is no cosmological constant.

Luckily, astronomers do not hold themselves to such high standards; the evidence for dark energy is very good and has convinced much of the community that it does exist. But there is desperate need for a better theoretical understanding of dark energy and for better observational determination of its properties.

On the observational front at least, we can guarantee that progress will be made (perhaps eventually even to the 20σ detection level!) With the projects currently underway and in development, observations of the microwave background, high redshift supernovae, weak lensing and large scale structure should continue to improve for some time; at the same time, observers continue to look for new ways of probing dark energy. These new data will be crucial if we are to solve the mystery that is dark energy.

Acknowledgements

I thank E. Papantonopoulos and the other organizers of the Third Aegean Summer School on The Invisible Universe for inviting me to be a part of a very stimulating school and also thank my collaborators for innumerable discussions related to these topics.

References

1. V. Sahni, Lect. Notes Phys. **653**, 141 (2004).
2. S. Weinberg, Rev. Mod. Phys. **61**, 1 (1989).
3. S.M. Carroll, W.H. Press and E.L. Turner, Ann. Rev. Astron. Astrophys. **30**, 499 (1992).
4. N. Straumann, arXiv:astro-ph/0203330.
5. P.J.E. Peebles and B. Ratra, Astrophys. J. **325**, L17 (1988).
6. R.R. Caldwell, R. Dave and P.J. Steinhardt, Phys. Rev. Lett. **80**, 1582 (1998).
7. S.M. Carroll, Phys. Rev. Lett. **81**, 3067 (1998).
8. P.J. Steinhardt, Quintessence and Cosmic Acceleration. In: *Structure Formation in the Universe*, ed by R.G. Crittenden, N. Turok (Kluwer, Dordrecht 2001) pp. 143–176.
9. M. Bucher and D.N. Spergel, Phys. Rev. D **60**, 043505 (1999).
10. R.A. Battye and A. Moss, JCAP **0506**, 001 (2005).
11. P. Pina Avelino, C.J.A. Martins, J. Menezes, R. Menezes and J.C.R. Oliveira, Phys. Rev. D **73**, 123519 (2006), arXiv:astro-ph/0602540.
12. A.Y. Kamenshchik, U. Moschella and V. Pasquier, Phys. Lett. B **511**, 265 (2001).
13. R.R. Caldwell, Phys. Lett. B **545**, 23 (2002).
14. R.R. Caldwell, M. Kamionkowski and N.N. Weinberg, Phys. Rev. Lett. **91**, 071301 (2003).
15. S.M. Carroll, M. Hoffman and M. Trodden, Phys. Rev. D **68**, 023509 (2003).

16. J.M. Cline, S. Jeon and G.D. Moore, Phys. Rev. D **70**, 043543 (2004).
17. R. Maartens, arXiv:astro-ph/0602415.
18. A.A. Penzias and R.W. Wilson, Astrophys. J. **142**, 419 (1965).
19. R.H. Dicke et al., Astrophys. J. **142**, 414 (1965).
20. D.J. Fixsen, E. Dwek, J.C. Mather, C.L. Bennett and R.A. Shafer, Astrophys. J. **508**, 123 (1998).
21. J.C. Mather, D.J. Fixsen, R.A. Shafer, C. Mosier and D.T. Wilkinson, Astrophys. J. **512**, 511 (1999).
22. G.F. Smoot et al., Astrophys. J. **396**, L1 (1992).
23. C.L. Bennett et al., Astrophys. J. Suppl. **148**, 1 (2003).
24. G. Hinshaw et al., arXiv:astro-ph/0603451.
25. K. Grainge et al., Mon. Not. Roy. Astron. Soc. **341**, L23 (2003).
26. C. Dickinson et al., Mon. Not. Roy. Astron. Soc. **353**, 732 (2004).
27. C.l. Kuo et al. [ACBAR collaboration], Astrophys. J. **600**, 32 (2004).
28. J.E. Ruhl et al., Astrophys. J. **599**, 786 (2003).
29. M.E. Abroe et al., Astrophys. J. **605**, 607 (2004).
30. A.C.S. Readhead et al., Astrophys. J. **609**, 498 (2004).
31. M. Kamionkowski, A. Kosowsky and A. Stebbins, Phys. Rev. Lett. **78**, 2058 (1997).
32. U. Seljak and M. Zaldarriaga, Phys. Rev. Lett. **78**, 2054 (1997).
33. D. Coulson, R.G. Crittenden and N.G. Turok, Phys. Rev. Lett. **73**, 2390 (1994).
34. J. Kovac, E.M. Leitch, C. Pryke, J.E. Carlstrom, N.W. Halverson and W.L. Holzapfel, Nature **420**, 772 (2002).
35. A.C.S. Readhead et al., Science, **306**, 836 (2004).
36. D. Barkats et al., Astrophys. J. **619**, L127 (2005).
37. T.E. Montroy et al., ApJ. **647**, 813 (2006), arXiv:astro-ph/0507514.
38. A. Kogut et al., Astrophys. J. Suppl. **148**, 161 (2003).
39. L. Page et al., arXiv:astro-ph/0603450.
40. P.J.E. Peebles and J.T. Yu, Astrophys. J. **162**, 815 (1970).
41. U. Seljak and M. Zaldarriaga, Astrophys. J. **469**, 437 (1996).
42. A. Lewis, A. Challinor and A. Lasenby, Astrophys. J. **538**, 473 (2000).
43. A. Challinor, arXiv:astro-ph/0403344.
44. C.P. Ma and E. Bertschinger, Astrophys. J. **455**, 7 (1995).
45. R.K. Sachs and A.M. Wolfe, Astrophys. J. **147**, 73 (1967).
46. W.J. Percival et al. [The 2dFGRS Team Collaboration], Mon. Not. Roy. Astron. Soc. **337**, 1068 (2002).
47. J. Silk, Astrophys. J. **151**, 459 (1968).
48. M.J. Rees and D.W. Sciama, Nature **217**, 511 (1968).
49. B.M. Schaefer and M. Bartelmann, Mon. Not. Roy. Astron. Soc. **373**, 1211 (2007), arXiv:astro-ph/0502208.
50. Y.B. Zeldovich and R.A. Sunyaev, Astrophys. Space Sci. **4**, 301 (1969).
51. R.A. Sunyaev and Y.B. Zeldovich, Mon. Not. Roy. Astron. Soc. **190**, 413 (1980).
52. J.P. Ostriker and E.T. Vishniac, Astrophys. J. **306**, L51 (1986).
53. E.T. Vishniac, Astrophys. J. **322**, 597 (1987).
54. R.B. Partridge: *3-K: The Cosmic microwave background radiation,* (Cambridge Univ. Pr., Cambridge 1995).
55. D.N. Spergel et al., arXiv:astro-ph/0603449.

56. M. Kaplinghat, L. Knox and Y.S. Song, Phys. Rev. Lett. **91**, 241301 (2003).
57. O. Elgaroy and O. Lahav, New J. Phys. **7**, 61 (2005).
58. W.L. Freedman et al., Astrophys. J. **553**, 47 (2001).
59. D. Huterer and M.S. Turner, Phys. Rev. D **64**, 123527 (2001).
60. L.M. Krauss and B. Chaboyer, Science **299**, 65 (2003).
61. J. Simon, L. Verde and R. Jimenez, Phys. Rev. D **71**, 123001 (2005).
62. A.G. Riess et al. [Supernova Search Team Collaboration], Astrophys. J. **607**, 665 (2004).
63. P. Astier et al., Astr. Astroph. **447**, 31 (2006), arXiv:astro-ph/0510447.
64. K.H. Chae et al., Phys. Rev. Lett. **89**, 151301 (2002).
65. C. Alcock and B. Paczynski, Nature **281**, 358:359 (1979).
66. D.J. Eisenstein et al., Astrophys. J. **633**, 560 (2005).
67. S. Cole et al. [The 2dFGRS Collaboration], Mon. Not. Roy. Astron. Soc. **362** 505 (2005).
68. W.J. Percival, arXiv:astro-ph/0601538.
69. A.R. Cooray and D. Huterer, Astrophys. J. **513**, L95 (1999).
70. W. Hu, Phys. Rev. D **66**, 083515 (2002).
71. B. Jain and A. Taylor, Phys. Rev. Lett. **91**, 141302 (2003).
72. W.H. Press and P. Schechter, Astrophys. J. **187** 425 (1974).
73. R.K. Sheth, H.J. Mo and G. Tormen, Mon. Not. Roy. Astron. Soc. **323**, 1 (2001).
74. A. Jenkins et al., Mon. Not. Roy. Astron. Soc. **321**, 372 (2001).
75. J. Weller, R. Battye and R. Kneissl, Phys. Rev. Lett. **88**, 231301 (2002).
76. P. Tozzi, arXiv:astro-ph/0602072.
77. R.G. Crittenden and N. Turok, Phys. Rev. Lett. **76**, 575 (1996).
78. A. Kinkhabwala and M. Kamionkowski, Phys. Rev. Lett. **82**, 4172 (1999).
79. H.V. Peiris and D.N. Spergel, Astrophys. J. **540**, 605 (2000).
80. N. Afshordi, Phys. Rev. D **70**, 083536 (2004).
81. S.P. Boughn, R.G. Crittenden and N.G. Turok, New Astron. **3**, 275 (1998).
82. S.P. Boughn and R.G. Crittenden, Phys. Rev. Lett. **88**, 021302 (2002).
83. S. Boughn and R. Crittenden, Nature **427**, 45 (2004).
84. M.R. Nolta et al., Astrophys. J. **608**, 10 (2004).
85. P. Fosalba and E. Gaztanaga, Mon. Not. Roy. Astron. Soc. **350**, L37 (2004).
86. P. Fosalba, E. Gaztanaga and F. Castander, Astrophys. J. **597**, L89 (2003).
87. R. Scranton et al. [SDSS Collaboration], arXiv:astro-ph/0307335.
88. N. Padmanabhan, C.M. Hirata, U. Seljak, D. Schlegel, J. Brinkmann and D.P. Schneider, Phys. Rev. D **72**, 043525 (2005).
89. N. Afshordi, Y.S. Loh and M.A. Strauss, Phys. Rev. D **69**, 083524 (2004).
90. P. Vielva, E. Martinez-Gonzalez and M. Tucci, arXiv:astro-ph/0408252.
91. J.D. McEwen, P. Vielva, M.P. Hobson, E. Martinez-Gonzalez and A.N. Lasenby, Mon. Not. Roy. Astron. Soc. **373**, 1211 (2007), arXiv:astro-ph/0602398.
92. A. Cabre, E. Gaztanaga, M. Manera, P. Fosalba and F. Castander, Mon. Not. Roy. Astron. Soc. Lett. **372**, L23 (2006), arXiv:astro-ph/0603690.
93. T. Giannantonio et al., Phys. Rev. D **74**, 063520 (2006), astro-ph/0607572.
94. E. Gaztanaga, M. Manera and T. Multamaki, Mon. Not. Roy. Astron. Soc. **365**, 171 (2006).
95. P.S. Corasaniti, T. Giannantonio and A. Melchiorri, Phys. Rev. D **71**, 123521 (2005).

96. J. Garriga, L. Pogosian and T. Vachaspati, Phys. Rev. D **69**, 063511 (2004).
97. L. Pogosian, JCAP **0504**, 015 (2005).
98. L. Pogosian, P.S. Corasaniti, C. Stephan-Otto, R. Crittenden and R. Nichol, Phys. Rev. D **72**, 103519 (2005).
99. R. Bean and O. Dore, Phys. Rev. D **69**, 083503 (2004).
100. J. Weller and A.M. Lewis, Mon. Not. Roy. Astron. Soc. **346**, 987 (2003).
101. W. Hu and R. Scranton, Phys. Rev. D **70**, 123002 (2004).
102. A. Blanchard, M. Douspis, M. Rowan-Robinson and S. Sarkar, Astron. Astrophys. **412**, 35 (2003).
103. A. Blanchard, M. Douspis, M. Rowan-Robinson and S. Sarkar, Astron. Astrophys. **449**, 925 (2006).
104. G. Starkman, private communication (2006).

8

Models of Dark Energy

M. Sami

Center for Theoretical Physics, Jamia Millia Islamia, New Delhi, India
sami@iucaa.ernet.in

Abstract. In this talk we present a pedagogical review of scalar field dynamics. The main emphasis is put on the underlying basic features rather than on concrete scalar field models. Cosmological dynamics of standard scalar fields, phantoms and tachyon fields is developed in detail. Scaling solutions are discussed emphasizing their importance in modelling dark energy. The developed concepts are implemented in an example of *quintessential* inflation. A brief discussion of scaling solutions for coupled quintessence is also included.

Accelerated expansion seems to have played an important role in the dynamical history of our universe. There is a firm belief, at present, that universe has passed through inflationary phase at early times and there have been growing evidences that it is accelerating at present. The recent measurement of the Wilkinson Microwave Anisotropy Probe (WMAP) in the Cosmic Microwave Background (CMB) made it clear that (i) the current state of the universe is very close to a critical and that (ii) primordial density perturbations that seeded large-scale structure in the universe are nearly scale-invariant and Gaussian, which are consistent with the inflationary paradigm. As for the current accelerating of universe, it is supported by observations of high redshift type Ia supernovae treated as standardized candles and, more indirectly, by observations of the cosmic microwave background and galaxy clustering. The criticality of universe supported by CMB observations fixes the total energy budget of universe. The study of large scale structure reveals that nearly 30 percent of the total cosmic budget is contributed by dark matter. Then there is a deficit of almost 70 percent; the supernovae observations tell us that the missing component is an exotic form of energy with large negative pressure dubbed *dark energy* [1, 2, 3, 4]. The recent observations on baryon oscillations provides yet another independent support to dark energy hypothesis. The idea that universe is in the state of acceleration is slowly establishing in modern cosmology.

The dynamics of our universe is described by Einstein equations in which the contribution of energy content of universe is represented by energy

M. Sami: *Models of Dark Energy*, Lect. Notes Phys. **720**, 219–256 (2007)
DOI 10.1007/978-3-540-71013-4_8   © Springer-Verlag Berlin Heidelberg 2007

momentum tensor appearing on RHS of these equations. The LHS represents pure geometry given by the curvature of space time. Einstein equations in their original form with energy momentum tensor of normal matter can not lead to acceleration. There are then two ways to obtain accelerated expansion, either by supplementing energy momentum tensor by dark energy component or by modifying the geometry itself. In the frame work of Dvali-Gabadadze-Porrati (DGP) brane worlds [5], the extra dimensional effects can lead to late time acceleration. The other alternative which is largely motivated by phenomenological considerations is related to the introduction of inverse powers of Ricci scalar in the Einstein Hilbert action [6]. The third intriguing possibility is provided by Bekenstein relativistic theory of modified gravity [7, 8, 9] which apart from spin two field contains a vector and a scalar field.

Due to the simplicity of the mechanism, most of the work in cosmology related late time acceleration is attributed to the assumption that within the framework of general relativity, cosmic acceleration is sourced by an energy-momentum tensor which has a large negative pressure. The simplest candidate of dark, yet most difficult from field theoretic point of view, is provided by cosmological constant. Due to its non evolving nature it is plagued with fine tuning problem which can be alleviate in dynamically evolving scalar field models. A variety of scalar field models have been conjectured for this purpose including quintessence [10, 11], phantoms [12, 13, 14], K-essence [15] and recently tachyonic scalar fields [16]. In this talk we present a review of cosmological dynamics of quintessence, phantoms and rolling tachyons. We describe in detail the concepts of field dynamics relevant to cosmic evolution with a special emphasis on scaling solutions. The example of quintessential inflation is worked out in detail.

We employ the metric signature (-,+,+,+) and use the reduced Planck mass $M_p^{-2} = 8\pi G \equiv \kappa^2$. In certain places we have adopted the unit $M_p = 1$. Finally we should mention that our list of references is restricted, in most of the places, we referred to reviews to help the readers.

8.1 Glimpses of FRW Cosmology

The Friedmann-Robertson-Walker (FRW) model is based on the assumption of homogeneity and isotropy which is approximately true at very large scales. The small deviation from homogeneity at early epochs played very important role in the dynamical history of our universe. The small density perturbations are believed to have grown via gravitational instability into the structure we see today in the universe. the origin of primordial perturbations is quantum mechanical and is out side the scope of standard big bang model. In what follows we shall review main features of FRW model necessary for the subsequent sections.

Homogeneity and isotropy forces the metric of space time to assume the form [17]

$$ds^2 = -dt^2 + a^2(t) \left(\frac{dr^2}{1 - kr^2} + r^2(d\theta^2 + \sin^2\theta d\phi^2) \right) ,$$

$$k = 0, \pm 1 , \tag{8.1}$$

where $a(t)$ is cosmic scale factor. Coordinates r, θ and ϕ are known as *comoving* coordinates. A freely moving particle comes to rest in these coordinates.

Equation (8.1) is purely a kinematic statement. In this problem the dynamics is associated with the scale factor$-$ a(t). Einstein equations allow to determine the scale factor provided the matter content of universe is specified. Constant k occurring in the metric (8.1) describes the geometry of spatial section of space-time. Its value is also determined once the matter distribution in the universe is known. Observations have repeatedly confirmed the spatially flat geometry ($k = 0$) in confirmation of the prediction of inflationary scenario.

8.1.1 Evolution Equations

The differential equations for the scale factor and the matter density follow from Einstein equations

$$G^\mu_\nu \equiv R^\mu_\nu - \frac{1}{2}\delta^\mu_\nu R = 8\pi G T^\mu_\nu , \tag{8.2}$$

where $G_{\mu,\nu}$ is the Einstein tensor, $R_{\mu\nu}$ is the Ricci tensor which depends on the metric and its derivatives and R is Ricci scalar. The energy momentum tensor $T_{\mu\nu}$ assumes a simplified form reminiscent of ideal perfect fluid in FRW background

$$T^\mu_\nu = Diag\left(-\rho, p, p, p\right) . \tag{8.3}$$

In this case the components of $G_{\mu\nu}$ can easily be computed

$$G^0_0 = -\frac{3}{a^2}\left(\dot{a}^2 + k\right) \tag{8.4}$$

$$G^j_i = \frac{1}{a^2}\left(2a\ddot{a} + \dot{a}^2 + k\right) \tag{8.5}$$

and all the other components of Einstein tensor are identically zero. Equations (8.2) then give the two independent equations

$$H^2 \equiv \frac{\dot{a}^2}{a^2} = \frac{8\pi G\rho}{3} - \frac{k}{a^2} \tag{8.6}$$

$$\frac{\ddot{a}}{a} = -\frac{4\pi G}{3}\left(\rho + 3p\right) . \tag{8.7}$$

The energy momentum tensor is conserved by virtue of the Bianchi identity $\Delta^\nu G^\mu_\nu = 0$ leading to the continuity equation

$$\dot{\rho} + 3H(\rho + p) = 0 . \tag{8.8}$$

Equations (8.6), (8.7) & (8.8) make a redundant set of equations convenient to use; one of the two (8.7) & (8.8) can be obtained using the other one and the Hubble (8.6). These equations supplemented with the equation of state $p(t) = p(\rho)$ uniquely determine $a(t)$, $p(t)$ and $\rho(t)$. Constant k also gets determined

$$\frac{k}{a^2} = H^2 \left(\Omega(t) - 1 \right) , \tag{8.9}$$

where $\Omega = \rho/\rho_c$ is the dimensionless density parameter and $\rho_c = 3H^2/8\pi G$ is critical density. The matter distribution clearly determines the spatial geometry of our universe namely

$$\Omega < 1 \quad or \ \rho < \rho_c \ \to k = -1 \tag{8.10}$$

$$\Omega = 1 \quad or \ \rho = \rho_c \ \to k = 0 \tag{8.11}$$

$$\Omega > 1 \quad or \ \rho > \rho_c \ \to k = 1 . \tag{8.12}$$

In case of $k = 0$, the value the scale factor at the present epoch a_0 can be normalized to a convenient value, say, $a_0 = 1$. In other cases it should be determined using the observed values of H_0 and $\Omega^{(0)}$ from the relation $a_0 H_0 = \left(|\Omega^{(0)} - 1| \right)^{-1/2}$. Observations on cosmic micro wave background radiation support the *critical* universe which is one of the predictions of inflation. We would therefore assume $k = 0$ in the subsequent description.

Acceleration

We now turn to the nature of expansion which is determined by the matter content in the universe. Equation (8.7) should be contrasted to the analogous situation in Newtonian gravity

$$\ddot{R} = -\frac{4\pi}{3} G\rho R \tag{8.13}$$

where R denotes the distance of the test particle from the center of a homogeneous sphere of mass density ρ. In general theory of relativity (GR), unlike the Newtonian case, pressure contributes to energy density and may qualitatively modify the dynamics. Indeed, from (8.7) we have

$$\ddot{a} > 0 \quad if \ p < -\frac{\rho}{3} \tag{8.14}$$

$$\ddot{a} < 0 \quad if \ p > -\frac{\rho}{3} . \tag{8.15}$$

Accelerated expansion, thus, is fuelled by an exotic form of matter of large negative pressure dubbed *dark energy* which turns gravity into a repulsive force. The simplest example of a perfect fluid of negative pressure is provided by cosmological constant associated with $\rho = constant$. In this case the continuity (8.8) yields the relation $p = -\rho$. A host of scalar field systems can also mimic negative pressure.

Assuming that the universe is filled with perfect barotropic fluid with constant equation of state parameter $w = p/\rho$ yields

$$\rho \propto a^{-3(1+w)} \tag{8.16}$$

$$a(t) \propto t^{\frac{2}{3(1+w)}} \quad (w > -1) \tag{8.17}$$

$$a(t) \propto e^{H\,t} \quad (w = -1) \,. \tag{8.18}$$

The last equation corresponds to cosmological constant which can be added to the energy momentum tensor of the perfect fluid. Interestingly, in four dimension and at the classical level, the only modification Einstein equations allow is associated with $T_{\mu\nu} \to +T_{\mu\nu} + \Lambda g_{\mu\nu}$. Historically such a modification was first proposed by Einstein to achieve a static solution which turns out to be unstable. It was later dropped by him after the Hubble's discovery. In presence of Λ, the evolution equations modify to

$$H^2 = \frac{8\pi G}{3} + \frac{\Lambda}{3} \,, \tag{8.19}$$

$$\frac{\ddot{a}}{a} = -\frac{4\pi G}{3}(\rho + 3p) + \frac{\Lambda}{3} \,. \tag{8.20}$$

From (8.20), it clearly follows that Λ term contributes negatively to the pressure term and hence exhibits repulsive effect.

Age Crisis and Cosmological Constant

Apart from the dark energy problem, cosmological constant has other important implications, in particular, in relation to the age problem. In any cosmological model with normal form of matter, the age of universe falls short as compared to the age of some well known old objects found in the universe. Remarkably, the presence of Λ can resolve the age problem. In order to appreciate the problem, let us first consider the case of flat dust dominated universe ($\Omega_m = 1$)

$$a(t) \propto t^{2/3} \to \quad H_0 = \frac{2}{3t_0} \,. \tag{8.21}$$

The present value of the Hubble parameter H_0 is not accurately know by the observations

$$H_0^{-1} = h^{-1}0.98 \times 10^{10} years \,, \tag{8.22}$$

$$0.8 < h < 0.64 \to \quad to = (8 - 10) \times 10^9 years \,. \tag{8.23}$$

This model is certainly in trouble as its prediction for age of universe fails to meet the solar age constraint $- t_0 > (11 - 12) \times 10^9 years$. One could try to improve the situation by invoking the open model with $\Omega_m^{(0)} < 1$. In this case the age of universe is expected to be larger than the flat dust dominated model— for less amount of matter, it would take longer for gravitational interaction to slow down the expansion rate to its present value. Indeed, in this case we have the exact expression

$$H_0 t_0 = \frac{1}{1 - \Omega_m^{(0)}} - \frac{\Omega_m^{(0)}}{2(1 - \Omega_m^{(0)})^{3/2}} \cosh^{-1}\left(\frac{2 - \Omega_m^{(0)}}{\Omega_m^{(0)}}\right) \tag{8.24}$$

from which follows that

$$H_0 t_0 = 1, \, for \ \ \Omega_m^{(0)} \to 0 \,, \tag{8.25}$$

$$H_0 t_0 = \frac{2}{3}, \, for \ \ \Omega_m^{(0)} \to 1 \,. \tag{8.26}$$

For obvious reasons, in case of the closed universe, the age would even be smaller than $2/3 H_0^{-1}$. Though the age of universe is larger than $2/3 H_0^{-1}$ for vanishingly small value of $\Omega_m^{(0)}$, such a model is note viable as $\Omega_m^{(0)} \simeq 0.3$ and universe is critical to a good accuracy. The problem can be solved in a flat universe dominated by cosmological constant. In fact, in a flat universe with two components $(\Omega_m^{(0)} + \Omega_\Lambda^{(0)} = 1)$, the Hubble equation

$$\left(\frac{\dot{a}}{a}\right)^2 = H_0^2 \left[\Omega_m^{(0)}\left(\frac{a_0}{a}\right)^2 + \Omega_\Lambda^{(0)}\right] \tag{8.27}$$

has the solution

$$\frac{a}{a_0} = \left(\frac{\Omega_m^{(0)}}{\Omega_\Lambda^{(0)}}\right)^{1/3} \sinh^{2/3}\left(\frac{3}{2}\Omega_m^{(0)\,1/2} H_0 t\right) \,, \tag{8.28}$$

which at $t = t_0$ yields the following expression for the age of universe

$$t_0 = \frac{2}{3}\frac{H_0^{-1}}{\Omega_\Lambda^{(0)\,1/2}} \ln\left(\frac{1 + \Omega_\Lambda^{(0)\,1/2}}{\Omega_m^{(0)\,1/2}}\right) \,. \tag{8.29}$$

In Fig. 8.1, we have plotted the age t_0 versus Ω_m. The age of universe is larger than H_0^{-1} for a Λ dominated universe. The numerical value of t_0 ($t_0 \simeq 0.96 H_0^{-1}$) is comfortable with observations for popular values of $\Omega_m^{(0)} = 0.3$ and $\Omega_\Lambda^{(0)} = 0.7$.

Super Acceleration

So far, we have restricted our attention to fluids with equation' of state parameter $w \geq -1$. The case of $w < -1$ corresponds to *phantom dark energy*

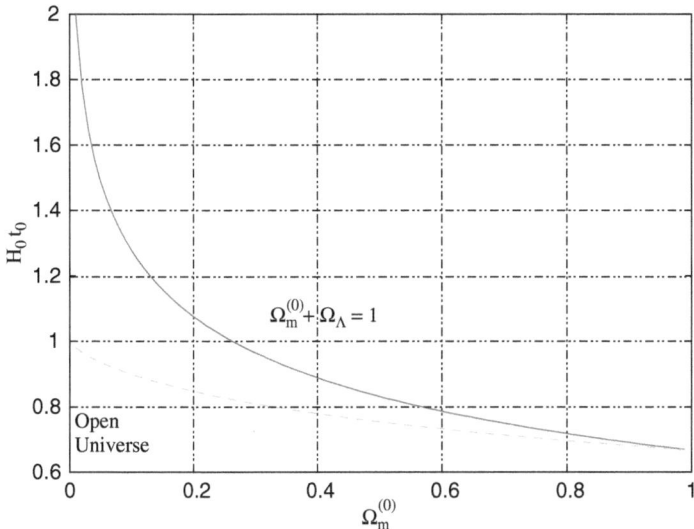

Fig. 8.1. Age of universe (in the units of H_0^{-1}) is plotted against $\Omega_m^{(0)}$ in a flat model (*solid line*) with $\Omega_m^{(0)} + \Omega_\Lambda^{(0)} = 1$ and matter dominated model (*dashed line*) with $\Omega_m^{(0)} = 1$

and requires separate considerations. The power law expansion $a(t) \sim t^n (n = 2/3(1+w))$ corresponds to shrinking universe for $n < 0$ ($w < -1$). The situation can easily be remedied by changing the sign of t and by introducing the origin of time t_s

$$a(t) = (t_s - t)^n , \qquad (8.30)$$

which is the generic solution of evolution equations for super-negative values of w and it gives rise to a very different future course of evolution

$$H = \frac{n}{t_s - t} \qquad (8.31)$$

$$R = 6\left[\frac{\ddot{a}}{a} + \left(\frac{\dot{a}}{a}\right)^2\right] = 6\frac{n(n-1) + n^2}{(t_s - t)^2} . \qquad (8.32)$$

The Hubble expansion rate diverges as $t \to t_s$ corresponding to infinitely large energy density after a finite time in future. The curvature also grows to infinity as $t \to t_s$. Such a situation is referred to Big Rip singularity. Big Rip can be avoided in specific models of phantom field with variable equation of state. It should also be emphasized that quantum effects become important in a situation when curvature becomes large. In that case one should take into account the higher order curvature corrections to Einstein Hilbert action which crucially modifies the structure of singularity.

8.1.2 Scalar Fields— As Perfect Fluids in FRW Background

Scalar fields naturally arise in unified models of interactions and also in string theory. Since the invent of inflation, they continue play an important role in cosmology. They are frequently used as candidates of dark energy. In the recent years a variety of scalar field models namely quintessence, phantoms, tachyons, K-essence, dilatonic ghosts and others have been investigated in the literature. In what follows we briefly describe some of these systems. Their dynamics will be dealt with in detail in Sect. III.

Standard Scalar Field

Let us consider the scalar field minimally coupled to gravity

$$S = - \int \left(\frac{1}{2} g^{\mu\nu} \partial_\mu \phi \partial_\nu \phi + V(\phi) \right) \sqrt{-g} d^4 x \ . \tag{8.33}$$

The Euler Lagrangian equation

$$\partial^\mu \frac{\delta(\sqrt{-g}\mathcal{L})}{\delta \partial^\mu \phi} - \frac{\delta(\sqrt{-g}\mathcal{L})}{\delta \phi} = 0 \ , \tag{8.34}$$

$$\sqrt{-g} = a^3(t)$$

for the action (8.33) in case of a homogeneous field acquires the form

$$\ddot{\phi} + 3H\dot{\phi} + V_\phi = 0 \ , \tag{8.35}$$

which is equivalent to the conservation equation

$$\frac{\dot{\rho}}{\rho} + 3H(1 + w) = 0 \ . \tag{8.36}$$

The energy momentum tensor

$$T_{\mu\nu} = -2 \frac{1}{\sqrt{-g}} \frac{\delta S}{\delta g^{\mu\nu}} \tag{8.37}$$

for the field ϕ which arises from the action (8.33) is given by

$$T_{\mu\nu} = \partial_\mu \phi \partial_\nu \phi - g_{\mu\nu} \left[\frac{1}{2} g^{\mu\nu} \partial_\mu \phi \partial_\nu \phi + V(\phi) \right] \ . \tag{8.38}$$

In the homogeneous and isotropic universe, the field energy density ρ_ϕ and pressure p_ϕ obtained from $T_{\mu\nu}$ are

$$T_{00} \equiv \rho = \frac{\dot{\phi}^2}{2} + V(\phi), \quad T^i_i \equiv p = \frac{\dot{\phi}^2}{2} - V(\phi) \ . \tag{8.39}$$

The field evolution equation (8.35) formally integrates to

$$\rho = \rho_0 e^{-6 \int \left(1 - \frac{2V}{\dot{\phi}^2 + 2V}\right) \frac{da}{a}} . \tag{8.40}$$

Thus the scaling of field energy density crucially depends upon the ratio of kinetic to potential energy. Depending upon the scalar field regime ρ can mimic a behavior ranging from cosmological constant to stiff matter

$$\rho \sim a^{-m} \qquad 0 < m < 6 . \tag{8.41}$$

This behavior is also clear intuitively namely the field ϕ rolling slowly along the flat wing of the potential gives rise to $p \simeq -\rho$ where as it gives $p \simeq \rho$ while dropping fast along the steep part of the potential. Interestingly, one can obtain the similar picture in the oscillatory regime for a power law type of potential.

Acceleration During Oscillations

As the scalar field evolves towards the minimum of its, the slow role ceases and a the scalar field enters into the regime of quasi periodic evolution with decaying amplitude. In what follows, we shall assume that the potential is even and has minimum at $\phi = 0$. When the field initially being displaced from the minimum of the potential, rolls below its slow roll value, the coherence oscillation regime, $\nu \gg H$, commences. The evolution equation can then be approximately solved by separating the two times scales namely the fast oscillation time scale and the longer expansion time scale. On the first time scale, the Hubble expansion can be neglected, and one obtain ϕ as a function of time;

$$t - t_0 = \pm \int \frac{1}{\sqrt{2(V_m - V(\phi))}} d\phi , \tag{8.42}$$

where $\rho \equiv V_m \equiv V(\phi_m)$; V_m being the maximum current value of the potential energy and ϕ_m being the field amplitude. On the longer time scale ρ and ϕ_m slowly decrease because Hubble damping term in (1). The average adiabatic index γ is defined as

$$\gamma = \left\langle \frac{\rho + p}{\rho} \right\rangle = \left\langle \frac{\dot{\phi}^2}{\rho} \right\rangle , \tag{8.43}$$

where$< . >$ denotes the time average over one oscillation. Equation (4) then tells that expansion during oscillations would continue ($\ddot{a} > 0$) if $\gamma < \frac{2}{3}$. The adiabatic evolution of $a(t)$ and ρ is given by,

$$a(t) \propto t^{\frac{2}{3\gamma}} , \tag{8.44}$$

$$\rho \equiv V(\phi_m) \propto t^{-2} . \tag{8.45}$$

As $\ddot{\phi} = \frac{-dV}{d\phi}$, the condition $\gamma < \frac{2}{3}$ can equivalently be written

$$\gamma = \left\langle \frac{\dot{\phi}^2}{\rho} \right\rangle = \frac{<\phi V_{,\phi}>}{V_m} = 2(1 - \frac{<V>}{V_m}) ,$$

$$= 2 \frac{\int_0^{\phi_m}(1 - V(\phi)/V_m)^{\frac{1}{2}}d\phi}{\int_0^{\phi_m}(1 - V(\phi)/V_m)^{-\frac{1}{2}}d\phi} = \frac{2p}{p+1} , \tag{8.46}$$

for a power law potential $V \sim \phi^{2p}$, which gives the average value of the equation of state parameter

$$\langle w \rangle = \frac{p-1}{p+1}, \qquad p < 1/2 \rightarrow \ acceleration . \tag{8.47}$$

Thus a quadratic potential, on the average, mimics dust where as the quartic potential exhibits radiation like behavior. It is really interesting that the scalar field in oscillatory regime can give rise to dark energy for $p < 1/2$.

While developing scalar fields models of dark energy, it is important to have some control on its dynamics. In what follows we show how to construct a field potential viable to desired cosmic evolution.

Construction of Field Potential for a Given Cosmological Evolution

After the invent of cosmological inflation, scalar field models have been frequently used in cosmology in various contexts; they play a central role specially in modelling dark energy. Our focus in the review will also be around these models. We should, however, caution the reader that the scalar field models have limited predictive power. The merits of these models should therefore be judged on the basis of generic features that might emerge in them. Indeed, for *a priori* given cosmological evolution, we can always construct a field potential that would produce it. We shall illustrate this simple fact in case of a power law expansion for a general cosmological background governed by the Friedmann equation

$$H^2 = \frac{\rho^q}{A} , \tag{8.48}$$

where $q = 2, 2/3$ correspond to Randall-Sundrum (RS) and Gauss-Bonnet (GB) brane worlds respectively; A is a constant which takes different values in different patches. We show below how to construct the field potential for ordinary scalar field propagating in a general background described by (8.48).

$$\ddot{\phi} + 3H\dot{\phi} + \frac{dV}{d\phi} . \tag{8.49}$$

Using (8.36) and (8.48) we obtain

$$1 + w = - \left(\frac{2}{3q} \right) \frac{\dot{H}}{H^2} . \tag{8.50}$$

From evolution equation (8.49) and the expression $\dot{\phi}^2 = V(1+w)(1-w)^{-1}$, we have the differential equation for the field potential V

$$\frac{\dot{V}}{V} = -\frac{\dot{f} + 6Hf}{1+f} ,$$

(8.51)

where $f = (1+w)(1-w)^{-1}$. Integrating (8.51) respecting (8.50), we get

$$V(t) = \frac{C}{3q} \left(\frac{3qH^2 + \dot{H}}{H^{2(q-1)/q}} \right) ,$$

(8.52)

where $C = A^{1/q}$ is an integrating constant. Expressing f in terms of H and its derivative through (8.50) and using (8.51), we obtain the $\phi(t)$

$$\phi(t) = \left(\frac{2C}{3q} \right)^{1/2} \int \left[-\frac{\dot{H}}{H^{2(q-1)/q}} \right]^{1/2} dt .$$

(8.53)

Equations (refpoteq) and (8.53) allow to find the field potential with a given expansion dynamics prescribed by $a(t)$. For $a(t) \sim t^n$, we are interested, we have

$$\phi - \phi_0 = D_{nq} t^{(q-1)/q} ,$$

(8.54)

$$V(t) = Cn^{2/q} \left(1 - \frac{n^{-(q+2)/2q}}{3q} \right) t^{-2/q} ,$$

(8.55)

where $D_{nq} = n^{(2-q)/q}(2C/3q)^{1/2}(q/q - 1)$ and $q \neq 1$. Combining (8.54) and (8.55) we get the expression for the potential as function of ϕ

$$V(\phi) = V_0 \phi^{-(2/q-1)} .$$

(8.56)

In case $q = 1$, the field logarithmically depends on time t and (8.56) leads to the well known exponential potential. For $q = 2$ corresponding to RS, we obtain $V(\phi) \sim 1/\phi^2$. The case of high energy GB regime ($q = 2/3$) leads to the power law behavior of $V(\phi)$

$$V(\phi) = V_0 \phi^6 .$$

(8.57)

Phantom Field

All these models of scalar field lead to the equation of state parameter w greater than or equal to minus one. However, the recent observations do not seem to exclude values of this parameter less than minus one. It is therefore important to look for theoretical possibilities to describe dark energy with $w < -1$ called phantom energy. In our opinion, the simplest alternative is provided by a phantom field, scalar field with negative kinetic energy. Such a field can be motivated from S-brane constructs in string theory. Historically, phantom fields were first introduced in Hoyle's version of the Steady State

Theory. In adherence to the Perfect Cosmological Principle, a creation field (C-field) was for the first time introduced [12] to reconcile with homogeneous density by creation of new matter in the voids caused by the expansion of Universe. It was further refined and reformulated in the Hoyle and Narlikar theory of gravitation [13]. Though the quantum theory of phantom fields is problematic, it is nevertheless interesting to examine the cosmological consequences of these fields at the classical level.

The Lagrangian of the phantom field minimally coupled to gravity is given by

$$\mathcal{L} = (16\pi G)^{-1}R + \frac{1}{2}g^{\mu\nu}\partial_{\mu}\phi\partial_{\nu} - V(\phi) , \tag{8.58}$$

where $V(\phi)$ is the phantom potential. The kinetic energy term of the phantom field in (8.58) enters with the opposite sign in contrast to the ordinary scalar field (we remind the reader that we use the metric signature, -,+,+,+). In a spatially flat FRW cosmology, the stress tensor that follows from (8.58) acquires the diagonal form $T_{\beta}^{\alpha} = diag\,(-\rho, p, p, p)$ where the pressure and energy density of field ϕ are given by

$$\rho = -\frac{\dot{\phi}^2}{2} + V(\phi) , \qquad p = -\frac{\dot{\phi}^2}{2} - V(\phi) . \tag{8.59}$$

The corresponding equation of state parameter is now given by

$$w \equiv \frac{p}{\rho} = \frac{\frac{\dot{\phi}^2}{2} + V(\phi)}{\frac{\dot{\phi}^2}{2} - V(\phi)} . \tag{8.60}$$

For $\rho > 0$, $w < -1$.

The equations of motion which follow from (8.58) are

$$\dot{H} = \frac{1}{2M_p^2}\dot{\phi}^2 \tag{8.61}$$

$$H^2 = \frac{1}{3M_p^2}\rho_{\phi} \tag{8.62}$$

$$\ddot{\phi} + 3H\dot{\phi} = V'(\phi) . \tag{8.63}$$

Note that the evolution equation (8.63) for the phantom field is same as that of the normal scalar field with inverted potential allowing the field with zero initial kinetic energy to roll up the hill; i.e. from lower value of potential to higher one. At the first look such a situation looks pathological. However, at present, the situation in cosmology is remarkably tolerant to any pathology if it can lead to a viable model.

As mentioned above the equation of state parameter with super negative values leads to Big Rip which can be avoided in a particular class of models. For instance, let us consider consider a model with

$$V(\phi) = V_0 \left[\cosh\left(\frac{\alpha\phi}{M_p} \right) \right]^{-1} .$$

(8.64)

Due to its peculiar properties, the phantom field, released at a distance from the origin with zero kinetic energy, moves to wards the top of the potential and crosses over to the other side and turns ba ck to execute the damped oscillation about the maximum of the potential (see Fig. 8.2). After a certain period of ti me the motion ceases and the field settles on the top of the potential permanently to mimic the de-Sitter like behavior ($w = -1$).

Rolling Tachyon

It was recently suggested that rolling tachyon condensate, in a class of string theories, might have interesting cosmological consequences. It was shown by Sen [16] that the decay of D-branes produces a pressure-less gas with finite energy density that resembles classical dust. Rolling tachyon has an interesting equation of state whose parameter smoothly interpolates between -1 and 0. Attempts have been made to construct viable cosmological model using rolling tachyon field as a suitable candidate for inflaton, dark matter or dark energy (see [3] and references therein for details). As for the inflation, the rolling tachyon models are faced with difficulties associated with reheating. In what follows we shall consider the tachyon potentials field to obtain viable models of dark energy.

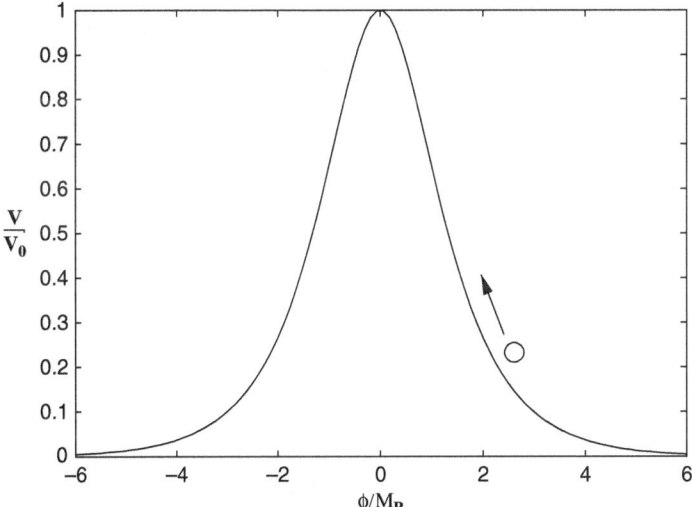

Fig. 8.2. Evolution of the phantom field is shown for the model described by (8.64). Due to the unusual behavior, the phantom field, released with zero kinetic energy away from the origin, moves towards the top of the potential. It sets into the damped oscillations about $\phi = 0$ and ultimately settles there permanently

The tachyon dynamics (on a non-BPS) D_3 brane can be described by an effective field theory with the following action

$$S = \int d^4x \left\{ \sqrt{-g} \left(\frac{R}{2\kappa^2} \right) - V(\phi)\sqrt{-\det(g_{ab} + \partial_a\phi\partial_b\phi)} \right\} .$$

(8.65)

The tachyon field measures the varying brane tension and is such that $V(\phi = \infty) = 0$ and $V(\phi = 0) = 1$. The effective potential obtained in open string theory has the form

$$V(\phi) = \frac{T_3}{\cosh\left(\frac{\phi}{\phi_0}\right)} ,$$

(8.66)

where $\phi_0 = \sqrt{2}$ for superstring and $\phi_0 = 2$ in case of bosonic string. We should note that the potential for the rolling scalar contains no free parameter to tune which is normally required for a viable cosmological evolution. For instance, the late time evolution of the scalar field with potential (8.66) can mimic the current accelerated expansion of universe provided the brane tension T_3 could be tuned to the critical energy density at the current epoch. However, this is absolutely out of scope from viewpoint of string theory as it leads to very small masses for massive string states. We are, therefore, led to think of another mechanism which would affect the D-brane tension and the slope of the scalar field potential without touching the string length and the string coupling constant. We shall hereafter show that these features are shared by the warped compactification. Consider the following warped metric

$$ds_{10}^2 = \beta(y_i)g_{ab}dx^a dx^b + \beta^{-1}(y_i)\hat{g}_{ij}dy^i dy^j ,$$

(8.67)

where the coordinates y_i represent the compact dimensions, and $\hat{g}_{ij}$ represent metric in the compact space. At some point in the y-space the factor β can be small. This corresponds to a scenario in which the brane moves in the compact dimensions reducing its tension. The tachyon action at a point y in the y-space becomes

$$S = -\int d^4x \beta^2 V(\phi)\sqrt{-\det(g_{ab} + \beta^{-1}\partial_a\phi\partial_b\phi)} .$$

(8.68)

Normalizing the scalar field as $\phi \rightarrow \sqrt{\beta}\phi$, one finds the standard Dirac-Born-Infeld (DBI) type action

$$S = -\int d^4x V(\phi)\sqrt{-\det(g_{ab} + \partial_a\phi\partial_b\phi)} ,$$

(8.69)

where now the potential is

$$V(\phi) = \frac{V_0}{\cosh\left(\frac{\sqrt{\beta}\phi}{\phi_0}\right)} , \quad \text{with } V_0 = \beta^2 T_3 .$$

(8.70)

The constant V_0 can be less than T_3 for small values of β with $\beta < 1$. In sections to follow, we shall also consider other forms of tachyon potential which can be inspired by string theory and others which are introduced by purely phenomenological considerations.

In a spatially flat Friedmann-Robertson-Walker (FRW) background, The energy density ρ and the pressure p which follow from action (8.65) are given by,

$$\rho = \frac{V(\phi)}{\sqrt{1 - \dot{\phi}^2}}, \tag{8.71}$$

$$p = -V(\phi)\sqrt{1 - \dot{\phi}^2} . \tag{8.72}$$

The equation of motion of the rolling scalar field follows from (8.8)

$$\frac{\ddot{\phi}}{1 - \dot{\phi}^2} + 3H\dot{\phi} + \frac{V_\phi}{V(\phi)} = 0 , \tag{8.73}$$

which is equivalent to the conservation equation

$$\frac{\dot{\rho}}{\rho} + 3H(1 + w) = 0 . \tag{8.74}$$

The tachyon dynamics is very different from the standard field case. Irrespective of steepness of tachyon potential, its equation of state parameter varies between 0 and -1. Thus reheating is impossible to achieve in this model, if tachyon field is to be an inflaton. However, it can be used as a candidate of dark energy as shown in one of the following sections.

We now look for the potential which can lead to power law type of expansion in case of tachyon field. In this case, the expression for $(1 + w)$ is also given by the (8.50) but the equation of state parameter w has a simple relation with $\dot{\phi}$

$$\dot{\phi}^2 = 1 + w . \tag{8.75}$$

Using (8.50) and (8.75) we get

$$\phi(t) = \int \left[-\frac{2\dot{H}}{3qH^2} \right] dt . \tag{8.76}$$

From (8.48), (8.50) and (8.72) we can express the potential $V(t)$ as

$$V(t) = (-w)^{1/2}\rho = H^{2/q}A^{1/q} \left(1 + \frac{2}{3q}\frac{\dot{H}}{H^2} \right) . \tag{8.77}$$

In case of Born-Infeld scalar field, (8.76) and (8.77) determine the field $\phi(t)$ and the potential $V(t)$ for given scale factor $a(t)$. In case of power law expansion $a(t) \propto t^n$, we obtain from (8.76) and (8.77)

$$\phi(t) - \phi_0 = \left(\frac{2}{3nq}\right)^{1/2} t \,, \tag{8.78}$$

$$V(t) = n^{2/q} A^{1/q} \left(1 - \frac{2}{3nq}\right)^{1/2} t^{-2/q} \,, \tag{8.79}$$

which finally lead to

$$V(\phi) = V_0 \phi^{-2/q} \,. \tag{8.80}$$

We should note that the power law expansion in the present case takes place with the constant velocity of the the field (see (8.78)) which is typical of Born-Infeld dynamics. For $q = 1$ corresponding to standard GR, (8.80) reduces to inverse square potential earlier obtained by Padmanabhan [18]. In case of RS which corresponds to $q = 2$, we get $V(\phi) \sim 1/\phi$. In case of high energy GB regime ($= 2/3$), the potential which can implement power law expansion turns out to be

$$V(\phi) = \frac{V_0}{\phi^3} \,. \tag{8.81}$$

This sort of hierarchy of potentials is understandable; in GR the required potential behaves as $1/\phi^2$ whereas in RS scenario due to the extra brane damping, the power law expansion can be achieved with the less inverse power of field. In the high energy GB regime, the Hubble damping is weaker than the standard FRW cosmology, thereby requiring larger inverse power of the field.

8.1.3 Current Acceleration and Observations in Brief

The direct evidence of current acceleration of universe is related to the observation of luminosity distance of high redshift supernovae by two groups independently in 1998 [1, 2]. The luminosity distance at high redshift is larger in dark energy dominated universe. Thus supernovae would appear fainter in case the universe is dominated by dark energy. The luminosity distance can be used to estimate the apparent magnitude m of the source given its absolute magnitude M. using the following relation often used in astronomy

$$m - M = 5 \log \left(\frac{D_L}{Mpc}\right) + 25 \,. \tag{8.82}$$

In order to get a feeling of the phenomenon (the reader is referred to excellent review of Perivolaropoulos [19] for details) let us consider two supernovae 1997ap at redshift $z = 0.83$ with $m = 24.32$ and 1992P at $z = 0.026$ with apparent magnitude $M = 16.08$. Since the supernovae are supposed to be the standard candles, their absolute magnitude is same. Secondly we shall use the fact that $D_L(z) \simeq z/H_0$ for small value of z. Equation (8.82) then yields the following estimate

$$H_0 D_L \simeq 1.16 \tag{8.83}$$

The theoretical estimate for the luminosity distance for flat universe tells us

$$D_L \simeq 0.95 H_0^{-1} , \qquad \Omega_m^{(0)} = 1 , \tag{8.84}$$

$$D_L \simeq 1.23 H_0^{-1} , \qquad \Omega_m^{(0)} = 0.3, \ \Omega_\Lambda^{(0)} = 0.7 . \tag{8.85}$$

The above estimate clearly lands a strong support to the case of dark energy dominated universe (see [19] for details).

An interesting proposal for visualizing acceleration in supernovae data was proposed in [20]. The authors displayed the data with error bars on the phase plane $(\dot{a}, a)$, see Fig. 8.3 for flat models with different values of Ω_m. The data at low red shift clearly confirms the presence of accelerated phase but due to large error bars it is not possible to choose a particular model. The later requires the interplay between the low redshift and the high redshift data [20]. The observations related to CMB and large scale structures independently support the dark energy scenario. The CMB anisotropies observed originally by COBE in 1992 and the recent WMAP data overwhelmingly support inflationary scenario. The location of the major peak around $l = 220$ tells us that $\Omega_{tot} \simeq 1$. Since the baryonic matter in the universe amounts to only 4%. Nearly 30% of the total energy content is contributed by non-luminous

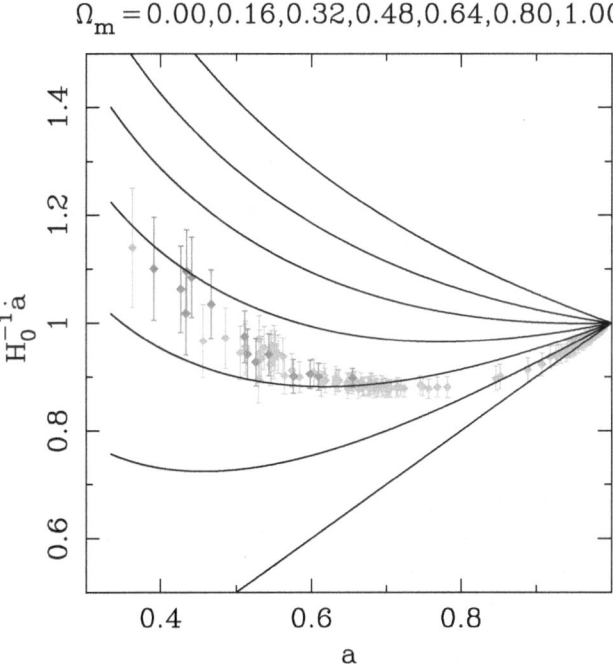

$\Omega_m = 0.00, 0.16, 0.32, 0.48, 0.64, 0.80, 1.00$

Fig. 8.3. The supernovae data points are displayed in the phase plane $(\dot{a}, a)$. The solid curves correspond to flat cosmological models for different values of Ω_m. The bottom and top curves corresponds to $\Omega_m = 0.0, 1.0$ respectively from [20]

component of non-barionic nature with dust like equation of state dubbed *cold dark matter*. There is then a deficit of about 70%–the missing component, known as dark energy. The CMB and the large scale galaxy clustering data is complimentary to supernova results; the combined analysis strongly points towards $\Omega_m = 0.3$ and $\Omega_\Lambda = 0.7$ universe.

However, in view of the fine tuning problem, it looks absolutely essential that dark energy be represented by a variable equation of state. At the same time, the quest for dark energy metamorphosis continues at the observational level.

8.2 Cosmological Constant Λ

Historically Λ was introduced by Einstein to achieve a static solution which turned out to be unstable. However, after the Hubble's redshift discovery in 1929, the motivation for having Λ was lost and it was dropped. Since then the cosmological constant was introduced time and again to remove the discrepancies between theory and observations and withdrawn when these discrepancies were resolved. It had come and gone several times making its come back finally, seemingly for ever!, in 1998 through supernova Ia observations. Recently much efforts have gone in understanding Λ in the frame work of quantum fields and string theory. In what follows we shall briefly mention these issues.

Λ as a Natural Free Parameter of Classical Gravity

It should be noted that a term proportional to the the metric $g_{\mu\nu}$ is missing on the right hand side of Einstein equation (8.2). Indeed the Bianch identity $\Delta^\nu G^\mu_\nu = 0$ implies that

$$G_{\mu\nu} = +\kappa T_{\mu\nu} - \Lambda g_{\mu\nu} \ , \tag{8.86}$$

with

$$\nabla_\nu T^{\mu\nu} = 0 \ , \tag{8.87}$$

where $T_{\mu\nu}$ is a symmetric tensor, and κ and Λ are constants. The demand that it should in the first approximation reduce to the Newtonian equation for gravitation will require $T_{\mu\nu}$ to represent the energy momentum tensor for matter and $\kappa = 8\pi G/c^2$ with Λ being negligible at the stellar scale. The Einstein equations should then read as

$$G_{\mu\nu} \equiv R_{\mu\nu} - \frac{1}{2}g_{\mu\nu}R = 8\pi G T_{\mu\nu} - \Lambda g_{\mu\nu} \ . \tag{8.88}$$

Note that the constant Λ enters into the equation naturally. It was introduced by Einstein in an ad-hoc manner to have a physically acceptable static model of the Universe and was subsequently withdrawn when Friedmann found the non-static model with acceptable physical properties. We would however like to maintain that it appears in the equation as naturally as the stress tensor $T_{\mu\nu}$ and hence should be considered on the same footing [21]. As for the classical physics, the cosmological constant is a free parameter of the theory and its numerical value should be determined from observations.

Λ Arising due to Vacuum Fluctuations

Cosmological constant can be associated with vacuum fluctuations in the quantum field theoretic context. Though the arguments are still at the level of numerology but may have far reaching consequences. Unlike the classical theory the cosmological constant Λ in this scheme is no longer a free parameter of the theory. Broadly the line of thinking takes the following route. The quantum effects in GR become important when the Einstein Hilbert action becomes of the order of Planck's constant; this happens at the Planck's length $L_p = \sqrt{(8\pi G)} \sim 10^{-32} cm$ corresponding to Planck energy which is of the order of $M_p^4 \sim 10^{72} GeV^4$. In the language of field theory, a system is described by a set of quantum fields. The ground state energy dubbed zero point energy or *vacuum* energy of a free quantum field is infinite.

This contribution is related the ordering ambiguity of fields in the classical Lagrangian and disappears when normal ordering is adopted. Since this procedure of throwing out the vacuum energy is ad hoc, one might try to cancel it by introducing the counter terms. The later, however requires fine tuning and may be regarded as unsatisfactory. Whether or not the zero point energy in field theory is realistic is still a debatable question. The divergence is related to the modes of very small wave length. As we are ignorant of physics around Planck scale we might be tempted to introduce a cut off at L_p and associate Λ with this fundamental scale. Thus we arrive at an estimate of vacuum energy $\rho_v \sim M_p^4$ (corresponding mass scale— $M_V \sim (\rho_V^{1/4})$ which is away by 120 orders of magnitudes from the observed value of this quantity. The vacuum energy may not be felt in the laboratory but plays important role in GR through its contribution to the energy momentum tensor as $< T_{\mu\nu} >_0 = -\rho_V g_{\mu\nu}$ and appears on the right hand side of Einstein equations

$$R_{\mu\nu} - \frac{1}{2} g_{\mu\nu} R = 8\pi G \left(T_{\mu\nu} + < T_{\mu\nu} >_0 \right) . \tag{8.89}$$

The problem of zero point energy is naturally resolved by invoking supersymmetry which has many other remarkable features. In the supersymmetric description, every bosonic degree of freedom has its Fermi counter part which contributes zero point energy with opposite sign compared to the bosonic degree of freedom thereby doing away with the vacuum energy. It is in this sense the supersymmetric theories do not admit a non-zero cosmological constant. However, we know that we do not leave in supersymmetric vacuum state and hence it should be broken. For a viable supersymmetric scenario, for instance if it is to be relevant to hierarchy problem, the suppersymmetry breaking scale should be around $M_{susy} \sim 10^3 GeV$. We are still away from the observed value by many orders of magnitudes. At present we do not know how Planck scale or SUSY breaking scales is related to the observed vacuum scale.

Λ from String Theory–de-Sitter Vacuua a la KKLT

In view of the observations related to supernova, large scale clustering and Micro wave background, the idea of late time acceleration has reached the level of general acceptability. It is, therefore, not surprising that tremendous efforts have recently been made in finding out de-Sitter solutions in supergravity and string theory. Using flux compactification, Kachru, Kallosh, Linde and Trivedi (KKLT) formulated a procedure to construct de-Sitter vacua of type IIB string theory [22]. They demonstrated that the life time of the vacua is larger that the age of universe and hence these solutions can be considered as stable for practical purposes. Although a fine-tuning problem of Λ still remains in this scenario, it is interesting that string theory gives rise to a stable de-Sitter vacua with all moduli fixed. We note that a vast number of different choices of fluxes leads to a complicated landscape with more than 10^{100} vacua. We should believe, if we can, that we live in one of them!.

8.2.1 Fine Tuning Problem

Inspite of the fact that introduction of Λ does not require an adhoc assumption and it is also not ruled out by observation as a candidate of dark energy; the scenario base upon Λ is faced with the worst type of fine tuning problem. The numerical value of Λ at early epochs should be tuned to a fantastic accuracy so as not to disturb todays physics. In order to appreciate the problem, let us consider the following ratio

$$\frac{\rho_\Lambda}{\frac{3H^2(t)}{8\pi G}} = \Omega_\Lambda \left(\frac{H_0}{H(t)} \right)^2 , \tag{8.90}$$

where $\Omega_\Lambda = (\rho_\Lambda/\rho_c) \simeq 0.7$. It will not disturb our estimate if we assume radiation domination today. In that case the ratio H/H_0 scales as a^{-2} and since the temperature is inversely proportional to the scale factor a, we find

$$\frac{\rho_\Lambda}{\frac{3H^2(t)}{8\pi G}} = 0.7 \left(\frac{T_0}{T} \right)^4 . \tag{8.91}$$

Since at the Planck ($T = T_p = M_p$) epoch $T_0/T \simeq 10^{-31}$, the ratio of ρ_Λ to $3H^2/8\pi G$ turns out to be of the order of 10^{-123}. On the theoretical ground, such a fine tuning related to the scale of cosmological constant is not acceptable. This problem led to the investigation of scalar field models of dark energy which can alleviate this problem to a considerable extent.

8.3 Dynamically Evolving Scalar Field Models of Dark Energy

Before entering into the detailed investigations of field dynamics, we shall first examine some of the general constraints on scalar field Lagrangian if it is to be relevant to cosmology.

8.3.1 Broad Features of Scalar Field Dynamics and Cosmological Relevance of Scaling Solutions

The scalar field aiming to describe dark energy is often imagined to be a relic of early universe physics. Depending upon the model, the scalar field energy density may be larger or smaller than the background (radiation/matter) energy density ρ_B. In case it is larger than the back ground density, the density ρ_ϕ should scale faster than ρ_B allowing radiation domination to commence which requires a steep scalar field potential. In this case the field energy density overshoots the background and becomes sub dominant to it. This leads to the locking regime for the scalar field. The field unlocks the moment its energy density becomes comparable to the background. Its further course of evolution crucially depend upon the form of field potential. In order to obtain viable dark energy models, we require that the energy density of the scalar field remains unimportant during radiation and matter dominant eras and emerges only at late times to give rise to the current acceleration of universe. It is then important to investigate cosmological scenarios in which the energy density of the scalar field mimics the background energy density. The cosmological solutions which satisfy this condition is called *scaling solutions* [23]. Namely scaling solutions are characterised by the relation

$$\rho_B/\rho_\phi = \text{const} . \tag{8.92}$$

We shall shortly demonstrate that exponential potentials give rise to scaling solutions for a minimally coupled scalar field, allowing the field energy density to mimic the background being sub-dominant during radiation and matter dominant eras. In this case, for any generic initial conditions, the field would sooner or later enter into the scaling regime (see Fig. 8.4). This allows to alleviate the fine tuning problem to a considerable extent. The same thing is true in case of the undershoot, i.e., when the field energy is smaller as compared to the background. In Fig. 8.5, we have displayed a cartoon depicting the field dynamics in absence of scaling solutions. For instance, we shall see later, scaling solutions, which could mimic realistic background, do not exist in case of phantom and tachyon fields. These models are plagued with additional fine tuning problem.

Scaling solutions exist in case of a steep exponential potential $V(\phi) \sim exp(\lambda\phi/M_p)$ with $\lambda^2 > 3(1 + w_m)$ (the field dominated case corresponds to $\lambda^2 < 3(1 + w_m)$ whereas $\lambda^2 < 2$ gives rise to ever accelerating universe).

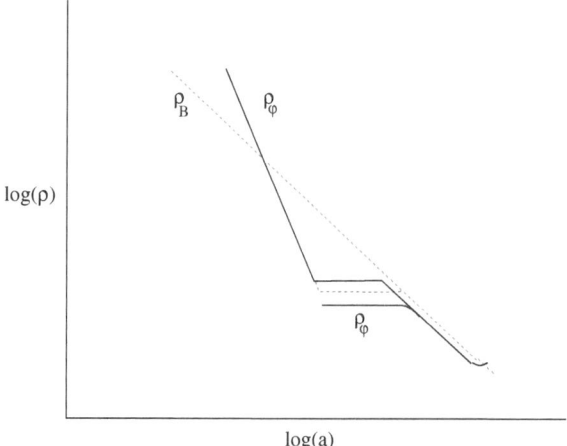

Fig. 8.4. Desired evolution of background and scalar field energy densities ρ_B and ρ_ϕ. In case of overshoot (*solid line*) and undershoot (*dotted line*), the field energy density (for different initial conditions) joins the attractor solution which mimics the background (*scaling solution*). At late times, the field energy density exits the scaling regime to become dominant

Nucleosynthesis puts stringent restriction on any additional degree of freedom which translates into a constraint on the slope of the exponential potential λ.

Late Time Evolution and Exit from Scaling Regime

Obviously, scaling solution is non-accelerating as the equation of state of the field ϕ equals to that of the background fluid ($w_\phi = w_m$) in this case. One then requires to introduce a late time feature in the potential allowing to exit from the scaling regime. Broadly there are two ways to get the required late time behavior for a minimally coupled scalar field:
(i) The potential changes into a power law type $V \sim \phi^{2q}$ which gives late time acceleration for $q < 1/2$ (e.g. $V(\phi) = V_0 \left[\cosh(\alpha\phi/M_p) - 1\right]^q$, $q > 0$ [24]).
(ii) The potential becomes shallow to support the slow-roll at large values of the field [25] allowing the field energy density to catch up with the background; such a solution is referred to a *tracker*.

The scalar field models in absence of the above described features suffer from the fine tuning problem similar to the case of cosmological constant.

Scalar fields should not interfere with the thermal history of universe, they are thus should satisfy certain constraints. An earlier constraint in the history of universe follows from nucleosynthesis which we briefly describe below [11].

Nucleosynthesis Constraint

The introduction of an extra degree of freedom (on the top of those already present in the standard model of particle physics) like a scalar field might effect

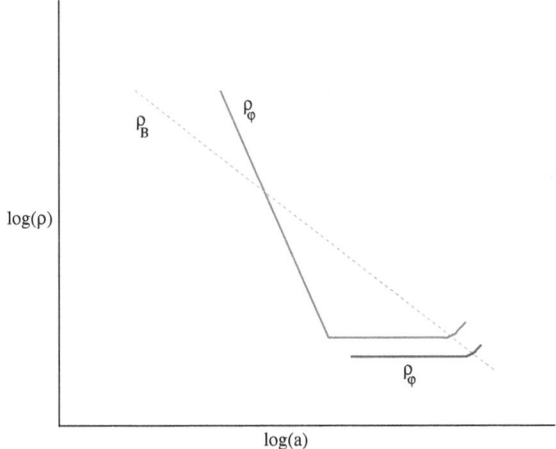

Fig. 8.5. Evolution of ρ_B, ρ_ϕ in absence of scaling solution. The scalar field after its energy density overshoots the background gets into locking regime where it mimics cosmological constant. It waits till its energy density becomes comparable to the background; it then begins evolving and takes over the background to account for the current acceleration

the abundance of light elements in the radiation dominated epoch. The presence of a minimally coupled scalar field effects the expansion rate at a given temperature. This effect becomes crucial at the nucleosynthesis epoch with temperature round $1\,MeV$ when the weak interactions (which keep neutrons and protons in equilibrium) freeze-out. The observationally allowed range of expansion rate at this temperature leads to a bound on the energy density of the scalar field

$$\Omega_\phi(T \sim 1MeV) < \frac{7\Delta N_{eff}/4}{10.75 + 7\Delta N_{eff}/4} , \qquad (8.93)$$

where ΔN_{eff} are the additional relativistic degrees of freedom and 10.75 is the effective number of standard model degrees of freedom. A conservative bound on the additional degrees of freedom used in the literature is given by $\Delta N_{eff} \simeq 1.5$. Equation (8.93) then yields a constraint

$$\Omega_\phi(T \sim 1MeV) < 0.2 , \qquad (8.94)$$

which results into a restriction on the slope of the potential (see Sect. V).

8.3.2 Autonomous Systems, Their Fixed Points and Stability

The dynamical systems which play an important role in cosmology belong to the class of the so called autonomous systems. In what follows we shall analyze the dynamics in great details of a variety of scalar field models. We first briefly record some basic definitions related to dynamical systems. Though, for simplicity we shall consider the system of two first order equations, the analysis can be extended to a system of any number of equations. Let us consider the system of two coupled differential equations for $x(t)$ and $y(t)$

$$\dot{x} = f(x, y, t) ,$$
$$\dot{y} = g(x, y, t) , \qquad (8.95)$$

where f and g are well behaved functions. System (8.95) is said to be autonomous if f and g do not contain explicit time dependent. The dynamics of these systems can be analysed in a standard way.

• Fixed or critical points

A point (x_c, y_c) is said to be a *fixed point* or *critical point* of the autonomous system if and only if

$$(f, g)|_{x_c, y_c} = 0 \qquad (8.96)$$

and a critical point (x_c, y_c) is called an *attractor* in case

$$(x(t), y(t)) \rightarrow (x_c, y_c) \ for \ t \rightarrow \infty . \qquad (8.97)$$

• Stability around the fixed points

The stability of each point can be studied by considering small perturbations δx and δy around the critical point (x_c, y_c), i.e.

$$x = x_c + \delta x , \quad y = y_c + \delta y . \qquad (8.98)$$

Substituting into (8.104) and (8.105), leads to the first-order differential equations:

$$\frac{\mathrm{d}}{\mathrm{d}N} \begin{pmatrix} \delta x \\ \delta y \end{pmatrix} = \mathcal{M} \begin{pmatrix} \delta x \\ \delta y \end{pmatrix} , \qquad (8.99)$$

where matrix $\mathcal{M}$ depends upon x_c and y_c $\left[\mathcal{M} = \begin{pmatrix} \frac{\partial f}{\partial x} & \frac{\partial f}{\partial y} \\ \frac{\partial g}{\partial x} & \frac{\partial g}{\partial y} \end{pmatrix}_{(x=x_c, y=y_c)} \right]$

The general solution for the evolution of linear perturbations can be written as

$$\delta x = C_1 exp(\mu_1 N) + C_2 exp(\mu_2 N) , \qquad (8.100)$$
$$\delta y = C_3 exp(\mu_1 N) + C_4 expp(\mu_2 N) , \qquad (8.101)$$

where μ_1 and μ_2 are the eigenvalues of matrix $\mathcal{M}$. Thus the stability around the fixed points depends upon the nature of eigenvalues. One generally uses the following classification:

- (i) Stable node: $\mu_1 < 0$ and $\mu_2 < 0$.
- (ii) Unstable node: $\mu_1 > 0$ and $\mu_2 > 0$.
- (iii) Saddle point: $\mu_1 < 0$ and $\mu_2 > 0$ (or $\mu_1 > 0$ and $\mu_2 < 0$).
- (iv) Stable spiral: The determinant of the matrix $\mathcal{M}$ is negative and the real parts of μ_1 and μ_2 are negative.

8.3.3 Quintessence

Let us consider a minimally coupled scalar field ϕ with a potential $V(\phi)$:

$$\mathcal{L} = \frac{1}{2}\epsilon\dot{\phi}^2 + V(\phi) , \qquad (8.102)$$

where $\epsilon = +1$ for an ordinary scalar field. Here we allow the possibility of phantom ($\epsilon = -1$) as we see in the next subsection.

In what follows we shall consider a cosmological evolution when the universe is filled by a scalar field ϕ and a barotropic fluid with an equation of state $w_m = p_m/\rho_m$. We introduce the following dimensionless quantities:

$$x \equiv \frac{\kappa\dot{\phi}}{\sqrt{6}H} , \quad y \equiv \frac{\kappa\sqrt{V}}{\sqrt{3}H} , \quad \lambda \equiv -\frac{V_\phi}{\kappa V} , \quad \Gamma \equiv \frac{VV_{\phi\phi}}{V_\phi^2} . \qquad (8.103)$$

For the Lagrangian density (8.102) the Einstein equations can be written in the following autonomous form (see [3] for details):

$$\frac{\mathrm{d}x}{\mathrm{d}N} = -3x + \frac{\sqrt{6}}{2}\epsilon\lambda y^2$$
$$+\frac{3}{2}x\left[(1-w_m)\epsilon x^2 + (1+w_m)(1-y^2)\right] , \qquad (8.104)$$

$$\frac{\mathrm{d}y}{\mathrm{d}N} = -\frac{\sqrt{6}}{2}\lambda xy$$
$$+\frac{3}{2}y\left[(1-w_m)\epsilon x^2 + (1+w_m)(1-y^2)\right] , \qquad (8.105)$$

$$\frac{\mathrm{d}\lambda}{\mathrm{d}N} = -\sqrt{6}\lambda^2(\Gamma - 1)x , \qquad (8.106)$$

together with a constraint equation

$$\epsilon x^2 + y^2 + \frac{\kappa^2\rho_m}{3H^2} = 1 , \qquad (8.107)$$

where $N \equiv \log(a)$. We note that the equation of state w and the fraction of the energy density Ω_ϕ for the field ϕ is

$$w_\phi \equiv \frac{p}{\rho} = \frac{\epsilon x^2 - y^2}{\epsilon x^2 + y^2} , \quad \Omega_\phi \equiv \frac{\kappa^2\rho}{3H^2} = \epsilon x^2 + y^2 . \qquad (8.108)$$

We also define the total effective equation of state:

$$w_{\mathrm{eff}} \equiv \frac{p+p_m}{\rho+\rho_m} = w_m + (1-w_m)\epsilon x^2 - (1+w_m)y^2 . \qquad (8.109)$$

An accelerated expansion occurs for $w_{\mathrm{eff}} < -1/3$. In this subsection we shall consider the normal scalar field ($\epsilon = +1$).

Constant λ

From (8.103) we find that the constant λ corresponds to an exponential potential [23]:

$$V(\phi) = V_0 e^{-\kappa\lambda\phi} . \tag{8.110}$$

In this case (8.106) is dropped from the dynamical system. One can obtain the fixed points by setting $dx/dN = 0$ and $dy/dN = 0$ in (8.104) and (8.105). This is summarized in Table 8.1.

In the next section we shall extend our analysis to the more general case in which dark energy is coupled to dark matter. The readers may refer to the next section in order to know precise values of the eigenvalues in a more general system. From Table 8.1 we find that there exists two stable fixed points (c) and (d). The point (c) is a stable node for $\lambda^2 < 3\gamma$. Since the effective equation of state is $w_{\rm eff} = w_\phi = -1+\lambda^2/3$, the accelerated expansion occurs for $\lambda^2 < 2$ in this case. The point (d) corresponds to a scaling solution in which the energy density of the field ϕ decreases proportionally to that of the barotropic fluid ($\gamma_\phi = \gamma$). Although this fixed point is stable for $\lambda^2 > 3\gamma$, we do not have an accelerated expansion in the case of non relativistic dark matter.

The above analysis of the critical points shows that one can obtain an accelerated expansion provided that the solutions approach the fixed point (c) with $\lambda^2 < 2$, in which case the final state of the universe is the scalar-field dominated one ($\Omega_\phi = 1$). The scaling solution (d) is not viable to explain the late-time acceleration. However this can be used to provide the cosmological evolution in which the scalar field decreases proportionally to that of the matter or radiation. If the slope of the exponential potential becomes shallow

Table 8.1. The properties of the critical points (s=saddle, p=point, un=unstable, n=node, st=stable, sp=spiral) from [3]. Here γ is defined by $\gamma \equiv 1 + w_m$

Name	x	y	Range	Stability	Ω_ϕ	γ_ϕ
(a)	0	0	$\forall\lambda, \gamma$	s. p. for $0 < \gamma < 2$	0	–
(b1)	1	0	$\forall\lambda,\gamma$	un. n. for $\lambda < \sqrt{6}$ s p for $\lambda > \sqrt{6}$	1	2
(b2)	-1	0	$\forall\lambda, \gamma$	un. n. for $\lambda > -\sqrt{6}$ s. p. for $\lambda < -\sqrt{6}$	1	2
(c)	$\lambda/\sqrt{6}$	$[1-\lambda^2/6]^{1/2}$	$\lambda^2 < 6$	st. n. for $\lambda^2 < 3\gamma$ st. n. for $3\gamma < \lambda^2 < 6$	1	$\lambda^2/3$
(d)	$(3/2)^{1/2}\gamma/\lambda$	$[3(2-\gamma)\gamma/2\lambda^2]^{1/2}$	$\lambda^2 > 3\gamma$	st. n. for $3\gamma < \lambda^2 < 24\gamma^2/(9\gamma-2)$ st. sp. for $\lambda^2 > 24\gamma^2/(9\gamma-2)$	$3\gamma/\lambda^2$	γ

to satisfy $\lambda^2 < 2$ near to the present, the universe exits from the scaling regime and approaches the fixed point (c) giving rise to an accelerated expansion.

Dynamically Changing λ

Exponential potentials correspond to constant λ and $\Gamma = 1$. Let us consider the potential $V(\phi)$ along which the field rolls down toward plus infinity ($\phi \to \infty$) This means that $x > 0$ in (8.106). If the condition

$$\Gamma > 1 , \tag{8.111}$$

is satisfied, λ decreases toward 0. Hence the slope of the potential becomes flat as $\lambda \to 0$, thereby giving rise to an accelerated expansion at late times. The condition (8.111) is regarded as the *tracking* condition under which the energy density of ϕ eventually catches up that of the fluid. In order to construct viable quintessence models, we require that the potential should satisfy the condition (8.111). For example, one has $\Gamma = (n+1)/n > 1$ for the inverse power-law potential $V(\phi) = V_0 \phi^{-n}$ with $n > 0$. This means that the tracking occurs for this potential.

When $\Gamma < 1$ the quantity λ increases toward infinity. Since the potential is steep in this case, the energy density of the scalar field becomes negligible compared to that of the fluid. Hence we do not have an accelerated expansion at late times.

In order to obtain the dynamical evolution of the system we need to solve (8.106) together with (8.104) and (8.105). Although λ is dynamically changing, one can exploit the discussion of constant λ by considering "instantaneous" critical points.

8.3.4 Phantoms

The phantom field corresponds to a negative kinematic sign, i.e $\epsilon = -1$ in (8.102). Let us consider the exponential potential given by (8.110). In this case (8.106) is dropped from the dynamical system. In Table 8.2 we show fixed points for the phantom field. The only stable solution is the scalar-field dominant solution (b), in which case the equation of the field ϕ is

$$w_\phi = -1 - \lambda^2/3 . \tag{8.112}$$

Hence w_ϕ is less than -1. The scaling solution (c) is unstable and exists only for $w_m < -1$. We note that the effective equation of state of the universe equals to w_ϕ, i.e., $w_{\text{eff}} = -1 - \lambda^2/3$. In this case the Hubble rate evolves as

$$H = \frac{2}{3(1 + w_{\text{eff}})(t - t_s)} , \tag{8.113}$$

where t_s is an integration constant. Hence H diverges for $t \to t_s$. This is so-called the Big Rip singularity at which the Hubble rate and the energy density

Table 8.2. The properties of the critical points (s=saddle, p=point, n=node, st=stable) for $\epsilon = -1$ (from [3])

Name	x	y	Range	Stab.	Ω_ϕ	w_ϕ
(a)	0	0	No for $0 \leq \Omega_\phi \leq 1$	s. p.	0	–
(b)	$-\lambda/\sqrt{6}$	$[1 + \lambda^2/6]^{1/2}$	All values	st. n.	1	$-1 - \lambda^2/3$
(c)	$\frac{\sqrt{6}(1+w_m)}{2\lambda}$	$[\frac{-3(1-w_m^2)}{2\lambda^2}]^{1/2}$	$w_m < -1$	s. p.	$\frac{-3(1+w_m)}{\lambda^2}$	w_m

of the universe exhibit divergence. We note that the phantom field rolls *up* the potential hill in order to lead to the increase of the energy density.

When the potential of the phantom is different from the exponential type, the quantity λ is dynamically changing in time. In this case the point (b) in Table 8.2 can be regarded as an instantaneous critical point. Then the equation of state w_ϕ varies in time, but the field behaves as a phantom since $w_\phi = -1 - \lambda^2/3 < -1$ is satisfied.

8.3.5 Tachyons

We shall take into account the contribution of a barotropic perfect fluid with an equation of state $p_B = (\gamma-1)\rho_B$. Then the background equations of motion are for rolling tachyon system are

$$\dot{H} = -\frac{\dot{\phi}^2 V(\phi)}{2M_p^2\sqrt{1 - \dot{\phi}^2}} - \frac{\gamma}{2}\frac{\rho_B}{M_p^2} \,, \tag{8.114}$$

$$\frac{\ddot{\phi}}{1 - \dot{\phi}^2} + 3H\dot{\phi} + \frac{V_\phi}{V} = 0 \,, \tag{8.115}$$

$$\dot{\rho}_B + 3\gamma H\rho_B = 0 \,, \tag{8.116}$$

together with a constraint equation:

$$3M_p^2 H^2 = \frac{V(\phi)}{\sqrt{1 - \dot{\phi}^2}} + \rho_B \,. \tag{8.117}$$

Defining the following dimensionless quantities:

$$x = \dot{\phi} \,, \quad y = \frac{\sqrt{V(\phi)}}{\sqrt{3}HM_p} \,, \tag{8.118}$$

we obtain the following autonomous equations

$$\frac{dx}{dN} = -(1 - x^2)(3x - \sqrt{3}\lambda y) \,, \tag{8.119}$$

$$\frac{dy}{dN} = \frac{y}{2}\left(-\sqrt{3}\lambda xy - \frac{3(\gamma - x^2)y^2}{\sqrt{1 - x^2}} + 3\gamma\right) , \tag{8.120}$$

$$\frac{d\lambda}{dN} = -\sqrt{3}\lambda^2 xy(\Gamma - 3/2) . \tag{8.121}$$

where

$$\lambda = -\frac{M_p V_\phi}{V^{3/2}} , \quad \Gamma = \frac{V V_{\phi\phi}}{V_\phi^2} . \tag{8.122}$$

We note that the allowed range of x and y is $0 \le x^2 + y^4 \le 1$ from the requirement: $0 \le \Omega_\phi \le 1$. Hence both x and y are finite in the range $0 \le x^2 \le 1$ and $0 \le y \le 1$. The effective equation of state for the field ϕ is

$$\gamma_\phi = \frac{\rho_\phi + p_\phi}{\rho_\phi} = \dot{\phi}^2 , \tag{8.123}$$

which means that $\gamma_\phi \ge 0$. The condition for inflation corresponds to $\dot{\phi}^2 < 2/3$.

Constant λ

From (8.121) we find that λ is a constant for $\Gamma = 3/2$. This case corresponds to an inverse square potential (For details, see [3])

$$V(\phi) = M^2\phi^{-2} . \tag{8.124}$$

The scalar-field dominated solution ($\Omega_\phi = 1$), in this case, corresponds to $\gamma_\phi = \lambda^2/3$ which can lead to an accelerated expansion for $\lambda^2 < 2$. No scaling solution which could mimic radiation or matter exist in this case (see [3]). Since λ is given by $\lambda = 2M_p/M$, the condition for an accelerated expansion gives a super-Planckian value of the mass scale, i.e., $M > \sqrt{2}M_p$. Such a large mass is problematic since this shows the breakdown of classical gravity. This problem can be alleviated for the inverse power-law potential $V(\phi) = M^{4-n}\phi^{-n}$, as we will see below.

Dynamically Changing λ

When the potential is different from the inverse square potential given in (8.124), λ is a dynamically changing quantity. As we have seen in the subsection of quintessence, there are basically two cases: (i) λ evolves toward zero, or (ii) $|\lambda|$ increases toward infinity. The case (i) is regarded as the tracking solution in which the energy density of the scalar field eventually dominates over that of the fluid. This situation is realized when the potential satisfies the condition

$$\Gamma > 3/2 , \tag{8.125}$$

as can be seen from (8.121). The case (ii) corresponds to the case in which the energy density of the scalar field becomes negligible compared to the fluid.

As an example let us consider the inverse power-law potential given by

$$V(\phi) = M^{4-n}\phi^{-n}, \quad n > 0 . \tag{8.126}$$

In this case one has $\Gamma = (n+1)/n$. Hence the scalar-field energy density dominates at late times for $n < 2$.

There exist a number of potentials that exhibit the behavior $|\lambda| \to \infty$ asymptotically. For example $V(\phi) = M^{4-n}\phi^{-n}$ with $n > 2$ and $V(\phi) = V_0 e^{-\mu\phi}$ with $\mu > 0$. In the latter case one has $\Gamma = 1$. In these cases, pressure less dust ia late time attractor where as the accelerated expansion can occur as a transient phenomenon. Extra fine tuning is needed in this case to obtain the current acceleration.

8.4 Scaling Solutions in Models of Coupled Quintessence

As we have already seen in the previous section, exponential potentials give rise to scaling solutions for a minimally coupled scalar field, allowing the field energy density to mimic the background being sub-dominant during radiation and matter dominant eras. In the previous section we found out the expression for Ω_ϕ for scaling solution which after combining with the nucleosynthesis constraint (8.94) gives

$$\Omega_\phi \equiv \frac{\rho_\phi}{\rho_\phi + \rho_m} = \frac{(1 + w_m)}{\lambda^2} < 0.2 \quad \to \quad \lambda > 5 . \tag{8.127}$$

In this case, however, one can not have an accelerated expansion at late times since ρ_ϕ mimics background. We briefly mentioned as how to exit the scaling regime, in models of minimally coupled scalar fields, to account for the current acceleration of universe.

If the scalar field ϕ is coupled to the background fluid, it is possible to obtain an accelerated expansion at late-times even in the case of steep exponential potentials. In this section we implement the coupling Q between the field and the barotropic fluid and show that scaling solutions can also account for accelerated expansion.

The evolution equations in presence of coupling acquire the form

$$\dot{\rho}_\phi + 3H(1 + w_\phi)\rho_\phi = -Q\rho_m\dot{\phi} \tag{8.128}$$

$$\dot{\rho}_m + 3H(1 + w_m)\rho_m = Q\rho_m\dot{\phi} , \tag{8.129}$$

$$\dot{H} = -\frac{1}{2}\Big[(1 + w_m)\rho_\phi + (1 + w_m)\rho_m\Big] . \tag{8.130}$$

Table 8.3. $Q \neq 0$, from [3]

x	y	Ω_ϕ	w_{eff}
$-\frac{\sqrt{6}Q}{3(1-w_m)}$	0	$\frac{2Q^2}{3(1-w_m)}$	1
1	0	1	1
-1	0	1	1
$\frac{\lambda}{\sqrt{6}}$	$[(1-\frac{\lambda^2}{6})]^{1/2}$	1	$-1+\frac{\lambda^2}{3}$
$\frac{\sqrt{6}(1+w_m)}{2(\lambda+Q)}$	$[\frac{2Q(\lambda+Q)+3\,(1-w_m^2)}{2(\lambda+Q)^2}]^{1/2}$	$\frac{Q(\lambda+Q)+3\,(1+w_m)}{(\lambda+Q)^2}$	$\frac{\lambda w_m-Q}{(\lambda+Q)}$

$$H^2 = \frac{\rho_\phi + \rho_m}{3} \,, \tag{8.131}$$

where coupling Q is field dependent in general. For simplicity, we shall assume constant coupling. The autonomous form of equations for exponential potential in presence of coupling takes the following form

$$\frac{dx}{dN} = -3x + \frac{\sqrt{6}}{2}\lambda y^2 + \frac{3}{2}x\Big[(1-w_m)\epsilon x^2 \tag{8.132}$$

$$+(1+w_m)(1-c_1 y^2)\Big] - \frac{\sqrt{6}Q}{2}(1-x^2-y^2) \,,$$

$$\frac{dy}{dN} = -\frac{\sqrt{6}}{2}\lambda xy + \frac{3}{2}y\left[(1-w_m)x^2 + (1+w_m)(1-y^2)\right] \,. \tag{8.133}$$

We display the critical points for coupled quintessence in the table 8.3 in which the last entry corresponds to scaling solution with effective equation of state $w_{eff} = 0$ for $Q = 0$ consistent with earlier analysis. It is remarkable that $w_{eff} \to -1$ for $Q >> \lambda$. Thus scaling solutions can account for acceleration in presence of coupling between field and the barotropic fluid. Unfortunately, they are not acceptable from CMB constraints. The general investigations of perturbations for coupled quintessence require further serious considerations.

8.5 Quintessential Inflation

In this section we shall work out the example of quintessential inflation which is an attempt to describe inflation and dark energy with a single scalar field. The description to follow would clearly demonstrate the utility of the tools developed in earlier sections. The problem was first addressed by Peebles and Vilenkin [26]. They introduced a potential for the field ϕ which allowed it to play the role of the inflaton in the early Universe and later to play the role of the quintessence field. To do this it was important that the potential did not have a minimum in which the inflaton field would completely decay at the end of the initial period of inflation. They proposed the following potential

$$V(\phi) = \begin{cases} \lambda(\phi^4 + M^4) & \text{for } \phi < 0 , \\ \frac{\lambda M^4}{1+(\phi/M)^\alpha} & \text{for } \phi \geq 0 . \end{cases} \tag{8.134}$$

For $\phi < 0$ we have ordinary chaotic inflation. Much later on, for $\phi > 0$ the universe once again begins to inflate but this time at the lower energy scale associated with quintessence. Reheating after inflation should have proceeded via gravitational particle production because of the absence of the potential minimum, but this mechanism is very inefficient and leads to an unwanted relic gravity wave background. The main difficulty for the realistic construction of quintessential inflation is that we need a flat potential during inflation but also require a steep potential during radiation and matter dominated periods. There are some nice resolutions of quintessential inflation in braneworld scenarios as we shall see below (see review. [27] and references therein on this theme). In these models, the scalar field exhibits the properties of tracker field. As a result it goes into hiding after the commencement of radiation domination; it emerges from the shadow only at late times to account for the observed accelerated expansion of universe. These models belong to the category of *non oscillating* models in which the standard reheating mechanism does not work. In this case, one can employ an alternative mechanism of reheating via quantum-mechanical particle production in time varying gravitational field at the end of inflation. However, then the inflaton energy density should red-shift faster than that of the produced particles so that radiation domination could commence. And this requires a steep field potential, which of course, cannot support inflation in the standard FRW cosmology. This is precisely where the brane [29] assisted inflation comes to our rescue. In the 4+1 dimensional brane scenario inspired by the Randall-Sundrum (RS) model, the standard Friedman equation is modified to

$$H^2 = \frac{1}{3M_p^2}\rho\left(1 + \frac{\rho}{2\lambda_b}\right) , \tag{8.135}$$

The presence of the quadratic density term ρ^2/λ_b (high energy corrections) in the Friedmann equation on the brane changes the expansion dynamics at early epochs (see [29] for details on the dynamics of brane worlds) Consequently, the field experiences greater damping and rolls down its potential slower than it would during the conventional inflation. This effect is reflected in the slow-roll parameters which have the form [29]

$$\epsilon = \epsilon_{\text{FRW}} \frac{1 + V/\lambda_b}{(1 + V/2\lambda_b)^2} ,$$

$$\eta = \eta_{\text{FRW}} (1 + V/2\lambda_b)^{-1} , \tag{8.136}$$

where

$$\epsilon_{\text{FRW}} = \frac{M_p^2}{2} \left(\frac{V'}{V}\right)^2 , \quad \eta_{\text{FRW}} = M_p^2 \left(\frac{V''}{V}\right) \tag{8.137}$$

are slow roll parameters in the absence of brane corrections. The influence of the brane term becomes important when $V/\lambda_b \gg 1$ and in this case we get

$$\epsilon \simeq 4\epsilon_{\text{FRW}}(V/\lambda_b)^{-1}, \quad \eta \simeq 2\eta_{\text{FRW}}(V/\lambda_b)^{-1} . \tag{8.138}$$

Clearly slow-roll ($\epsilon, \eta \ll 1$) is easier to achieve when $V/\lambda_b \gg 1$ and on this basis one can expect inflation to occur even for relatively steep potentials, such the exponential and the inverse power-law. The model of quintessential inflation [27] based upon reheating via gravitational particle production is faced with difficulties associated with excessive production of gravity waves. Indeed the reheating mechanism based upon this process is extremely inefficient. The energy density of so produced radiation sis typically one part in 10^{16} to the scalar-field energy density at the end of inflation. As a result, these models have prolonged kinetic regime during which the amplitude of primordial gravity waves enhances and violates the nucleosynthesis constraint. Hence, it is necessary to look for alternative mechanisms more efficient than the gravitational particle production to address the problem. However this problem may be alleviated in instant preheating scenario [28] in the presence of an interaction $g^2\phi^2\chi^2$ between inflaton ϕ and another field χ. This mechanism is quite efficient and robust, and is well suited to non-oscillating models. It describes a new method of realizing quintessential inflation on the brane in which inflation is followed by 'instant preheating'. The larger reheating temperature in this model results in a smaller amplitude of relic gravity waves which is consistent with the nucleosynthesis bounds [27]. Figure 8.6 shows the post inflationary evolution of scalar field energy density for the potential given by

$$V(\phi) = V_0 \left[\cosh(\kappa\lambda\phi) - 1\right]^n . \tag{8.139}$$

This potential has following asymptotic forms:

$$V(\phi) \simeq \begin{cases} \widetilde{V}_0 e^{-n\kappa\lambda\phi} & (|\lambda\phi| \gg 1, \ \phi < 0) , \\ \widetilde{V}_0 (\kappa\lambda\phi)^{2n} & (|\lambda\phi| \ll 1) , \end{cases} \tag{8.140}$$

where $\widetilde{V}_0 = V_0/2^n$. The existence of scaling solution for exponential potential ($V \sim exp(\kappa\lambda\phi)$) tells us that $\lambda^2 > 3\gamma$ where as nucleosynthesis constraint makes the potential further steeper as $\Omega_\phi = 3\gamma/\lambda^2 < 0.2 \rightarrow \lambda > 5$. Potential (8.140) is suitable for unification of inflation and quintessence. In this case, for a given number of e-foldings, the COBE normalization allows to estimate the brane tension λ_b and the field potential at the end of inflation. Tuning the model parameters (λ – slope of the potential and V_0), we can account for the current acceleration with $\Omega_\phi^{(0)} \simeq 0.7$ and $\Omega_m^{(0)} \simeq 0.3$ [27]. However, the recent

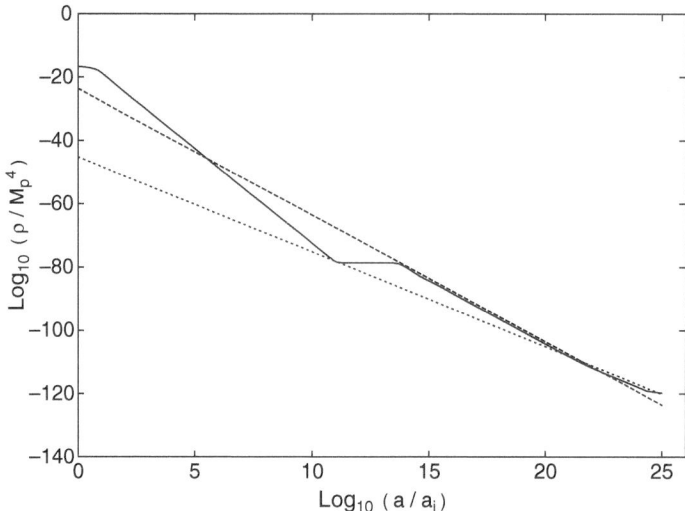

Fig. 8.6. The post-inflationary evolution of the scalar field energy density (*solid line*), radiation (*dashed line*) and cold dark matter (*dotted line*) is shown as a function of the scale factor for the quintessential inflation model described by (8.140) with $V_0^{1/4} \simeq 10^{-30} M_p$, $\lambda = 50$ and $n = 0.1$. After brane effects have ended, the field energy density ρ_ϕ enters the kinetic regime and soon drops below the radiation density. After a brief interval during which $< w_\phi > \simeq -1$, the scalar field begins to track first radiation and then matter. At very late times (present epoch) the scalar field plays the role of quintessence and makes the universe accelerate. The evolution of the energy density is shown from the end of inflation until the present epoch. From [32]

measurement of CMB anisotropies by WMAP places fairly strong constraints on inflationary models. The ratio of tensor perturbations to scalar perturbations turns out to be large in case of steep exponential potential pushing the model outside the 2σ observational bound [30]. However, the model can be rescued in case a Gauss-Bonnet term is present in five dimensional bulk [31, 32].

In order to see how it comes about, let us consider Einstein-Gauss-Bonnet action for five dimensional bulk containing a 4D brane

$$S = \frac{1}{2\kappa_5^2} \int d^5x \sqrt{-g} \{ R - 2\Lambda_5 + \alpha_{\mathrm{GB}}[R^2 - 4R_{AB}R^{AB}$$
$$+ R_{ABCD}R^{ABCD}]\} + \int d^4x \sqrt{-h}(L_m - \lambda_b) , \qquad (8.141)$$

R refers to the Ricci scalars in the bulk metric g_{AB} and h_{AB} is the induced metric on the brane; α_{GB} has dimensions of $(length)^2$ and is the Gauss-Bonnet coupling, while λ_b is the brane tension and $\Lambda_5 (< 0)$ is the bulk cosmological constant. The constant κ_5 contains the M_5, the 5D fundamental energy scale $(\kappa_5^2 = M_5^{-3})$.

The analysis of modified Friedmann [34] equation which follows from the above action shows that there is a characteristic GB energy scale M_{GB} [34] such that,

$$\rho \gg M_{\mathrm{GB}}^4 \;\Rightarrow\; H^2 \approx \left[\frac{\kappa_5^2}{16\alpha_{\mathrm{GB}}}\,\rho\right]^{2/3}, \tag{8.142}$$

$$M_{\mathrm{GB}}^4 \gg \rho \gg \lambda_b \;\Rightarrow\; H^2 \approx \frac{\kappa^2}{6\lambda_b}\,\rho^2, \tag{8.143}$$

$$\rho \ll \lambda_b \;\Rightarrow\; H^2 \approx \frac{\kappa^2}{3}\,\rho. \tag{8.144}$$

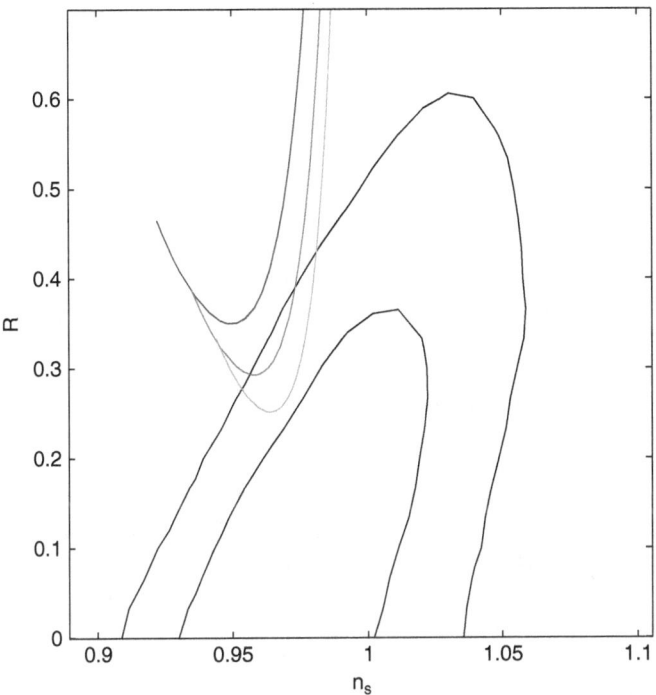

Fig. 8.7. Plot of R ($R \equiv 16A_T^2/A_S^2$ – according to the normalization used here [31]) versus the spectral index n_S in case of the exponential potential for the number of inflationary e-foldings $\mathcal{N} = 50, 60, 70$ (from top to bottom) along with the 1σ and 2σ observational contours. These curves exhibit a minimum in the intermediate region between GB (*extreme right*) and the RS (*extreme left*) regimes. The upper limit on n_S is dictated by the quantum gravity limit where as the lower bound is fixed by the requirement of ending inflation in the RS regime [31]. For a larger value of the number of e-folds $\mathcal{N}$, more points are seen to be within the 2σ bound. Clearly, steep inflation in the deep GB regime is not favored due to the large value of R in spite the spectral index being very close to 1 there. From [32]

It should be noted that Hubble law acquires an unusual form for energies higher than the GB scale. Interestingly, for an exponential potential, the modified (8.142) leads to exactly scale invariant spectrum for primordial density perturbations. Inflation continues below GB scale and terminates in the RS regime leading to the spectral index very close to one. However, as shown in [31, 43], the tensor to scalar ratio of perturbations(R) also increases towards the high energy GB regime. It is known that the value of R is larger in case of RS brane world as compared to the standard GR. While moving from the RS regime characterized by $H^2 \propto \rho^2$ to GB regime described by $H^2 \propto \rho^{2/3}$, we pass through an intermediate region which mimics GR like behavior. It is not surprising that the ratio R has minimum at an intermediate energy scale between RS and GB, see Fig. 8.7. We conclude that a successful scenario of quintessential inflation on the Gauss-Bonnet braneworld can be constructed which agrees with CMB+LSS observations.

8.6 Conclusions

In this talk we have reviewed the general features of scalar field dynamics. Our discussion has been mainly pedagogical in nature. we tried to present the basic features of standard scalar field, phantoms and rolling tachyon. Introducing the basic definitions and concepts, we have shown as how to find the critical pints and investigate stability around them. This is a standard technique needed for building the scalar field models desired for a viable cosmic evolution. The two often used mechanisms for the exit from scaling regime are also described in detail. In case of phantoms and rolling tachyon, we have shown that there exits no scaling solutions which would mimic the realistic background fluid (radiation/matter). Thus, in these case, there will be dependency on the initial conditions of the field leading to fine tuning problems. These models should therefore be judged on the basis of generic features which might arise in them. The rolling tachyon is inspired by string theory whereas as phantoms might be supported by observations!.

After developing the basic techniques of scalar field dynamics, we worked out the example of quintessential inflation. we have shown in detail how to implement the techniques for building a unified model of inflation and quintessence with a single scalar field.

In this talk we have not touched upon the observational status of dark energy models. We have also not discussed the alternatives to dark energy. The interested reader is refereed to other talks on these topics in the same proceedings. The supernovae observations are not yet sufficient to decide the metamorphosis of dark energy. There have been claims and anti-claims for dynamically evolving dark energy using supernovae, CMB and large scale studies. Given the present observational status of cosmology, it would be fair to say that the nature of dark energy remains to be a mystery of the millennium. It could be any thing or it could be nothing!

Acknowledgements

I thank, G. Agelika, E. J. Copeland, Naresh Dadhich, Sergei Odintsov, T. Padmanabhan, Varun Sahni, N. Savchenko, Parampreet Singh and Shinji Tsujikawa for useful discussions. I also thank Gunma National College of Technology (Japan) for hospitality where the part of the talk was written. I am extremely thankful to the organisers of Third Aegean Summer school for giving me opportunity to present the review on dark energy models.

References

1. S. Perlmutter et al., (1999). Astrophys. J **517**, 565
2. A. Riess, et al. (1999). Astrophys. J,**117**, 707
3. E.J. Copeland, M. Sami and Shinji Tsujikawa, *Dynamics of dark energy*, hep-th/0603057.
4. V. Sahni and A.A. Starobinsky, Int. J. Mod. Phys. D **9**, 373 (2000); S.M. Carroll, Living Rev. Rel. **4**, 1 (2001); T. Padmanabhan, Phys. Rept. **380**, 235 (2003); P.J.E. Peebles and B. Ratra, Rev. Mod. Phys. **75**, 559 (2003).
5. A. Lue, Phys. Rept. **423**, 1 (2006) astro-ph/0510068.
6. S. Nojiri, S.D. Odintsov, hep-th/0601213.
7. Jacob D. Bekenstein, Phys. Rev. D **70**, 083509 (2004); Erratum-Lect. Notes phys. D71 069901 (2005).
8. Robert H. Sanders, astro-ph/0601431.
9. Luz Maria Diaz-Rivera, Lado Samushia, B. Ratra, astro-ph/0601153.
10. C. Wetterich, Nucl. Phys. B **302**, 668 (1988); J.A. Frieman, C.T. Hill, A. Stebbins and I. Waga, Phys. Rev. Lett. **75**, 2077 (1995) [arXiv:astro-ph/9505060]; I. Zlatev, L.M. Wang and P.J. Steinhardt, Phys. Rev. Lett. **82**, 896 (1999) [arXiv:astro-ph/9807002]; P. Brax and J. Martin, Phys. Rev. D **61**, 103502 (2000) [arXiv:astro-ph/9912046]; T. Barreiro, E.J. Copeland and N.J. Nunes, Phys. Rev. D **61**, 127301 (2000) [arXiv:astro-ph/9910214]; A. Albrecht and C. Skordis.
11. P.G. Ferreira and M. Joyce, Phys. Rev. Lett. **79**, 4740 (1997) [arXiv:astro-ph/9707286].
12. F. Hoyle, Mon. Not. R. Astr. Soc. **108**, 372 (1948); **109**, 365 (1949).
13. F. Hoyle and J.V. Narlikar, Proc. Roy. Soc. A **282**, 191 (1964); Mon. Not. R. Astr. Soc. **155**, 305 (1972); J.V. Narlikar and T. Padmanabhan, Phys. Rev. D **32**, 1928 (1985).
14. R.R. Caldwell, Phys. Lett. B **545**, 23–29 (2002).
15. C. Armendáriz-Picón, T. Damour and V. Mukhanov, Phys. Lett. B **458**, 219 (1999), [hep-th/9904075]; J. Garriga and V. Mukhanov, Phys. Lett. B **458**, 219 (1999), [hep-th/9904176]; T. Chiba, T. Okabe and M. Yamaguchi, Phys. Rev. D**62**, 023511 (2000) [astro-ph/9912463]; C. Armendáriz-Picón, V. Mukhanov and P.J. Steinhardt, Phys. Rev. Lett. **85**, 4438 (2000) [astro-ph/0004134]; C. Armendáriz-Picón, V. Mukhanov and P.J. Steinhardt, Phys. Rev. D **63**, 103510 (2001) [astro-ph/0006373]; M. Malquarti and A.R. Liddle, Phys. Rev. D **66**, 023524 (2002) [astro-ph/0203232].

16. A. Sen, JHEP **0204**, 048 (2002); JHEP **0207**, 065 (2002); A. Sen, JHEP **9910**, 008 (1999); M.R. Garousi, Nucl. Phys. B **584**, 284 (2000); Nucl. Phys. B **647**, 117 (2002); JHEP **0305**, 058 (2003); E.A. Bergshoeff, M. de Roo, T.C. de Wit, E. Eyras, S. Panda, JHEP **0005**, 009 (2000); J. Kluson, Phys. Rev. D **62**, 126003 (2000); D. Kutasov and V. Niarchos, Nucl. Phys. B **666**, 56 (2003).
17. T. Padmanabhan, astro-ph/0602117.
18. T. Padmanabhan, Phys. Rev. D **66** (2002) 021301.
19. L. Perivolaropoulos, astro-ph/0601014.
20. T. Padmanabhan and T. Roy Choudhury, Mon. Not. Roy. Astron. Soc. **344**, 823 (2003).
21. N. Dadhich, gr-qc/0405115.
22. Shamit Kachru, Renata Kallosh andrei Linde, Sandip P. Trivedi, Phys. Rev. D **68** 046005 (2003).
23. E.J. Copeland, A.R. Liddle and D. Wands, "Phys. Rev. D **57**, 4686 (1998).
24. V. Sahni and L.M. Wang, Phys. Rev. D **62**, 103517 (2000).
25. T. Barreiro, E.J. Copeland and N.J. Nunes, Phys. Rev. D **61**, 127301 (2000).
26. P.J.E. Peebles and A. Vilenkin, Phys. Rev. D **59**, 063505 (1999).
27. M. Sami and N. Dadhich, hep-th/04050.
28. G.N. Felder, L. Kofman and A.D. Linde, Phys. Rev. D **60**, 103505 (1999); G.N. Felder, L. Kofman and A.D. Linde, Phys. Rev. D **59**, 123523 (1999).
29. R. Maartens, Living Rev. Rel. **7**, 7 (2004).
30. A.R. Liddle and A.J. Smith, Phys. Rev. D **68**, 061301 (2003); S. Tsujikawa and A.R. Liddle, JCAP **0403**, 001 (2004).
31. S. Tsujikawa, M. Sami and R. Maartens, Phys. Rev. D **70**, 063525 (2004).
32. M. Sami and V. Sahni, Phys. Rev. D **70**, 083513 (2004).
33. J.-F. Dufaux, J. Lidsey, R. Maartens, M. Sami, Phys. Rev. D **70**, 083525 (2004) hep-th/0404161.
34. S.L. Dubovsky and V.A. Rubakov, Phys. Rev. D **67**, 104014 (2003) [arXiv:hep-th/0212222].

9

Accelerating Universe: Observational Status and Theoretical Implications

Leandros Perivolaropoulos

Department of Physics, University of Ioannina, Greece
leandros@cc.uoi.gr

Abstract. This is a pedagogical review of the recent observational data obtained from type Ia supernova surveys that support the accelerating expansion of the universe. The methods for the analysis of the data are reviewed and the theoretical implications obtained from their analysis are discussed.

9.1 Introduction

Recent distance-redshift surveys [1, 2, 3, 4, 5, 6] of cosmologically distant Type Ia supernovae (SnIa) have indicated that the universe has recently (at redshift $z \simeq 0.5$) entered a phase of accelerating expansion. This expansion has been attributed to a dark energy [7] component with negative pressure which can induce repulsive gravity and thus cause accelerated expansion. The evidence for dark energy has been indirectly verified by Cosmic Microwave Background (CMB) [8] and large scale structure [9] observations.

The simplest and most obvious candidate for this dark energy is the cosmological constant [10] with equation of state $w = p/\rho = -1$. The extremely fine tuned value of the cosmological constant required to induce the observed accelerated expansion has led to a variety of alternative models where the dark energy component varies with time. Many of these models make use of a homogeneous, time dependent minimally coupled scalar field ϕ (quintessence [11, 12]) whose dynamics is determined by a specially designed potential $V(\phi)$ inducing the appropriate time dependence of the field equation of state $w(z) = p(\phi)/\rho(\phi)$. Given the observed $w(z)$, the quintessence potential can in principle be determined. Other physically motivated models predicting late accelerated expansion include modified gravity [13, 14, 15, 16], Chaplygin gas [17], Cardassian cosmology [18], theories with compactified extra dimensions [19, 20], braneworld models [21] etc. Such cosmological models predict specific forms of the Hubble parameter $H(z)$ as a function of redshift z. The observational determination of the recent expansion history $H(z)$ is therefore important for the identification of the viable cosmological models.

L. Perivolaropoulos: *Accelerating Universe*, Lect. Notes Phys. **720**, 257–290 (2007)
DOI 10.1007/978-3-540-71013-4_9   © Springer-Verlag Berlin Heidelberg 2007

The most direct and reliable method to observationally determine the recent expansion history of the universe $H(z)$ is to measure the redshift z and the apparent luminosity of cosmological distant indicators (standard candles) whose absolute luminosity is known. The luminosity distance vs. redshift is thus obtained which in turn leads to the Hubble expansion history $H(z)$.

The goal of this review is to present the methods used to construct the recent expansion history $H(z)$ from SnIa data and discuss the most recent observational results and their theoretical implications. In the next section I review the method used to determine $H(z)$ from cosmological distance indicators and discuss SnIa as the most suitable cosmological standard candles. In Sect. 9.3 I show the most recent observational results for $H(z)$ and discuss their possible interpretations other than accelerating expansion. In Sect. 9.4 I discuss some of the main theoretical implications of the observed $H(z)$ with emphasis on the various parametrizations of dark energy (the simplest being the cosmological constant). The best fit parametrizations are shown and their common features are pointed out. The physical origin of models predicting the best fit form of $H(z)$ is discussed in Sect. 9.5 where I distinguish between minimally coupled scalar fields (quintessence) and modified gravity theories. An equation of state of dark energy with $w < -1$ is obtained by a specific type of dark energy called *phantom energy* [22]. This type of dark energy is faced with theoretical challenges related to the stability of the theories that predict it. Since however the SnIa data are consistent with phantom energy it is interesting to investigate the implications of such an energy. These implications are reviewed in Sect. 9.6 with emphasis to the Big Rip future singularity implied by such models as the potential death of the universe. Finally, in Sect. 9.7 I review the future observational and theoretical prospects related to the investigation of the physical origin of dark energy and summarize the main conclusions of this review.

9.2 Expansion History
from the Luminosity Distances of SnIa

Consider a luminous cosmological object emitting at total power L (absolute luminosity) in radiation within a particular wavelength band. Consider also an observer (see Fig. 9.1) at a distance d_L from the luminous object. In a static cosmological setup, the power radiated by the luminous object is distributed in the spherical surface with radius d_L and therefore the intensity l (apparent luminosity) detected by the observer is

$$l = \frac{L}{4\pi d_L^2} \, . \tag{9.1}$$

The quantity

$$d_L \equiv \sqrt{\frac{L}{4\pi l}} \tag{9.2}$$

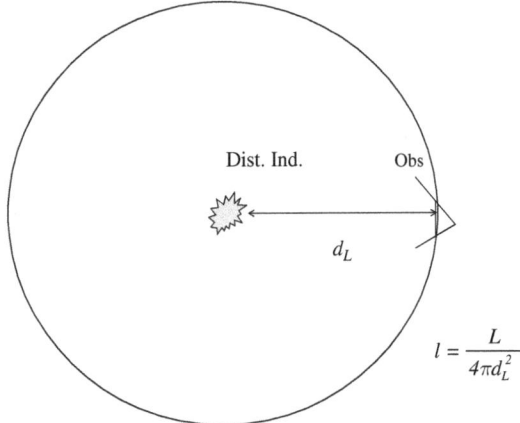

Fig. 9.1. The luminosity distance obtained from the apparent and absolute luminosities

is known as the luminosity distance to the luminous object and in a static universe it coincides with the actual distance. In an expanding universe however, the energy of the radiation detected by the observer has been reduced not only because of the distribution of photons on the spherical surface but also because the energy of the photons has been redshifted while their detection rate is reduced compared to their emission rate due to the cosmological expansion [23]. Both of these expansion effects give a reduction of the detected energy by a factor $\frac{a(t_0)}{a(t)} = (1+z)$ where $a(t)$ is the scale factor of the universe at cosmic time t and t_0 is the present time. Usually a is normalized so that $a(t_0) = 1$. Thus the detected apparent luminosity in an expanding background may be written as

$$l = \frac{L}{4\pi a(t_0)^2 x(z)^2 (1+z)^2} , \tag{9.3}$$

where $x(z)$ is the comoving distance to the luminus object emitting with redshift z. This implies that in an expanding universe the luminosity distance $d_L(z)$ is related to the comoving distance $x(z)$ by the relation

$$d_L(z) = x(z)(1+z) . \tag{9.4}$$

Using (9.4) and the fact that light geodesics in a flat expanding background obey

$$c\,dt = a(z)\,dx(z) \tag{9.5}$$

it is straightforward to eliminate $x(z)$ and express the expansion rate of the universe $H(z) \equiv \frac{\dot{a}}{a}(z)$ at a redshift z (scale factor $a = \frac{1}{1+z}$) in terms of the observable luminosity distance as

$$H(z) = c[\frac{d}{dz}(\frac{d_L(z)}{1+z})]^{-1} . \tag{9.6}$$

This is an important relation that connects the theoretically predictable Hubble expansion history $H(z)$ with the observable luminosity distance $d_L(z)$ in the context of a spatially flat universe. Therefore, if the absolute luminosity of cosmologically distant objects is known and their apparent luminosity is measured as a function of redshift, (9.2) can be used to calculate their luminosity distance $d_L(z)$ as a function of redshift. The expansion history $H(z)$ can then be deduced by differentiation with respect to the redshift using (9.6). Reversely, if a theoretically predicted $H(z)$ is given, the corresponding predicted $d_L(z)$ is obtained from (9.6) by integrating $H(z)$ as

$$d_L(z) = c\,(1+z) \int_0^z \frac{dz'}{H(z')} \; . \tag{9.7}$$

This predicted $d_L(z)$ can be compared with the observed $d_L(z)$ to test the consistency of the theoretical model with observations. In practice astronomers do not refer to the ratio of absolute over apparent luminosity. Instead they use the difference between apparent magnitude m and absolute magnitude M which is connected to the above ratio by the relation

$$m - M = 2.5\,log_{10}(\frac{L}{l}) \; . \tag{9.8}$$

A particularly useful diagram which illustrates the expansion history of the Universe is the *Hubble diagram*. The x-axis of a Hubble diagram (see Fig. 9.2) shows the redshift z of cosmological luminous objects while the y-axis shows the physical distance Δr to these objects.

In the context of a cosmological setup the redshift z is connected to the scale factor $a(t)$ at the time of emission of radiation by $1 + z = a(t_0)/a(t)$ where t_0 is the present time. On the other hand, the distance to the luminous object is related to the time in the past t_{past} when the radiation emission was made. Therefore, the Hubble diagram contains information about the time dependence of the scale factor $a(t)$. The slope of this diagram at a given redshift denotes the inverse of the expansion rate $\frac{\dot{a}}{a}(z) \equiv H(z)$ ie

$$\Delta r = \frac{1}{H(z)}c\,z \; . \tag{9.9}$$

In an accelerating universe the expansion rate $H(z)$ was smaller in the past (high redshift) and therefore the slope H^{-1} of the Hubble diagram is larger at high redshift. Thus, at given redshift, luminous objects appear to be further away (dimmer) compared to an empty universe expanding with a constant rate (see Fig. 9.2).

The luminous objects used in the construction of the Hubble diagram are objects whose absolute luminosity is known and therefore their distance can be evaluated from their apparent luminosity along the lines discussed above. Such objects are known as *distance indicators* or *standard candles*. A list of common distance indicators used in astrophysics and cosmology is shown in

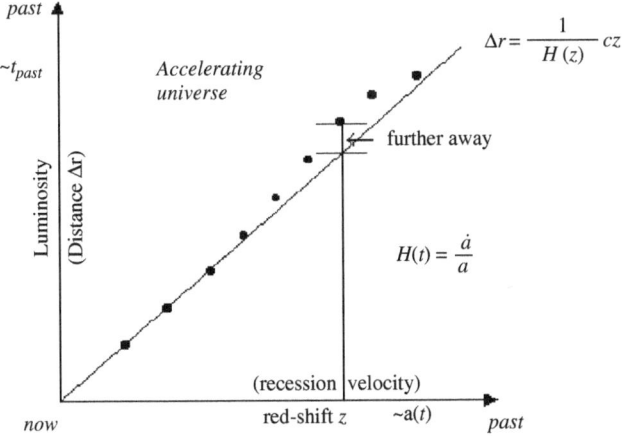

Fig. 9.2. The Hubble diagram. In an accelerating universe luminous objects at a given redshift appear to be dimmer

Table 9.1 along with the range of distances where these objects are visible and the corresponding accuracy in the determination of their absolute magnitude. As shown in Table 9.1 the best choice distance indicators for cosmology are SnIa not only because they are extremely luminous (at their peak they are as luminous as a bright galaxy) but also because their absolute magnitude can be determined at a high accuracy.

Type Ia supernovae emerge in binary star systems where one of the companion stars has a mass below the Chandrasekhar limit $1.4M_\odot$ and therefore ends up (after hydrogen and helium burning) as white dwarf supported by degeneracy pressure. Once the other companion reaches its red giant phase the white dwarf begins gravitational striping of the outer envelop of the red giant thus accreting matter from the companion star. Once the white dwarf reaches a mass equal to the Chandrasekhar limit, the degeneracy pressure is unable

Table 9.1. Extragalactic distance indicators (from [24])

Technique	Range of distance	Accuracy (1σ)
Cepheids	< LMC to 25 Mpc	0.15 mag
SNIa	4 Mpc to > 2 Gpc	0.2 mag
Expand. Phot. Meth./SnII	LMC to 200 Mpc	0.4 mag
Planetary Nebulae	LMC to 20 Mpc	0.1 mag
Surf. Brightness Fluct	1 Mpc to 100 Mpc	0.1 mag
Tully Fisher	1 Mpc to 100 Mpc	0.3 mag
Brightest Cluster Gal.	50 Mpc to 1 Gpc	0.3 mag
Glob. Cluster Lum. Fun.	1 Mpc to 100 Mpc	0.4 mag
Sunyaev-Zeldovich	100 Mpc to > 1 Gpc	0.4 mag
Gravitational Lensing	5 Gpc	0.4 mag

to support the gravitational pressure, the white dwarf shrinks and increases its temperature igniting carbon fussion. This leads to violent explosion which is detected by a light curve which rapidly increases luminosity in a time scale of less than a month, reaches a maximum and disappears in a timescale of 1-2 months (see Fig. 9.3). Type Ia are the preferred distance indicators for cosmology for several reasons:

1. They are exceedingly luminous. At their peak luminosity they reach an absolute magnitude of $M \simeq -19$ which corresponds to about $10^{10} M_\odot$.
2. They have a relatively small dispersion of peak absolute magnitude.
3. Their explosion mechanism is fairly uniform and well understood.
4. There is no cosmic evolution of their explosion mechanism according to known physics.
5. There are several local SnIa to be used for testing SnIa physics and for calibrating the absolute magnitude of distant SnIa.

On the other hand, the main problem for using SnIa as standard candles is that they are not easy to detect and it is impossible to predict a SnIa explosion. In fact the expected number of SnIa exploding per galaxy is 1-2 per millenium. It is therefore important to develop a *search strategy* in order to efficiently search for SnIa at an early stage of their light curve. The method used (with minor variations) to discover and follow up photometrically and spectroscopically SnIa consists of the following steps [1, 2, 3, 4]:

1. Observe a number of wide fields of apparently empty sky out of the plane of our Galaxy. Tens of thousands of galaxies are observed in a few patches of sky.
2. Come back three weeks later (next new moon) to observe the same galaxies over again.

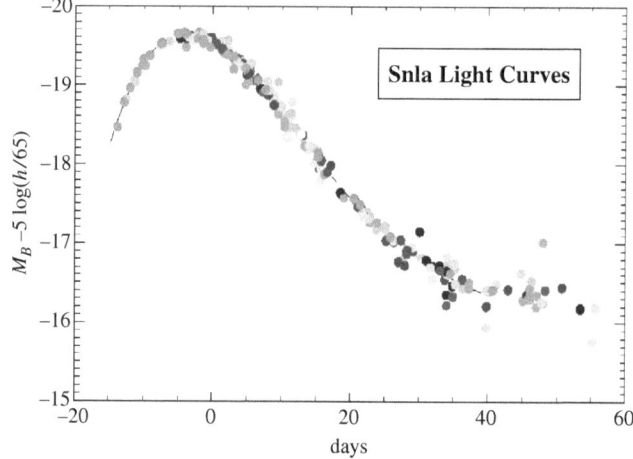

Fig. 9.3. Typical SnIa light-curve

3. Subtract images to identify on average 12–14 SnIa.
4. Schedule in advance follow up photometry and spectroscopy on these SnIa as they brighten to peak and fade away.

Given the relatively short time difference (three weeks) between first and second observation, most SnIa do not have time to reach peak brightness so almost all the discoveries are pre-maximum. This strategy turns a rare, random event into something that can be studied in a systematic way. This strategy is illustrated in Figs. 9.4 and 9.5 (from [25]). The outcome of this observation strategy is a set of SnIa light curves in various bands of the spectrum (see Fig. 9.6). These light curves are very similar to each other and their peak apparent luminosity could be used to construct the Hubble diagram assuming a common absolute luminosity.

Before this is done however a few corrections must be made to take into account the minor intrinsic absolute luminosity differences (due to composition differences) among SnIa as well as the radiation extinction due to the intergalactic medium. Using samples of closeby SnIa it has been empirically observed that the minor differences of SnIa absolute luminosity are connected with differences in the shape of their light curves. Broad slowly declining light curves (stretch factor $s > 1$) correspond to brighter SnIa while narrower rapidly declining light curves (stretch factor $s < 1$) correspond to intrinsically fainter SnIa.

This stretch factor dependence of the SnIa absolute luminosity has been verified using closeby SnIa [27] It was shown that contraction of broad light

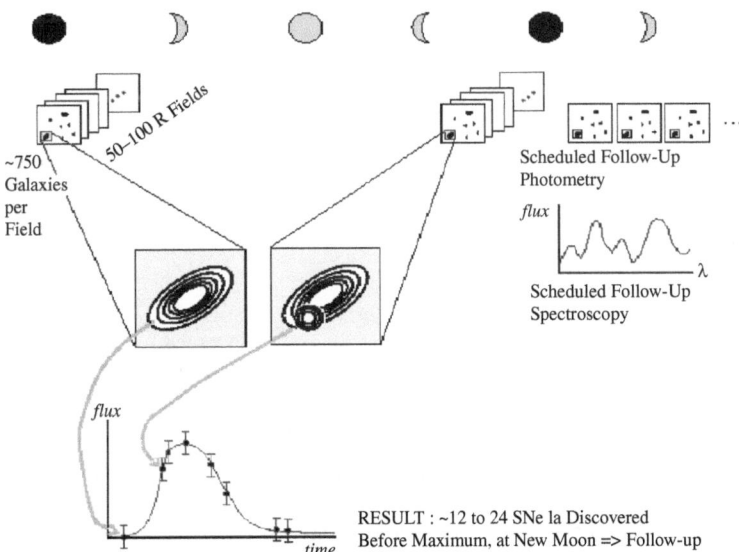

Fig. 9.4. Search strategy to discover of supernovae in a scheduled, systematic procedure [25]

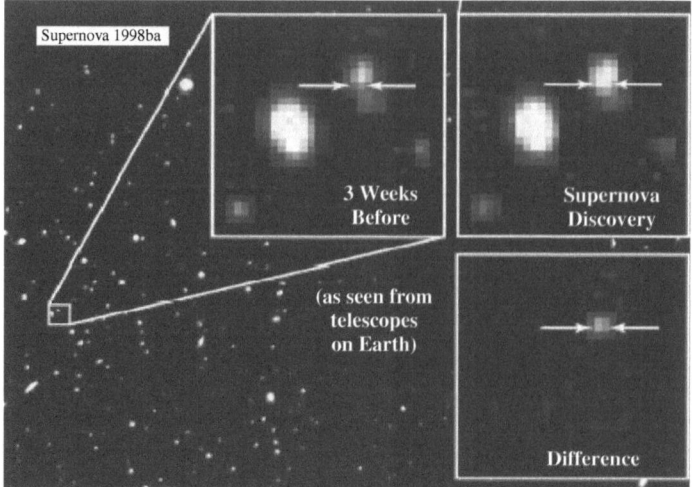

Fig. 9.5. Supernova 1997cj, an example of a supernova discovery using the search strategy described in the text involving subtraction of images

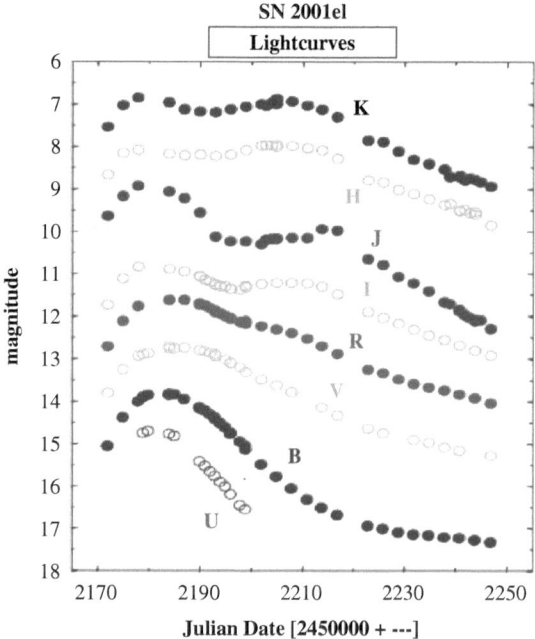

Fig. 9.6. A set of light curves from SN2001el in various bands of the spectrum

curves while reducing peak luminosity and stretching narrow light curves while increasing peak luminosity makes these light curves coincide (see Fig. 9.7).

In addition to the stretch factor correction an additional correction must be made in order to compare the light curves of high redshift SnIa with those

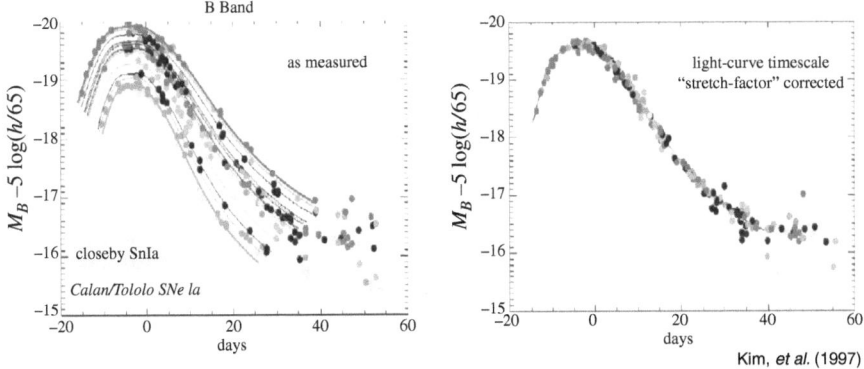

Fig. 9.7. Left: The range of lightcurve for low-redshift supernovae discovered by the Calan/Tololo Supernova Survey. At these redshifts, the relative distances can be determined (from redshift), so their relative brightnesses are known. **Right**: The same lightcurves after calibrating the supernova brightness using the stretch of the timescale of the lightcurve as an indicator of brightness (and the color at peak as an indicator of dust absorption)

of lower redshift. In particular all light curves must be transformed to the same reference frame and in particular the rest frame of the SnIa. For example a low redshift light curve of the blue B band of the spectrum should be compared with the *appropriate* red R band light curve of a high redshift SnIa. The transformation also includes correction for the cosmic time dilation (events at redshift z last $1 + z$ times longer than events at $z \simeq 0$).

These corrections consist the *K-correction* and is used in addition to the stretch factor correction discussed above. The K-correction transformation is illustrated in Fig. 9.8.

9.3 Observational Results

The first project in which SnIa were used to determine the cosmological constant energy was the research from Perlmutter et al. in 1997 [27]. The project was known as the Supernova Cosmology Project (SCP). Applying the above described methods they discovered seven distant SnIa at redshift $0.35 < z < 0.65$. When discovered, the supernovae were followed for a year by different telescopes on earth to obtain good photometry data in different bands, in order to measure good magnitudes. The Hubble diagram they constructed was consistent with standard Friedman cosmology without dark energy or cosmological constant.

A year after their first publication, Perlmutter et al. published in Nature [1] an update on their initial results. They had included the measurements of a very high-redshifted z = 0.83 Supernova Ia. This dramatically changed their

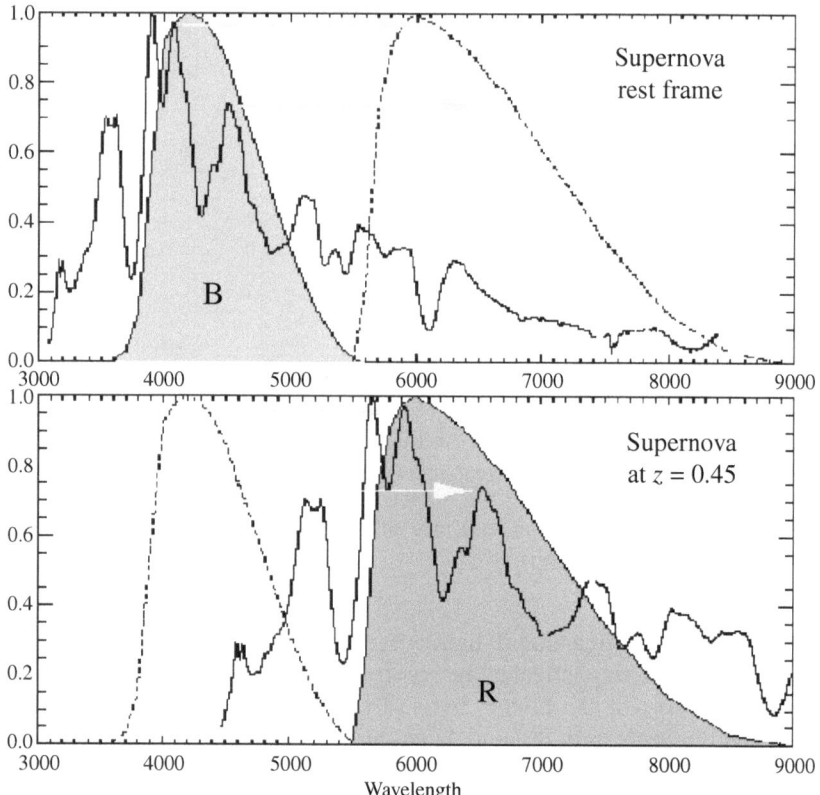

Fig. 9.8. Slightly different parts of the supernova spectrum are observed through the B filter transmission function at low redshift (*upper panel*) and through the R filter transmission function at high redshift (*lower panel*). This small difference is accounted for by the "cross-filter K-correction"[26]

conlusions. The standard decelerating Friedman cosmology was rulled out at about 99% confidence level. The newly discovered Supernova indicated a universe with accelerating expansion dominated by dark energy. These results were confirmed independently by another pioneer group (High-z Supernova Search Team (HSST)) searching for SnIa and measuring the expansion history $H(z)$ (Riess et al. in 1998 [2]). They had discovered 16 SnIa at $0.16 < z < 0.62$ and their $H(z)$ also indicated accelerating expansion ruling out for a flat universe. Their data also permitted them to definitely rule out decelerating Friedman cosmology at about 99% confidence level.

In 2003 Tonry et al. [3] reported the results of their observations of eight newly discovered SnIa. These SnIa were found in the region $0.3 < z < 1.2$. Together with previously acquired SnIa data they had a data set of more than 100 SnIa. This dataset confirmed the previous findings of accelerated expansion and gave the first hints of decelerated expansion at redshifts $z \overset{>}{\sim} 0.6$ when matter is expected to begin dominating over dark energy. This transition from decelerating to accelerating expansion was confirmed and pinpointed accurately by Riess et al. in 2004 [5] who included in the analysis 16 new high-redshift SnIa obtained with HST and reanalyzed all the available data in a uniform and robust manner constructing a robust and reliable dataset consisting of 157 points known as the Gold dataset. These SnIa included 6 of the 7 highest redshift SnIa known with $z > 1.25$. With these new observations, they could clearly identify the transition from a decelerating towards an accelerating universe to be at $z = 0.46 \pm 0.13$. It was also possible to rule out the effect of dust on the dimming of distant SnIa, since the accelerating/decelerating transition makes the effect of dimming inverse. The Hubble diagram obtained from the Gold dataset is shown in Fig. 9.9 where the corrected apparent magnitude $m(z)$ of the 157 SnIa is plotted versus the redshift z. The apparent magnitude $m(z)$ is related to the corresponding luminosity distance d_L of the SnIa by

$$m(z) = M + 5 log_{10}[\frac{d_L(z)}{Mpc}] + 25 , \qquad (9.10)$$

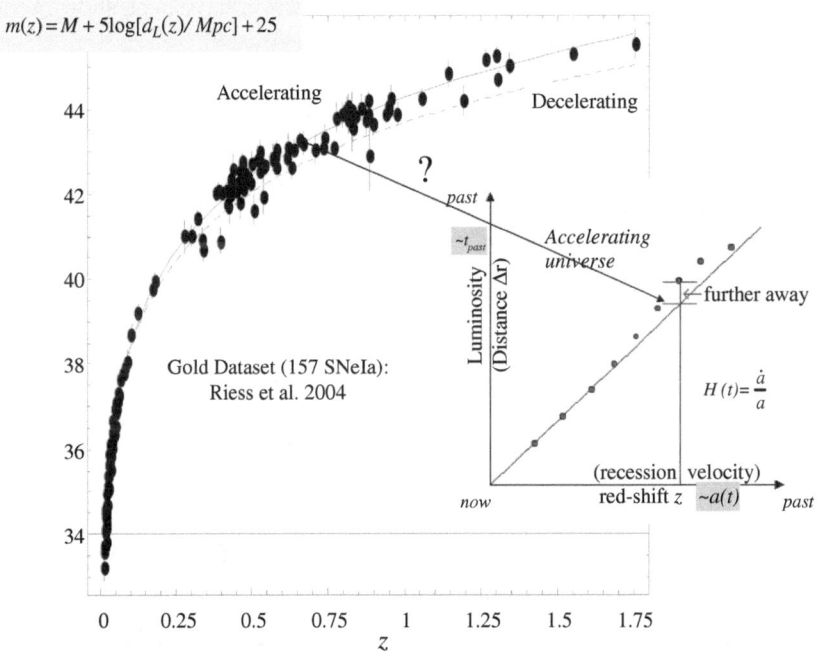

Fig. 9.9. The apparent magnitude $m(z)$ vs redshift as obtained from the Gold dataset. It is not easy to distinguish between accelerating and decelerating expansion in such a diagram

where M is the absolute magnitude which is assumed to be constant for standard candles like SnIa after the corrections discussed in Sect. 9.2 are implemented.

A potential problem of plots like the one of Fig. 9.9 is that it is not easy to tell immediately if the data favor an accelerating or decelerating universe. This would be easy to tell in the Hubble diagram of Fig. 9.2 where the distance is plotted vs redshift and is superposed with the distance-redshift relation $d_L^{empty}(z)$ of an empty universe with $H(z)$ constant. An even more efficient plot for such a purpose would be the plot of the ratio $d_L(z)/d_L^{empty}(z)$ (or its log_{10}) which can immediately distinguish accelerating from decelerating expansion by comparing with the $d_L(z)/d_L^{empty}(z) = 1$ line. Such a plot is shown in Fig. 9.10 [5] using both the raw Gold sample data and the same data binned in redshift bins.

The lines of zero acceleration, constant acceleration and constant deceleration are also shown for comparison. Clearly the best fit is obtained by an expansion which is accelerating at recent times ($z \lesssim 0.5$) and decelerating at earlier times ($z \gtrsim 0.5$) when matter is expected to dominate.

The interpretation of the data assuming that the observed dimming at high redshift is due to larger distance may not be the only possible interpretation. The most natural alternative interpretations however have been shown to lead

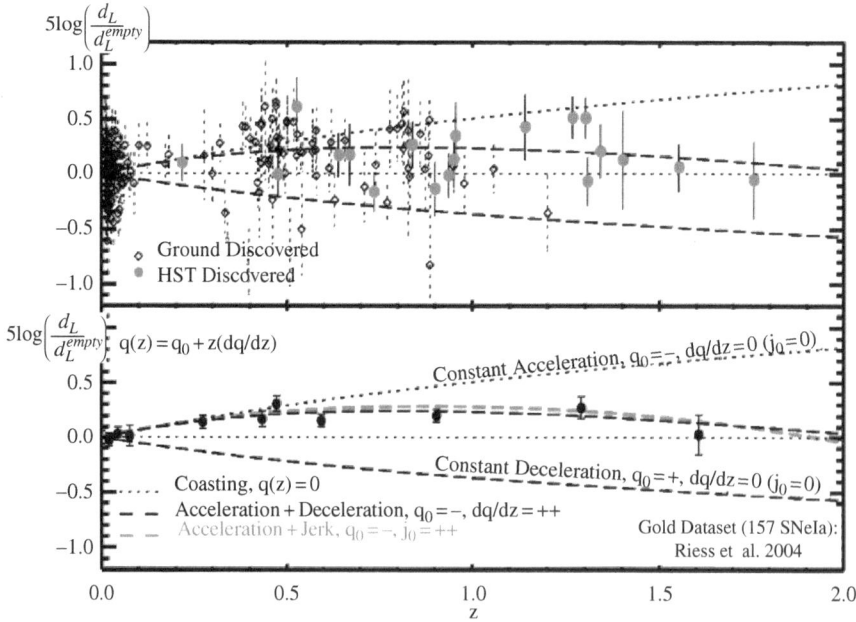

Fig. 9.10. The reduced Hubble diagram used to distinguish between accelerating and decelerating expansion [5]

to inconsistencies and none of them has been favored as a viable alternative at present. These alternative interpretations include the following:

- **Intergalactic Dust:** Ordinary astrophysical dust does not obscure equally at all wavelengths, but scatters blue light preferentially, leading to the well-known phenomenon of "reddening". Spectral measurements [5] reveal a negligible amount of reddening, implying that any hypothetical dust must be a novel "grey" variety inducing no spectral distortions [28].
- **Grey Dust:** Grey dust is highly constrained by observations: first, it predicts further increase of dimming at higher redshifts $z \gtrsim 0.5$ which is not observed; and second, intergalactic dust would absorb ultraviolet/optical radiation and re-emit it at far infrared wavelengths, leading to stringent constraints from observations of the cosmological far-infrared background. Thus, while the possibility of obscuration has not been entirely eliminated, it requires a novel kind of dust which is already highly constrained (and may be convincingly ruled out by further observations).
- **Evolution of SnIa:** The supernova search teams have found consistency in the spectral and photometric properties of SnIa over a variety of redshifts and environments [5] (e.g. in elliptical vs. spiral galaxies). Thus despite the relevant tests there is currently no evidence that the observed dimming can be attributed to evolution of SnIa.

According to the best of our current understanding, the supernova results indicating an accelerating universe seem likely to be trustworthy. Needless to say, however, the possibility of a neglected systematic effect can not be definitively excluded. Future experiments, discussed in Sect. 9.7 will both help us improve our understanding of the physics of supernovae and allow a determination of the distance/redshift relation to sufficient precision to distinguish between the effects of an accelerating universe and those of possible astrophysical phenomena.

9.4 Dark Energy and Negative Pressure

Our current knowledge of the expansion history of the universe can be summarized as follows: The universe originated at an initial state that was very close to a density singularity known as the Big Bang. Soon after that it entered a phase of superluminal accelerating expansion known as inflation. During inflation causally connected regions of the universe exited out of the horizon, the universe approached spatial flatness and the primordial fluctuations that gave rise to structure were generated. At the end of inflation the universe was initially dominated by radiation and later by matter whose attractive gravitational properties induced a decelerating expansion.

The SnIa data discussed in Sect. 9.3 (along with other less direct cosmological observations [8, 9]) strongly suggest that the universe has recently entered a phase of accelerating expansion at a redshift $z \simeq 0.5$. This accelerating expansion can not be supported by the attractive gravitational properties of regular matter. The obvious question to address is therefore 'What are the properties of the additional component required to support this acceleration?'. To address this question we must consider the dynamical equation that determines the evolution of the scale factor $a(t)$. This equation is the Friedman equation which is obtained by combining General Relativity with the cosmological principle of homogeneity and isotropy of the universe. It may be written as

$$\frac{\ddot{a}}{a} = -\frac{4\pi G}{3} \sum_i (\rho_i + 3p_i) = -\frac{4\pi G}{3} [\rho_m + (\rho_X + 3p_X)], \qquad (9.11)$$

where ρ_i and p_i are the densities and pressures of the contents of the universe assumed to behave as ideal fluids. The only directly detected fluids in the universe are matter $(\rho_m, p_m = 0)$ and the subdominant radiation $(\rho_r, p_r = \rho_r/3)$. Both of these fluids are unable to cancel the minus sign on the rhs of the Friedman equation and can therefore only lead to decelerating expansion. Accelerating expansion in the context of general relativity can only be obtained by assuming the existence of an additional component $(\rho_X, p_X = w\rho_X)$ termed 'dark energy' which could potentially change the minus sign of (9.11) and thus lead to accelerating expansion. Assuming a positive energy density for dark energy (required to achieve flatness) it becomes clear that negative pressure is required for accelerating expansion. In fact, writing the Friedman (9.11) in terms of the dark energy equation of state parameter w as

$$\frac{\ddot{a}}{a} = -\frac{4\pi G}{3} [\rho_m + \rho_X(1 + 3w)] \qquad (9.12)$$

it becomes clear that a $w < -\frac{1}{3}$ is required for accelerating expansion implying repulsive gravitational properties for dark energy.

The redshift dependence of the dark energy can be easily connected to the equation of state parameter w by combining the energy conservation $d(\rho_X a^3) = -p_x d(a^3)$ with the equation of state $p_X = w\rho_X$ as

$$\rho_X \sim a^{-3(1+w)} = (1 + z)^{3(1+w)}. \qquad (9.13)$$

This redshift dependence is related to the observable expansion history $H(z)$ through the Friedman equation

$$H(z)^2 = \frac{\dot{a}^2}{a^2} = \frac{8\pi G}{3} [\rho_{0m}(\frac{a_0}{a})^3 + \rho_X(a)] = H_0^2 [\Omega_{0m}(1 + z)^3 + \Omega_X(z)] \quad (9.14)$$

where the density parameter $\Omega \equiv \rho/\rho_{0crit}$ for matter is constrained by large scale structure observations to a value (prior) $\Omega_{0m} \simeq 0.3$. Using this prior, the dark energy density parameter $\Omega_X(z) \equiv \rho_X(z)/\rho_{0crit}$ and the corresponding equation of state parameter w may be constrained from the observed $H(z)$.

In addition to $\Omega_X(z)$, the luminosity distance-redshift relation $d_L(z)$ obtained from SnIa observations can constrain other cosmological parameters. The only parameter however obtained directly from $d_L(z)$ (using (9.6)) is the Hubble parameter $H(z)$. Other cosmological parameters can be obtained from $H(z)$ as follows:

– The age of the universe t_0 is obtained as:

$$t_0 = \int_0^\infty \frac{dz}{(1+z)H(z)} \, . \tag{9.15}$$

– The present Hubble parameter $H_0 = H(z = 0)$.
– The deceleration parameter $q(z) \equiv \ddot{a}a/\dot{a}^2$

$$q(z) = (1+z)\frac{dlnH}{dz} - 1 \tag{9.16}$$

and its present value $q_0 \equiv q(z = 0)$.
– The density parameters for matter and dark energy are related to $H(z)$ through the Friedman (9.14).
– The equation of state parameter $w(z)$ obtained as [29, 30]

$$w(z) = \frac{p_X(z)}{\rho_X(z)} = \frac{\frac{2}{3}(1+z)\frac{d \ln H}{dz} - 1}{1 - (\frac{H_0}{H})^2\Omega_{0m}(1+z)^3} \tag{9.17}$$

obtained using the Friedman (9.12) and (9.14).

The most interesting parameter from the theoretical point of view (apart from $H(z)$ itself) is the dark energy equation of state parameter $w(z)$. This parameter probes directly the gravitational properties of dark energy which are predicted by theoretical models. The downside of it is that it requires two differentiations of the observable $d_L(z)$ to be obtained and is therefore very sensitive to observational errors.

The simplest form of dark energy corresponds to a time independent energy density obtained when $w = -1$ (see (9.13)). This is the well known cosmological constant which was first introduced by Einstein in 1917 two years after the publication of the General Relativity (GR) equation

$$G_{\mu\nu} = \kappa T_{\mu\nu} \, , \tag{9.18}$$

where $\kappa = 8\pi G/c^2$. At the time the 'standard' cosmological model was a static universe because the observed stars of the Milky Way were found to have negligible velocities. The goal of Einstein was to apply GR in cosmology and obtain a static universe using matter only. It became clear that the attractive gravitational properties of matter made it impossible to obtain a static cosmology from (9.18). A repulsive component was required and at the time of major revolutions in the forms of physical laws it seemed more natural to obtain it by modifying the gravitational law than by adding new forms of energy density. The simplest generalization of (9.18) involves the introduction of a term proportional to the metric $g_{\mu\nu}$. The GR equation becomes

$$G_{\mu\nu} - \Lambda g_{\mu\nu} = \kappa T_{\mu\nu} \, , \tag{9.19}$$

where Λ is the cosmological constant. The repulsive nature of the cosmological constant becomes clear by the metric of a point mass (Schwarschild-de Sitter metric) which, in the Newtonian limit leads to a gravitational potential

$$V(r) = -\frac{GM}{r} - \frac{\Lambda r^2}{6} \, , \tag{9.20}$$

which in addition to the usual attractive gravitational term has a repulsive term proportional to the cosmological constant Λ. This repulsive gravitational force can lead to a static (but unstable) universe in a cosmological setup and in the presence of a matter fluid. A few years after the introduction of the cosmological constant by Einstein came Hubble's discovery that the universe is expanding and it became clear that the cosmological constant was an unnecessary complication of GR. It was then that Einstein (according to Gamow's autobiography) called the introduction of the cosmological constant *'the biggest blunder of my life'*. In a letter to Lemaitre in 1947 Einstein wrote: *'Since I introduced this term I had always had a bad conscience. I am unable to believe that such an ugly thing is actually realized in nature'*. As discussed below, there is better reason than ever before to believe that the cosmological constant may be non-zero, and Einstein may not have blundered after all.

If the cosmological constant is moved to the right hand side of (9.19) it may be incorporated in the energy momentum tensor as an ideal fluid with $\rho_\Lambda = \Lambda/8\pi G$ and $w = -1$. In the context of field theory such an energy momentum tensor is obtained by a scalar field ϕ with potential $V(\phi)$ at its vacuum state ϕ_0 i.e. $\partial_\mu\phi = 0$ and $T_{\mu\nu} = -V(\phi_0)g_{\mu\nu}$. Even though the cosmological constant may be physically motivated in the context of field theory and consistent with cosmological observation there are two important problems associated with it:

– *Why is it so incredibly small?* Observationally, the cosmological constant density is 120 orders of magnitude smaller than the energy density associated with the Planck scale – the obvious cut off. Furthermore, the standard model of cosmology posits that very early on the universe experienced a period of inflation: A brief period of very rapid acceleration, during which the Hubble constant was about 52 orders of magnitude larger than the value observed today. How could the cosmological constant have been so large then, and so small now? This is sometimes called *the cosmological constant problem*.
– *The 'coincidence problem':* Why is the energy density of matter nearly equal to the dark energy density today?

Despite the above problems and given that the cosmological constant is the simplest dark energy model, it is important to investigate the degree to which it is consistent with the SnIa data. I will now describe the main steps involved in this analysis. According to the Friedman equation the predicted Hubble expansion in a flat universe and in the presence of matter and a cosmological constant is

$$H(z)^2 = \frac{\dot{a}^2}{a^2} = \frac{8\pi G}{3}\rho_{0m}(\frac{a_0}{a})^3 + \frac{\Lambda}{3} = H_0^2[\Omega_{0m}(1+z)^3 + \Omega_\Lambda] , \quad (9.21)$$

where $\Omega_\Lambda = \rho_\Lambda / \rho_{0crit}$ and

$$\Omega_{0m} + \Omega_\Lambda = 1 . \quad (9.22)$$

This is the LCDM (Λ+Cold Dark Matter) which is currently the minimal standard model of cosmology. The predicted $H(z)$ has a single free parameter which we wish to constrain by fitting to the SnIa luminosity distance-redshift data.

Observations measure the apparent luminosity vs redshift $l(z)$ or equivalently the apparent magnitude vs redshift $m(z)$ which are related to the luminosity distance by

$$2.5log_{10}(\frac{L}{l(z)}) = m(z) - M - 25 = 5log_{10}(\frac{d_L(z)_{obs}}{Mpc}) . \quad (9.23)$$

From the theory point of view the predicted observable is the Hubble parameter (9.21) which is related to the theoretically predicted luminosity distance $d_L(z)$ by (9.7). In this case $d_L(z)$ depends on the single parameter Ω_{0m} and takes the form

$$d_L(z; \Omega_{0m})_{th} = c (1+z) \int_0^z \frac{dz'}{H(z'; \Omega_{0m})} . \quad (9.24)$$

Constraints on the parameter Ω_{0m} are obtained by the maximum likelihood method [31] which involves the minimization of the $\chi^2(\Omega_{0m})$ defined as

$$\chi^2(\Omega_{0m}) = \sum_{i=1}^N \frac{[d_L(z)_{obs} - d_L(z; \Omega_{0m})_{th}]^2}{\sigma_i^2} , \quad (9.25)$$

where N is the number of the observed SnIa luminosity distances and σ_i are the corresponding 1σ errors which include errors due to flux uncertainties, internal dispersion of SnIa absolute magnitude and peculiar velocity dispersion. If flatness is not imposed as a prior through (9.22) then $d_L(z)_{th}$ depends on two parameters (Ω_{0m} and Ω_Λ) and the relation between $d_L(z; \Omega_{0m}, \Omega_\Lambda)_{th}$ and $H(z; \Omega_{0m}, \Omega_\Lambda)$ takes the form

$$d_L(z)_{th} = \frac{c(1+z)}{\sqrt{\Omega_{0m} + \Omega_\Lambda - 1}} sin[\sqrt{\Omega_{0m} + \Omega_\Lambda - 1} \int_0^z dz' \frac{1}{H(z)}] . \quad (9.26)$$

In this case the minimization of (9.25) leads to constraints on both Ω_{0m} and Ω_Λ. This is the only direct and precise observational probe that can place constraints directly on Ω_Λ. Most other observational probes based on large scale structure observations place constraints on Ω_{0m} which are indirectly related to Ω_Λ in the context of a flatness prior.

As discussed in Sect. 9.2 the acceleration of the universe has been confirmed using the above maximum likelihood method since 1998 [1, 2]. Even the early datasets of 1998 [1, 2] were able to rule out the flat matter dominated universe (SCDM: $\Omega_{0m} = 1$, $\Omega_\Lambda = 0$) at 99% confidence level. The latest datasets are the Gold dataset ($N = 157$ in the redshift range $0 < z < 1.75$) discussed in Sect. 9.2 and the first year SNLS (Supernova Legacy Survey) dataset which consists of 71 datapoints in the range $0 < z < 1$ plus 44 previously published closeby SnIa. The 68% and 95% χ^2 contours in the (Ω_{0m} and Ω_Λ) parameter space obtained using the maximum likelihood method are shown in Fig. 9.11 for the SNLS dataset, a truncated version of the Gold dataset (TG) with $0 < z < 1$ and the Full Gold (FG) dataset. The following comments can be made on these plots:

– The two versions of the Gold dataset favor a closed universe instead of a flat universe ($\Omega_{tot}^{TG} = 2.16 \pm 0.59$, $\Omega_{tot}^{FG} = 1.44 \pm 0.44$). This trend is not realized by the SNLS dataset which gives $\Omega_{tot}^{SNLS} = 1.07 \pm 0.52$.
– The point corresponding to SCDM ($\Omega_{0m}, \Omega_\Lambda) = (1, 0)$ is ruled out by all datasets at a confidence level more than 10σ.
– If we use a prior constraint of flatness $\Omega_{0m} + \Omega_\Lambda = 1$ thus restricting on the corresponding dotted line of Fig. 9.1 and using the parametrization

$$H(z)^2 = H_0^2[\Omega_{0m}(1 + z)^2 + (1 - \Omega_{0m})] \tag{9.27}$$

we find minimizing $\chi^2(\Omega_{0m})$ of (9.25)

$$\Omega_{0m}^{SNLS} = 0.26 \pm 0.04 , \tag{9.28}$$

$$\Omega_{0m}^{TG} = 0.30 \pm 0.05 , \tag{9.29}$$

$$\Omega_{0m}^{FG} = 0.31 \pm 0.04 . \tag{9.30}$$

These values of Ω_{0m} are consistent with corresponding constraints from the CMB [8] and large scale structure observations [9].

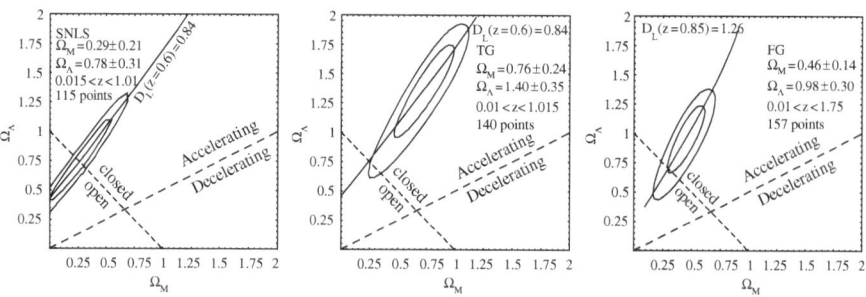

Fig. 9.11. The 68% and 95% χ^2 contours in the (Ω_{0m} and Ω_Λ) parameter space obtained using the SNLS, TG and FG datasets (from [32])

Even though LCDM is the simplest dark energy model and is currently consistent with all cosmological observations (especially with the SNLS dataset) the question that may still be address is the following: 'Is it possible to get better fits (lowering χ^2 further) with different $H(z)$ parametrizations and if yes what are the common features of there better fits?' The strategy towards addressing this question involves the following steps:

- Consider a physical model and extract the predicted recent expansion history $H(z; a_1, a_2, ..., a_n)$ as a function of the model parameters $a_1, a_2, ..., a_n$. Alternatively a model independent parametrization for $H(z; a_1, a_2, ..., a_n)$ (or equivalently $w(z; a_1, a_2, ..., a_n)$) may be constructed aiming at the best possible fit to the data with a small number of parameters (usually 3 or less).
- Use (9.7) to obtain the theoretically predicted luminosity distance as a function of z, $d_L(z; a_1, a_2, ..., a_n)_{th}$.
- Use the observed luminosity distances $d_L(z_i)_{obs}$ to construct χ^2 along the lines of (9.25) and minimize it with respect to the parameters $a_1, a_2, ..., a_n$.
- From the resulting best fit parameter values $\bar{a}_1, \bar{a}_2, ..., \bar{a}_n$ (and their error bars) construct the best fit $H(z; \bar{a}_1, \bar{a}_2, ..., \bar{a}_n)$, $d_L(z; \bar{a}_1, \bar{a}_2, ..., \bar{a}_n)$ and $w(z; \bar{a}_1, ..., \bar{a}_n)$. The quality of fit is measured by the depth of the minimum of $\chi^2_{min}(\bar{a}_1, ..., \bar{a}_n)$.

Most useful parametrizations reduce to LCDM of (9.21) for specific parameter values giving a χ^2_{LCDM} for these parameter values. Let

$$\Delta\chi^2_{LCDM} \equiv \chi^2_{min}(\bar{a}_1, \bar{a}_2, ..., \bar{a}_n) - \chi^2_{LCDM} . \tag{9.31}$$

The value of $\Delta\chi^2_{LCDM}$ is usually negative since χ^2 is usually further reduced due to the larger number of parameters compared to LCDM. For a given number of parameters the value of $\Delta\chi^2_{LCDM}$ gives a measure of the probability of having LCDM physically realized in the context of a given parametrization [33]. The smaller this probability is, the more 'superior' this parametrization is compared to LCDM. For example for a two parameter parametrization and $|\Delta\chi^2_{LCDM}| > 2.3$ the parameters of LCDM are more than 1σ away from the best fit parameter values of the given parametrization. This statistical test has been quantified in [33] and applied to several $H(z)$ parametrizations.

As an example let us consider the two parameter polynomial parametrization allowing for dark energy evolution

$$H(z)^2 = H_0^2[\Omega_{0m}(1+z)^3 + a_2(1+z)^2 + a_1(1+z) + (1-a_2-a_1-\Omega_{0m})] \tag{9.32}$$

in the context of the Full Gold dataset. Applying the above described χ^2 minimization leads to the best fit parameter values $a_1 = 1.67 \pm 1.03$ and $a_2 = -4.16 \pm 2.53$. The corresponding $|\Delta\chi^2_{LCDM}|$ is found to be 2.9 which implies that the LCDM parameters values ($a_1 = a_2 = 0$) are in the range of $1\sigma - 2\sigma$ away from the best fit values.

The same analysis can be repeated for various different parametrizations in an effort to identify the common features of the best fit parametrizations. For example two other dynamical dark energy parametrizations used commonly in the literature are defined in terms of $w(z)$ as

– Parametrization A:

$$w(z) \quad = \quad w_0 + w_1 \, z \,, \tag{9.33}$$
$$H^2(z) \quad = \quad H_0^2 [\Omega_{0m}(1+z)^3 + \tag{9.34}$$
$$+ (1 - \Omega_{0m})(1+z)^{3(1+w_0-w_1)} e^{3w_1 z}] \,.$$

– Parametrization B:

$$w(z) = w_0 + w_1 \frac{z}{1+z} \,, \tag{9.35}$$
$$H^2(z) = H_0^2 [\Omega_{0m}(1+z)^3 +$$
$$+(1 - \Omega_{0m})(1+z)^{3(1+w_0+w_1)} e^{3w_1[1/(1+z)-1]}] \,. \tag{9.36}$$

where the corresponding forms of $H(z)$ are derived using (9.17). The best fit forms of $w(z)$ obtained from a variety of these and other parametrizations [33] in the context of the Full Gold dataset are shown in Fig. 9.12.

Even though these best fit forms appear very different at redshifts $z > 0.5$ (mainly due to the two derivatives involved in obtaining $w(z)$ from $d_L(z)$), in the range $0 < z < 0.5$ they appear to have an interesting common feature: they all cross the line $w = -1$ also known as the Phantom Divide Line (PDL). As discussed in the next section this feature is difficult to reproduce in most theoretical models based on minimally coupled scalar fields and therefore if

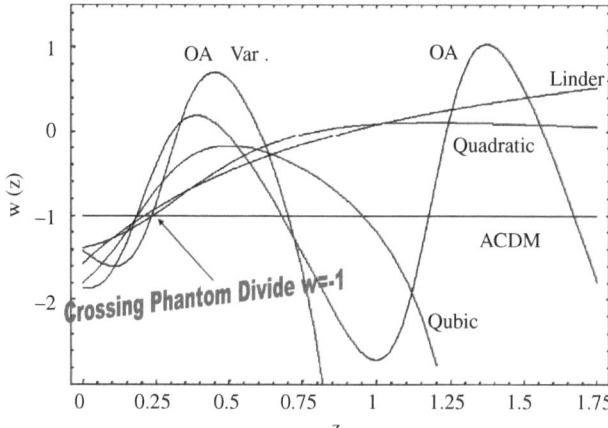

Fig. 9.12. The best fit forms of $w(z)$ obtained from a variety of parametrizations [33] in the context of the Full Gold dataset. Notice that they all cross the line $w = -1$ also known as the Phantom Divide Line (PDL)

it persisted in other independent datasets it could be a very useful tool in discriminating among theoretical models. Unfortunately if the same analysis is made in the context of the more recent SNLS dataset it seems that this common feature does not persist. In Fig. 9.13 the best fit $w(z)$ (along with the 1σ error region) is shown in the context of three different datasets (in analogy with Fig. 9.11) for the there different parametrizations (A, B and polynomial of (9.32) (called C in Fig. 9.13)). Even though the crossing of the PDL is realized at best fit for both the FG and TG datasets it is not realized at best fit when the SNLS is used. Thus we must wait until further SnIa datasets are released before the issue is settled. In Fig. 9.14 I show the 1σ and 2σ χ^2 contours corresponding to parametrizations A and B with a prior of $\Omega_{0m} = 0.24$ confirming the fact that the SNLS dataset provides best fit

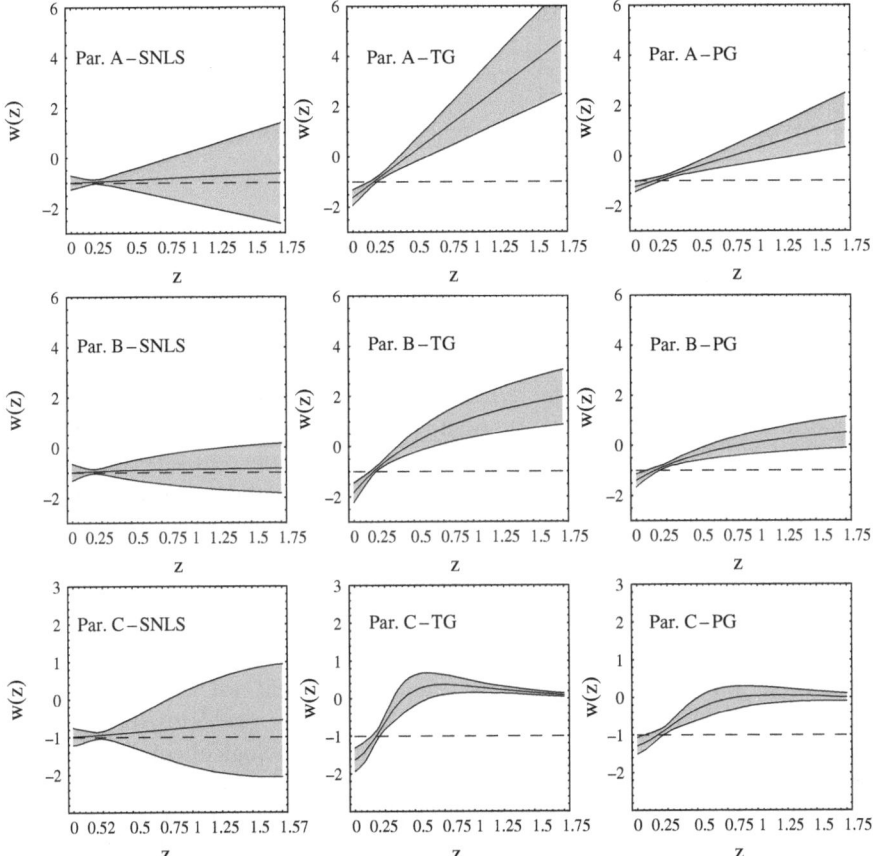

Fig. 9.13. The best fit $w(z)$ (along with the 1σ error (*shaded region*)) is shown in the context of three different datasets (in analogy with Fig. 11) for there different parametrizations (A, B and C) [32]

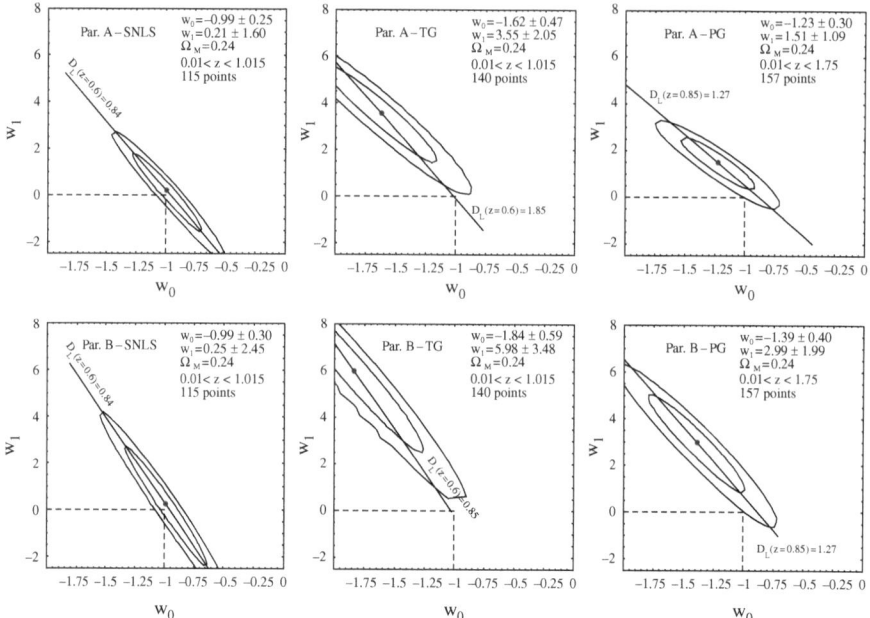

Fig. 9.14. The 1σ and 2σ χ^2 contours corresponding to parametrizations A and B with a prior of $\Omega_{0m} = 0.24$. Notice that the SNLS dataset provides best fit parameter values that are almost identical to those corresponding to LCDM ($w_0 = -1$, $w_1 = 0$)

parameter values that are almost identical to those corresponding to LCDM ($w_0 = -1$, $w_1 = 0$) despite the dynamical degrees of freedom incorporated in the parametrizations A and B. It should be pointed out however that despite the differences in the best fit parametrizations, the three datasets (SNLS, TG and FG) are consistent with each other at the 95% confidence range (see e.g. Fig. 9.14) and they are all consistent with flat LCDM with $\Omega_{0m} \simeq 0.3$.

9.5 Dynamical Evolution of Dark Energy

Even though LCDM is the simplest model consistent with current cosmological data it is plagued with theoretical fine tuning problems discussed in the previous section (the 'coincidence' and the 'cosmological constant' problems). In additions dynamical dark energy parametrizations of $H(z)$ provide in certain cases significantly better fits to the SnIa data. Therefore the investigation of physically motivated models that predict a dynamical evolution of dark energy is an interesting and challenging problem (see also M. Sami's contribution in this volume of models of dark energy).

The role of dark energy can be played by any physical field with positive energy and negative pressure which violates the strong energy condition

$\rho + 3p > 0$ $(w > -\frac{1}{3})$. Quintessence scalar fields [34] with small positive kinetic term $(-1 < w < -\frac{1}{3})$ violate the strong energy condition but not the
dominant energy condition $\rho + p > 0$. Their energy density scales down with
the cosmic expansion and so does the cosmic acceleration rate. Phantom fields
[35] with negative kinetic term $(w < -1)$ violate the strong energy condition,
the dominant energy condition and maybe physically unstable. However, they
are also consistent with current cosmological data and according to recent
studies [30, 33, 36] they maybe favored over their quintessence counterparts.

Homogeneous quintessence or phantom scalar fields are described by Lagrangians of the form

$$\mathcal{L} = \pm\frac{1}{2}\dot{\phi}^2 - V(\phi) , \qquad (9.37)$$

where the upper (lower) sign corresponds to a quintessence (phantom) field in
(9.37) and in what follows. The corresponding equation of state parameter is

$$w = \frac{p}{\rho} = \frac{\pm\frac{1}{2}\dot{\phi}^2 - V(\phi)}{\pm\frac{1}{2}\dot{\phi}^2 + V(\phi)} . \qquad (9.38)$$

For quintessence (phantom) models with $V(\phi) > 0$ $(V(\phi) < 0)$ the parameter
w remains in the range $-1 < w < 1$. For an arbitrary sign of $V(\phi)$ the
above restriction does not apply but it is still impossible for w to cross the
PDL $w = -1$ in a continuous manner. The reason is that for $w = -1$ a zero
kinetic term $\pm\dot{\phi}^2$ is required and the continuous transition from $w < -1$ to
$w > -1$ (or vice versa) would require a change of sign of the kinetic term. The
sign of this term however is fixed in both quintessence and phantom models.
This difficulty in crossing the PDL $w = -1$ could play an important role in
identifying the correct model for dark energy in view of the fact that data
favor $w \simeq -1$ and furthermore parametrizations of $w(z)$ where the PDL is
crossed appear to be favored over the cosmological constant $w = -1$ according
to the Gold dataset as discussed in the previous section.

It is therefore interesting to consider the available quintessence and phantom scalar field models and compare the consistency with data of the predicted
forms of $w(z)$ among themselves and with arbitrary parametrizations of $w(z)$
that cross the PDL. This task has been recently undertaken by several authors
in the context of testing the predictions of phantom and quintessence scalar
field models [36, 37].

As an example we may consider a particular class of scalar field potentials
of the form

$$V(\phi) = s\,\phi , \qquad (9.39)$$

where I have followed [38] and set $\phi = 0$ at $V = 0$. As discussed in Sect. 9.2
(see also [38]) the field may be assumed to be frozen ($\dot{\phi} = 0$) at early times due
to the large cosmic friction $H(t)$. It has been argued [39] that such a potential
is favored by anthropic principle considerations because galaxy formation is
possible only in regions where $V(\phi)$ is in a narrow range around $V = 0$

and in such a range any potential is well approximated by a linear function. In addition such a potential can provide a potential solution to the cosmic coincidence problem [40].

The cosmological evolution in the context of such a model [41] is obtained by solving the coupled Friedman-Robertson-Walker (FRW) and the scalar field equation

$$\frac{\ddot{a}}{a} = M_{Pl}\frac{1}{3M_p^2}(\dot{\phi}^2 + s\,\phi) - \frac{\Omega_{0m}H_0^2}{2a^3} \ , \tag{9.40}$$

$$\ddot{\phi} + 3\frac{\dot{a}}{a}\dot{\phi} - s = 0 \ . \tag{9.41}$$

where $M_p = (8\pi G)^{-1/2}$ is the Planck mass and I have assumed a potential of the form

$$V(\phi) = M_{Pl}s\,\phi \tag{9.42}$$

where the upper (lower) sign corresponds to quintessence (phantom) models. The solution of the system (9.40)–(9.41) for both positive and negative values of the single parameter of the model s, is a straightforward numerical problem [41] which leads to the predicted forms of $H(z; s)$ and $w(z; s)$. These forms may then be fit to the SnIa datasets for the determination of the best fit value of the parameter s. This task has been undertaken in [41] using the Full Gold dataset. The best fit value of s was found to be practically indistinguishable from zero which corresponds to the cosmological constant for both the quintessence and the phantom cases. The predicted forms of $w(z)$ for a phantom and a quintessence case and $s \simeq 2$ is shown in Fig. 9.15. The value of $\Delta\chi^2_{LCDM}$ is positive in both cases which implies that the fit is worse compared to LCDM. The main reason for this is that both the quintessence and

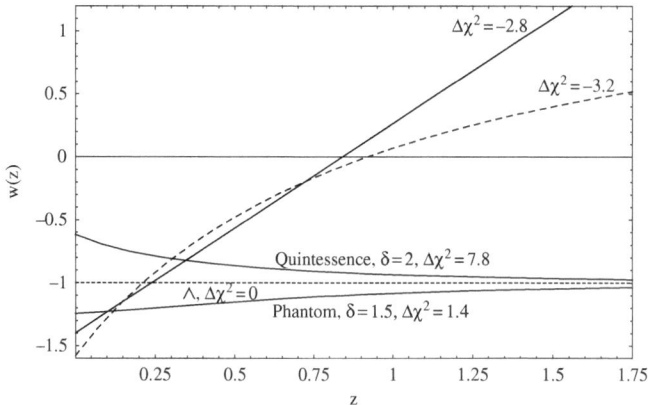

Fig. 9.15. The predicted forms of $w(z)$ for a phantom and a quintessence case and $s \simeq 2$ provide worse fits to the Gold dataset than LCDM and even worse compared to best fit parametrizations that cross the PDL [41]

phantom minimally coupled scalar field models do not allow for crossing of the PDL line for any parameter value as discussed above. In contrast, the best fit $w(z)$ parametrizations A and B of (9.33)–(9.35) which allow for PDL crossing have a negative $\Delta\chi^2_{LCDM}$ in the context of the Gold dataset as shown in Fig. 9.15 and therefore provide better fits than the field theory models. It should be stressed however that in the context of the SNLS dataset, parametrizations that allow for crossing of the PDL do not seem to have a similar advantage as discussed in the previous section.

The difficulty in crossing the PDL $w = -1$ described above could play an important role in identifying the correct model for dark energy in view of the fact that data favor $w \simeq -1$ and furthermore parametrizations of $w(z)$ where the PDL is crossed appear to be favored over the cosmological constant $w = -1$ in the context of the Gold dataset. Even for generalized k-essence Lagrangians [42, 43] of a minimally coupled scalar field e.g.

$$\mathcal{L} = \frac{1}{2}f(\phi)\dot{\phi}^2 - V(\phi) \tag{9.43}$$

it has been shown [44] to be impossible to obtain crossing of the PDL. Multiple field Lagrangians (combinations of phantom with quintessence fields [45, 46, 47, 48]) have been shown to in principle achieve PDL crossing but such models are complicated and without clear physical motivation (but see [49] for an interesting physically motivated model).

The obvious class of theories that could lead to a solution of the above described problem is the non-minimally coupled scalar fields. Such theories are realized in a universe where gravity is described by a scalar-tensor theory and their study is well motivated for two reasons:

1. A scalar-tensor theory of gravity is predicted by all fundamental quantum theories that involve extra dimensions. Such are all known theories that attempt to unify gravity with the other interactions (e.g. supergravity (SUGRA), M-theory etc.).
2. Scalar fields emerging from scalar tensor theories (extended quintessence) can predict an expansion rate $H(z)$ that violates the inequality

$$\frac{d(H(z)^2/H_0^2)}{dz} \geq 3\Omega_{0m}(1+z)^2 , \tag{9.44}$$

which is equivalent to crossing the PDL $w = -1$ (see e.g. [50]).

In fact it has been shown in [50] that in contrast to minimally coupled quintessence, scalar tensor theories can reproduce the main features of the best fit Hubble expansion history obtained from the Gold dataset. However, the precise determination of the scalar tensor theory potentials requires more accurate SnIa data and additional cosmological observational input.

9.6 The Fate of a Phantom Dominated Universe: Big Rip

As discussed in Sect. 9.4 the Gold dataset favors a dynamical dark energy with present value of the equation of state parameter w in the phantom regime. If this trend is verified by future datasets and if w remains in the phantom regime in the future then the fate of the universe acquires novel interesting features. The energy density of phantom fields increases with time and so does the predicted expansion acceleration rate $\frac{\ddot{a}}{a}$. This monotonically increasing acceleration rate of the expansion may be shown to lead to a novel kind of singularity which occurs at a finite future time and is characterized by divergences of the scale factor a, the Hubble parameter H its derivative $\dot{H}$ and the scalar curvature. This singularity has been called 'Big Smash' [51] the first time it was discussed and 'Big Rip' [52] in a more recent study. An immediate consequence of the very rapid expansion rate as the Big Rip singularity is approached is the dissociation of bound systems due to the buildup of repulsive negative pressure in the interior of these systems.

This dissociation of bound systems can be studied by considering the spacetime in the vicinity of a point mass M placed in an expanding background in order to study the effects of the cosmic expansion on bound systems. Such a metric should interpolate between a static Schwarzschild metric at small distances from M and a time dependent Friedmann spacetime at large distances. In the Newtonian limit (weak field, low velocities) such an interpolating metric takes the form [53]:

$$ds^2 = (1 - \frac{2GM}{a(t)\rho}) \cdot dt^2 - a(t)^2 \cdot (d\rho^2 + \rho^2 \cdot (d\theta^2 + sin^2\theta d\varphi^2)) , \qquad (9.45)$$

where ρ is the comoving radial coordinate. Using

$$r = a(t) \cdot \rho \qquad (9.46)$$

the geodesics corresponding to the line element (9.45) take the form

$$-(\ddot{r} - \frac{\ddot{a}}{a}r) - \frac{GM}{r^2} + r\dot{\varphi}^2 = 0 \qquad (9.47)$$

and

$$r^2\dot{\varphi} = L , \qquad (9.48)$$

where L is the constant angular momentum per unit mass. Therefore the radial equation of motion for a test particle in the Newtonian limit considered is

$$\ddot{r} = \frac{\ddot{a}}{a}r + \frac{L^2}{r^3} - \frac{GM}{r^2} . \qquad (9.49)$$

The first term on the rhs proportional to the cosmic acceleration is a time dependent repulsive term which is increasing with time for $w < -1$. This is easy to see by considering the Friedman (9.12) combined with the dark energy evolution $\rho_X \sim a^{-3(1+w)}$ where the scale factor obtained from the Friedman equation is

$$a(t) = \frac{a(t_m)}{[-w + (1+w)t/t_m]^{-\frac{2}{3(1+w)}}} \quad for \ \ t > t_m \quad (9.50)$$

and t_m is the transition time from decelerating to accelerating expansion. For phantom energy ($w < -1$) the scale factor diverges at a finite time

$$t_* = \frac{w}{1+w}t_m > 0 \quad (9.51)$$

leading to the Big Rip singularity. Clearly, the time dependent repulsive term of (9.49) diverges at the Big Rip singularity.

A quantitative analysis [54] shows that the geodesic (9.49) is equivalent to a Newtonian equation with a time-dependent effective potential that determines the dynamics of the bound system which in dimesionless form is [54]

$$V_{eff} = -\frac{\omega_0^2}{r} + \frac{\omega_0^2}{2r^2} - \frac{1}{2}\lambda(t)^2 r^2 \ , \quad (9.52)$$

where

$$\lambda(t) = \frac{\sqrt{2|1+3w|}}{3(-w + (1+w)t)} \ , \quad (9.53)$$

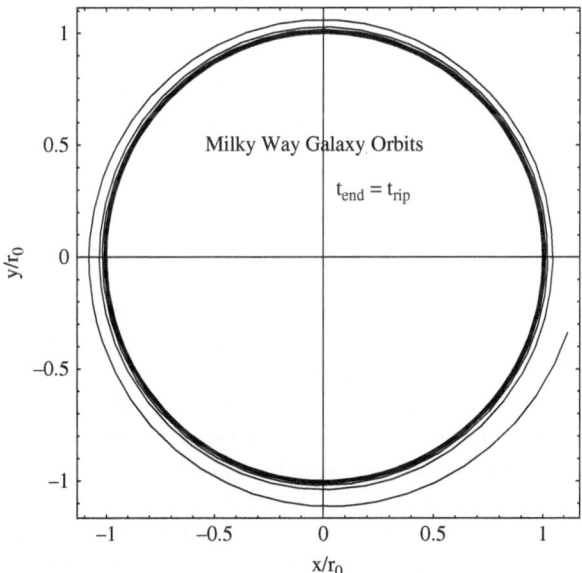

Fig. 9.16. The numerically obtained evolution of a galactic size two body system at times close to the predicted dissociation time t_{rip} [54]

Table 9.2. The difference between dissociation times t_{rip} and the big rip time t_* for three bound systems in years as predicted by (9.55). The dissociation times t_{rip} for the three bound systems in units of t_m are also shown in column 3. The value $w = -1.2$ was assumed [54]

System	$t_* - t_{rip}$ (yrs)	t_{rip}/t_m
Solar System	$1.88 \cdot 10^4$	6.00
Milky Way	$3.59 \cdot 10^8$	5.94
Coma Cluster	$1.58 \cdot 10^{10}$	3.19

with $w < -1$ and ω_0 is defined as

$$\omega_0^2 = \frac{GM}{r_0^3} t_m^2 \ . \tag{9.54}$$

At $t = 1$ the system is assumed to be in circular orbit with radius given by the minimum $r_{min}(t)$ of the effective potential of (9.52). It is easy to show that the minimum of the effective potential (9.52) disappears at a time t_{rip} which obeys

$$t_* - t_{rip} = \frac{16\sqrt{3}}{9} \frac{T\sqrt{2|1+3w|}}{6\pi|1+w|} \ . \tag{9.55}$$

The value of the bound system dissociation time t_{rip} may be verified by numerically solving the geodesic Newtonian equation of a test particle with the effective potential (9.52). The resulting evolution close to the predicted dissociation time t_{rip} is shown in Fig. 9.16 for $w = -1.2$ and verifies the dissociation time predicted by (9.55). Using the appropriate values for the bound system masses M the dissociation times of cosmological bound systems may be obtained. These are shown in Table 9.2.

9.7 Future Prospects-Conclusion

The question of the physical origin and dynamical evolution properties of dark energy is the central question currently in cosmology. Since the most sensitive and direct probes towards the answer of this question are distance-redshift surveys of SnIa there has been intense activity during the recent years towards designing and implementing such projects using ground based and satellite observatories. Large arrays of CCDs such as MOSAIC camera at Cerro Tololo Inter-American Obsrevatory, the SUPRIME camera at Subaru or the MEGA-CAM at the Canada-France-Hawaii Telescope (CFHT) are some of the best ground based tools for supernova searches. These devices work well in the reddest bands (800–900nm) where the ultraviolet and visible light of redshifted high-z SnIa is detected. Searches from the ground have the advantages of large telescope apertures (Subaru for example has 10 times the collecting area of the Hubble Space Telescope (HST)) and large CCD arrays (the CFHT has a 378-milion pixel camera compared to the Advanced Camera for Surveys on HST which has 16 million pixels). On the other hand the advantage of space satellite observatories like the HST include avoiding the bright and variable night-sky encountered in the near infrared, the potential for much sharper imaging for point sources like supernovae to distinguish them from galaxies in which they reside and better control over the observing conditions which need not factor in weather and moonlight.

The original two SnIa search teams (the Supernova Cosmology Project and the High-z Supernova Search Team) have evolved to a number of ongoing and proposed search projects both satellite and ground based. These projects (see Fig. 9.17) include the following:

– **The GOODS [56], the Higher-z Supernova Search Team(HZT)** [55]. This has originated from the High-z Supernova Search Team and has A. Riess of Space Telescope Sci. Inst. as its team leader. This team is in collaboration with the GOODS program (Great Observatories Origin Deep Survey) using the ACS of the HST to detect and analyze high redshift $(0.5 < z < 2)$ SnIa. Successive GOODS observations are spaced by 45 days providing 5 epochs of data on two fields: the Hubble Deep Field (HDF) north and south. Whereas the GOODS team adds these images to build a superdeep field, the HZT subtracts the accumulated template image from each incoming frame. Thus the HZT has already detected more than 42 supernovae in the above redshift range.

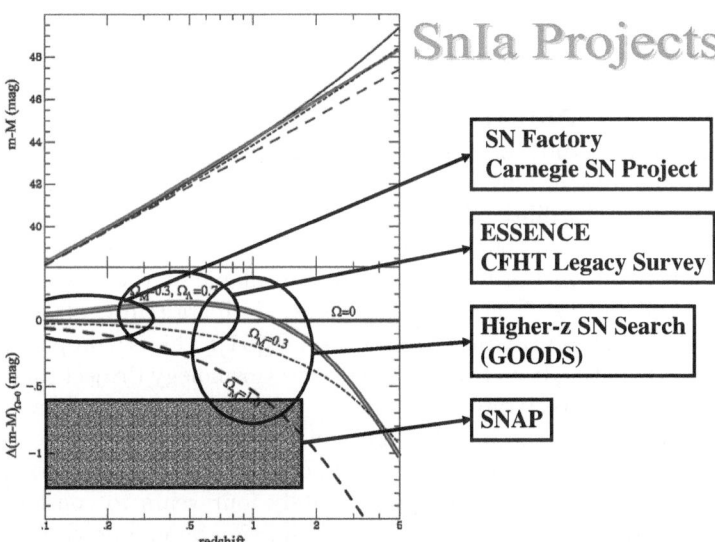

Fig. 9.17. Ongoing and proposed SnIa search projects with the corresponding redshift ranges

- **Equation of State:(ESSENCE)[57], SupErNovae Trace Cosmic Expansion**. This has also originated from the High-z Supernova Search Team and has C. Stubbs of the Univ. of Washington, C. Smith and N. Suntzeff of Cerro Tololo as its team leaders. This ongoing program aims to find and measure 200 SnIa's in the redshift range of $0.15 < z < 0.7$ where the transition from decelerating to accelerating expansion occurs. Spectroscopic backup to the program comes from the ground based Gemini, Magellan, VLT, Keck and MMT Obsevatory. The ESSENCE project is a five-year endeavor, with the goal of tightly constraining the time average of the equation-of-state parameter $w = p/\rho$ of the dark energy. To help minimize systematic errors, all of their ground-based photometry is obtained with the same telescope and instrument. In 2003 the highest-redshift subset of ESSENCE supernovae was selected for detailed study with HST.
- **The Supenova Legacy Survey (SNLS)[58]:** The CFHT Legacy Survey aims at detecting and monitoring about 1000 supernovae in the redshift range $0 < z < 1$ with Megaprime at the Canada-France-Hawaii telescope between 2003 and 2008. High-z spectroscopy of SnIa is being carried on 8m class telescopes (Gemini, VLT, Keck). Team representatives are: C. Pritchet (Univ. Victoria), P. Astier (CNRS/IN2P3), S. Basa (CNRS/INSU) et al. The SNLS has recently released the first year dataset [6].
- **Nearby Supernova Factory (SNF)[59]:** The Nearby Supernova Factory (SNF) is an international collaboration based at Lawrence Berkeley National Laboratory. Greg Aldering of Berkeley Lab's Physics Division is the principal investigator of the SNF. The goal of the SNF is to discover and carefully study 300 to 600 nearby Type Ia supernovae in the redshift range $0 < z < 0.3$.
- **Carnegie SN Project (CSP)[60]:** The goal of the project is the comprehensive study of both Type Ia and II Supernovae in the local ($z < 0.07$) universe. This is a long-term program with the goal of obtaining exceedingly-well calibrated optical/near-infrared light curves and optical spectroscopy of over 200 Type nnIa and Type nII supernovae. The CSP takes advantage of the unique resources available at the Las Campanas Observatory (LCO). The team leader is R. Carlberg (Univ. of Toronto).
- **Supernova Acceleration Probe (SNAP)[61]:** This is a proposed space mission originating from LBNL's Supernova Cosmology Project that would increase the discovery rate for SnIa's to about 2000 per year. The satellite called SNAP (Supernova / Acceleration Probe) would be a space based telescope with a one square degree field of view with 1 billion pixels. The project schedule would take approximately four years to construct and launch SNAP, and another three years of mission observations. SNAP has a 2 meter telescope with a large field of view: 600 times the sky area of the Hubble Space Telescopes Wide Field Camera. By repeatedly imaging 15 square degrees of the sky, SNAP will accurately measure the energy spectra

and brightness over time for over 2,000 Type Ia supernovae, discovering them just after they explode.

These projects aim at addressing important questions related to the physical origin and dynamical properties of dark energy. In particular these questions can be structured as follows:

- Can the accelerating expansion be attributed to a dark energy ideal fluid with negative pressure or is it necessary to implement extensions of GR to understand the origin of the accelerating expansion?
- Is w evolving with redshift and crossing the PDL? If the crossing of the PDL by $w(z)$ is confirmed then it is quite likely that extensions of GR will be required to explain observations.
- Is the cosmological constant consistent with data? If it remains consistent with future more detailed data then the theoretical efforts should be focused on resolving the coincidence and the cosmological constant problems which may require anthropic principle arguments.

The main points of this brief review may be summarized as follows:

- *Dark energy* with *negative pressure* can explain SnIa cosmological data indicating accelerating expansion of the universe.
- The existence of a *cosmological constant* is consistent with SnIa data but other *evolving* forms of dark energy *crossing the $w = -1$* line may provide better fits to some of the recent data (Gold dataset).
- New *observational projects* are underway and are expected to lead to significant progress in the understanding of the properties of *dark energy*.

References

1. S. Perlmutter et al., Nature (London) **391**, 51(1998).
2. A. Riess et al., Astron. J. **116** 1009 (1998).
3. J.L. Tonry et al., (2003). Astroph. J. **594**, 1.
4. S.J. Perlmutter et al., Astroph. J. **517**, 565 (1999); B. Barris et al. (2004). Astroph. J. **602**, 571; R. Knop et al., (2003) Astroph. J. **598**, 102
5. A.G. Riess et al. [Supernova Search Team Collaboration], Astrophys. J. **607**, 665 (2004).
6. P. Astier et al., Astron. Astrophys. **447**, 31 (2006) [arXiv:astro-ph/0510447].
7. V. Sahni, Lect. Notes Phys. **653**, 141 (2004) [arXiv:astro-ph/0403324].
8. D. Spergel et al., Astrophys. J. Suppl. **148**, 175 (2003) [arXiv:astro-ph/0302209].
9. M. Tegmark et al., Phys. Rev. D69 (2004) 103501,astro-ph/0310723
10. V. Sahni and A.A. Starobinsky, Int. J. Mod. Phys. D **9**, 373 (2000) [arXiv:astro-ph/9904398].
11. P.J. Peebles and B. Ratra, Astrophys. J. **325**, L17 (1988); R.R. Caldwell, R. Dave and P.J. Steinhardt, Phys. Rev. Lett. **80**, 1582 (1998) [arXiv:astro-ph/9708069]; I. Zlatev, L.M. Wang and P.J. Steinhardt, Phys. Rev. Lett. **82**, 896 (1999).

12. V. Sahni and L.M. Wang, Phys. Rev. D **62**, 103517 (2000) [arXiv:astro-ph/9910097].

13. F. Perrotta, C. Baccigalupi and S. Matarrese, Phys. Rev. D **61**, 023507 (2000) [arXiv:astro-ph/9906066]; A. Riazuelo and J.P. Uzan, Phys. Rev. D **66**, 023525 (2002) [arXiv:astro-ph/0107386]; G. Esposito-Farese and D. Polarski, Phys. Rev. D **63**, 063504 (2001) [arXiv:gr-qc/0009034].

14. D.F. Torres, Phys. Rev. D **66**, 043522 (2002) [arXiv:astro-ph/0204504].

15. M. Axenides and K. Dimopoulos, JCAP **0407**, 010 (2004) [arXiv:hep-ph/0401238].

16. S. Nojiri and S.D. Odintsov, Gen. Rel. Grav. **36**, 1765 (2004) [arXiv:hep-th/0308176].

17. A.Y. Kamenshchik, U. Moschella and V. Pasquier, Phys. Lett. B **511**, 265 (2001) [arXiv:gr-qc/0103004].

18. K. Freese and M. Lewis, Phys. Lett. B **540**, 1 (2002) [arXiv:astro-ph/0201229].

19. L. Perivolaropoulos and C. Sourdis, Phys. Rev. D **66**, 084018 (2002) [arXiv:hep-ph/0204155]; M. Sami, N. Savchenko and A. Toporensky, Phys. Rev. D **70**, 123528 (2004) [arXiv:hep-th/0408140]; T.R. Mongan, Gen. Rel. Grav. **33**, 1415 (2001) [arXiv:gr-qc/0103021].

20. L. Perivolaropoulos, Phys. Rev. D **67**, 123516 (2003) [arXiv:hep-ph/0301237].

21. V. Sahni and Y. Shtanov, JCAP **0311**, 014 (2003) [arXiv:astro-ph/0202346].

22. R.R. Caldwell, Phys. Lett. B **545**, 23 (2002) [arXiv:astro-ph/9908168].

23. Edward W. Kolb, Michael S. Turner, *The Early Universe*, (Frontiers in Physics 1990).

24. G.H. Jacoby et al., Pub. Astron. Soc. Pac. **104**, 599 (1992); S. van den Bergh, Pub. Astron. Soc. Pac. **104**, 861 (1992); S. van den Bergh, Pub. Astron. Soc. Pac. **106**, 1113 (1994); M. Fukugita, C.J. Hogan and P.J.E. Peebles, Nature **366**, 309 (1993); A. Sandage and G. Tammann, in Critical Dialogues in Cosmology, ed. N. Turok (Princeton, 1997), astro-ph 9611170; W. Freedman, ibid., astro-ph 9612024; C.J. Hogan, http://pdg.lbl.gov/1998/hubblerpp.pdf (1997).

25. S. Perlmutter, et al., 1995. in Presentations at the NATO ASI in Aiguablava, Spain, LBL-38400, page I.1 and in Thermonuclear Supernova, P. Ruiz-Lapuente, R. Canal and J. Isern, eds. (Dordrecht: Kluwer, 1997).

26. A. Kim, A. Goobar and S. Perlmutter, Publ. Astron. Soc. Pac. **108**, 190 (1996) [arXiv:astro-ph/9505024].

27. S. Perlmutter et al., Astrophys.J. **483** 565 (1997).

28. A.N. Aguirre, Astrophys. J. **512**, L19 (1999) [arXiv:astro-ph/9811316]; A. N. Aguirre, Astrophys. J., 512, L19–L22, (1999).

29. D. Huterer and M.S. Turner, Phys. Rev. D **64**, 123527 (2001) [arXiv:astro-ph/0012510].

30. S. Nesseris and L. Perivolaropoulos, Phys. Rev. D **70**, 043531 (2004) [arXiv:astro-ph/0401556].

31. W.H. Press et al., 'Numerical Recipes', Cambridge University Press (1994).

32. S. Nesseris and L. Perivolaropoulos, Phys. Rev. D **72**, 123519 (2005) [arXiv:astro-ph/0511040].

33. R. Lazkoz, S. Nesseris and L. Perivolaropoulos, JCAP **0511**, 010 (2005) [arXiv:astro-ph/0503230].

34. B. Ratra and P.J.E. Peebles, Phys. Rev. D**37**, 3406 (1988); Rev. Mod. Phys. **75**, 559 (2003) [arXiv:astro-ph/0207347]; C. Wetterich, Nucl. Phys. B **302**, 668(1988); P.G. Ferreira and M. Joyce, Phys. Rev. D.**58**, 023503 (1998).

35. R.R. Caldwell, Phys. Lett. B **545**, 23 (2002) [arXiv:astro-ph/9908168]; P. Singh, M. Sami and N. Dadhich, Phys. Rev. D **68**, 023522 (2003) [arXiv:hep-th/0305110]; V.B. Johri, Phys. Rev. D **70**, 041303 (2004) [arXiv:astro-ph/0311293]; M. Sami and A. Toporensky, Mod. Phys. Lett. A **19**, 1509 (2004) [arXiv:gr-qc/0312009].

36. U. Alam, V. Sahni, T.D. Saini and A.A. Starobinsky, Mon. Not. Roy. Astron. Soc. **354**, 275 (2004) [arXiv:astro-ph/0311364]; Y. Wang and P. Mukherjee, Astrophys. J. **606**, 654 (2004) [arXiv:astro-ph/0312192]; D. Huterer and A. Cooray, Phys. Rev. D **71**, 023506 (2005) [arXiv:astro-ph/0404062]; R.A. Daly and S.G. Djorgovski, Astrophys. J. **597**, 9 (2003) [arXiv:astro-ph/0305197]; R.A. Daly and S.G. Djorgovski, arXiv:astro-ph/0512576.

37. M.C. Bento, O. Bertolami, N.M.C. Santos and A.A. Sen, Phys. Rev. D **71**, 063501 (2005) [arXiv:astro-ph/0412638]; J.Q. Xia, G.B. Zhao, B. Feng, H. Li and X. Zhang, Phys. Rev. D **73**, 063521 (2006) [arXiv:astro-ph/0511625]; U. Alam and V. Sahni, Phys. Rev. D **73**, 084024 (2006) [astro-ph/0511473]; H.K. Jassal, J.S. Bagla and T. Padmanabhan, Phys. Rev. D **72**, 103503 (2005) [arXiv:astro-ph/0506748]; V.B. Johri and P.K. Rath [arXiv:astro-ph/0510017]; H. Li, B. Feng, J.Q. Xia and X. Zhang, Phys. Rev. D **73**, 103503 (2006) [arXiv:astro-ph/0509272]; X. Zhang and F.Q. Wu, Phys. Rev. D **72**, 043524 (2005) [arXiv:astro-ph/0506310]; R. Opher and A. Pelinson, arXiv:astro-ph/0505476; D. Jain, J.S. Alcaniz and A. Dev, Nucl. Phys. B **732**, 379 (2006) [arXiv:astro-ph/0409431].

38. J. Garriga, L. Pogosian and T. Vachaspati, Phys. Rev. D **69**, 063511 (2004) [arXiv:astro-ph/0311412].

39. J. Garriga and A. Vilenkin, Phys. Rev. D **61**, 083502 (2000) [arXiv:astro-ph/9908115]; J. Garriga and A. Vilenkin, Phys. Rev. D **67**, 043503 (2003) [arXiv:astro-ph/0210358]; J. Garriga, A. Linde and A. Vilenkin, Phys. Rev. D **69**, 063521 (2004) [arXiv:hep-th/0310034].

40. P.P. Avelino, Phys. Lett. B **611**, 15 (2005) [arXiv:astro-ph/0411033].

41. L. Perivolaropoulos, Phys. Rev. D **71**, 063503 (2005) [arXiv:astro-ph/0412308].

42. C. Armendariz-Picon, V. Mukhanov and P.J. Steinhardt, Phys. Rev. D **63**, 103510 (2001) [arXiv:astro-ph/0006373]; R.J. Scherrer, Phys. Rev. Lett. **93**, 011301 (2004) [arXiv:astro-ph/0402316].

43. A. Melchiorri, L. Mersini, C.J. Odman and M. Trodden, Phys. Rev. D **68**, 043509 (2003) [arXiv:astro-ph/0211522].

44. A. Vikman, Phys. Rev. D **71**, 023515 (2005) [arXiv:astro-ph/0407107].

45. Z.K. Guo, Y.S. Piao, X.M. Zhang and Y.Z. Zhang, Phys. Lett. B **608**, 177 (2005) [arXiv:astro-ph/0410654]; B. Feng, X.L. Wang and X.M. Zhang, Phys. Lett. B **607**, 35 (2005) [arXiv:astro-ph/0404224]; B. Feng, M. Li, Y.S. Piao and X. Zhang, Phys. Lett. B **634**, 101 (2006) [arXiv:astro-ph/0407432]; X.F. Zhang, H. Li, Y.S. Piao and X.M. Zhang, Mod. Phys. Lett. A **21**, 231 (2006) [arXiv:astro-ph/0501652].

46. R.R. Caldwell and M. Doran, Phys. Rev. D **72**, 043527 (2005) [arXiv:astro-ph/0501104].

47. W. Hu, Phys. Rev. D **71**, 047301 (2005) [arXiv:astro-ph/0410680].

48. H. Stefancic, Phys. Rev. D **71**, 124036 (2005) [arXiv:astro-ph/0504518].

49. S. Nojiri, S.D. Odintsov and M. Sasaki, Phys. Rev. D **71**, 123509 (2005) [arXiv:hep-th/0504052]; B.M.N. Carter and I.P. Neupane, Phys. Lett. B **638**, 94 (2006) [arXiv:hep-th/0510109].

50. L. Perivolaropoulos, JCAP **0510**, 001 (2005) [arXiv:astro-ph/0504582]; S. Tsujikawa, Phys. Rev. D **72**, 083512 (2005) [arXiv:astro-ph/0508542].

51. B. McInnes, JHEP **0208**, 029 (2002) [arXiv:hep-th/0112066].

52. R.R. Caldwell, M. Kamionkowski and N.N. Weinberg, Phys. Rev. Lett. **91**, 071301 (2003) [arXiv:astro-ph/0302506].

53. G.C. McVittie, MNRAS **93**, 325 (1933).

54. S. Nesseris and L. Perivolaropoulos, Phys. Rev. D **70**, 123529 (2004) [arXiv:astro-ph/0410309].

55. L.G. Strolger et al., Astrophys. J. **613**, 200 (2004) [arXiv:astro-ph/0406546].

56. M. Giavalisco [The GOODS Team Collaboration], Astrophys. J. **600**, L93 (2004) [arXiv:astro-ph/0309105].

57. J. Sollerman et al., ESA Spec. Publ. **637**, 141 (2006) arXiv:astro-ph/0510026.

58. N. Palanque-Delabrouille [SNLS Collaboration], arXiv:astro-ph/0509425.

59. W.M. Wood-Vasey et al., New Astron. Rev. **48**, 637 (2004) [arXiv:astro-ph/0401513].

60. W.L. Freedman [The Carnegie Supernova Project Collaboration], arXiv:astro-ph/0411176.

61. G. Aldering [SNAP Collaboration], arXiv:astro-ph/0209550.

Part III

Dark Matter and Dark Energy Beyond the Standard Theory of General Relativity

10

The Physics of Extra Dimensions

Ignatios Antoniadis

Department of Physics, CERN - Theory Division, 1211 Geneva 23, Switzerland[**]
ignatios.antoniadis@cern.ch

Abstract. Lowering the string scale in the TeV region provides a theoretical framework for solving the mass hierarchy problem and unifying all interactions. The apparent weakness of gravity can then be accounted by the existence of large internal dimensions, in the submillimeter region, and transverse to a braneworld where our universe must be confined. I review the main properties of this scenario and its implications for observations at both particle colliders, and in non-accelerator gravity experiments. Such effects are for instance the production of Kaluza-Klein resonances, graviton emission in the bulk of extra dimensions, and a radical change of gravitational forces in the submillimeter range. I also discuss the warped case and localization of gravity in the presence of infinite size extra dimensions.

10.1 Introduction

During the last few decades, physics beyond the Standard Model (SM) was guided from the problem of mass hierarchy. This can be formulated as the question of why gravity appears to us so weak compared to the other three known fundamental interactions corresponding to the electromagnetic, weak and strong nuclear forces. Indeed, gravitational interactions are suppressed by a very high energy scale, the Planck mass $M_P \sim 10^{19}$ GeV, associated to a length $l_P \sim 10^{-35}$ m, where they are expected to become important. In a quantum theory, the hierarchy implies a severe fine tuning of the fundamental parameters in more than 30 decimal places in order to keep the masses of elementary particles at their observed values. The reason is that quantum radiative corrections to all masses generated by the Higgs vacuum expectation value (VEV) are proportional to the ultraviolet cutoff which in the presence of gravity is fixed by the Planck mass. As a result, all masses are "attracted" to become about 10^{16} times heavier than their observed values.

[**] On leave from CPHT (UMR CNRS 7644) Ecole Polytechnique, F-91128 Palaiseau.

I. Antoniadis: *The Physics of Extra Dimensions*, Lect. Notes Phys. **720**, 293–321 (2007)
DOI 10.1007/978-3-540-71013-4_10 © Springer-Verlag Berlin Heidelberg 2007

Besides compositeness, there are three main theories that have been proposed and studied extensively during the last years, corresponding to different approaches of dealing with the mass hierarchy problem. (1) Low energy supersymmetry with all superparticle masses in the TeV region. Indeed, in the limit of exact supersymmetry, quadratically divergent corrections to the Higgs self-energy are exactly cancelled, while in the softly broken case, they are cut-off by the supersymmetry breaking mass splittings. (2) TeV scale strings, in which quadratic divergences are cutoff by the string scale and low energy supersymmetry is not needed. (3) Split supersymmetry, where scalar masses are heavy while fermions (gauginos and higgsinos) are light. Thus, gauge coupling unification and dark matter candidate are preserved but the mass hierarchy should be stabilized by a different way and the low energy world appears to be fine-tuned. All these ideas are experimentally testable at high-energy particle colliders and in particular at LHC. Below, I discuss their implementation in string theory.

The appropriate and most convenient framework for low energy supersymmetry and grand unification is the perturbative heterotic string. Indeed, in this theory, gravity and gauge interactions have the same origin, as massless modes of the closed heterotic string, and they are unified at the string scale M_s. As a result, the Planck mass M_P is predicted to be proportional to M_s:

$$M_P = M_s/g \,, \tag{10.1}$$

where g is the gauge coupling. In the simplest constructions all gauge couplings are the same at the string scale, given by the four-dimensional (4d) string coupling, and thus no grand unified group is needed for unification. In our conventions $\alpha_{\mathrm{GUT}} = g^2 \simeq 0.04$, leading to a discrepancy between the string and grand unification scale M_{GUT} by almost two orders of magnitude. Explaining this gap introduces in general new parameters or a new scale, and the predictive power is essentially lost. This is the main defect of this framework, which remains though an open and interesting possibility.

The other two ideas have both as natural framework of realization type I string theory with D-branes. Unlike in the heterotic string, gauge and gravitational interactions have now different origin. The latter are described again by closed strings, while the former emerge as excitations of open strings with endpoints confined on D-branes [1]. This leads to a braneworld description of our universe, which should be localized on a hypersurface, i.e. a membrane extended in p spatial dimensions, called p-brane (see Fig. 10.1). Closed strings propagate in all nine dimensions of string theory: in those extended along the p-brane, called parallel, as well as in the transverse ones. On the contrary, open strings are attached on the p-brane. Obviously, our p-brane world must have at least the three known dimensions of space. But it may contain more: the extra $d_\parallel = p - 3$ parallel dimensions must have a finite size, in order to be unobservable at present energies, and can be as large as $\mathrm{TeV}^{-1} \sim 10^{-18}$ m [2]. On the other hand, transverse dimensions interact with us only gravitationally and experimental bounds are much

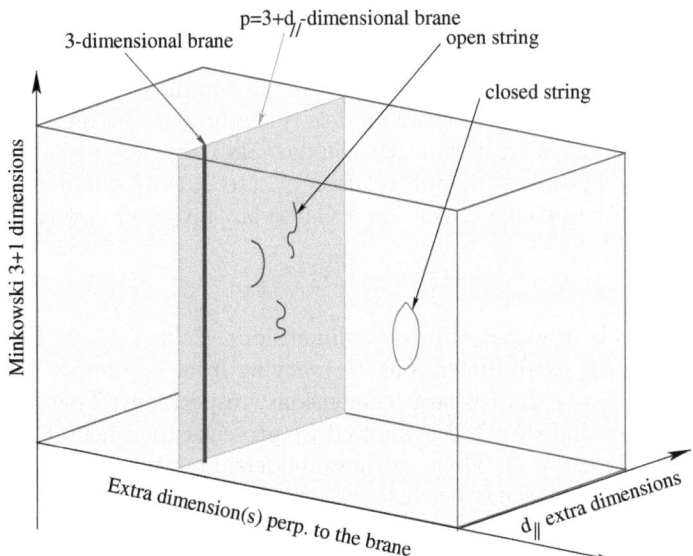

Fig. 10.1. In the type I string framework, our Universe contains, besides the three known spatial dimensions (*denoted by a single line*), some extra dimensions ($d_\parallel = p - 3$) parallel to our world p-brane where endpoints of open strings are confined, as well as some transverse dimensions where only gravity described by closed strings can propagate

weaker: their size should be less than about 0.1 mm [3]. In the following, I review the main properties and experimental signatures of low string scale models [4, 5].

10.2 Framework

In type I theory, the different origin of gauge and gravitational interactions implies that the relation between the Planck and string scales is not linear as (10.1) of the heterotic string. The requirement that string theory should be weakly coupled, constrain the size of all parallel dimensions to be of order of the string length, while transverse dimensions remain unrestricted. Assuming an isotropic transverse space of $n = 9 - p$ compact dimensions of common radius $R_\perp$, one finds:

$$M_P^2 = \frac{1}{g^4} M_s^{2+n} R_\perp^n, \qquad g_s \simeq g^2 . \tag{10.2}$$

where g_s is the string coupling. It follows that the type I string scale can be chosen hierarchically smaller than the Planck mass [4, 6] at the expense of introducing extra large transverse dimensions felt only by gravity, while keeping the string coupling small [4]. The weakness of 4d gravity compared to gauge interactions (ratio M_W/M_P) is then attributed to the largeness of the transverse space $R_\perp$ compared to the string length $l_s = M_s^{-1}$.

An important property of these models is that gravity becomes effectively $(4 + n)$-dimensional with a strength comparable to those of gauge interactions at the string scale. The first relation of (10.2) can be understood as a consequence of the $(4 + n)$-dimensional Gauss law for gravity, with

$$M_*^{(4+n)} = M_s^{2+n}/g^4 \qquad (10.3)$$

the effective scale of gravity in $4 + n$ dimensions. Taking $M_s \simeq 1$ TeV, one finds a size for the extra dimensions $R_\perp$ varying from 10^8 km, .1 mm, down to a Fermi for $n = 1, 2,$ or 6 large dimensions, respectively. This shows that while $n = 1$ is excluded, $n \geq 2$ is allowed by present experimental bounds on gravitational forces [3, 7]. Thus, in these models, gravity appears to us very weak at macroscopic scales because its intensity is spread in the "hidden" extra dimensions. At distances shorter than $R_\perp$, it should deviate from Newton's law, which may be possible to explore in laboratory experiments (see Fig. 10.2).

The main experimental implications of TeV scale strings in particle accelerators are of three types, in correspondence with the three different sectors that are generally present: (i) new compactified parallel dimensions, (ii) new extra large transverse dimensions and low scale quantum gravity, and (iii) genuine string and quantum gravity effects. On the other hand, there exist interesting implications in non accelerator table-top experiments due to the exchange of gravitons or other possible states living in the bulk.

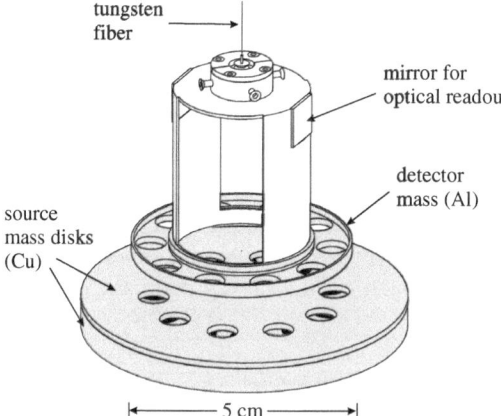

Fig. 10.2. Torsion pendulum that tested Newton's law at 130 nm. Several sources of background noise were eliminated using appropriate devices

10.3 Experimental Implications in Accelerators

10.3.1 World-brane Extra Dimensions

In this case $RM_s \gtrsim 1$, and the associated compactification scale $R_\parallel^{-1}$ would be the first scale of new physics that should be found increasing the beam energy [2, 8]. There are several reasons for the existence of such dimensions. It is a logical possibility, since out of the six extra dimensions of string theory only two are needed for lowering the string scale, and thus the effective p-brane of our world has in general $d_\parallel \equiv p - 3 \leq 4$. Moreover, they can be used to address several physical problems in braneworld models, such as obtaining different SM gauge couplings, explaining fermion mass hierarchies due to different localization points of quarks and leptons in the extra dimensions, providing calculable mechanisms of supersymmetry breaking, etc.

The main consequence is the existence of Kaluza-Klein (KK) excitations for all SM particles that propagate along the extra parallel dimensions. Their masses are given by:

$$M_m^2 = M_0^2 + \frac{m^2}{R_\parallel^2}; \quad m = 0, \pm 1, \pm 2, \ldots \tag{10.4}$$

where we used $d_\parallel = 1$, and M_0 is the higher dimensional mass. The zero-mode $m = 0$ is identified with the 4d state, while the higher modes have the same quantum numbers with the lowest one, except for their mass given in (10.4). There are two types of experimental signatures of such dimensions [8, 9, 10]: (i) virtual exchange of KK excitations, leading to deviations in cross-sections compared to the SM prediction, that can be used to extract bounds on the compactification scale; (ii) direct production of KK modes.

On general grounds, there can be two different kinds of models with qualitatively different signatures depending on the localization properties of matter fermion fields. If the latter are localized in 3d brane intersections, they do not have excitations and KK momentum is not conserved because of the breaking of translation invariance in the extra dimension(s). KK modes of gauge bosons are then singly produced giving rise to generally strong bounds on the compactification scale and new resonances that can be observed in experiments. Otherwise, they can be produced only in pairs due to the KK momentum conservation, making the bounds weaker but the resonances difficult to observe.

When the internal momentum is conserved, the interaction vertex involving KK modes has the same 4d tree-level gauge coupling. On the other hand, their couplings to localized matter have an exponential form factor suppressing the interactions of heavy modes. This form factor can be viewed as the fact that the branes intersection has a finite thickness. For instance, the coupling of the KK excitations of gauge fields $A^\mu(x, y) = \sum_m A_m^\mu \exp i\frac{my}{R_\parallel}$ to the charge density $j_\mu(x)$ of massless localized fermions is described by the effective action [11]:

$$\int d^4x \sum_m e^{-\ln 16 \frac{m^2 l_s^2}{2R_\parallel^2}} \, j_\mu(x) \, A_m^\mu(x) \,. \tag{10.5}$$

After Fourier transform in position space, it becomes:

$$\int d^4x \, dy \, \frac{1}{(2\pi \ln 16)^2} e^{-\frac{y^2 M_s^2}{2\ln 16}} \, j_\mu(x) \, A^\mu(x,y) \,, \tag{10.6}$$

from which we see that localized fermions form a Gaussian distribution of charge with a width $\sigma = \sqrt{\ln 16} \, l_s \sim 1.66 \, l_s$.

To simplify the analysis, let us consider first the case $d_\parallel = 1$ where some of the gauge fields arise from an effective 4-brane, while fermions are localized states on brane intersections. Since the corresponding gauge couplings are reduced by the size of the large dimension $R_\parallel M_s$ compared to the others, one can account for the ratio of the weak to strong interactions strengths if the $SU(2)$ brane extends along the extra dimension, while $SU(3)$ does not. As a result, there are 3 distinct cases to study [10], denoted by (t,l,l), (t,l,t) and (t,t,l), where the three positions in the brackets correspond to the three SM gauge group factors $SU(3) \times SU(2) \times U(1)$ and those with l (longitudinal) feel the extra dimension, while those with t (transverse) do not.

In the (t,l,l) case, there are KK excitations of $SU(2) \times U(1)$ gauge bosons: $W_\pm^{(m)}$, $\gamma^{(m)}$ and $Z^{(m)}$. Performing a χ^2 fit of the electroweak observables, one finds that if the Higgs is a bulk state (l), $R_\parallel^{-1} \gtrsim 3.5$ TeV [12]. This implies that LHC can produce at most the first KK mode. Different choices for localization of matter and Higgs fields lead to bounds, lying in the range $1 - 5$ TeV [12].

In addition to virtual effects, KK excitations can be produced on-shell at LHC as new resonances [9] (see Fig. 10.3). There are two different channels, neutral Drell–Yan processes $pp \to l^+ l^- X$ and the charged channel $l^\pm \nu$, corresponding to the production of the KK modes $\gamma^{(1)}$, $Z^{(1)}$ and $W_\pm^{(1)}$, respectively. The discovery limits are about 6 TeV, while the exclusion bounds 15 TeV. An interesting observation in the case of $\gamma^{(1)} + Z^{(1)}$ is that interferences can lead to a "dip" just before the resonance. There are some ways to distinguish the corresponding signals from other possible origin of new physics, such as models with new gauge bosons. In fact, in the (t,l,l) and (t,l,t) cases, one expects two resonances located practically at the same mass value. This property is not shared by most of other new gauge boson models. Moreover, the heights and widths of the resonances are directly related to those of SM gauge bosons in the corresponding channels.

In the (t,l,t) case, only the $SU(2)$ factor feels the extra dimension and the limits set by the KK states of $W^\pm$ remain the same. On the other hand, in the (t,t,l) case where only $U(1)_Y$ feels the extra dimension, the limits are weaker and the exclusion bound is around 8 TeV. In addition to these simple possibilities, brane constructions lead often to cases where part of $U(1)_Y$ is t

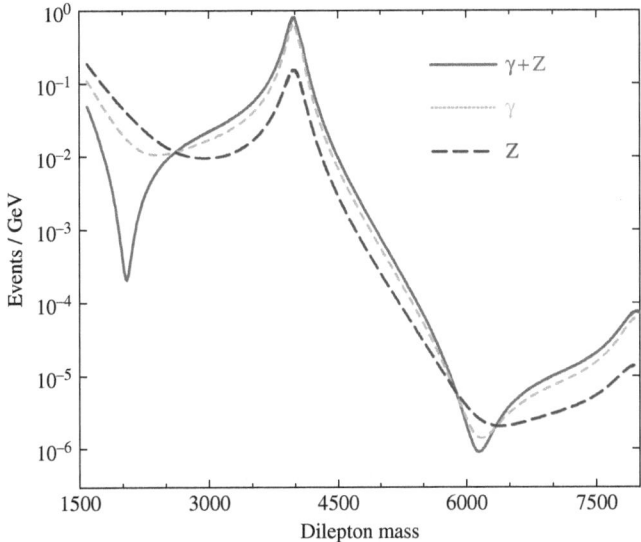

Fig. 10.3. Production of the first KK modes of the photon and of the Z boson at LHC, decaying to electron-positron pairs. The number of expected events is plotted as a function of the energy of the pair in GeV. From highest to lowest: excitation of $\gamma + Z$, γ and Z

and part is l. If $SU(2)$ is l the limits come again from $W^{\pm}$, while if it is t then it will be difficult to distinguish this case from a generic extra $U(1)'$. A good statistics would be needed to see the deviation in the tail of the resonance as being due to effects additional to those of a generic $U(1)'$ resonance. Finally, in the case of two or more parallel dimensions, the sum in the exchange of the KK modes diverges in the limit $R_{\parallel}M_s \gg 1$ and needs to be regularized using the form factor (10.5). Cross-sections become bigger yielding stronger bounds, while resonances are closer implying that more of them could be reached by LHC.

On the other hand, if all SM particles propagate in the extra dimension (called universal)[1], KK modes can only be produced in pairs and the lower bound on the compactification scale becomes weaker, of order of 300–500 GeV. Moreover, no resonances can be observed at LHC, so that this scenario appears very similar to low energy supersymmetry. In fact, KK parity can even play the role of R-parity, implying that the lightest KK mode is stable and can be a dark matter candidate in analogy to the LSP [13].

---

[1] Although interesting, this scenario seems difficult to be realized, since 4d chirality requires non-trivial action of orbifold twists with localized chiral states at the fixed points.

10.3.2 Extra Large Transverse Dimensions

The main experimental signal is gravitational radiation in the bulk from any physical process on the world-brane. In fact, the very existence of branes breaks translation invariance in the transverse dimensions and gravitons can be emitted from the brane into the bulk. During a collision of center of mass energy $\sqrt{s}$, there are $\sim (\sqrt{s}R_\perp)^n$ KK excitations of gravitons with tiny masses, that can be emitted. Each of these states looks from the 4d point of view as a massive, quasi-stable, extremely weakly coupled (s/M_P^2 suppressed) particle that escapes from the detector. The total effect is a missing-energy cross-section roughly of order:

$$\frac{(\sqrt{s}R_\perp)^n}{M_P^2} \sim \frac{1}{s}\left(\frac{\sqrt{s}}{M_s}\right)^{n+2}. \tag{10.7}$$

Explicit computation of these effects leads to the bounds given in Table 10.1.

However, larger radii are allowed if one relaxes the assumption of isotropy, by taking for instance two large dimensions with different radii.

Figure 10.4 shows the cross-section for graviton emission in the bulk, corresponding to the process $pp \rightarrow jet + graviton$ at LHC, together with the SM background [14]. For a given value of M_s, the cross-section for graviton emission decreases with the number of large transverse dimensions, in contrast to the case of parallel dimensions. The reason is that gravity becomes weaker if there are more dimensions because there is more space for the gravitational field to escape. There is a particular energy and angular distribution of the produced gravitons that arise from the distribution in mass of KK states of spin-2. This can be contrasted to other sources of missing energy and might be a smoking gun for the extra dimensional nature of such a signal.

In Table 10.1, there are also included astrophysical and cosmological bounds. Astrophysical bounds [15, 16] arise from the requirement that the radiation of gravitons should not carry on too much of the gravitational

Table 10.1. Limits on $R_\perp$ in mm

Experiment	$n = 2$	$n = 4$	$n = 6$
	Collider bounds		
LEP 2	5×10^{-1}	2×10^{-8}	7×10^{-11}
Tevatron	5×10^{-1}	10^{-8}	4×10^{-11}
LHC	4×10^{-3}	6×10^{-10}	3×10^{-12}
NLC	10^{-2}	10^{-9}	6×10^{-12}
	Present non-collider bounds		
SN1987A	3×10^{-4}	10^{-8}	6×10^{-10}
COMPTEL	5×10^{-5}	-	-

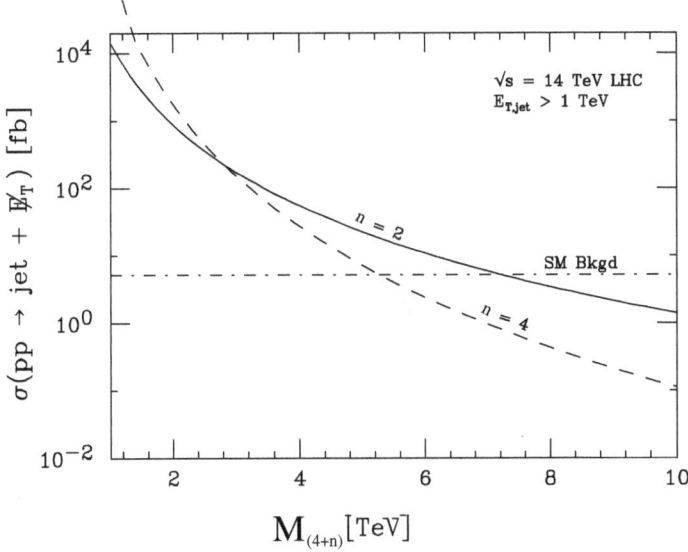

Fig. 10.4. Missing energy due to graviton emission at LHC, as a function of the higher-dimensional gravity scale M_*, produced together with a hadronic jet. The expected cross-section is shown for $n = 2$ and $n = 4$ extra dimensions, together with the SM background

binding energy released during core collapse of supernovae. In fact, the measurements of Kamiokande and IMB for SN1987A suggest that the main channel is neutrino fluxes. The best cosmological bound [17] is obtained from requiring that decay of bulk gravitons to photons do not generate a spike in the energy spectrum of the photon background measured by the COMPTEL instrument. Bulk gravitons are expected to be produced just before nucleosynthesis due to thermal radiation from the brane. The limits assume that the temperature was at most 1 MeV as nucleosynthesis begins, and become stronger if temperature is increased.

10.3.3 String Effects

At low energies, the interaction of light (string) states is described by an effective field theory. Their exchange generates in particular four-fermion operators that can be used to extract independent bounds on the string scale. In analogy with the bounds on longitudinal extra dimensions, there are two cases depending on the localization properties of matter fermions. If they come from open strings with both ends on the same stack of branes, exchange of massive open string modes gives rise to dimension eight effective operators, involving four fermions and two space-time derivatives [11, 18]. The corresponding bounds on the string scale are then around 500 GeV. On the other hand, if matter fermions are localized on non-trivial brane intersections, one obtains dimension six four-fermion operators and the bounds become stronger: $M_s \gtrsim 2 - 3$ TeV [5, 11]. At energies higher than the string scale, new spectacular phenomena are expected to occur, related to string physics and quantum gravity effects, such as possible micro-black hole production [19]. Particle accelerators would then become the best tools for studying quantum gravity and string theory.

10.4 Supersymmetry in the Bulk and Short Range Forces

10.4.1 Sub-millimeter Forces

Besides the spectacular predictions in accelerators, there are also modifications of gravitation in the sub-millimeter range, which can be tested in "table-top" experiments that measure gravity at short distances. There are three categories of such predictions:

(i) Deviations from the Newton's law $1/r^2$ behavior to $1/r^{2+n}$, which can be observable for $n = 2$ large transverse dimensions of sub-millimeter size. This case is particularly attractive on theoretical grounds because of the logarithmic sensitivity of SM couplings on the size of transverse space [20], that allows to determine the hierarchy [21].

(ii) New scalar forces in the sub-millimeter range, related to the mechanism of supersymmetry breaking, and mediated by light scalar fields φ with masses [4, 22]:

$$m_\varphi \simeq \frac{m_{susy}^2}{M_P} \simeq 10^{-4} - 10^{-6} \text{ eV} ,\qquad (10.8)$$

for a supersymmetry breaking scale $m_{susy} \simeq 1 - 10$ TeV. They correspond to Compton wavelengths of 1 mm to 10 μm. m_{susy} can be either $1/R_\parallel$ if supersymmetry is broken by compactification [22], or the string scale if it is broken "maximally" on our world-brane [4]. A universal attractive scalar force is mediated by the radion modulus $\varphi \equiv M_P \ln R$, with R the radius of the longitudinal or transverse dimension(s). In the former case, the result (10.8) follows from the behavior of the vacuum energy density $\Lambda \sim 1/R_\parallel^4$ for large $R_\parallel$ (up to logarithmic corrections). In the latter, supersymmetry is broken primarily on the brane, and thus its transmission to the bulk is gravitationally suppressed, leading to (10.8). For $n = 2$, there may be an enhancement factor of the radion mass by $\ln R_\perp M_s \simeq 30$ decreasing its wavelength by an order of magnitude [21].

The coupling of the radius modulus to matter relative to gravity can be easily computed and is given by:

$$\sqrt{\alpha_\varphi} = \frac{1}{M} \frac{\partial M}{\partial \varphi} \; ; \quad \alpha_\varphi = \begin{cases} \frac{\partial \ln \Lambda_{\text{QCD}}}{\partial \ln R} \simeq \frac{1}{3} \text{ for } R_\parallel \\[2ex] \frac{2n}{n+2} = 1 - 1.5 \text{ for } R_\perp \end{cases} \qquad (10.9)$$

where M denotes a generic physical mass. In the longitudinal case, the coupling arises dominantly through the radius dependence of the QCD gauge coupling [22], while in the case of transverse dimension, it can be deduced from the rescaling of the metric which changes the string to the Einstein frame and depends slightly on the bulk dimensionality ($\alpha = 1 - 1.5$ for $n = 2 - 6$) [21]. Such a force can be tested in microgravity experiments and should be contrasted with the change of Newton's law due the presence of extra dimensions that is observable only for $n = 2$ [3, 7]. The resulting bounds from an analysis of the radion effects are [3]:

$$M_* \gtrsim 3 - 4.5 \, \text{TeV} \quad \text{for} \quad n = 2 - 6 \,. \tag{10.10}$$

In principle there can be other light moduli which couple with even larger strengths. For example the dilaton, whose VEV determines the string coupling, if it does not acquire large mass from some dynamical supersymmetric mechanism, can lead to a force of strength 2000 times bigger than gravity [23]. (iii) Non universal repulsive forces much stronger than gravity, mediated by possible abelian gauge fields in the bulk [15, 24]. Such fields acquire tiny masses of the order of M_s^2/M_P, as in (10.8), due to brane localized anomalies [24]. Although their gauge coupling is infinitesimally small, $g_A \sim M_s/M_P \simeq 10^{-16}$, it is still bigger that the gravitational coupling E/M_P for typical energies $E \sim 1$ GeV, and the strength of the new force would be $10^6 - 10^8$ stronger than gravity. This is an interesting region which will be soon explored in micro-gravity experiments (see Fig. 10.5). Note that in this case supernova constraints impose that there should be at least four large extra dimensions in the bulk [15].

In Fig. 10.5 we depict the actual information from previous, present and upcoming experiments [7, 21]. The solid lines indicate the present limits from the experiments indicated. The excluded regions lie above these solid lines. Measuring gravitational strength forces at short distances is challenging. The dashed thick lines give the expected sensitivity of the various experiments, which will improve the actual limits by roughly two orders of magnitude, while the horizontal dashed lines correspond to the theoretical predictions for the graviton in the case $n = 2$ and for the radion in the transverse case. These limits are compared to those obtained from particle accelerator experiments in Table 10.1. Finally, in Figs. 10.6 and 10.7, we display recent improved bounds for new forces at very short distances by focusing on the right hand side of Fig. 10.5, near the origin [7].

10.4.2 Brane Non-linear Supersymmetry

When the closed string sector is supersymmetric, supersymmetry on a generic brane configuration is non-linearly realized even if the spectrum is not supersymmetric and brane fields have no superpartners. The reason is that the gravitino must couple to a conserved current locally, implying the existence

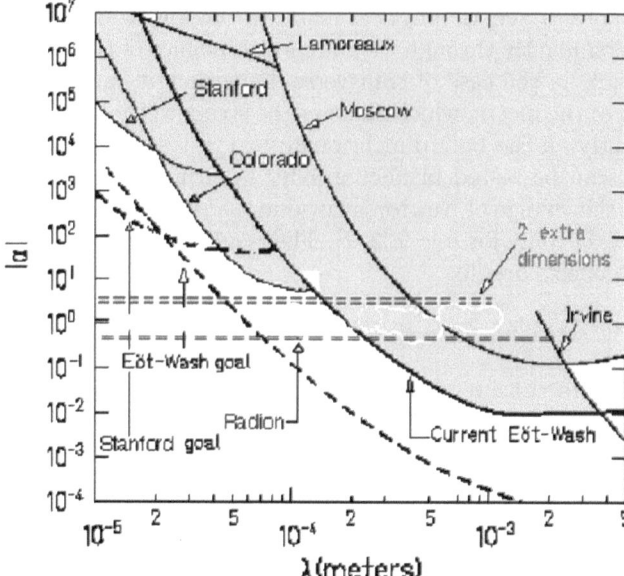

Fig. 10.5. Present limits on non-Newtonian forces at short distances (*regions above dotted lines*), as a function of their range λ and their strength relative to gravity α. The limits are compared to new forces mediated by the graviton in the case of two large extra dimensions, and by the radion

of a goldstino on the brane world-volume. The goldstino is exactly massless in the infinite (transverse) volume limit and is expected to acquire a small mass suppressed by the volume, of order (10.8). In the standard realization, its coupling to matter is given via the energy momentum tensor [25], while in

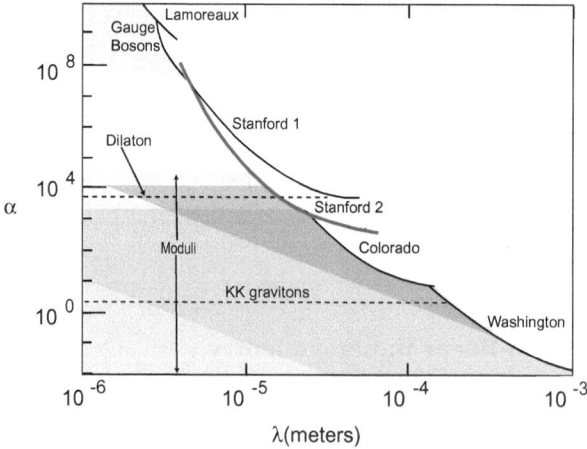

Fig. 10.6. Bounds on non-Newtonian forces in the range 6-20 μm (see S. J. Smullin et al., in [7])

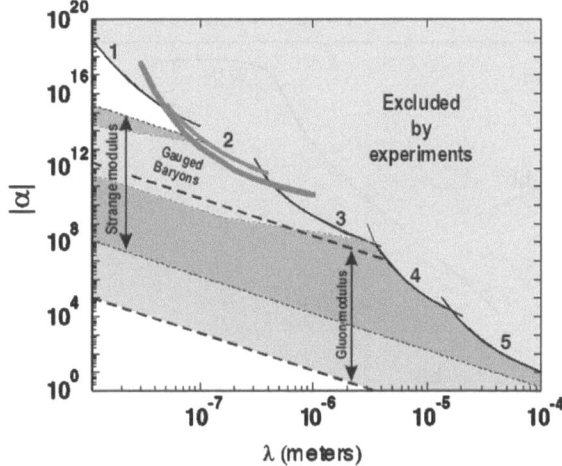

Fig. 10.7. Bounds on non-Newtonian forces in the range around 200 nm (see R. S. Decca et al., in [7]). Curves 4 and 5 correspond to Stanford and Colorado experiments, respectively, of Fig. 10.6 (see also J C. Long and J. C. Price of [7])

general there are more terms invariant under non-linear supersymmetry that have been classified, up to dimension eight [26, 27].

An explicit computation was performed for a generic intersection of two brane stacks, leading to three irreducible couplings, besides the standard one [27]: two of dimension six involving the goldstino, a matter fermion and a scalar or gauge field, and one four-fermion operator of dimension eight. Their strength is set by the goldstino decay constant κ, up to model-independent numerical coefficients which are independent of the brane angles. Obviously, at low energies the dominant operators are those of dimension six. In the minimal case of (non-supersymmetric) SM, only one of these two operators may exist, that couples the goldstino χ with the Higgs H and a lepton doublet L:

$$\mathcal{L}_{\chi}^{int} = 2\kappa(D_{\mu}H)(LD^{\mu}\chi) + h.c., \tag{10.11}$$

where the goldstino decay constant is given by the total brane tension

$$\frac{1}{2\,\kappa^2} = N_1\,T_1 + N_2\,T_2\,; \quad T_i = \frac{M_s^4}{4\pi^2 g_i^2}\,, \tag{10.12}$$

with N_i the number of branes in each stack. It is important to notice that the effective interaction (10.11) conserves the total lepton number L, as long as we assign to the goldstino a total lepton number $L(\chi) = -1$ [28]. To simplify the analysis, we will consider the simplest case where (10.11) exists only for the first generation and L is the electron doublet [28].

The effective interaction (10.11) gives rise mainly to the decays $W^{\pm} \to e^{\pm}\chi$ and $Z, H \to \nu\chi$. It turns out that the invisible Z width gives the strongest limit on κ which can be translated to a bound on the string scale $M_s \gtrsim$

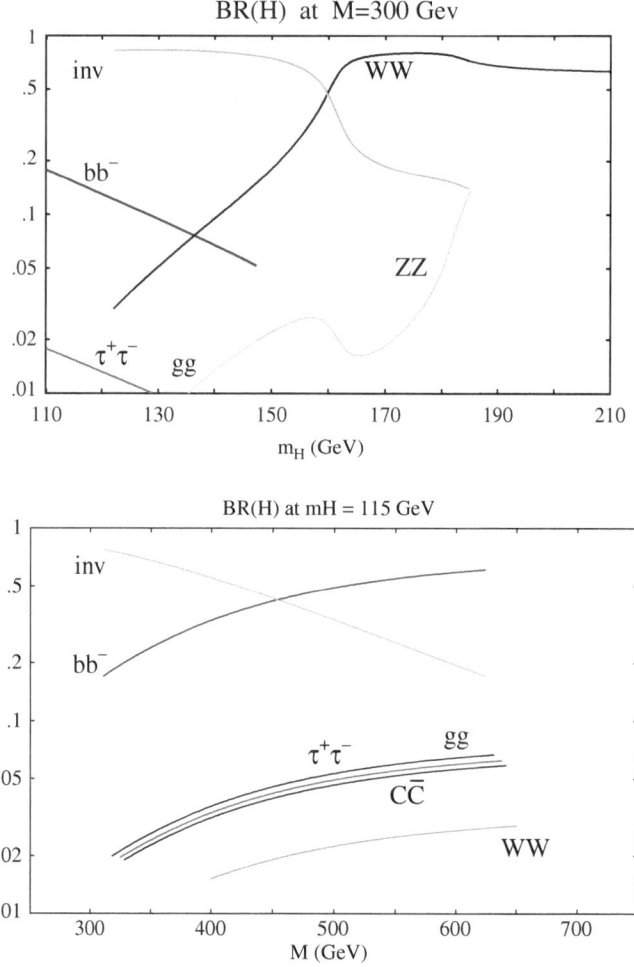

Fig. 10.8. Higgs branching rations, as functions either of the Higgs mass m_H for a fixed value of the string scale $M_s \simeq 2M = 600$ GeV, or of $M \simeq M_s/2$ for $m_H = 115$ GeV

500 GeV, comparable to other collider bounds. This allows for the striking possibility of a Higgs boson decaying dominantly, or at least with a sizable branching ratio, via such an invisible mode, for a wide range of the parameter space (M_s, m_H), as seen in Fig. 10.8.

10.5 Electroweak Symmetry Breaking

Non-supersymmetric TeV strings offer also a framework to realize gauge symmetry breaking radiatively. Indeed, from the effective field theory point of view, one expects quadratically divergent one-loop contributions to the masses of scalar fields. The divergences are cut off by M_s and if the corrections are negative, they can induce electroweak symmetry breaking and explain the mild hierarchy between the weak and a string scale at a few TeV, in terms of a loop factor [29]. More precisely, in the minimal case of one Higgs doublet H, the scalar potential is:

$$V = \lambda (H^\dagger H)^2 + \mu^2 (H^\dagger H) , \qquad (10.13)$$

where λ arises at tree-level. Moreover, in any model where the Higgs field comes from an open string with both ends fixed on the same brane stack, it is given by an appropriate truncation of a supersymmetric theory. Within the minimal spectrum of the SM, $\lambda = (g_2^2 + g'^2)/8$, with g_2 and g' the $SU(2)$ and $U(1)_Y$ gauge couplings. On the other hand, μ^2 is generated at one loop:

$$\mu^2 = -\varepsilon^2 g^2 M_s^2 , \qquad (10.14)$$

where ε is a loop factor that can be estimated from a toy model computation and varies in the region $\epsilon \sim 10^{-1} - 10^{-3}$.

Indeed, consider for illustration a simple case where the whole one-loop effective potential of a scalar field can be computed. We assume for instance one extra dimension compactified on a circle of radius $R > 1$ (in string units). An interesting situation is provided by a class of models where a non-vanishing VEV for a scalar (Higgs) field ϕ results in shifting the mass of each KK excitation by a constant $a(\phi)$:

$$M_m^2 = \left(\frac{m + a(\phi)}{R} \right)^2 , \qquad (10.15)$$

with m the KK integer momentum number. Such mass shifts arise for instance in the presence of a Wilson line, $a = q \oint \frac{dy}{2\pi} g A$, where A is the internal component of a gauge field with gauge coupling g, and q is the charge of a given state under the corresponding generator. A straightforward computation shows that the ϕ-dependent part of the one-loop effective potential is given by [30]:

$$V_{eff} = -Tr(-)^F \frac{R}{32\,\pi^{3/2}} \sum_n e^{2\pi i n a} \int_0^\infty dl\; l^{3/2} f_s(l)\; e^{-\pi^2 n^2 R^2 l} \qquad (10.16)$$

where $F = 0,1$ for bosons and fermions, respectively. We have included a regulating function $f_s(l)$ which contains for example the effects of string oscillators. To understand its role we will consider the two limits $R >> 1$ and $R << 1$. In the first case only the $l \to 0$ region contributes to the integral. This means that the effective potential receives sizable contributions only from the infrared (field theory) degrees of freedom. In this limit we would have $f_s(l) \to 1$. For example, in the string model considered in [29]:

$$f_s(l) = \left[\frac{1}{4l} \frac{\theta_2}{\eta^3} (il + \frac{1}{2}) \right]^4 \to 1 \qquad \text{for} \qquad l \to 0 \qquad (10.17)$$

and the field theory result is finite and can be explicitly computed. As a result of the Taylor expansion around $a = 0$, we are able to extract the one-loop contribution to the coefficient of the term of the potential quadratic in the Higgs field. It is given by a loop factor times the compactification scale [30]. One thus obtains $\mu^2 \sim g^2/R^2$ up to a proportionality constant which is calculable in the effective field theory. On the other hand, if we consider $R \to 0$, which by T-duality corresponds to taking the extra dimension as transverse and very large, the one-loop effective potential receives contributions from the whole tower of string oscillators as appearing in $f_s(l)$, leading to squared masses given by a loop factor times M_s^2, according to (10.14).

More precisely, from the expression (10.16), one finds:

$$\varepsilon^2(R) = \frac{1}{2\pi^2} \int_0^\infty \frac{dl}{(2\,l)^{5/2}} \frac{\theta_2^4}{4\eta^{12}} \left(il + \frac{1}{2} \right) R^3 \sum_n n^2 e^{-2\pi n^2 R^2 l} \qquad (10.18)$$

which is plotted in Fig. 10.9. For the asymptotic value $R \to 0$ (corresponding upon T-duality to a large transverse dimension of radius $1/R$), $\varepsilon(0) \simeq 0.14$, and the effective cut-off for the mass term is M_s, as can be seen from(10.14). At large R, $\mu^2(R)$ falls off as $1/R^2$, which is the effective cut-off in the limit $R \to \infty$, as we argued above, in agreement with field theory results in the presence of a compactified extra dimension [22, 31]. In fact, in the limit $R \to \infty$, an analytic approximation to $\varepsilon(R)$ gives:

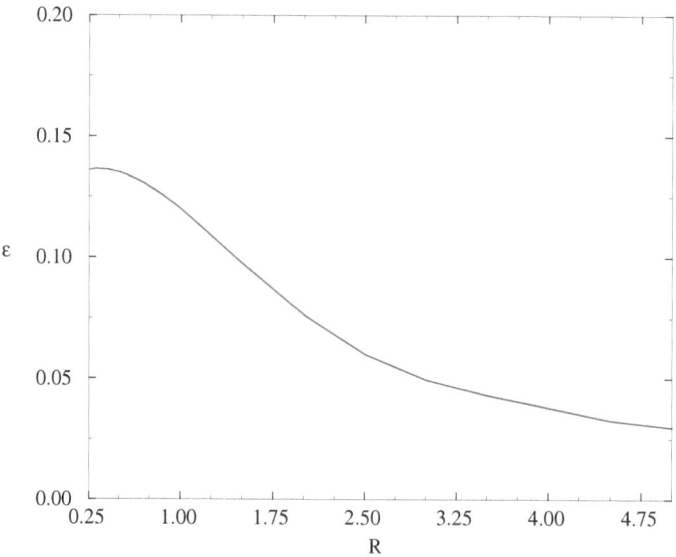

Fig. 10.9. The coefficient ε of the one loop Higgs mass (10.14)

$$\varepsilon(R) \simeq \frac{\varepsilon_\infty}{M_s R}, \qquad \varepsilon_\infty^2 = \frac{3\,\zeta(5)}{4\,\pi^4} \simeq 0.008 \qquad (10.19)$$

The potential (10.13) has the usual minimum, given by the VEV of the neutral component of the Higgs doublet $v = \sqrt{-\mu^2/\lambda}$. Using the relation of v with the Z gauge boson mass, $M_Z^2 = (g_2^2 + g'^2)v^2/4$, and the expression of the quartic coupling λ, one obtains for the Higgs mass a prediction which is the MSSM value for $\tan\beta \to \infty$ and $m_A \to \infty$: $m_H = M_Z$. The tree level Higgs mass is known to receive important radiative corrections from the top-quark sector and rises to values around 120 GeV. Furthermore, from (10.14), one can compute M_s in terms of the Higgs mass $m_H^2 = -2\mu^2$:

$$M_s = \frac{m_H}{\sqrt{2}\,g\varepsilon} \qquad (10.20)$$

yielding naturally values in the TeV range.

10.6 Standard Model on D-branes

The gauge group closest to the Standard Model one can easily obtain with D-branes is $U(3) \times U(2) \times U(1)$. The first factor arises from three coincident "color" D-branes. An open string with one end on them is a triplet under $SU(3)$ and carries the same $U(1)$ charge for all three components. Thus, the $U(1)$ factor of $U(3)$ has to be identified with *gauged* baryon number. Similarly, $U(2)$ arises from two coincident "weak" D-branes and the corresponding abelian factor is identified with *gauged* weak-doublet number. Finally, an extra $U(1)$ D-brane is necessary in order to accommodate the Standard Model without breaking the baryon number [32]. In principle this $U(1)$ brane can be chosen to be independent of the other two collections with its own gauge coupling. To improve the predictability of the model, we choose to put it on top of either the color or the weak D-branes [33]. In either case, the model has two independent gauge couplings g_3 and g_2 corresponding, respectively, to the gauge groups $U(3)$ and $U(2)$. The $U(1)$ gauge coupling g_1 is equal to either g_3 or g_2.

Let us denote by Q_3, Q_2 and Q_1 the three $U(1)$ charges of $U(3) \times U(2) \times U(1)$, in a self explanatory notation. Under $SU(3) \times SU(2) \times U(1)_3 \times U(1)_2 \times U(1)_1$, the members of a family of quarks and leptons have the following quantum numbers:

$$
\begin{aligned}
&Q \;\; (\mathbf{3}, \mathbf{2}; 1, w, 0)_{1/6} \\
&u^c \;\; (\bar{\mathbf{3}}, \mathbf{1}; -1, 0, x)_{-2/3} \\
&d^c \;\; (\bar{\mathbf{3}}, \mathbf{1}; -1, 0, y)_{1/3} \qquad (10.21)\\
&L \;\; (\mathbf{1}, \mathbf{2}; 0, 1, z)_{-1/2} \\
&l^c \;\; (\mathbf{1}, \mathbf{1}; 0, 0, 1)_1
\end{aligned}
$$

The values of the $U(1)$ charges x, y, z, w will be fixed below so that they lead to the right hypercharges, shown for completeness as subscripts.

It turns out that there are two possible ways of embedding the Standard Model particle spectrum on these stacks of branes [32], which are shown pictorially in Fig. 10.10. The quark doublet Q corresponds necessarily to a massless excitation of an open string with its two ends on the two different collections of branes (color and weak). As seen from the figure, a fourth brane stack is needed for a complete embedding, which is chosen to be a $U(1)_b$ extended in the bulk. This is welcome since one can accommodate right handed neutrinos as open string states on the bulk with sufficiently small Yukawa couplings suppressed by the large volume of the bulk [34]. The two models are obtained by an exchange of the up and down antiquarks, u^c and d^c, which correspond to open strings with one end on the color branes and the other either on the $U(1)$ brane, or on the $U(1)_b$ in the bulk. The lepton doublet L arises from an open string stretched between the weak branes and $U(1)_b$, while the antilepton l^c corresponds to a string with one end on the $U(1)$ brane and the other in the bulk. For completeness, we also show the two possible Higgs states H_u and H_d that are both necessary in order to give tree-level masses to all quarks and leptons of the heaviest generation.

The weak hypercharge Y is a linear combination of the three $U(1)$'s:

$$Y = Q_1 + \frac{1}{2}Q_2 + c_3 Q_3 \quad ; \quad c_3 = -1/3 \text{ or } 2/3 , \qquad (10.22)$$

where Q_N denotes the $U(1)$ generator of $U(N)$ normalized so that the fundamental representation of $SU(N)$ has unit charge. The corresponding $U(1)$ charges appearing in (10.21) are $x = -1$ or 0, $y = 0$ or 1, $z = -1$, and $w = 1$ or -1, for $c_3 = -1/3$ or $2/3$, respectively. The hypercharge coupling g_Y is

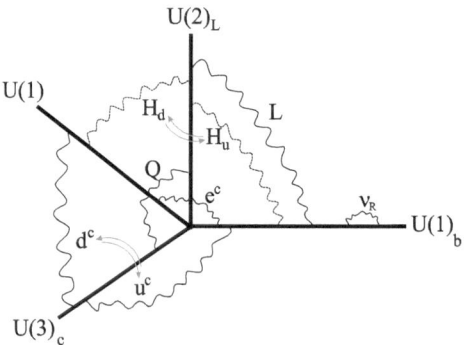

Fig. 10.10. A minimal Standard Model embedding on D-branes

given by[2]:

$$\frac{1}{g_Y^2} = \frac{2}{g_1^2} + \frac{4c_2^2}{g_2^2} + \frac{6c_3^2}{g_3^2} \,.$$

(10.23)

It follows that the weak angle $\sin^2 \theta_W$, is given by:

$$\sin^2 \theta_W \equiv \frac{g_Y^2}{g_2^2 + g_Y^2} = \frac{1}{2 + 2g_2^2/g_1^2 + 6c_3^2 g_2^2/g_3^2} \,,$$

(10.24)

where g_N is the gauge coupling of $SU(N)$ and $g_1 = g_2$ or $g_1 = g_3$ at the string scale. In order to compare the theoretical predictions with the experimental value of $\sin^2 \theta_W$ at M_s, we plot in Fig. 10.11 the corresponding curves as functions of M_s.

The solid line is the experimental curve. The dashed line is the plot of the function (10.24) for $g_1 = g_2$ with $c_3 = -1/3$ while the dotted-dashed line corresponds to $g_1 = g_3$ with $c_3 = 2/3$. The other two possibilities are not shown because they lead to a value of M_s which is too high to protect the hierarchy. Thus, the second case, where the $U(1)$ brane is on top of the color branes, is compatible with low energy data for $M_s \sim 6 - 8$ TeV and $g_s \simeq 0.9$.

From (10.24) and Fig. 10.11, we find the ratio of the $SU(2)$ and $SU(3)$ gauge couplings at the string scale to be $\alpha_2/\alpha_3 \sim 0.4$. This ratio can be arranged by an appropriate choice of the relevant moduli. For instance, one may choose the color and $U(1)$ branes to be D3 branes while the weak branes to be D7 branes. Then, the ratio of couplings above can be explained by choosing the volume of the four compact dimensions of the seven branes to be $V_4 = 2.5$ in string units. This being larger than one is consistent with the picture above. Moreover it predicts an interesting spectrum of KK states

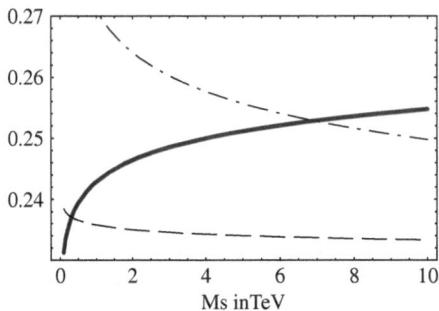

Fig. 10.11. The experimental value of $\sin^2 \theta_W$ (*thick curve*), and the theoretical predictions (10.24)

---

[2] The gauge couplings $g_{2,3}$ are determined at the tree-level by the string coupling and other moduli, like radii of longitudinal dimensions. In higher orders, they also receive string threshold corrections.

for the Standard model, different from the naive choices that have appeared hitherto: the only Standard Model particles that have KK descendants are the W bosons as well as the hypercharge gauge boson. However, since the hypercharge is a linear combination of the three $U(1)$'s, the massive $U(1)$ KK gauge bosons do not couple to the hypercharge but to the weak doublet number.

10.7 Non-compact Extra Dimensions and Localized Gravity

There are several motivations to study localization of gravity in non-compact extra dimensions: (i) it avoids the problem of fixing the moduli associated to the size of the compactification manifold; (ii) it provides a new approach to the mass hierarchy problem; (iii) there are modifications of gravity at large distances that may have interesting observational consequences. Two types of models have been studied: warped metrics in curved space [35], and in-finite size extra dimensions in flat space [36]. The former, although largely inspired by stringy developments and having used many string-theoretic tech-niques, have not yet a clear and calculable string theory realization [37]. In any case, since curved space is always difficult to handle in string theory, in the following we concentrate mainly on the latter, formulated in flat space with gravity localized on a subspace of the bulk. It turns out that these models of induced gravity have an interesting string theory realization [38] that we describe below, after presenting first a brief overview of the warped case [39].

10.7.1 Warped Spaces

In these models, space-time is a slice of anti de Sitter space (AdS) in $d = 5$ di-mensions while our universe forms a four-dimensional (4d) flat boundary [35]. The corresponding line element is:

$$ds^2 = e^{-2k|y|}\eta_{\mu\nu}dx^\mu dx^\nu + dy^2 \quad ; \quad \Lambda = -24M^3k^2 , \qquad (10.25)$$

where M, Λ are the 5d Planck mass and cosmological constant, respectively, and the parameter k is the curvature of AdS_5. The fifth coordinate y is re-stricted on the interval $[0, \pi r_c]$. Thus, this model requires two 'branes', a UV and an IR, located at the two end-points of the interval, $y = 0$ and $y = \pi r_c$, respectively. The vanishing of the 4d cosmological constant requires to fine tune the two tensions: $T = -T' = 24M^3k^2$. The 4d Planck mass is given by:

$$M_P^2 = \frac{1}{k}(1 - e^{-2\pi kr_c})M^3 . \qquad (10.26)$$

Note that the IR brane can move to infinity by taking the limit $r_c \to \infty$, while M_P is kept finite and thus 4d gravity is always present on the brane. The reason is that the internal volume remains finite in the non-compact limit along the positive y axis. As a result, gravity is kept localized on the UV brane, while the Newtonian potential gets corrections, $1/r + 1/k^2 r^3$, which are identical with those arising in the compact case of two flat extra dimensions. Using the experimental limit $k^{-1} \lesssim 0.1$ mm and the relation (10.26), one finds a bound for the 5d gravity scale $M \gtrsim 10^8$ GeV, corresponding to a brane tension $T \gtrsim 1$ TeV. Notice that this bound is not valid in the compact case of six extra dimensions, because their size is in the fermi range and thus the $1/r^3$ deviations of Newton's law are cutoff at shorter distances.

10.7.2 The Induced Gravity Model

The DGP model and its generalizations are specified by a bulk Einstein-Hilbert (EH) term and a four-dimensional EH term [36]:

$$M^{2+n} \int_{\mathcal{M}_{4+n}} d^4x d^n y \sqrt{G}\, \mathcal{R}_{(4+n)} + M_P^2 \int_{\mathcal{M}_4} d^4x \sqrt{g}\, \mathcal{R}_{(4)} \; ; \quad M_P^2 \equiv r_c^n M^{2+n}$$

(10.27)

with M and M_P the (possibly independent) respective Planck scales. The scale $M \geq 1$ TeV would be related to the short-distance scale below which UV quantum gravity or stringy effects are important. The four-dimensional metric is the restriction of the bulk metric $g_{\mu\nu} = G_{\mu\nu}|$ and we assume the WORLD[3] rigid, allowing the gauge $G_{i\mu}| = 0$ with $i \geq 5$. Finally, only intrinsic curvature terms are omitted but no Gibbons–Hawking term is needed.

Co-dimension One

In the case of co-dimension one bulk ($n = 1$) and δ-function localization, it is easy to see that r_c is a crossover scale where gravity changes behavior on the WORLD. Indeed, by Fourier transform the quadratic part of the action (10.27) with respect to the 4d position x, at the WORLD position $y = 0$, one obtains $M^{2+n}(p^{2-n} + r_c^n p^2)$, where p is the 4d momentum. It follows that for distances smaller than r_c (large momenta), the first term becomes irrelevant and the graviton propagator on the "brane" exhibits four-dimensional behavior $(1/p^2)$ with Planck constant $M_P = M^3 r_c$. On the contrary, at large distances, the first term becomes dominant and the graviton propagator acquires a five-dimensional fall-off $(1/p)$ with Planck constant M. Imposing r_c to be larger than the size of the universe, $r_c \gtrsim 10^{28}$ cm, one finds $M \lesssim 100$

---

[3] We avoid calling $\mathcal{M}_4$ a brane because, as we will see below, gravity localizes on singularities of the internal manifold, such as orbifold fixed points. Branes with localized matter can be introduced independently of gravity localization.

MeV, which seems to be in conflict with experimental bounds. However, there were arguments that these bounds can be evaded, even for values of the fundamental scale $M^{-1} \sim 1$ mm that one may need for suppressing the quantum corrections of the cosmological constant [36].

On the other hand, in the presence of non-zero brane thickness w, a new crossover length-scale seems to appear, $R_c \sim (wr_c)^{1/2}$ [40] or $r_c^{3/5} w^{2/5}$ [41].

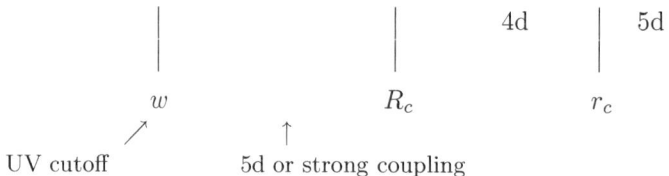

Below this scale, the theory acquires either again a five-dimensional behavior, or a strong coupling regime. For $r_c \sim 10^{28}$ cm, the new crossover scale is of order $R_c \sim 10^{-4} - 10$ m.

Higher Co-dimension

The situation changes drastically for more than one non-compact bulk dimensions, $n > 1$, due to the ultraviolet properties of the higher-dimensional theories. Indeed, from the action (10.27), the effective potential between two test masses in four dimensions

$$\int [d^3 x]\, e^{-ip\cdot x}\, V(x) = \frac{D(p)}{1 + r_c^n\, p^2\, D(p)} \left[\tilde{T}_{\mu\nu} T^{\mu\nu} - \frac{1}{2+n} \tilde{T}_\mu^\mu\, T_\nu^\nu \right] \quad (10.28)$$

$$D(p) = \int [d^n q]\, \frac{f_w(q)}{p^2 + q^2} \quad (10.29)$$

is a function of the bulk graviton retarded Green's function $G(x,0;0,0) = \int [d^4 p] \times e^{ip\cdot x}\, D(p)$ evaluated for two points localized on the WORLD ($y = y' = 0$). The integral (10.29) is UV-divergent for $n > 1$ unless there is a non-trivial brane thickness profile $f_w(q)$ of width w. If the four-dimensional WORLD has zero thickness, $f_w(q) \sim 1$, the bulk graviton does not have a normalizable wave function. It therefore cannot contribute to the induced potential, which always takes the form $V(p) \sim 1/p^2$ and Newton's law remains four-dimensional at all distances.

For a non-zero thickness w, there is only one crossover length scale, R_c:

$$R_c = w \left(\frac{r_c}{w} \right)^{\frac{n}{2}}, \quad (10.30)$$

above which one obtains a higher-dimensional behaviour [42]. Therefore the effective potential presents two regimes: (i) at short distances ($w \ll r \ll R_c$)

the gravitational interactions are mediated by the localized four-dimensional graviton and Newton's potential on the WORLD is given by $V(r) \sim 1/r$ and, (ii) at large distances $(r \gg R_c)$ the modes of the bulk graviton dominate, changing the potential. Note that for $n = 1$ the expressions (10.28) and (10.29) are finite and unambiguously give $V(r) \sim 1/r$ for $r \gg r_c$. For a co-dimension bigger than 1, the precise behavior for large-distance interactions depends *crucially* on the UV completion of the theory.

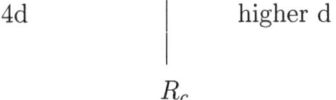

At this point we stress a fundamental difference with the *finite extra dimensions* scenarios. In these cases Newton's law gets higher-dimensional at distances smaller than the characteristic size of the extra dimensions. This is precisely the opposite of the case of infinite volume extra dimensions that we discuss here.

As mentioned above, for higher co-dimension, there is an interplay between UV regularization and IR behavior of the theory. Indeed, several works in the literature raised unitarity [43] and strong coupling problems [44] which depend crucially on the UV completion of the theory. A unitary UV regularization for the higher co-dimension version of the model has been proposed in [45]. It would be interesting to address these questions in a precise string theory context. Actually, using for UV cutoff on the "brane" the 4d Planck length $w \sim l_P$, one gets for the crossover scale (10.30): $R_c \sim M^{-1}(M_P/M)^{n/2}$. Putting $M \gtrsim 1$ TeV leads to $R_c \lesssim 10^{8(n-2)}$ cm. Imposing $R_c \gtrsim 10^{28}$ cm, one then finds that the number of extra dimensions must be at least six, $n \geq 6$, which is realized nicely in string theory and provides an additional motivation for studying possible string theory realizations.

10.7.3 String Theory Realization

In the following, we explain how to realize the gravity induced model (10.27) with $n \geq 6$ as the low-energy effective action of string theory on a noncompact six-dimensional manifold $\mathcal{M}_6$ [38]. We work in the context of $\mathcal{N} = 2$ supergravities in four dimensions but the mechanism for localizing gravity is independent of the number of supersymmetries. Of course for $\mathcal{N} \geq 3$ supersymmetries, there is no localization. We also start with the compact case and take the decompactification limit. The localized properties are then encoded in the different volume dependences.

In string perturbation, corrections to the four-dimensional Planck mass are in general very restrictive. In the heterotic string, they vanish to all orders in perturbation theory [46]; in type I theory, there are moduli-dependent corrections generated by open strings [47], but they vanish when the manifold $\mathcal{M}_6$ is decompactified; in type II theories, they are constant, independent of the moduli of the manifold $\mathcal{M}_6$, and receive contributions only from tree and one-loop levels that we describe below (at least for supersymmetric backgrounds) [38, 48]. Finally, in the context of M-theory, one obtains a similar localized action of gravity kinetic terms in five dimensions, corresponding to the strong coupling limit of type IIA string [38].

The origin of the two EH terms in (10.27) can be traced back to the perturbative corrections to the eight-derivative effective action of type II strings in ten dimensions. These corrections include the tree-level and one-loop terms given by:[4]

$$\frac{1}{l_s^8} \int_{M_{10}} \frac{1}{g_s^2} \mathcal{R}_{(10)} + \frac{1}{l_s^2} \int_{M_{10}} \left(\frac{2\zeta(3)}{g_s^2} + 4\zeta(2) \right) t_8 t_8 R^4 \qquad (10.31)$$
$$- \frac{1}{l_s^2} \int_{M_{10}} \left(\frac{2\zeta(3)}{g_s^2} M_{\mathrm{Pl}} 4\zeta(2) \right) R \wedge R \wedge R \wedge R \wedge e \wedge e + \cdots$$

where ϕ is the dilaton field determining the string coupling $g_s = e^{\langle \phi \rangle}$, and the $\pm$ sign corresponds to the type IIA/B theory.

On a direct product space-time $\mathcal{M}_6 \times \mathbb{R}^4$, the $t_8 t_8 R^4$ contribute in four dimensions to R^2 and R^4 terms [48]. At the level of zero modes, the second R^4 term in (10.31) splits as:

$$\int_{M_6} R \wedge R \wedge R \times \int_{M_4} \mathcal{R}_{(4)} = \chi \int_{M_4} \mathcal{R}_{(4)} , \qquad (10.32)$$

where χ is the Euler number of the M_6 compactification manifold. We thus obtain the action terms:

$$\frac{1}{l_s^8} \int_{M_4 \times M_6} \frac{1}{g_s^2} \mathcal{R}_{(10)} + \frac{\chi}{l_s^2} \int_{M_4} \left(-\frac{2\zeta(3)}{g_s^2} \pm 4\zeta(2) \right) \mathcal{R}_{(4)} , \qquad (10.33)$$

which gives the expressions for the Planck masses M and M_p:

$$M^2 \sim M_s^2 / g_s^{1/2} \quad ; \quad M_P^2 \sim \chi (\frac{c_0}{g_s^2} + c_1) M_s^2 , \qquad (10.34)$$

with $c_0 = -2\zeta(3)$ and $c_1 = \pm 4\zeta(2) = \pm 2\pi^2/3$.

It is interesting that the appearance of the induced 4d localized term preserves $\mathcal{N} = 2$ supersymmetry and is independent of the localization mechanism of matter fields (for instance on D-branes). Localization requires the internal space $\mathcal{M}_6$ to have a non-zero Euler characteristic $\chi \neq 0$. Actually, in type IIA/B compactified on a Calabi-Yau manifold, χ counts the difference between the numbers of $\mathcal{N} = 2$ vector multiplets and hypermultiplets:

---

[4] The rank-eight tensor t_8 is defined as $t_8 M^4 \equiv -6(\mathrm{tr}M^2)^2 + 24\mathrm{tr}M^4$. See [49] for more details.

$\chi = \pm 4(n_V - n_H)$ (where the graviton multiplet counts as one vector). Moreover, in the non-compact limit, the Euler number can in general split in different singular points of the internal space, $\chi = \sum_I \chi_I$, giving rise to different
localized terms at various points y_I of the internal space. A number of conclusions (confirmed by string calculations in [38]) can be reached by looking
closely at (10.33):

▷ $M_p \gg M$ requires a large non-zero Euler characteristic for M_6, and/or
a weak string coupling constant $g_s \to 0$.

▷ Since χ is a topological invariant the localized $\mathcal{R}_{(4)}$ term coming from the
closed string sector is universal, independent of the background geometry and
dependent only on the internal topology[5]. It is a matter of simple inspection to
see that if one wants to have a localized EH term in less than ten dimensions,
namely something linear in curvature, with non-compact internal space in all
directions, *the only possible dimension is four* (or five in the strong coupling
M-theory limit).

▷ In order to find the width w of the localized term, one has to do a
separate analysis. On general grounds, using dimensional analysis in the limit
$M_P \to \infty$, one expects the effective width to vanish as a power of $l_P \equiv M_P^{-1}$:
$w \sim l_P^\nu/l_s^{\nu-1}$ with $\nu > 0$. The computation of ν for a general Calabi-Yau space,
besides its technical difficulty, presents an additional important complication:
from the expression (10.34), $l_P \sim g_s l_s$ in the weak coupling limit. Thus, w
vanishes in perturbation theory and one has to perform a non-perturbative
analysis to extract its behavior. Alternatively, one can examine the case of
orbifolds. In this limit, $c_0 = 0$, $l_P \sim l_s$, and the hierarchy $M_P > M$ is achieved
only in the limit of large χ.

The one-loop graviton amplitude for the supersymmetric orbifold $T^6/\mathbb{Z}_N$,
takes the form of a sum of quasi-localized contributions at the positions of the
fixed points x_f of the orbifold [38]:

$$\langle V_g^3 \rangle \sim \frac{1}{N} \sum_{(h,g)} \sum_{x_f} \int_\mathcal{F} \frac{d^2\tau}{\tau_2^2} \int \prod_{i=1}^3 \frac{d^2 z_i}{\tau_2} \frac{1}{F_{(h,g)}(\tau, z_i)^3} e^{-\frac{(y-x_f)^2}{\alpha' F_{(h,g)}(\tau, z_i)}} , \quad (10.35)$$

where (h, g) denote the orbifold twists and $\tau = \tau_1 + i\tau_2$ is the complex modulus of the world-sheet torus, integrated over its fundamental domain $\mathcal{F}$. The
above expression (10.35) gives the three-point amplitude involving three 4d
gravitons on-shell. Focusing on one particular fixed point $x_f = 0$ and sending
the radii to infinity, we obtain the effective action for the quasi-localized EH
term

$$\chi \int d^4x d^6y \sqrt{g} f_w(y) \mathcal{R}_{(4)} \quad (10.36)$$

---

[5] Field theory computations of [50] show that the Planck mass renormalization
depends on the UV behavior of the matter fields coupling to the external metric.
But, even in the supersymmetric case, the corrections are not obviously given by
an index.

with a width given by the four-dimensional induced Planck mass

$$w \simeq l_P = l_s \, \chi^{-1/2} \, , \tag{10.37}$$

and the power $\nu = 1$.

Summary of the Results

Using $w \sim l_P$ and the relations (10.34) in the weak coupling limit (with $c_0 \neq 0$), the crossover radius of (10.30) is given by the string parameters $(n = 6)$

$$R_c = \frac{r_c^3}{w^2} \sim g_s \frac{l_s^4}{l_P^3} \simeq g_s \times 10^{32} \text{ cm} \, , \tag{10.38}$$

for $M_s \simeq 1$ TeV. Because R_c has to be of cosmological size, the string coupling can be relatively small, and the Euler number $|\chi| \simeq g_s^2 l_P \sim g_s^2 \times 10^{32}$ must be very large. The hierarchy is obtained mainly thanks to the large value of χ, so that lowering the bound on R_c lowers the value of χ. Our actual knowledge of gravity at very large distances indicates [51] that R_c should be of the order of the Hubble radius $R_c \simeq 10^{28}$ cm, which implies $g_s \geq 10^{-4}$ and $|\chi| \gtrsim 10^{24}$. A large Euler number implies only a large number of closed string massless particles with no a-priori constraint on the observable gauge and matter sectors, which can be introduced for instance on D3-branes placed at the position where gravity localization occurs. All these particles are localized at the orbifold fixed points (or where the Euler number is concentrated in the general case), and should have sufficiently suppressed gravitational-type couplings, so that their presence with such a huge multiplicity does not contradict observations. Note that these results depend crucially on the scaling of the width w in terms of the Planck length: $w \sim l_P^\nu$, implying $R_c \sim 1/l_P^{2\nu+1}$ in string units. If there are models with $\nu > 1$, the required value of χ will be much lower, becoming $\mathcal{O}(1)$ for $\nu \geq 3/2$. In this case, the hierarchy could be determined by tuning the string coupling to infinitesimal values, $g_s \sim 10^{-16}$.

The explicit string realization of localized induced gravity models offers a consistent framework that allows to address a certain number of interesting physics problems. In particular, the effective UV cutoff and the study of the gravity force among matter sources localized on D-branes. It would be also interesting to perform explicit model building and study in detail the phenomenological consequences of these models and compare to other realizations of TeV strings with compact dimensions.

Acknowledgments

This work was supported in part by the European Commission under the RTN contract MRTN-CT-2004-503369, and in part by the INTAS contract 03-51-6346.

References

1. C. Angelantonj and A. Sagnotti, *Phys. Rept.* **371**, 1 (2002) [Erratum-ibid. **376**, 339 (2003)] [arXiv:hep-th/0204089].
2. I. Antoniadis, Phys. Lett. B **246**, 377 (1990).
3. C.D. Hoyle, D.J. Kapner, B.R. Heckel, E.G. Adelberger, J.H. Gundlach, U. Schmidt and H.E. Swanson, *Phys. Rev.* D **70**, 042004 (2004).
4. N. Arkani-Hamed, S. Dimopoulos and G.R. Dvali, *Phys. Lett.* B **429**, 263 (1998) [arXiv:hep-ph/9803315]; I. Antoniadis, N. Arkani-Hamed, S. Dimopoulos and G.R. Dvali, *Phys. Lett.* B **436**, 257 (1998) [arXiv:hep-ph/9804398].
5. For a review see e.g. I. Antoniadis, *Prepared for NATO Advanced Study Institute and EC Summer School on Progress in String, Field and Particle Theory, Cargese, Corsica, France (2002)*; and references therein.
6. J.D. Lykken, *Phys. Rev.* D **54**, 3693 (1996) [arXiv:hep-th/9603133].
7. J.C. Long and J.C. Price, Comptes Rendus Physique **4**, 337 (2003); R.S. Decca, D. Lopez, H.B. Chan, E. Fischbach, D.E. Krause and C.R. Jamell, *Phys. Rev. Lett.* **94**, 240401 (2005); S.J. Smullin, A.A. Geraci, D.M. Weld, J. Chiaverini, S. Holmes and A. Kapitulnik, arXiv:hep-ph/0508204; H. Abele, S. Haeßler and A. Westphal, in 271th WE-Heraeus-Seminar, Bad Honnef (2002).
8. I. Antoniadis and K. Benakli, Phys. Lett. B **326**, 69 (1994).
9. I. Antoniadis, K. Benakli and M. nnQuirós, Phys. Lett. B **331**, 313 (1994); Phys. Lett. B **460**, 176 (1999); P. Nath, Y. Yamada and M. Yamaguchi, Phys. Lett. B **466**, 100 (1999); T.G. Rizzo and J.D. Wells, Phys. Rev. D **61**, 016007 (2000); T.G. Rizzo, Phys. Rev. D **61**, 055005 (2000); A. De Rujula, A. Donini, M.B. Gavela and S. Rigolin, Phys. Lett. B **482**, 195 (2000).
10. E. Accomando, I. Antoniadis and K. Benakli, Nucl. Phys. B **579**, 3 (2000).
11. I. Antoniadis, K. Benakli and A. Laugier, JHEP **0105**, 044 (2001).
12. P. Nath and M. Yamaguchi, Phys. Rev. D **60**, 116004 (1999); Phys. Rev. D **60**, 116006 (1999); M. Masip and A. Pomarol, Phys. Rev. D **60**, 096005 (1999); W.J. Marciano, Phys. Rev. D **60**, 093006 (1999); A. Strumia, Phys. Lett. B **466**, 107 (1999); R. Casalbuoni, S. De Curtis, D. Dominici and R. Gatto, Phys. Lett. B **462**, 48 (1999); C.D. Carone, Phys. Rev. D **61**, 015008 (2000); A. Delgado, A. Pomarol and M. Quirós, JHEP **1**, 30 (2000).
13. G. Servant and T.M.P. Tait, Nucl. Phys. B **650**, 391 (2003).
14. G.F. Giudice, R. Rattazzi and J.D. Wells, Nucl. Phys. B **544**, 3 (1999); E.A. Mirabelli, M. Perelstein and M.E. Peskin, Phys. Rev. Lett. **82**, 2236 (1999); T. Han, J.D. Lykken and R. Zhang, Phys. Rev. D **59**, 105006 (1999); K. Cheung and W.-Y. Keung, Phys. Rev. D **60**, 112003 (1999); C. Balázs et al., Phys. Rev. Lett. **83**, 2112 (1999); L3 Collaboration (M. Acciarri et al.), Phys. Lett. B **464**, 135 (1999) and **470**, 281 (1999); J.L. Hewett, Phys. Rev. Lett. **82**, 4765 (1999).
15. N. Arkani-Hamed, S. Dimopoulos and G. Dvali, Phys. Rev. D **59**, 086004 (1999).
16. S. Cullen and M. Perelstein, Phys. Rev. Lett. **83**, 268 (1999); V. Barger, T. Han, C. Kao and R.J. Zhang, Phys. Lett. B **461**, 34 (1999).
17. K. Benakli and S. Davidson, Phys. Rev. D **60**, 025004 (1999); L.J. Hall and D. Smith, Phys. Rev. D **60**, 085008 (1999).
18. S. Cullen, M. Perelstein and M.E. Peskin, Phys. Rev. D **62**, 055012 (2000); D. Bourilkov, Phys. Rev. D **62**, 076005 (2000); L3 Collaboration (M. Acciarri et al.), Phys. Lett. B **489**, 81 (2000).

19. S.B. Giddings and S. Thomas, Phys. Rev. D **65**, 056010 (2002); S. Dimopoulos and G. Landsberg, Phys. Rev. Lett. **87**, 161602 (2001).
20. I. Antoniadis, C. Bachas, Phys. Lett. B **450**, 83 (1999).
21. I. Antoniadis, K. Benakli, A. Laugier and T. Maillard, Nucl. Phys. B **662**, 40 (2003) [arXiv:hep-ph/0211409].
22. I. Antoniadis, S. Dimopoulos and G. Dvali, Nucl. Phys. B **516**, 70 (1998); S. Ferrara, C. Kounnas and F. Zwirner, Nucl. Phys. B **429**, 589 (1994).
23. T.R. Taylor and G. Veneziano, Phys. Lett. B **213**, 450 (1988).
24. I. Antoniadis, E. Kiritsis and J. Rizos, Nucl. Phys. B **637**, 92 (2002).
25. D.V. Volkov and V.P. Akulov, JETP Lett. **16**, 438 (1972); Phys. Lett. B **46**, 109 (1973).
26. A. Brignole, F. Feruglio and F. Zwirner, JHEP **9711**, 001 (1997); T.E. Clark, T. Lee, S.T. Love and G. Wu, Phys. Rev. D **57**, 5912 (1998); M.A. Luty and E. Ponton, Phys. Rev. D **57**, 4167 (1998); I. Antoniadis, K. Benakli and A. Laugier, Nucl. Phys. B **631**, 3 (2002).
27. I. Antoniadis and M. Tuckmantel, Nucl. Phys. B **697**, 3 (2004).
28. I. Antoniadis, M. Tuckmantel and F. Zwirner, Nucl. Phys. B **707**, 215 (2005) [arXiv:hep-ph/0410165].
29. I. Antoniadis, K. Benakli and M. Quirós, Nucl. Phys. B **583**, 35 (2000).
30. I. Antoniadis, K. Benakli and M. Quiros, New Jour. Phys. **3**, 20 (2001).
31. I. Antoniadis, C. Muñoz and M. Quirós, Nucl. Phys. B **397**, 515 (1993); I. Antoniadis and M. Quirós, Phys. Lett. B **392**, 61 (1997); A. Pomarol and M. Quirós, Phys. Lett. B **438**, 225 (1998); I. Antoniadis, S. Dimopoulos, A. Pomarol and M. Quirós, Nucl. Phys. B **544**, 503 (1999); A. Delgado, A. Pomarol and M. Quirós, Phys. Rev. D **60**, 095008 (1999); R. Barbieri, L.J. Hall and Y. Nomura, Phys. Rev. D **63**, 105007 (2001).
32. I. Antoniadis, E. Kiritsis and T.N. Tomaras, Phys. Lett. B **486**, 186 (2000); I. Antoniadis, E. Kiritsis, J. Rizos and T.N. Tomaras, Nucl. Phys. B **660**, 81 (2003).
33. G. Shiu and S.-H.H. Tye, Phys. Rev. D **58**, 106007 (1998); Z. Kakushadze and S.-H.H. Tye, Nucl. Phys. B **548**, 180 (1999); L.E. Ibáñez, C. Muñoz and S. Rigolin, Nucl. Phys. B **553**, 43 (1999).
34. K.R. Dienes, E. Dudas and T. Gherghetta, Nucl. Phys. B **557**, 25 (1999) [arXiv:hep-ph/9811428]; N. Arkani-Hamed, S. Dimopoulos, G.R. Dvali and J. March-Russell, Phys. Rev. D **65**, 024032 (2002) [arXiv:hep-ph/9811448]; G.R. Dvali and A.Y. Smirnov, Nucl. Phys. B **563**, 63 (1999).
35. L. Randall and R. Sundrum, Phys. Rev. Lett. **83**, 4690 (1999); Phys. Rev. Lett. **83**, 3370 (1999).
36. G.R. Dvali, G. Gabadadze and M. Porrati, Phys. Lett. B **485**, 208 (2000).
37. H. Verlinde, Nucl. Phys. B **580**, 264 (2000); S.B. Giddings, S. Kachru and J. Polchinski, Phys. Rev. D **66**, 106006 (2002).
38. I. Antoniadis, R. Minasian and P. Vanhove, Nucl. Phys. B **648**, 69 (2003) [arXiv:hep-th/0209030].
39. For a recent review see e.g. R. Maartens, Living Rev. Rel. **7**, 7 (2004) [arXiv:gr-qc/0312059]; same proceedings and references therein.
40. E. Kiritsis, N. Tetradis and T.N. Tomaras, JHEP **0108**, 012 (2001).
41. M.A. Luty, M. Porrati and R. Rattazzi, [arXiv:hep-th/0303116].
42. G.R. Dvali and G. Gabadadze, Phys. Rev. D **63**, 065007 (2001); G.R. Dvali, G. Gabadadze, M. Kolanovic and F. N itti, Phys. Rev. D **64**, 084004 (2001).

43. S.L. Dubovsky and V.A. Rubakov, Phys. Rev. D **67**, 104014 (2003) [arXiv:hep-th/0212222].
44. V.A. Rubakov, [arXiv:hep-th/0303125].
45. M. Kolanovic, M. Porrati and J.W. Rombouts, Phys. Rev. D **68**, 064018 (2003) [arXiv:hep-th/0304148].
46. I. Antoniadis, E. Gava and K.S. N arain, Phys. Lett. B **283**, 209 (1992).
47. I. Antoniadis, C. Bachas, C. Fabre, H. Partouche and T.R. Taylor, Nucl. Phys. B **489**, 160 (1997); I. Antoniadis, H. Partouche and T.R. Taylor, Nucl. Phys. B **499**, 29 (1997).
48. I. Antoniadis, S. Ferrara, R. Minasian and K.S. Narain, Nucl. Phys. B **507**, 571 (1997).
49. K. Peeters, P. Vanhove and A. Westerberg, the associated superalgebras and their formulation in superspace," Class. Quant. Grav. **18**, 843 (2001).
50. S.L. Adler, Rev. Mod. Phys. **54**, 729 (1982) [Erratum-ibid. **55** (1983) 837].
51. A. Lue and G. Starkman, Phys. Rev. D **67**, 064002 (2003).

11

Dark Energy from Brane-world Gravity

Roy Maartens

Institute of Cosmology & Gravitation, Portsmouth University, Portsmouth PO1 2EG, UK
Roy.Maartens@port.ac.uk

Abstract. Recent observations provide strong evidence that the universe is accelerating. This confronts theory with a severe challenge. Explanations of the acceleration within the framework of general relativity are plagued by difficulties. General relativistic models require a "dark energy" field with effectively negative pressure. An alternative to dark energy is that gravity itself may behave differently from general relativity on the largest scales, in such a way as to produce acceleration. The alternative approach of modified gravity also faces severe difficulties, but does provide a new angle on the problem. This review considers an example of modified gravity, provided by brane-world models that self-accelerate at late times.

11.1 Introduction

The current "standard model" of cosmology – the inflationary cold dark matter model with cosmological constant (LCDM), based on general relativity and particle physics (the minimal supersymmetric extension of the Standard Model) – provides an excellent fit to the wealth of high-precision observational data [1]. In particular, independent data sets from CMB anisotropies, galaxy surveys and SNe redshifts, provide a consistent set of model parameters. For the fundamental energy density parameters, this is shown in Fig. 11.1. The data indicates that the cosmic energy budget is given by

$$\Omega_\Lambda \approx 0.7, \quad \Omega_M \approx 0.3 , \tag{11.1}$$

leading to the dramatic conclusion that the universe is undergoing a late-time acceleration. The data further indicates that the universe is (nearly) spatially flat, and that the primordial perturbations are (nearly) scale-invariant, adiabatic and Gaussian.

This standard model is remarkably successful, but we know that its theoretical foundation, general relativity, breaks down at high enough energies, usually taken to be at the Planck scale,

$$E \gtrsim M_p \sim 10^{16} \, \text{TeV} . \tag{11.2}$$

R. Maartens: *Dark Energy from Brane-world Gravity*, Lect. Notes Phys. **720**, 323–332 (2007)
DOI 10.1007/978-3-540-71013-4_11
© Springer-Verlag Berlin Heidelberg 2007

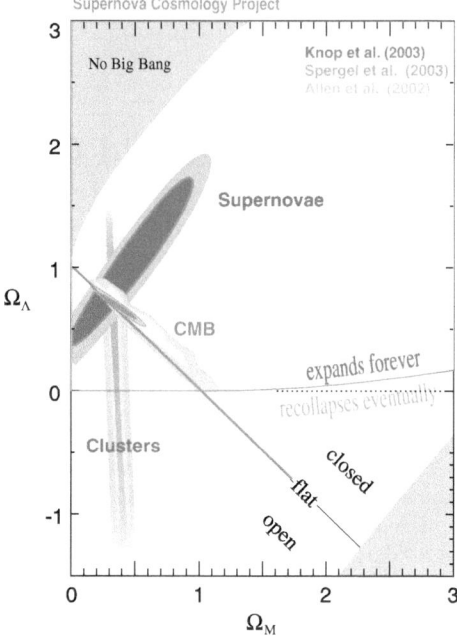

Fig. 11.1. Observational constraints in the $(\Omega_\Lambda, \Omega_M)$ plane (from [2])

The LCDM model can only provide limited insight into the very early universe. Indeed, the crucial role played by inflation belies the fact that inflation remains an effective theory without yet a basis in fundamental theory. A quantum gravity theory will be able to probe higher energies and earlier times, and should provide a consistent basis for inflation, or an alternative that replaces inflation within the standard cosmological model.

An even bigger theoretical problem than inflation is that of the recent accelerated expansion of the universe. Within the framework of general relativity, the acceleration must originate from a dark energy field with effectively negative pressure ($w \equiv p/\rho < -\frac{1}{3}$), such as vacuum energy ($w = -1$) or a slow-rolling scalar field ("quintessence", $w > -1$). So far, none of the available models has a natural explanation.

For the simplest option of vacuum energy, i.e. the LCDM model, the incredibly small value of the cosmological constant

$$\rho_{\Lambda,\,obs} = \frac{\Lambda}{(8\pi G)} \sim H_0^2 \, M_P^2 \sim (10^{-33}\,\mathrm{eV})^2 (10^{19}\,\mathrm{GeV})^2 = 10^{-57}\,\mathrm{GeV}^4 \,,$$

$$\rho_{\Lambda,\,theory} \sim M_{fundamental}^4 > 1\,\mathrm{TeV}^4 \gg \rho_{\Lambda,\,obs} \,, \tag{11.3}$$

cannot be explained by current particle physics. In addition, the value needs to be incredibly fine-tuned,

$$\Omega_\Lambda \sim \Omega_M \, , \tag{11.4}$$

which also has no natural explanation. Quintessence models attempt to address the fine-tuning problem, but do not succeed fully – and also cannot address the problem of how Λ is set exactly to 0. Quantum gravity will hopefully provide a solution to the problems of vacuum energy and fine-tuning.

Alternatively, it is possible that there is no dark energy, but instead a low-energy/ large-scale (i.e. "infrared") modification to general relativity that accounts for late-time acceleration. Schematically, we are modifying the geometric side of the field equations,

$$G_{\mu\nu} + G_{\mu\nu}^{\text{dark}} = 8\pi G T_{\mu\nu} \, , \tag{11.5}$$

rather than the matter side,

$$G_{\mu\nu} = 8\pi G \left(T_{\mu\nu} + T_{\mu\nu}^{\text{dark}} \right) \, , \tag{11.6}$$

as in general relativity.

It is important to stress that a consistent modification of general relativity requires a covariant formulation of the field equations in the general case, i.e. including inhomogeneities and anisotropies. It is not sufficient to propose ad hoc modifications of the Friedman equation, of the form

$$f(H^2) = \frac{8\pi G}{3}\rho \ \text{ or } \ H^2 = \frac{8\pi G}{3}g(\rho) \, , \tag{11.7}$$

for some functions f or g. We can compute the SNe redshifts using this equation – but we *cannot* compute the density perturbations without knowing the covariant parent theory that leads to such a modified Friedmann equation.

An infra-red modification to general relativity could emerge within the framework of quantum gravity, in addition to the ultraviolet modification that must arise at high energies in the very early universe. The leading candidate for a quantum gravity theory, string theory, is able to remove the infinities of quantum field theory and unify the fundamental interactions, including gravity. But there is a price – the theory is only consistent in 9 space dimensions. Branes are extended objects of higher dimension than strings, and play a fundamental role in the theory, especially D-branes, on which open strings can end. Roughly speaking, open strings, which describe the non-gravitational sector, are attached at their endpoints to branes, while the closed strings of the gravitational sector can move freely in the higher-dimensional "bulk" spacetime. Classically, this is realised via the localization of matter and radiation fields on the brane, with gravity propagating in the bulk (see Fig. 11.2).

The implementation of string theory in cosmology is extremely difficult, given the complexity of the theory. This motivates the development of phenomenology, as an intermediary between observations and fundamental theory. (Indeed, the development of inflationary cosmology has been a very valuable exercise in phenomenology.) Brane-world cosmological models inherit

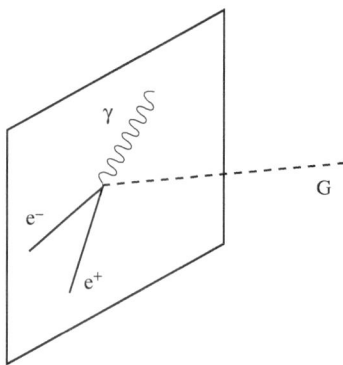

Fig. 11.2. The confinement of matter to the brane, while gravity propagates in the bulk (from [3])

key aspects of string theory, but do not attempt to impose the full machinery of the theory. Instead, drastic simplifications are introduced in order to be able to construct cosmological models that can be used to compute observational predictions (see [4] for reviews in this spirit). Cosmological data can then be used to constrain the brane-world models, and hopefully thus provide constraints on string theory, as well as pointers for the further development of string theory.

It turns out that even the simplest brane-world models are remarkably rich – and the computation of their cosmological perturbations is remarkably complicated, and still incomplete. Here I will describe brane-world cosmologies of Dvali-Gabadadze-Porrati (DGP) type [5]. These are 5-dimensional models, with an infinite extra dimension. (We effectively assume that 5 of the extra dimensions in the "parent" string theory may be ignored at low energies.)

11.2 KK Modes of the Graviton

The brane-world mechanism, whereby matter is confined to the brane while gravity accesses the bulk, means that extra dimensions can be much larger than in the conventional Kaluza-Klein (KK) mechanism, where matter and gravity both access all dimensions. The dilution of gravity via the bulk effectively weakens gravity on the brane, so that the true, higher-dimensional Planck scale can be significantly lower than the effective 4D Planck scale M_p.

The higher-dimensional graviton has massive 4D modes felt on the brane, known as KK modes, in addition to the massless mode of 4D gravity. From a geometric viewpoint, the KK modes can also be understood via the fact that the projection of the null graviton 5-momentum $p_a^{(5)}$ onto the brane is timelike. If the unit normal to the brane is n_a, then the induced metric on the brane is

$$g_{ab} = g_{ab}^{(5)} - n_a n_b, \; g_{ab}^{(5)} n^a n^b = 1, \; g_{ab} n^b = 0 \,, \tag{11.8}$$

and the 5-momentum may be decomposed as

$$p_a^{(5)} = m n_a + p_a, \; p_a n^a = 0, \; m = p_a^{(5)} n^a \,, \tag{11.9}$$

where $p^a = g^{ab} p_b^{(5)}$ is the projection along the brane, depending on the orientation of the 5-momentum relative to the brane. The effective 4-momentum of the 5D graviton is thus p_a. Expanding $g_{ab}^{(5)} p_{(5)}^a p_{(5)}^b = 0$, we find that

$$g_{ab} p^a p^b = -m^2 \,. \tag{11.10}$$

It follows that the 5D graviton has an effective mass m on the brane. The usual 4D graviton corresponds to the zero mode, $m = 0$, when $p_a^{(5)}$ is tangent to the brane.

The extra dimensions lead to new scalar and vector degrees of freedom on the brane. The spin-2 5D graviton is represented by a metric perturbation $h_{ab}^{(5)}$ that is transverse traceless:

$$g_{ab}^{(5)} \to g_{ab}^{(5)} + h_{ab}^{(5)}, \; h^{(5)a}{}_a = 0 = \partial_b h^{(5)b}{}_a \,. \tag{11.11}$$

In a suitable gauge, $h_{ab}^{(5)}$ contains a 3D transverse traceless perturbation h_{ij}, a 3D transverse vector perturbation Σ_i and a scalar perturbation β, which each satisfy the 5D wave equation:

$$h^i{}_i = 0 = \partial_j h^{ij}, \; \partial_i \Sigma^i = 0 \,, \tag{11.12}$$

$$(\Box + \partial_y^2) \begin{pmatrix} \beta \\ \Sigma_i \\ h_{ij} \end{pmatrix} = 0 \,. \tag{11.13}$$

The 5 degrees of freedom (polarizations) in the 5D graviton are felt on the brane as:

- a 4D spin-2 graviton h_{ij} (2 polarizations)
- a 4D spin-1 gravi-vector (gravi-photon) Σ_i (2 polarizations)
- a 4D spin-0 gravi-scalar β.

The massive modes of the 5D graviton are represented via massive modes in all 3 of these fields on the brane. The standard 4D graviton corresponds to the massless zero-mode of h_{ij}.

11.3 DGP Type Brane-worlds: Self-accelerating Cosmologies

Could the late-time acceleration of the universe be a gravitational effect?[1] An historical precedent is provided by attempts to explain the anomalous

---

[1] Note that this would not remove the problem of explaining why the vacuum energy does not gravitate.

precession of Mercury's perihelion by a "dark planet". In the end, it was
discovered that a modification to Newtonian gravity was needed.

An alternative to dark energy plus general relativity is provided by models where the acceleration is due to modifications of gravity on very large scales, $r \gtrsim H_0^{-1}$. It is very difficult to produce infrared corrections to general relativity by modifying the 4D Einstein-Hilbert action,

$$\int d^4 x \sqrt{-g}\, R \;\; \rightarrow \;\; \int d^4 x \sqrt{-g}\, f(R, R_{\mu\nu} R^{\mu\nu}, \ldots) \,. \tag{11.14}$$

Typically, instabilities arise or the action has no natural motivation. The DGP brane-world offers a higher-dimensional approach to the problem, which effectively has infinite extra degrees of freedom from a 4D viewpoint.

Most brane-world models modify general relativity at high energies. The main examples are those of Randall-Sundrum (RS) type [6], where a Friedman-Robertson-Walker brane is embedded in an anti de Sitter bulk, with curvature radius ℓ. At low energies $H\ell \ll 1$, the zero-mode of the graviton dominates on the brane, and general relativity is recovered to a good approximation. At high energies, $H\ell \gg 1$, the massive modes of the graviton dominate over the zero mode, and gravity on the brane behaves increasingly in a 5D way. On the brane, the standard conservation equation holds,

$$\dot{\rho} + 3H(\rho + p) = 0 \,, \tag{11.15}$$

but the Friedmann equation is modified by an ultraviolet correction:

$$H^2 = \frac{8\pi G}{3} \rho \left(1 + \frac{2\pi G \ell^2}{3} \rho \right) + \frac{\Lambda}{3} \,. \tag{11.16}$$

The ρ^2 term is the ultraviolet term. At low energies, this term is negligible, and we recover $H^2 \propto \rho + \Lambda/8\pi G$. At high energies, gravity "leaks" off the brane and $H^2 \propto \rho^2$. This 5D behaviour means that a given energy density produces a greater rate of expansion than it would in general relativity. As a consequence, inflation in the early universe is modified in interesting ways.

In the DGP case the bulk is 5D Minkowski spacetime. Unlike the AdS bulk of the RS model, the Minkowski bulk has infinite volume. Consequently, there is no normalizable zero-mode of the graviton in the DGP brane-world. Gravity leaks off the 4D brane into the bulk at large scales. At small scales, gravity is effectively bound to the brane and 4D dynamics is recovered to a good approximation. The transition from 4- to 5D behaviour is governed by a crossover scale r_c; the weak-field gravitational potential behaves as

$$\Psi \sim \begin{cases} r^{-1} \text{ for } r \ll r_c \\ r^{-2} \text{ for } r \gg r_c \end{cases} \tag{11.17}$$

Gravity leakage at late times initiates acceleration – not due to any negative pressure field, but due to the weakening of gravity on the brane. 4D gravity is recovered at high energy via the lightest KK modes of the graviton, effectively via an ultralight metastable graviton.

The energy conservation equation remains the same as in general relativity, but the Friedman equation is modified:

$$\dot{\rho} + 3H(\rho + p) = 0 \ , \tag{11.18}$$

$$H^2 - \frac{H}{r_c} = \frac{8\pi G}{3}\rho \ . \tag{11.19}$$

This shows that at early times, $Hr_c \gg 1$, the general relativistic Friedman equation is recovered. By contrast, at late times in a CDM universe, with $\rho \propto a^{-3} \to 0$, we have

$$H \to H_\infty = \frac{1}{r_c} \ . \tag{11.20}$$

Since $H_0 > H_\infty$, in order to achieve self-acceleration at late times, we require

$$r_c \gtrsim H_0^{-1} \ , \tag{11.21}$$

and this is confirmed by fitting SNe observations, as shown in Fig. 11.3. This comparison is aided by introducing a dimensionless cross-over parameter,

$$\Omega_{r_c} = \frac{1}{4(H_0 r_c)^2} \ . \tag{11.22}$$

It should be emphasized that the DGP Friedman (11.19) is derived covariantly from a 5D gravitational action,

$$\int d^5x \, \sqrt{-g^{(5)}} \, R^{(5)} + r_c \int d^4x \, \sqrt{-g} \, R \ . \tag{11.23}$$

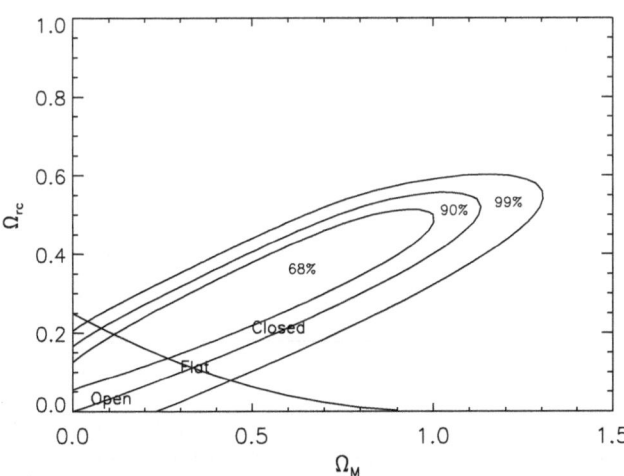

Fig. 11.3. Constraints from SNe redshifts on DGP models. (From [7])

LCDM and DGP can both account for the SNe observations, with the fine-tuned values $\Lambda \sim H_0^2$ and $r_c \sim H_0^{-1}$ respectively. This degeneracy may be broken by observations based on structure formation, since the two models suppress the growth of density perturbations in different ways [8, 9]. The distance-based SNe observations draw only upon the background 4D Friedman equation (11.19) in DGP models, and therefore there are quintessence models in general relativity that can produce precisely the same SNe redshifts as DGP [10]. By contrast, structure formation observations require the 5D perturbations in DGP, and one cannot find equivalent general relativity models [11].

For LCDM, the analysis of density perturbations is well understood. For DGP it is much more subtle and complicated. Although matter is confined to the 4D brane, gravity is fundamentally 5D, and the bulk gravitational field responds to and backreacts on density perturbations. The evolution of density perturbations requires an analysis based on the 5D nature of gravity. In particular, the 5D gravitational field produces an anisotropic stress on the 4D universe. Some previous results are based on inappropriately neglecting this stress and all 5D effects – as a consequence, the 4D Bianchi identity on the brane is violated, i.e. $\nabla^\nu G_{\mu\nu} \neq 0$, and the results are inconsistent.

When the 5D effects are incorporated [11], the 4D Bianchi identity is satisfied. (The results of [11] confirm and generalize those of [8].) The consistent modified evolution equation for density perturbations is

$$\ddot{\Delta} + 2H\dot{\Delta} = 4\pi G \left\{ 1 - \frac{(2Hr_c - 1)}{3[2(Hr_c)^2 - 2Hr_c + 1]} \right\} \rho\Delta , \qquad (11.24)$$

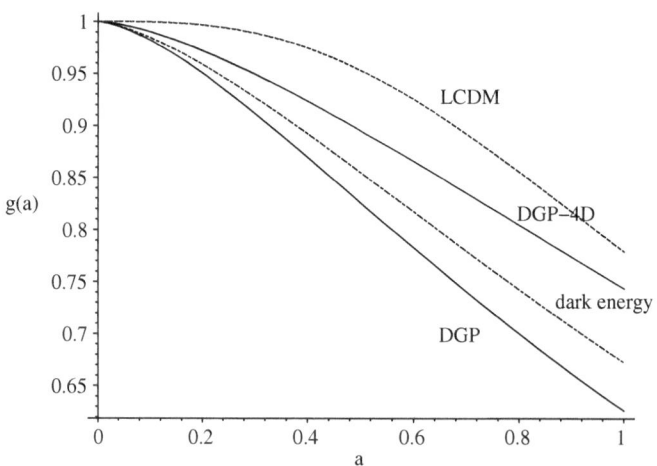

Fig. 11.4. The growth factor $g(a) = \Delta(a)/a$ for LCDM (*long dashed*) and DGP (*solid, thick*), as well as for a dark energy model with the same expansion history as DGP (*solid, thick*). DGP-4D (*solid, thin*) shows the incorrect result in which the 5D effects are set to zero. (From [11])

where the term in braces encodes the 5D correction. The linear growth factor, $g(a) = \Delta(a)/a$ (i.e. normalized to the flat CDM case, $\Delta \propto a$), is shown in Fig. 11.4.

It must be emphasized that these results apply on subhorizon scales. On superhorizon scales, where the 5D effects are strongest, the problem has yet to be solved. This solution is necessary before one can compute the large-angle CMB anisotropies. It should also be remarked that the late-time asymptotic de Sitter solution in DGP cosmological models has a ghost problem [12], which may have implications for the analysis of density perturbations.

11.4 Conclusion

In conclusion, DGP brane-world models, which are inspired by ideas from string theory, provide a rich and interesting phenomenology for modified gravity. These models can account for the late-time acceleration without the need for dark energy – gravity leakage from the 4D brane at large scales leads to self-acceleration. The 5D graviton, i.e. its KK modes, plays a crucial role, which has been emphasized in this article.

Acknowledgements

I thank the organizers for the invitation to present this work, which was supported by PPARC.

References

1. See, e.g. D. Scott, [arXiv:astro-ph/0510731].
2. R.A. Knop et al. [The Supernova Cosmology Project Collaboration], Astrophys. J. **598**, 102 (2003) [arXiv:astro-ph/0309368].
3. M. Cavaglia, Int. J. Mod. Phys. A**18**, 1843 (2003) [arXiv:hep-ph/0210296].
4. R. Maartens, Living Rev. Rel. **7**, 7 (2004) [arXiv:gr-qc/0312059];
   P. Brax, C. van de Bruck and A.C. Davis, Rept. Prog. Phys. **67**, 2183 (2004) [arXiv:hep-th/0404011];
   V. Sahni, [arXiv:astro-ph/0502032];
   R. Durrer, AIP Conf. Proc. **782**, 202 (2005) [arXiv:hep-th/0507006];
   D. Langlois, [arXiv:hep-th/0509231[;
   A. Lue, Phys. Rept. **423**, 1 (2006) [arXiv:astro-ph/0510068];
   D. Wands, [arXiv:gr-qc/0601078].
5. G.R. Dvali, G. Gabadadze and M. Porrati, Phys. Lett. B**484**, 112 (2000) [arXiv:hep-th/0002190];
   C. Deffayet, Phys. Lett. B**502**, 199 (2001) [arXiv:hep-th/0010186].
6. L. Randall and R. Sundrum, Phys. Rev. Lett. **83**, 4690 (1999) [arXiv:hep-th/9906064];
   P. Binetruy, C. Deffayet, U. Ellwanger and D. Langlois, Phys. Lett. B **477**, 285 (2000) [arXiv:hep-th/9910219].

7. C. Deffayet, S.J. Landau, J. Raux, M. Zaldarriaga and P. Astier, Phys. Rev. D **66**, 024019 (2002) [arXiv:astro-ph/0201164].

8. A. Lue, R. Scoccimarro and G.D. Starkman, Phys. Rev. D **69**, 124015 (2004) [arXiv:astro-ph/0401515];
A. Lue and G. Starkman, Phys. Rev. D **67**, 064002 (2003) [arXiv:astro-ph/0212083].

9. Y.S. Song, Phys. Rev. D **71**, 024026 (2005) [arXiv:astro-ph/0407489];
L. Knox, Y.S. Song and J.A. Tyson, [arXiv:astro-ph/0503644];
M. Ishak, A. Upadhye and D.N. Spergel, Phys. Rev. D **74**, 043513 (2006) [arXiv:astro-ph/0507184];
I. Sawicki and S.M. Carroll, Phys. Scripta T **121**, 119 (2005) [arXiv:astro-ph/0510364].

10. E.V. Linder, Phys. Rev. D **72**, 043529 (2005) [arXiv:astro-ph/0507263].

11. K. Koyama and R. Maartens, JCAP **0610**, 016 (2006) [arXiv:astro-ph/0511634].

12. D. Gorbunov, K. Koyama and S. Sibiryakov, Phys. Rev. D, to appear [arXiv:hep-th/0512097].

12

The Issue of Dark Energy in String Theory

Nick Mavromatos

King's College London, Department of Physics, Theoretical Physics, Strand,
London WC2R 2LS, UK
Nikolaos.Mavromatos@kcl.ac.uk

Abstract. Recent astrophysical observations, pertaining to either high-redshift
supernovae or cosmic microwave background temperature fluctuations, as those measured recently by the WMAP satellite, provide us with data of unprecedented accuracy, pointing towards two (related) facts: (i) our Universe is accelerated at present,
and (ii) more than 70% of its energy content consists of an unknown substance,
termed dark energy, which is believed responsible for its current acceleration. Both
of these facts are a challenge to String theory. In this review I outline briefly the
challenges, the problems and possible avenues for research towards a resolution of
the Dark Energy issue in string theory.

12.1 Introduction

Recent Astrophysical Data, from either studies of distant supernovae type
Ia [1], or precision measurements of temperature fluctuations in the cosmic
microwave background radiation from the WMAP satellite [2], point towards a
current-era acceleration of our Universe, as well as a very peculiar energy budget for it: 70% of its energy density consists of an unknown energy substance,
termed *Dark Energy*. In fact, global best-fit models from a compilation of all
the presently available data are based on simple Einstein-Friedman Universes
with a (four space-time dimensional) positive *cosmological constant* Λ, whose
value saturate the Newtonian upper limit obtained from galactic dynamics.
In order of magnitude,

$$\Lambda \sim 10^{-122} M_P^4 \qquad (M_P^4 = 10^{19} \text{ GeV}) . \qquad (12.1)$$

Although, as a classical (general relativistic) field theory, such a model is fairly
simple, from a quantum theory view point it appears to be the less understood
at present. The reason is simple: Since in cosmology [3] the radiation and
matter energy densities scale with inverse powers of the scale factor, a^{-4}
and a^{-3} respectively, in a Universe with a positive cosmological constant Λ,
the vacuum energy density remains constant and positive, and eventually

N. E. Mavromatos: *The Issue of Dark Energy in String Theory*, Lect. Notes Phys. **720**,
333–374 (2007)
DOI 10.1007/978-3-540-71013-4_12   © Springer-Verlag Berlin Heidelberg 2007

dominates the energy budget. The asymptotic (in time) Universe becomes a *de Sitter* one, and in such a Universe the scale factor will increase exponentially,

$$a(t) = a_0 e^{\sqrt{\frac{A}{3}}t} . \tag{12.2}$$

This in turn implies that the Universe will eventually enter an inflationary phase again, and in fact it will accelerate eternally, since $\ddot{a} > 0$, where the overdot denotes derivative with respect to the Robertson-Walker cosmic time, t, defined by:

$$ds^2_{RW} = -dt^2 + a^2(t)ds^2_{\text{spatial}} . \tag{12.3}$$

In such de Sitter Universes there is unfortunately a *cosmic horizon*

$$\delta \propto \int_{t_0}^{t_{End}} \frac{cdt}{a(t)} < \infty , \tag{12.4}$$

where t_{End} indicates the end of time. For a closed Universe $t_{End} < \infty$, but for an open or flat Universe $t_{End} \to \infty$. The Cosmic Microwave (CMB) data of WMAP and other experiments at present indicate that our Universe is spatially *flat*, and hence $t_{End} \to \infty$.

The presence of a cosmic horizon implies that it is not possible to define pure state vectors of quantum asymptotic (in time) states. Therefore, the entire concept of a well-defined and gauge invariant Scattering matrix S breaks down in quantum field theories defined on such de Sitter space-time backgrounds. For string theory this is bad news, because, by construction [4], perturbative string theory is based on well-defined scattering amplitudes for the various excitations, and hence on a well-defined S-matrix [5]. The accommodation of de Sitter space-times as consistent backgrounds is, therefore, a challenge for string theory, and certainly one of the most important issues I would like to discuss in this brief review.

A straightforward way out, would be *quintessence*-like scenarios for dark energy [6], according to which the latter is due to a potential of a time dependent scalar field, which has not yet reached its equilibrium point. If then the asymptotic value of the dark energy vanishes in such a way that there is no cosmic horizon, then the model could be accommodated within string-inspired effective field theories, given that an asymptotic S-matrix could be defined in such a case.

However, this does not mean that de Sitter Universes *per se* cannot be accommodated somehow into a (possibly non perturbative) string theory framework. Their anti-de-Sitter (AdS, negative cosmological constant) counterparts certainly do, and in fact there have been important development towards a holographic property of quantum field theories in such Universes, due to the celebrated Maldacena conjecture [7], concerning quantum properties of (supersymmetric) conformal field theories on the boundary of AdS space-time. As we shall discuss in the next section, similar conjectures [8] may characterize their de Sitter counterparts, and this could be a way forward to accommodate such a space-time into string theory.

Finally, a more straightforward (perturbative) approach to discuss de Sitter and inflationary models in string theories, would be to use the so-called *non-critical* (or *Liouville*) string framework [9], dealing with a mathematically consistent way of discussing strings propagating in non-conformal backgrounds, of which de Sitter space-time is one example. This theory, however, at least as far as computation of the pertinent correlation functions are concerned, has not been developed to the same level of mathematical understanding as the critical strings. A crucial ingredient in this approach is the identification of the Liouville mode with the target time [10], which allows for some non-conformal backgrounds in string theory, including de Sitter space-times and accelerated Universes, to be accommodated in a mathematically consistent manner. We shall cover this approach in some detail in Sect. 12.4.

We should stress at this point, that the above considerations, regarding S-matrix amplitudes in de Sitter Universes, refer to pure perturbative string theories. In the modern approach to strings, where membrane (D-brane) structures [11] also appear as mathematically consistent entities, the presence of a dark energy on the brane is unavoidable, unless extreme conditions of (a hight number of) unbroken supersymmetries and a static nature of brane worlds are imposed. However, in brane cosmology one needs moving branes, in order to obtain a cosmological space-time [12], and in this case, target space-time supersymmetry breaks down, due to the brane motion, resulting in non-trivial vacuum energy contributions on the brane (see Maartens' contribution in this volume). If one accommodates brane models within a string theory framework, then, the important question arises as to how one can formulate the string theory of the excitations on the brane in the presence of such vacuum energy contributions.

The structure of the article will be the following: in Sect. 12.2, I will deal with mathematical properties of de Sitter space-times: after reviewing briefly basic features of this geometry, I will describe modern approaches to the issue of placing a quantum field theory in de Sitter space-times, by discussing briefly a holographic conjecture, put forward by Strominger [8], according to which a quantum field theory on the single boundary of de Sitter space can be related to a classical theory in the bulk, in a way not dissimilar to the celebrated Maldacena conjecture [7] for anti-de Sitter spaces (negative cosmological constant space-times). In Sect. 12.3, I will discuss the issue of cosmic horizons in perturbative string theory, and give further arguments that consistent perturbative strings cannot be characterized by such horizons. In Sect. 12.4, I will discuss quintessence scenarios in strings, where the dilaton behaves as the quintessence field, responsible for the current acceleration of the Universe. I will discuss two opposite examples, a pre Big-Bang scenario [13], in which the string coupling increases at late times, with string loop corrections playing a dominant rôle, and another scenario [10, 14], in which the string coupling becomes more and more perturbative as the time passes, leading asymptotically to a vanishing dark energy, in such a way that S-matrix states can be defined. In this second scenario the current-era acceleration parameter turns out to be proportional to the square of the string coupling, which at present enjoys perturbative values compatible with particle physics phenomenology. I will briefly discuss predictions of such models in the context of recent data, but also unresolved problems. I will not discuss the issue of dark energy in brane cosmologies in this article, as this is a topic covered by other contributions in this volume. Conclusions and directions for future research in the issue of Dark Energy in Strings will be presented in Sect. 12.5.

12.2 De Sitter (dS) Universes from a Modern Perspective

In this section I shall give a very brief overview of the most important properties of de Sitter space, relevant for the purposes of this lecture. For more details the reader is referred to [8], and references therein, where a concise exposition of the most important properties of classical and quantum theories of de Sitter space is given.

12.2.1 Classical Properties

The classical Geometrical picture of a de Sitter space-time is that of a single-sheet hyperboloid, depicted in Fig. 12.1. This hypersurface can be constructed from a flat (d+1)-dimensional Minkowski space-time, with coordinates (X^0, X^i), $i = 1, \ldots d$, by means of the equation:

$$-(X^0)^2 + (X^1)^2 + \ldots + (X^d)^2 = \ell^2 , \tag{12.5}$$

where the parameter ℓ has units of length, and is called the *de Sitter radius*.

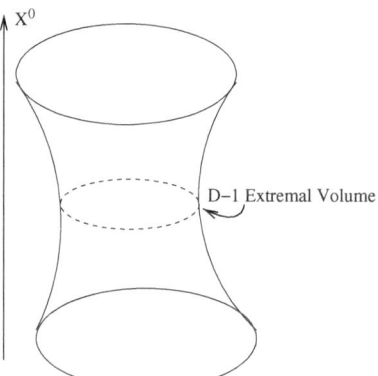

Fig. 12.1. A de Sitter space is a single-sheet hyperboloid, obtained from a (d+1)-dimensional Minkowski space-time by an appropriate embedding of the hypersurface $-(X^0)^2 + (X^1)^2 + \ldots + (X^d)^2 = \ell^2$, where ℓ is the de Sitter radius

The classical Einstein equations, which yield as a solution this space-time, involve a *positive cosmological constant*

$$R_{\mu\nu} - \frac{1}{2}g_{\mu\nu}R + \Lambda g_{\mu\nu} = 0 , \qquad \Lambda = \frac{(d-2)(d-1)}{2\ell^2} \qquad (12.6)$$

There are various coordinates one can use for the description of such space-times, whose detailed description is given in [8]. The most useful one, which is also most relevant for our purposes, and also helps us to understand more clearly the causal properties of the de Sitter space-time, is the *conformal coordinate system*, (T, θ_i), $i = 1, \ldots d$, in terms of which the line element reads:

$$ds^2 = \frac{1}{\cos^2 T} \left(-dT^2 + d\Omega_{d-1}^2 \right) , \qquad (12.7)$$

where Ω is the usual angular part, expressed in terms of θ_i's.

In terms of these coordinates, one arrives easily at the Penrose diagram for the de Sitter space, depicted in Fig. 12.2(a), which contains all the information about the causal structure.

A peculiar feature of this space, but quite important for the development of a consistent quantum gravity theory, is the fact that *no single observer can access the entire space-time* (see Figs. 12.2(b),(c)). As we see from the figure,

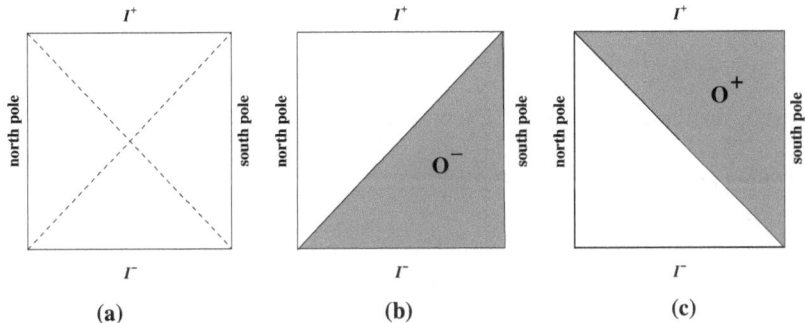

(a)   (b)   (c)

Fig. 12.2. (a) The Penrose diagram for the de Sitter space, constructed by means of conformal coordinates. Horizontal slices represent the extremal volume S^{d-1} (cf. *dashed line* in Fig. 12.1), whilst every point in the interior represents a S^{d-2}. The vertical slices marked as north and south pole are time-like surfaces. The I^+ (I^-) surfaces correspond to the future (past) infinity, and they are the surfaces where all the null geodesics ($ds^2 = 0$) originate and terminate. The diagonal dotted lines represent the past and future horizons of an observer at the south pole. Due to the existence of an horizon in this geometry, a light that starts at, say, the north pole at I^- will reach the south pole by the time it reaches I^+ *infinitely far* in the future. (b) A classical observer sitting at the south pole will never be able to observe anything past the dotted line that stretches from the north pole at I^- to the south pole at I^+ (causal past region $\mathcal{O}^-$). (c) Similarly, the south pole observer will only be able to send a message *only* to the causal future region $\mathcal{O}^+$

the causal past and future regions of an observer sitting, say, at the south pole will only be the portions $\mathcal{O}^-$ and $\mathcal{O}^+$, respectively. Their intersection (called *causal diamond*) is the only *region* of the de Sitter space that is fully accessible to an observer at the south pole.

12.2.2 Quantum Field Theory on dS: Thermodynamical Properties

Due to the aforementioned fact of inaccessibility of the entire region of dS space to a classical observer, a consistent formulation of quantum field theory in such space-times is still an open issue, and certainly it is expected to be rather different from the corresponding one in Minkowskian space-times, where such inaccessibility problems are absent. Essentially, the presence of an horizon will cause problems in defining appropriately asymptotic quantum states, and hence a scattering matrix, as we discussed in the introduction (cf. (12.4)) for the case of cosmological de Sitter space-times. This issue is still wide open, and below we outline some more consequences of this fact.

The situation somewhat resembles that of a Black Hole (BH). In that problem, there is an horizon, for an asymptotic observer, who lies far away from it and makes his/her measurements locally. The entropy S associated with an event horizon in the BH case is given by the Bekenstein-Hawking area law [15, 16]

$$S = \frac{1}{4G_N}A ,\tag{12.8}$$

where G_N is the gravitational (Newton) constant, and A is the area of the horizon. This is a macroscopic formula, which essentially describes how properties of event horizons in General Relativity change as their parameters (area, in this case) are varied.

The quantum BH Hawking radiate, and from this type of particle creation one comes to the conclusion that there is a temperature T_H characterizing the exterior space-time ('Hawking' temperature), measured at infinity. For the case of a Schwarzschild BH

$$T_H = \frac{\hbar}{8\pi M} ,\tag{12.9}$$

where M is the mass (energy) of the BH. In the BH case this temperature is found by integrating the thermodynamic relation $dS_{BH}/dM = 1/T_H$.

From a quantum field theory point of view, we can understand such formulae from the fact that they describe some effective "loss" of information, associated with modes that go beyond the horizon, and hence are lost for ever for the classical asymptotic observer.

Indeed, if one considers the exterior portion of space-time of the BH as an open system, and the interior as constituting the "environment", with which the physical world interacts, then the Bekenstein-area law formula may be derived simply even in flat Minkowski space-times with a boundary of area A. For instance, Srednicki [17] has demonstrated that by tracing the density matrix of a massless scalar field (taken as a toy, but illustrative, example) over degrees of freedom residing inside an imaginary sphere, embedded in a flat Minkowski space-time, the result leads to an entropy for the scalar field which is proportional to the area, and *not* the volume of the sphere.

In view of this analogy, one would therefore expect that, in all cases where there is a region in space-time inaccessible to an asymptotic observer of a quantum field theory on such geometries, there should be an entropy associated with the area of the region. This should also be expected in the dS case, in view of the existence of a de Sitter horizon. If one postulates some thermodynamic properties associated with the entropy, one arrives also at an effective temperature concept.

The deep issue in the black hole case is to understand the precise coefficient $1/4G_N$ in (12.8), in other words develop a sufficiently correct quantum theory for such space-times, in which it will be possible to count the *microstates* of the BH exactly. If the latter are associated to a *Von Neumann* entropy, $S_{VN} = -k_B \mathrm{Tr}\rho\ln\rho$, where ρ is the density matrix of the system under consideration (the quantum BH in our case), and the Tr is taken over all microstates, then one should show that $S_{VN} = S$ given by (12.8).

At present this is one of the most important issues in theoretical physics. A precise counting of microstates, however, leading to a relation of the form (12.8) has become possible for certain highly supersymmetric black hole backgrounds in string theory, saturating the so-called Bogomolnyi-Prasad-Sommerfield (BPS) bound [18]. It is, though, still unproven for the general case of non supersymmetric black holes, which are the likely types to be encountered in our physical world.

We now come to the dS case. Indeed, as one would expect from the above generic arguments, there should be an entropy associated with the horizon. In fact it is, and there is also a temperature ('Gibbons-Hawking temperature') [19], in complete analogy with the BH case. In fact, the temperature is given in terms of the de Sitter radius by:

$$T_{GH} = \frac{1}{2\pi\ell} \qquad (12.10)$$

and the entropy, associated with the de Sitter horizon of area A, is given exactly by the formula (12.8).

These properties can be proven by considering a quantum field on the dS background and evaluating its Green functions. Such an analysis shows that, in the case of massive quantum fields, an observer, moving along a time-like geodesic of dS space, observes a thermal bath of particles, when the massive field is in its vacuum state $|0>$. It turns out that the correct type of Green functions to be used in this case are the thermal ones. For details we refer the reader to the lectures by Strominger [8], and references therein. Such an analysis allows also for the computation of the effective temperature the dS space is associated with.

The entropy of the de Sitter space S_{dS}, then, is found following the argument suggested by Gibbons and Hawking [19], according to which

$$\frac{dS_{dS}}{d(-E_{dS})} = \frac{1}{T_{GH}} , \qquad (12.11)$$

where E_{dS} is the energy of the dS space. Notice the minus sign in front of E_{dS}. This stems from the fact that what we call energy in dS space is not as simple as the mass of the BH case. To understand qualitatively what might happen in the dS case, we should first start from the principle outlined above, that the entropy of the space is associated with "stuff" behind the horizon. We do not, at present, have any idea what the "microstates" of the dS vacuum are, but let us suppose for the sake of the argument, that an entropy is associated with them (this assumption is probably correct).

In general relativity energy on a surface is defined as an integral of a total derivative, which therefore reduces to a surface integral on the boundary of the surface, and hence vanishes for a closed surface. Because of this vanishing result, if we consider a closed surface on de Sitter space, and we put, say, positive energy on the south pole, then there must be necessarily some negative energy at the north pole to compensate, and yield a zero result. One can see this explicitly in the case of a Schwarzschild-de Sitter spacetime, where the singularity at the north pole, behind the dS horizon, carries negative energy [8]. From the BH analogy, it is therefore more sensible to vary with respect to this negative energy, and this explains the relative minus sign in (12.11) yielding the correct expression for the area law in the dS case.

The important point to notice, however, is that, despite the formal similarity of the dS with the BH, in the former case no one understands, at present, the precise *microscopic* origin of the entropy and temperature. It is not clear what the microstates behind the dS horizon are. Certainly they constitute an "environment" with which the quantum field theory on dS space interacts.

This question acquires much bigger importance in cosmologies with a positive cosmological constant, which are currently favoured by the astrophysical data [1, 2]. Indeed in such cases, the asymptotic (in time) Universe will enter a pure de-Sitter-space phase, since all the matter energy density will be diluted, scaling with the scale factor as a^{-3}, thereby leaving us only with the constant vacuum energy contribution Λ. As discussed in the beginning of the lecture, the cosmological horizon will be given by (12.4), and in this case the dS radius ℓ, in terms of which the entropy and temperature are expressed, is associated with Λ by (12.6), essentially its square root.

12.2.3 Lack of Scattering Matrix and Intrinsic CPT Violation in dS?

The important question, therefore, from a quantum field theory viewpoint on such cosmologies and in general dS-like space-times, concerns the kind of quantum field theories one can define consistently in such a situation. In this respect, the situation is dual to the BH case in the following sense: in a BH, there is an horizon which defines a space-time boundary for an asymptotic observer who lies far away from it. In a full quantum theory the BH evaporates due to Hawking radiation. Although the above thermodynamics arguments are valid for large semi-classical BH, one expects the Hawking evaporation process to continue until the BH acquires a size comparable to the characteristic scale of quantum gravity (QG), the Planck length $\ell_P = 1/M_P$, with $M_P \sim 10^{19}$ GeV. Such *microscopic* BH may either evaporate completely, leaving behind a naked singularity, or, better -thus satisfying the cosmic censorship hypothesis, according to which there are no unshielded space-time singularities in the physical world - disappear in a space-time "foam", namely in a QG ground state, consisting of dynamical "flashing on and off" microscopic BH. In such a case, an initially pure quantum state will in principle be observed as mixed by the asymptotic observer, given that "part of the state quantum numbers" will be kept inside the foamy black holes ("effective information loss"), and hence these will constitute degrees of freedom inaccessible to the observer.

Barring the important concept of *holographic* properties, which we shall come to later on, which may indeed characterize such singular space-times in QG, a situation like this will imply an *effective non-unitary evolution* of quantum states of matter in such backgrounds, and hence gravitational *decoherence*.

A similar situation will characterize the dS space, which is dual to the BH analogue, in the sense that the observer is inside the (cosmological) horizon, in contrast to the BH where he/she was lying outside. However the situation concerning the inability to define asymptotically pure state vectors for the quantum state of matter fields remains in this case.

The lack of a proper definition of pure "out" state asymptotic vectors in *both* situations, implies that a gauge invariant scattering matrix is also ill defined in the dS case. By a theorem due to Wald then [20], one cannot define in such quantum field theories a quantum mechanical CPT operator. This leads to quantum decoherence of matter propagating in such de Sitter space-times. For more details I refer the interested reader to [21], where possible phenomenological consequences of such a decoherence are discussed in detail.

For the purposes of this lecture, the reader should bear in mind that CPT invariant quantum field theories is the cornerstone of modern particle physics phenomenology. Hence, if the issue of CPT symmetry needs to be modified or violated in a dS space for the above reasons, then this brings up immediately the question as to how one can formulate consistent particle physics models in such space-times.

12.2.4 Holographic Properties of dS? Towards a Quantum Graviy Theory

A final, but important aspect, that might characterize a quantum theory in de Sitter space-times, is the aforementioned property of *holography*. If this happens, then the above-mentioned information loss paradox will not occur, since in that case all the information that would otherwise be lost behind the horizon surface would somehow be *reflected* back to (or tunnel through) the surface, and thus could be accessible to an outside observer. Thus, a mathematically consistent quantum mechanical picture of gravity in the presence of space-time boundaries could be in place.

I must stress, at this point, an important issue for which there is often confusion in the literature. If quantum gravity turns out to lead to open-system quantum mechanics for matter theories, this is not necessarily a mathematical inconsistency. It simply means that there is information carried out by the quantum-gravitational degrees of freedom, which however may not be easy to retrieve in a perturbative treatment. Of course, even in such situations, the complete system, "gravity plus matter", is mathematically a closed quantum system. On the other hand, if holography is valid, then one simply does not have to worry about any effective loss of information due to the space-time boundary, and hence the situation becomes much cleaner.

Holographic properties of anti-de-Sitter (AdS) spaces (negative cosmological constant) are encoded in the celebrated Maldacena conjecture [7], according to which the quantum correlators of a conformal quantum field theory on the boundary of the AdS space can be evaluated by means of classical gravity in the bulk of this space. This conjecture, known with the abbreviation AdS/CFT correspondence, has been verified to a number of highly super-symmetric backgrounds in string theory, but of course it may not be valid in (realistic) non conformal, non supersymmetric cases. The issue for such cases is still open.

A similar conjecture in de Sitter space-times has been put forward by Strominger [8]. The conjecture, which is not proven at present, can be formulated as follows:

Consider an operator $\phi(x_i)$ of quantum gravity in a de Sitter space, inserted at points x_i on the hypersurfaces I^- or I^+. The dS/CFT conjecture states that correlation functions of this operator at the points x_i can be generated by an appropriate Euclidean conformal field theory

$$\langle \phi(x_1) \ldots \phi(x_i) \rangle_{dS^{d+1}} \longleftrightarrow \langle \mathcal{O}_\phi(x_1) \ldots \mathcal{O}_\phi(x_i) \rangle_{S^d} , \qquad (12.12)$$

where $\mathcal{O}_\phi(x)$ is an operator of the CFT associated with the operator ϕ.

For the simple, but quite instructive case, of a three dimensional dS_3 space, a proof of this correspondence has been given in [8], making appropriate use of properties of the asymptotic symmetry group of gravity for dS_3. I refer the interested reader to that work, and references therein, for more details.

Before closing this section, I would like to stress that the dS/CFT conjecture may not be valid in realistic cosmologies, in which the quantum field theories of relevance are certainly not conformal. If, however, this conjecture is valid, then this is a very big step towards a CPT invariant, non-perturbative, construction of a quantum theory of gravity.

The holographic principle [22] will basically allow for any possible information loss associated with the presence of the cosmological horizon to decay with the cosmic time, in such a way that an asymptotic observer will not eventually loose any information. This will allow for a consistent CPT operator to be defined, then, according to the above-mentioned theorem of Wald [20].

If true for the dS case, one expects a similar holographic property to be valid for the BH case as well. In fact recently, Hawking argued [23] this to be the case in a BH quantum theory of gravity, but in my opinion his arguments are not supported by any rigorous calculation. Hawking's argument is based on the fact that any consistent theory of gravity should involve an appropriate sum over topologies, including the Minkowskian one (trivial). In Hawking's argument, then, the Euclidean path integrals over the non-trivial topologies, that would give non-unitary contributions, and hence information loss, lead to expressions in scattering amplitudes that decay exponentially with time, thereby leaving only the trivial topology contributions, which are unitary. As we said, however, there is no rigorous computation involved to support this argument, at least at present, not withstanding the fact that the Euclidean formalism seems crucial to the result (although, arguably we know of no other way of performing a proper quantum gravity path integral). Hence, the issue of unitarity in effective low-energy theories of quantum gravity is still wide open in my opinion, and constitutes a challenge for both theory and phenomenology of quantum gravity [21].

12.3 No Horizons in Perturbative (Critical) String Theory

As discussed above, if holography is valid, there should, in principle, be no issue regarding string theory, and hence CPT would be a good symmetry of the theory, as seems desirable from a modern M-theory point of view [24].

If, however, holography is not valid for realistic non-supersymmetric, non-conformal theories, then such a situation is most problematic in string theory, which, as mentioned in the beginning, at least in its perturbative treatment is based on a formalism with *well-defined* scattering amplitudes [5].

Apart from the scattering-matrix and CPT-based issues, there are other arguments that exclude the existence of horizons in perturbative string theory [25]. These arguments derive from considerations of the shape of the potentials arising from supersymmetry breaking scenarios in perturbative string theory, whose coupling (before compactification) is defined by the exponential of the dilaton field $g_s = e^{\Phi}$.

The situation becomes cleanest if we consider, for simplicity and definiteness, the case of a single scalar, canonically normalized, field ϕ, playing the rôle of the quintessence field in a Robertson-Walker space-time with scale factor $a(t)$, with t the cosmic time. The field depends only on time, since we assume homogeneity. Such a field could be the dilaton, or other modulus field from the string multiplet [4].

Consider the lowest order Friedmann equation, as well as the equation of motion of the field ϕ in $D+1$ dimensions (the overdot denotes cosmic-time derivative), which are (formally) derived from the σ-model β-functions of a perturbative string theory

$$
H^2 \equiv \left(\frac{\dot{a}}{a}\right)^2 = \frac{2\kappa^2}{D(D-1)}E , \qquad E = \frac{(\dot{\phi})^2}{2} + V(\phi) ,
$$
$$
\ddot{\phi} + DH\dot{\phi} + V'(\phi) = 0 , \tag{12.13}
$$

with E the total energy of the scalar field, V its potential, and a prime indicating variations with respect to the field ϕ. We obtain the following expressions for the scale factor $a(t)$ and the cosmic horizon δ:

$$
a(t) = \exp\left(\int d\phi \sqrt{\frac{E}{D(D-1)(E-V)}}\right) ,
$$
$$
\delta = \int^\infty \frac{dt}{a} = \int d\phi \frac{1}{a\dot{\phi}} = \int d\phi \frac{1}{a\sqrt{2(E-V)}} . \tag{12.14}
$$

The condition for the existence of a cosmic horizon is of course the convergence of the integral on the right-hand-side of the expression for δ. This depends on the asymptotic behaviour of the potential V as compared to the total energy E. This behaviour can be studied in a generic perturbative string theory, based on the form of low energy potentials of possible quintessence candidates, such as dilaton, moduli etc. Because realistic string theories involve at a certain stage supersymmetry in target space, which is broken as we go down to the four dimensional world after compactification, or as we lower the energy from the string (Planck) scale, such arguments depend on the form of the potential, dictated by supersymmetry breaking considerations. The form is such that $\delta \to \infty$ in (12.14), and hence there are no horizons. In this lecture I will not give further details [25] on the form of the supersymmetry breaking string theory potentials, because the above-mentioned CPT/scattering-matrix based argument is more general, and encompasses such cases, and it is the most fundamental reason for the incompatibility of perturbative strings with space-time backgrounds with horizons.

I would like to stress, however, that these arguments refer to the traditional critical strings, without branes, where a low-energy field theory derives from conformal invariance conditions. From this latter point of view it is straightforward to understand the problem of incorporating cosmologies with horizons, such as inflation or in general de Sitter space-times, in perturbative strings. A tree-world-sheet σ-model on, say, graviton backgrounds, whose conformal invariance conditions would normally yield the target-space geometry, reads to order α' (α' denotes henceforth the Regge slope) [4]:

$$\beta_{\mu\nu} = R_{\mu\nu} + \dots \tag{12.15}$$

where the ... indicate contributions from other background fields, such as dilaton *etc.*.

Ignoring the other fields, conformal invariance of the perttrurbative stringy σ-model would require a Ricci flat $R_{\mu\nu} = 0$ background, which is not the case of a dS space, for which (cf. (12.6))

$$R_{\mu\nu} = \Lambda g_{\mu\nu} . \tag{12.16}$$

To generate such corrections in the early days of string theory, Fischler and Susskind [26] had to invoke renormalization group corrections to the above-tree level β-function (12.15), induced by higher string loops, i.e. higher topologies of the σ-model world sheet. Tadpoles $\mathcal{J}$ of dilatons at one string loop order (torus topologies) yielded a dS (or AdS depending on the sign of $\mathcal{J}$) type contribution to the graviton β-function, $\mathcal{J} g_{\mu\nu}$. The basic idea behind this approach is to accept that world-sheet surfaces of higher topologies with handles whose size is smaller than the short-distance cutoff of the world-sheet theory, will not be 'seen' as higher- topologies but appear 'effectively' as tree level ones. They will, therefore, lead to loop corrections to the traditional tree-level β-functions of the various background fields, which cannot be discovered at tree level. Conformal invariance implies of course that tori with such small handles are equivalent to world-sheet spheres but with a long thin tube connected to them. For more details on this I refer the interested reader in my lectures in the First Aegean School [3].

Nevertheless, this approach does not solve the problem, despite its formal simplicity and elegance. The reason is two fold: first, string-loop perturbation theory is not Borel-resummable, and as such, the expansion in powers of genus of closed Riemann surfaces with handles (and holes if open strings are included), does not converge mathematically, hence it cannot give sensible answers for strong or intermediate string couplings. It is indeed, expected, that the dark energy is a property of a full theory of quantum gravity, and as such, an explanation of it should *not* be restricted only to perturbative string theory. Second, as already mentioned several times above, a string propagating in a space-time with a loop-induced cosmological constant will not be characterized by a well-defined scattering matrix, which by definition is a 'must' for perturbative string theory.

Thus, the issue remains as to what kind of dark energy one is likely to encounter in string theory. In the next section I will discuss stringy scenarios for time-dependent dark energy, relaxing to zero asymptotically in time, in an attempt to accommodate well-defined string scattering amplitudes.

12.4 Dilaton Quintessence in String Theory

In this section we shall be concerned with the propagation of strings in homogeneous cosmological backgrounds, consisting of dilaton and graviton fields that depend only on the target time. In particular, we shall discuss Robertson-Walker (RW) homogeneous cosmologies. The incorporation of time-dependent backgrounds in string theory is not a straightforward issue: the basic problem is the proper implementation of the conformal invariance conditions of the pertinent σ-model, describing perturbative stringy excitations [4]. The problem arises because such backgrounds are not vacuum solutions of Einstein's equations, and as such they require non-trivial "matter" or "dark-energy" contributions to the stress tensor. It then becomes clear that such a situation cannot involve simply the propagation of strings in gravitational backgrounds alone. Extra fields from the (gravitational) string multiplet, such as the dilaton, should be considered in order to ensure that the appropriate conformal invariance conditions are satisfied.

In this section we shall discuss several cases of such backgrounds: we shall start from the simplest case of a linear (in time) dilaton background [27], satisfying the appropriate conformal invariance conditions, which are known to be equivalent to on-shell dynamical equations of motion of a low-energy effective action [4]. Next we shall proceed to discuss more complicated cosmological (time-dependent only) dilaton and graviton backgrounds, associated with pre-Big-Bang scenarios for the Universe [13], characterized by dilaton-driven acceleration at the current epoch. We shall consider backgrounds that still satisfy on-shell target-space dynamical equations. Finally, in the last part of the lecture, we shall discuss *off equilibrium* situations [10, 36], arising from cosmically catastrophic events in string cosmologies, such as brane world collisions, which are responsible for deviations from conformal invariance of the associated stringy σ-models. In this case, Liouville dressing [9] is required in order to restore conformal invariance. Upon the identification of the Liouville mode with the target time [10], which as we shall discuss below is forced by the dynamics, this procedure results to dynamical-dark-energy scenarios, in which the corresponding dilaton-driven acceleration of the universe diminishes with the cosmic time in such a way that, asymptotically in time, both the Universe acceleration and the (dilaton) dark energy contributions decay to zero.

12.4.1 An Expanding Universe in String Theory

One of the simplest, and most natural quintessence fields, to generate a dynamical dark energy component for the string Universe is the dilaton Φ, a scalar field that appears in the basic gravitational multiplet of any (super)string theory [4]. Dilaton cosmology has been originated by Antoniadis, Bachas, Ellis and Nanopoulos in [27], where the basic steps for a correct formulation of an expanding Robertson-Walker Universe in string theory have been taken, consistently with conformal invariance conditions[1]. The crucial rôle of a time dependent dilaton field had been emphasized.

In [27] a time-dependent dilaton background, with a linear dependence on time in the so-called σ-model frame was assumed. Such backgrounds, even when the σ-model metric is flat, lead to exact solutions (to all orders in α') of the conformal invariance conditions of the pertinent stringy σ-model, and so are acceptable solutions from a perturbative viewpoint. It was argued in [27] that such backgrounds describe linearly-expanding Robertson-Walker Universes, which were shown to be exact conformal-invariant solutions, corresponding to Wess-Zumino models on appropriate group manifolds.

The pertinent σ-model action in a cosmological (time-dependent only) background of graviton $G(t)$, antisymmetric tensor $B(t)$ and dilaton $\Phi(t)$ fields reads [4]:

$$S_\sigma = \frac{1}{4\pi\alpha'} \int_\Sigma d^2\xi [\sqrt{-\gamma} G_{\mu\nu} \partial_\alpha X^\mu \partial^\alpha X^\nu + i\epsilon^{\alpha\beta} B_{\mu\nu} \partial_\alpha X^\mu \partial_\beta X^\nu + \alpha' \sqrt{-\gamma} R^{(2)} \Phi] ,$$
(12.17)

where Σ denotes the world-sheet, with metric γ and the topology of a sphere, α are world-sheet indices, and μ, ν are target space-time indices. The important point of [27] was the rôle of target time t as a specific dilaton background, linear in that coordinate, of the form

$$\Phi = \text{const} - \frac{1}{2} Q t , \qquad (12.18)$$

where Q is a constant and $Q^2 > 0$ is the σ-model central-charge deficit, allowing this *supercritical* string theory to be formulated in some number of dimensions different from the critical number. Consistency of the underlying world-sheet conformal field theory, as well as modular invariance of the string scattering amplitudes, required *discrete* values of Q^2, when expressed in units of the string length M_s [27]. This was the first example of a non-critical string cosmology, with the spatial target-space coordinates X^i, $i = 1, \ldots D - 1$, playing the rôle of σ-model fields. This non-critical string was not conformal invariant, and hence required Liouville dressing [9]. The Liouville field had time-like signature in target space, since the central charge deficit $Q^2 > 0$ in the model of [27], and its zero mode played the rôle of target time.

As a result of the non-trivial dilaton field, the Einstein term in the effective D-dimensional low-energy field theory action is conformally rescaled by $e^{-2\Phi}$. This requires a redefinition of the σ-model frame space-time metric $G_{\mu\nu}$ to the 'physical' Einstein metric $g^E_{\mu\nu}$:

---

[1] In fact that work was actually the first work on Liouville supercritical strings [9], with the Liouville mode identified with the target time, although this had not been recognized in the original work, but later [10].

$$g_{\mu\nu}^{E} = e^{-\frac{4\Phi}{D-2}} G_{\mu\nu} \ . \tag{12.19}$$

Target time must also be rescaled, so that the metric acquires the standard Robertson-Walker (RW) form in the normalized Einstein frame for the effective action:

$$ds_{E}^{2} = -dt_{E}^{2} + a_{E}^{2}(t_{E}) \left(dr^{2} + r^{2}d\Omega^{2}\right) \ , \tag{12.20}$$

where we show the example of a spatially-flat RW metric for definiteness, and $a_E(t_E)$ is an appropriate scale factor, which is a function of t_E alone in the homogeneous cosmological backgrounds we assume throughout.

The Einstein-frame time is related to the time in the σ-model frame [27] by:

$$dt_E = e^{-2\Phi/(D-2)}dt \qquad \rightarrow \qquad t_E = \int^t e^{-2\Phi(t')/(D-2)}dt' \ . \tag{12.21}$$

The linear dilaton background (12.18) yields the following relation between the Einstein and σ-model frame times:

$$t_E = c_1 + \frac{D-2}{Q}e^{\frac{Q}{D-2}t} \ , \tag{12.22}$$

where c_1 is a constant, which can be set to zero by an appropriate shift of the origin of time. Thus, a dilaton background (12.18) that is linear in the σ-model time scales logarithmically with the Einstein time (Robertson-Walker cosmic time) t_E:

$$\Phi(t_E) = \text{const.} - \frac{D-2}{2}\ln(\frac{Q}{D-2}t_E) \ . \tag{12.23}$$

In this regime, the string coupling [4]:

$$g_s = \exp\left(\Phi(t)\right) \tag{12.24}$$

varies with the cosmic time t_E as $g_s^2(t_E) \equiv e^{2\Phi} \propto \frac{1}{t_E^{D-2}}$, thereby implying a vanishing effective string coupling asymptotically in cosmic time. In the linear dilaton background of [27], the asymptotic space-time metric in the Einstein frame reads:

$$ds^2 = -dt_E^2 + a_0^2 t_E^2 \left(dr^2 + r^2 d\Omega^2\right) \ , \tag{12.25}$$

where a_0 is a constant. Clearly, there is no acceleration in the expansion of the Universe (12.25).

The effective low-energy action on the four-dimensional brane world for the gravitational multiplet of the string in the Einstein frame reads [27]:

$$S_{\text{eff}}^{\text{brane}} = \int d^4x \sqrt{-g}\{R - 2(\partial_\mu\Phi)^2 - \frac{1}{2}e^{4\Phi}(\partial_\mu b)^2 - \frac{2}{3}e^{2\Phi}\delta c\} \ , \tag{12.26}$$

where b is the four-dimensional axion field associated with a four-dimensional representation of the antisymmetric tensor, and $\delta c = C_{\text{int}} - c^*$, where C_{int}

is the central charge of the conformal world-sheet theory corresponding to the transverse (internal) string dimensions, and $c^* = 22(6)$ is the critical value of this internal central charge of the (super)string theory for flat four-dimensional space-times. The linear dilaton configuration (12.18) corresponds, in this language, to a background charge Q of the conformal theory, which contributes a term $-3Q^2$ (in our normalization) to the total central charge. The latter includes the contributions from the four uncompactified dimensions of our world. In the case of a flat four-dimensional Minkowski space-time, one has $C_{\text{total}} = 4 - 3Q^2 + C_{\text{int}} = 4 - 3Q^2 + c^* + \delta c$, which should equal 26 (10). This implies that $C_{\text{int}} = 22 + 3Q^2$ $(6 + 3Q^2)$ for bosonic (supersymmetric) strings.

An important result in [27] was the discovery of an exact conformal field theory corresponding to the dilaton background (12.23) and a constant-curvature (Milne) static metric in the σ-model frame (or, equivalently, a linearly-expanding Robertson-Walker Universe in the Einstein frame). The conformal field theory corresponds to a Wess-Zumino-Witten two-dimensional world-sheet model on a group manifold $O(3)$ with appropriate constant curvature, whose coordinates correspond to the spatial components of the four-dimensional metric and antisymmetric tensor fields, together with a free world-sheet field corresponding to the target time coordinate. The total central charge in this more general case reads $C_{\text{total}} = 4 - 3Q^2 - \frac{6}{k+2} + C_{\text{int}}$, where k is a positive integer corresponding to the level of the Kac-Moody algebra associated with the WZW model on the group manifold. The value of Q is chosen in such a way that the overall central charge to be $c = 26$ and the theory is conformally invariant. Since such unitary conformal field theories have *discrete* values of their central charges, which accumulate to integers or half-integers from *below*, it follows that the values of the central charge deficit δc are *discrete* and *finite* in number. From a physical point of view, this implies that the linear-dilaton Universe may either stay in such a state for ever, for a given δc, or tunnel between the various discrete levels before relaxing to a critical $\delta c = 0$ theory. It was argued in [27] that, due to the above-mentioned finiteness of the set of allowed discrete values of the central charge deficit δc, the Universe could reach flat four-dimensional Minkowski space-time, and thus exit from the expanding phase, after a finite number of phase transitions.

The analysis in [27] also showed that there are tachyonic mass shifts of order $-Q^2$ in the bosonic string excitations, but not in the fermionic ones. This implies the appearance of tachyonic instabilities and the breaking of target-space supersymmetry in such backgrounds, as far as the excitation spectrum is concerned. The instabilities could trigger the cosmological phase transitions, since they correspond to relevant renormalization group world-sheet operators, and hence initiate the flow of the internal unitary conformal field theory towards minimization of its central charge, in accordance with the Zamolodchikov c-theorem [28]. In semi-realistic cosmological models [14] such tachyons decouple from the spectrum relatively quickly. On the other hand, as a result of the form of the dilaton in the Einstein frame (12.23), we observe that the dark-energy density for this (four-dimensional) Universe, $\Lambda \equiv e^{2\Phi}\delta c$, is relaxing to zero with a $1/t_E^{(D-2)}$ dependence on the Einstein-frame time for each of the equilibrium values of δc. Therefore, the breaking of supersymmetry induced by the linear dilaton is only an obstruction [29], rather than a spontaneous breaking, in the sense that it appears only temporarily in the boson-fermion mass splittings between the excitations, whilst the vacuum energy of the asymptotic equilibrium theory vanishes.

12.4.2 Pre Big Bang Scenaria

After the work of [27], dilaton cosmology has been discussed in a plethora of interesting works, most of them associated with the so-called 'pre-Big-Bang' (pBB) cosmologies [13], suggested by Veneziano, and pursued further by Gasperini, Veneziano and collaborators. For the interested reader, this type of cosmology has been reviewed by the author in the First Aegean School [3].

The basic feature behind the approach, is the fact that the dilaton has such time dependence in these models that, as the cosmic time elapses, the string coupling $g_s = e^{\Phi}$ grows stronger at late stages of the Universe. The dilaton potential in the pre Big-Bang approach, which may be generated by higher string loop corrections, has the generic form depicted in Fig. 12.3 [13]. The situation is opposite that of [27], where as we have seen the string coupling becomes weaker with the cosmic time, and perturbative strings are sufficient for a description of the Universe at late epochs.

I will not discuss in great detail the pBB theories, since there are excellent reviews on the subject [13], where the interested reader is referred to for more details. For our purposes here, I would like to emphasize the basic predictions

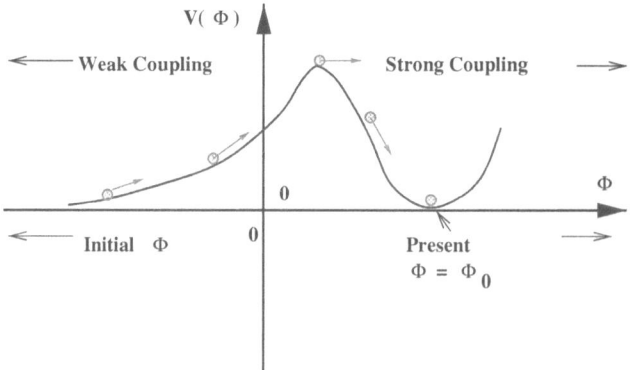

Fig. 12.3. The dilaton potential in the pre Big-Bang scenario of string cosmology. The string coupling grows strong at late times, and hence current-era is described by strongly-coupled strings, where higher string loop corrections matter

of this model regarding the rôle of dilaton as a a quintessence field, responsible for late-time acceleration of the string Universe.

The starting point is the string frame, low-energy, string-inspired effective action with graviton and dilaton backgrounds [4], to lowest order in the α' expansion, but including dilaton-dependent loop (and non-perturbative) corrections, which are essential given that at late epochs the dilaton grows strong in pBB scenarios. Such corrections are encoded in a few "form factors" [13] $\psi(\Phi)$, $Z(\Phi)$, $\alpha(\Phi)$, ..., and in an effective dilaton potential $V(\Phi)$. The effective action reads:

$$S = -\frac{M_s^2}{2} \int d^4x \sqrt{-\tilde{g}} \left[e^{-\psi(\Phi)}\tilde{R} + Z(\Phi)\left(\tilde{\nabla}\Phi\right)^2 + \frac{2}{M_s^2}V(\Phi) \right]$$

$$- \frac{1}{16\pi} \int d^4x \frac{\sqrt{-\tilde{g}}}{\alpha(\Phi)} F_{\mu\nu}^2 + \Gamma_m(\Phi, \tilde{g}, \text{matter}) , \tag{12.27}$$

where we follow the conventions of [13].

The four dimensional action above is the result of compactification. It is also assumed that the corresponding moduli have been frozen at the string scale. In the approach of [13] it is assumed that the form factors $\psi(\Phi)$, $Z(\Phi)$, $\alpha(\Phi)$ approach a finite limit as $\Phi \to +\infty$ while, in the same limit, $V \to 0$. The fields appearing in the matter action Γ_m are in general non-minimally and non-universally coupled to the dilaton (also because of the loop corrections).

In the Einstein frame the action (12.27) becomes

$$S = -\frac{M_P^2}{2} \int d^4x \sqrt{-g} \left[R - \frac{k(\Phi)^2}{2}\left(\nabla\Phi\right)^2 + \frac{2}{M_P^2}\hat{V}(\Phi) \right]$$

$$- \frac{1}{16\pi} \int d^4x \frac{\sqrt{-g}}{\alpha(\Phi)} F_{\mu\nu}^2 + \Gamma_m(\Phi, c_1^2 g_{\mu\nu}e^{\psi}, \text{matter}) , \tag{12.28}$$

where

$$k^2(\Phi) = 3\psi'^2 - 2e^{\psi}Z , \qquad \hat{V} = c_1^4 e^{2\psi}V . \tag{12.29}$$

The pertinent equations of motion for the graviton field read (in units where $M_P^2 = (8\pi G_N)^{-1} = 2$):

$$6H^2 = \rho + \rho_\Phi , \tag{12.30}$$

$$4\dot{H} + 6H^2 = -p - p_\Phi , \tag{12.31}$$

while the dilaton equation is:

$$k^2(\Phi)\left(\ddot{\Phi} + 3H\dot{\Phi}\right) + k(\Phi)k'(\Phi)\dot{\Phi}^2 + \hat{V}'(\Phi) + \frac{1}{2}[\psi'(\Phi)(\rho - 3p) + \sigma] = 0 . \tag{12.32}$$

In the above equations $H = \dot{a}/a$, a dot denotes differentiation with respect to the Einstein cosmic time, and we have used the definitions:

$$\rho_\Phi = \frac{1}{2}k^2(\Phi)\dot{\Phi}^2 + \hat{V}(\Phi) , \qquad p_\Phi = \frac{1}{2}k^2(\Phi)\dot{\Phi}^2 - \hat{V}(\Phi) . \tag{12.33}$$

After some manipulations the pertinent equations of motion, describing the dynamics of the system, read (all the quantities refer to the Einstein frame):

$$2\,H^2\,k^2\,\frac{d^2\Phi}{d\chi^2} + k^2\left(\frac{1}{2}\rho_m + \frac{1}{3}\rho_r + \hat{V}\right)\frac{d\Phi}{d\chi} + 2H^2\,k\,k'\left(\frac{d\Phi}{d\chi}\right)^2 +$$

$$2\hat{V}' + \psi'\rho_m + \sigma = 0\,,$$

$$H^2\left[6 - \frac{k^2}{2}\left(\frac{d\Phi}{d\chi}\right)^2\right] = \rho_m + \rho_r + \hat{V}\,,$$

$$\sigma \equiv -\frac{1}{c_1^2}\frac{2}{\sqrt{-g}}\frac{\delta(\Gamma_m + \frac{F_{\mu\nu}^2}{\alpha(\Phi)} - \text{terms})}{\delta\Phi} = \sigma_m + \sigma_r\,,$$

$$c_1^2 \equiv \lim_{\Phi\to+\infty}e^{-\psi(\Phi)}\,, \tag{12.34}$$

where $\chi = \ln a$, with a the scale factor in units of the present day scale, and the suffix $r(m)$ stands for radiation (matter) components.

The matter evolution equation, on the other hand, can be split into the various components (radiation (r), baryonic (b) and dark matter(d)):

$$\frac{d\rho_r}{d\chi} + 4\rho_r - \frac{\sigma_r}{2}\frac{d\Phi}{d\chi} = 0\,,$$

$$\frac{d\rho_b}{d\chi} + 3\rho_b - \frac{1}{2}\left(\psi'\rho_b + \sigma_b\right)\frac{d\Phi}{d\chi} = 0$$

$$\frac{d\rho_d}{d\chi} + 3\rho_d - \frac{1}{2}\left(\psi'\rho_d + \sigma_d\right)\frac{d\Phi}{d\chi} = 0\,, \tag{12.35}$$

and for the dilaton energy density ρ_Φ one can obtain the equation

$$\frac{d\rho_\Phi}{d\chi} + 6\rho_\Phi - 6\hat{V}(\Phi) + \frac{1}{2}\left(\psi'\rho_m + \sigma\right)\frac{d\Phi}{d\chi} = 0\,. \tag{12.36}$$

It is important to notice that one of the basic assumptions of the pBB scenarios is that, as the dilaton $\Phi \to +\infty$, there is a finite limit of the corresponding form factors [13]

$$e^{-\psi(\Phi)} = c_1^2 + b_1 e^{-\Phi} + \mathcal{O}(e^{-2\Phi})\,, \qquad Z(\Phi) = -c_2^2 + b_2 e^{-\Phi} + \mathcal{O}(e^{-2\Phi})\,,$$

$$\alpha(\Phi)^{-1} = a_0^{-1} + b_3 e^{-\Phi} + \mathcal{O}(e^{-2\Phi})\,. \tag{12.37}$$

where $c_i, i = 1,2\ b_i,\ i = 1,2,3$ and a_0 appropriate constants. Furthermore, one assumes that the effective dilaton potential originates purely from non perturbative effects, and thus has the form $\hat{V} = V_0 e^{-\Phi} + \mathcal{O}(e^{-2\Phi})$, tending to 0 as $\Phi \to +\infty$.

It remains to be seen whether the above are true in a complete string theory model, where (non-perturbative) summation over world-sheet genera is not understood at present. This is one of the reasons why, personally, I would prefer to use string models with weak string couplings at late eras, where perturbation theory is applicable and thus reliable predictions can be made. We shall consider such cases in the next chapter, however the pertinent string theories we shall employ are non-critical, as resulting from non equilibrium situations in the Early stages of the Universe.

The analysis of [13], based on (12.34), (12.35) and (12.36), leads to predictions regarding the behaviour of the various cosmological parameters of the pBB dilaton cosmology. Under various approximations and assumptions, which I will not go through, but I would stress that they are due to the fact that the various form factors and the dilaton couplings to matter are not known in this approach due to the (uncontrolled) loop corrections, one can solve the above equations to obtain the asymptotic evolution of the Hubble factor and of the dominant energy density in this approach,

$$H \sim a^{-3/(2+q)} , \qquad\qquad \rho \sim a^{-6/(2+q)} . \qquad (12.38)$$

where $q = \mathcal{O}(1) = \sigma_d/\rho_d$ is related to dark matter components, assumed dominant asymptotically, and is expressed [13] in terms of the various energy densities in the model $q = 2\frac{\Omega_V - \Omega_k}{1+\Omega_k-\Omega_V}$. We have defined $\Omega_i = \rho_i/6H^2$, with the suffix k denoting terms pertaining to the form-factor-$k(\Phi)$ (12.29) contributions to the vacuum energy density and the suffix V the dilaton potential contributions to it, $\rho_\Phi = \rho_k + \rho_V$. The dilaton equation of state in these models is given by $w_\Phi = \frac{\Omega_k - \Omega_V}{\Omega_k + \Omega_V}$. Some simple models of dark matter, assumed dominant in the asymptotic time regime, have been invoked in order to arrive at the behaviour (12.38). Their respective energy density is such that $\Omega_d + \Omega_k + \Omega_V = 1$. The dilatonic charges of such models, that is the appropriate dilaton factors that couple to the kinetic and interaction terms of the dark matter fields, play a crucial rôle in determining the late-time behaviour of q. The resulting asymptotic deceleration parameter of this Universe is given by:

$$q_{\text{decel}} \equiv -\frac{\dot{H} + H^2}{H^2} = -\frac{1}{H}\frac{dH}{d\chi} - 1 \sim \frac{1-q}{2+q} \qquad (12.39)$$

which implies that the universe would be accelerating asymptotically if $q > 1$. As we have seen above, this information relies heavily on the properties of the dark matter in this approach.

The evolution of the various cosmological parameters in a typical of such pBB models is given in Figs. 12.4 and 12.5, taken from the second in [13]. As we see, current and/or late-eras acceleration of the Universe can be arranged in these simplified models. However, what remains to be done in this context is to discuss detailed supersymmetric low-energy models obtained from realistic string theory, something which may not be feasible until one obtains control of the full non-perturbative regime of strings.

The reader should bear in mind that the above approach involves string theory on background fields which satisfy their equations of motion, and hence it is a case of critical strings. However, in occasions such as Early Universe cosmology, on-shell situations might not always be in place. An initial cosmically catastrophic event, such as the collision of two brane worlds or a Big

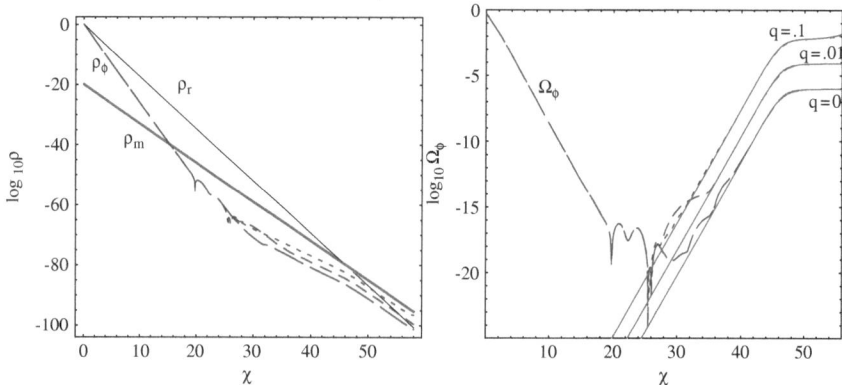

Fig. 12.4. Time evolution of ρ_ϕ for $q = 0$ (*dash-dotted curve*), $q = 0.01$ (*dashed curve*) and $q = 0.1$ (*dotted curve*) in the pre Big Bang cosmology [13]. The initial scale is $a_i = 10^{-20} a_{\rm eq}$, and the epoch of matter-radiation equality corresponds to $\chi \simeq 46$. **Left panel**: the dilaton energy density is compared with the radiation (*thin solid curve*) and matter (*bold solid curve*) energy density. **Right panel**: the dilaton energy density (in critical units) is compared with the analytical estimates for the focusing and dragging phases

Bang, certainly takes the theory way out of equilibrium. It might therefore be that the currently observed acceleration of the Universe is due to some relaxation process from an early-Universe cosmic catastrophe.

This is the point of view we shall discuss next, namely we shall attempt to formulate such off-equilibrium scenarios within the context of (non-critical) strings propagating in off-shell backgrounds. We shall associate the notion of non equilibrium in strings with that of deviations from conformal invariance of the pertinent σ-models describing perturbative stringy excitations at times long enough after the initial collision so that the σ-model approach suffices, but such that relaxation, non-critical-string effects are still important.

12.4.3 Non Critical Strings and Dark Energy

The General Idea

Pre Big Bang scenaria, as we have just discussed, involve strong string couplings at late times, and hence the various form factors appearing in the effective actions are essentially unknown for the present era.

An alternative approach, is to invoke the weak coupling late-era dilaton cosmology of [27], which has the advantage that at late eras perturbative σ-model calculations are reliable, and hence one can perform concrete computations and predictions. The analysis of [27] however has to be generalized to include inflationary and other backgrounds with horizons, if the dark matter issue and accelerating Universes are to be tackled. This cannot be achieved with the simple linear dilaton backgrounds of [27].

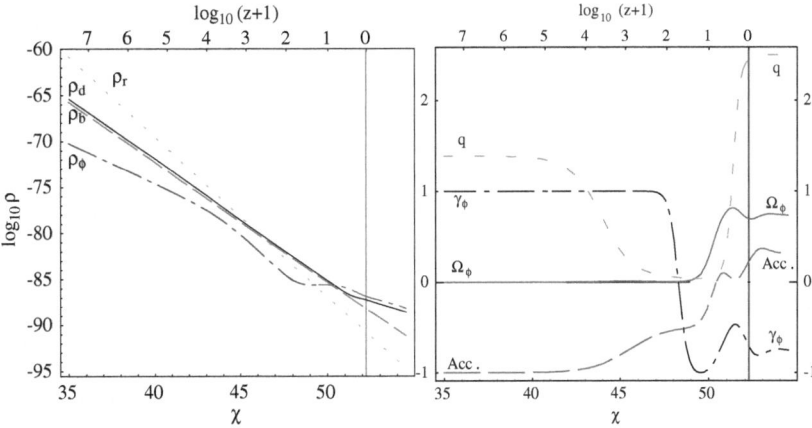

Fig. 12.5. Left panel: Late-time evolution of the dark matter (*solid curve*), baryonic matter (*dashed curve*), radiation (*dotted curve*) and the dilaton (*dash-dotted curve*) energy densities, for the pBB string cosmology model of [13]. The upper horizontal axis gives the $\log_{10}$ of the redshift parameter. **Right panel**: for the same model, the late-time evolution of q (*fine-dashed curve*), w_Φ (*dash-dotted curve*), Ω_Φ (*solid curve*) and of the acceleration parameter $\ddot{a}a/\dot{a}^2$ (*dashed curve*)

In [10] we went one step beyond the analysis in [27], and considered more complicated σ-model metric backgrounds that did not satisfy the σ-model conformal-invariance conditions, and therefore needed Liouville dressing [9] to restore conformal invariance. Such backgrounds could even be time-dependent, living in $(d + 1)$-dimensional target space-times. Various mathematically consistent forms of non-criticality can be considered, for instance cosmic catastrophes such as the collision of brane worlds [30, 35]. Such models lead to supercriticality of the associated σ models describing stringy excitations on the brane worlds. The Liouville dressing of such non-critical models results in $(d+2)$-dimensional target spaces with two time directions. An important point in [10] was the identification of the (world-sheet zero mode of the) Liouville field with the target time, thereby restricting the Liouville-dressed σ model to a $(d+1)$-dimensional hypersurface of the $(d+2)$-dimensional target space, thus maintaining the initial target space-time dimensionality. We stress that this identification is possible only in cases where the initial σ model is supercritical, so that the Liouville mode has time-like signature [9, 27]. In certain models [30, 35], such an identification was proven to be energetically preferable from a target-space viewpoint, since it minimized certain effective potentials in the low-energy field theory corresponding to the string theory at hand.

All such cosmologies require some physical reason for the initial departure from the conformal invariance of the underlying σ model that describes string excitations in such Universes. The reason could be an initial quantum fluctuation, or, in brane models, a catastrophic cosmic event such as the collision of two or more brane worlds. Such non-critical σ models relax asymptotically to conformal σ models, which may be viewed as equilibrium points in string theory space, as illustrated in Fig. 12.6. In some interesting cases of relevance to cosmology [14], which are particularly generic, the asymptotic conformal field theory is that of [27] with a linear dilaton and a flat Minkowski target-space metric in the σ-model frame. In others, the asymptotic theory is characterized by a constant dilaton and a Minkowskian space-time [30]. Since, as we discussed in [10] and review briefly below, the evolution of the central-charge deficit of such a non-critical σ model, $Q^2(t)$, plays a crucial rôle in inducing the various phases of the Universe, including an inflationary phase, graceful exit from it, thermalization and a contemporary phase of accelerating expansion, we term such Liouville string-based cosmologies Q-*Cosmologies*.

The use of Liouville strings to describe the evolution of our Universe has a broad motivation, since non-critical strings are associated with non-equilibrium situations, as are likely to have occurred in the early Universe. The space of non-critical string theories is much larger than that of critical strings. It is therefore remarkable that the departure from criticality may enhance the predictability of string theory to the extent that a purely stringy quantity such

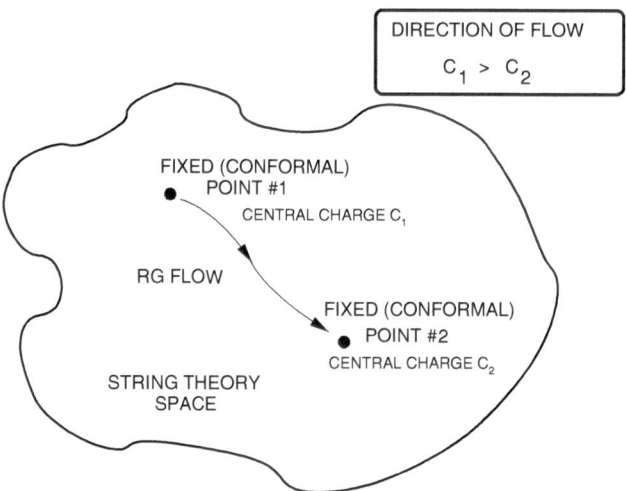

Fig. 12.6. A schematic view of string theory space, which is an infinite-dimensional manifold endowed with a (Zamolodchikov) metric [28]. The dots denote conformal string backgrounds. A non-conformal string flows (in a two-dimensional renormalization-group sense) from one fixed point to another, either of which could be a hypersurface in theory space. The direction of the flow is irreversible, and is directed towards the fixed point with a lesser value of the central charge, for unitary theories, or, for general theories, towards minimization of the degrees of freedom of the system

as the string coupling g_s may become accessible to experiment via its relation to the present-era cosmic acceleration parameter: $g_s^2 = -q^0$ [36]. Another example arises in a non-critical string approach to inflation, if the Big Bang is identified with the collision of two D-branes [35]. In such a scenario, astrophysical observations may place important bounds on the recoil velocity of the brane worlds after the collision, and lead to an estimate of the separation of the branes at the end of the inflationary period [33].

In such a framework, the identification of target time with a world-sheet renormalization group scale, the zero mode of the Liouville field [10], provides a novel way of selecting the ground state of the string theory. This is not necessarily associated with minimization of energy, but could simply be a result of cosmic chance. It may be a random global event that the initial state of our cosmos corresponds to a certain Gaussian fixed point in the space of string theories, which is then perturbed into a Big Bang by some relevant (in a world-sheet sense) deformation, which makes the theory non-critical, and hence out of equilibrium from a target space-time viewpoint. The theory then flows, as indicated in Fig. 12.6, along some specific renormalization group trajectory, heading asymptotically to some ground state that is a local extremum corresponding to an infrared fixed point of this perturbed world-sheet σ-model theory. This approach allows for many 'parallel universes' to be implemented, and our world might be just one of these. Each Universe may flow between different fixed points, its trajectory following a perturbation by a different operator. It seems to us that this scenario is more attractive and specific than the landscape scenario [26], which has recently been advocated as a framework for parametrizing our ignorance of the true nature of string/M theory.

Liouville Strings: A Brief Review of the Formalism

We commence our analysis with a brief review of the Liouville dressing procedure for non-critical strings, with the Liouville mode viewed as a local world-sheet renormalization group scale [10]. Consider a conformal σ-model, described by an action S^* on the world-sheet Σ, which is deformed by (non conformal) deformations $\int_\Sigma g^i V_i d^2\sigma$, with V_i appropriate vertex operators.

$$S_g = S^* + \int_\Sigma g^i V_i d^2\sigma \,. \tag{12.40}$$

The non-conformal nature of the couplings g^i implies that their (flat)world sheet renormalization group β-functions, β^i, are non-vanishing. The generic structure of such β-functions, close to a fixed point, $\{g^i = 0\}$ reads:

$$\beta^i = (h_i - 2)g^i + c^i_{jk}g^j g^k + O(g^3) \,, \tag{12.41}$$

where h_i are the appropriate conformal dimensions. In the context of Liouville strings, world-sheet gravitational dressing is required. The "gravitationally" dressed couplings, $\lambda^i(g,\phi)$, which from our point of view correspond to renormalized couplings in a curved space, read to $O(g^2)$ [9], in a weak field g^i-expansion we assume throughout:

$$\lambda^i(g,\phi) = g^i e^{\alpha_i \phi} + \frac{\pi}{Q + 2\alpha_i} c^i_{jk} g^j g^k \phi e^{\alpha_i \phi} + O(g^3), \qquad Q^2 = \frac{1}{3}(c - c^*),$$

(12.42)

where ϕ is the (world-sheet zero mode) of the Liouville field, and Q^2 is the central charge deficit, with $c = c[g]$ the ('running') central charge of the deformed theory [28], and c^* one of its critical values (conformal point) about which the theory is perturbed by means of the operators V^i. Close to a fixed point Q^2 may be considered as independent of g, but this is not true in general. Finally, α_i are the gravitational anomalous dimensions:

$$\alpha_i(\alpha_i + Q) = 2 - h_i \qquad \text{for} \qquad c \geq c^*.$$

(12.43)

Below we shall concentrate exclusively to the supercritical string case, $Q^2 \geq 0$, which from the point of view of identifying the Liouville mode with target time, corresponds to a Minkowskian signature spacetime manifold.

Due to the renormalization (12.42), the critical-string conformal invariance conditions, amounting to the vanishing of flat-space β-functions, are now substituted by:

$$\ddot{\lambda}^i + Q\dot{\lambda}^i = -\beta^i(\lambda) + \dots \qquad \text{for } c \geq c^*,$$

(12.44)

where the minus sign in front of the β-function on the right-hand-side is due to the supercriticality ($c > c^*$) of the string, the overdot denotes derivative with respect to the Liouville mode ϕ, and the $\dots$ denote higher-order terms, quadratic in $\dot{\lambda}^i$, $\mathcal{O}\left((\dot{\lambda}^i)^2\right)$. As we argued in [10], such terms can either be removed by field redefinitions, or alternatively are negligible if one works in the neighbourhood of a world-sheet renormalization-group fixed point, which is the case we shall consider in this work. The notation $\beta^i(\lambda)$ denotes flat-world-sheet β-functions but with the formal substitution $g^i \to \lambda^i(g,\phi)$. Note the minus sign in front of the flat world sheet β-functions β^i in (12.44), which is characteristic of the supercriticality of the string [9]. Notice that upon the identification of the Liouville mode ϕ with the target time t the overdot denotes temporal derivative.

Unless otherwise stated, for notational brevity from now on we shall use the notation

$$\lambda^i \to g^i$$

(12.45)

since we shall only be dealing with Liouville renormalized background fields $g^i(\phi, X^\mu)$ (with μ a target-space-time index).

We now mention that, in the case of stringy σ models, where the couplings g^i are background fields, depending on the coordinates of the target space-time, the diffeomorphism invariance of the target space results in the replacement of (12.44) by:

$$\ddot{g}^i + Q(t)\dot{g}^i = -\tilde{\beta}^i,$$

(12.46)

where the $\tilde{\beta}^i$ are the Weyl anomaly coefficients of the stringy σ model in the background $\{g^i\}$, which differ from the ordinary world-sheet renormalization-group β^i functions by terms of the form:

$$\tilde{\beta}^i = \beta^i + \delta g^i \ . \tag{12.47}$$

In the above formula δg^i denote transformations of the background field g^i under infinitesimal general coordinate transformations, e.g. for gravitons [4] $\tilde{\beta}^G_{\mu\nu} = \beta^G_{\mu\nu} + \nabla_{(\mu}W_{\nu)}$, with $W_\mu = \nabla_\mu\Phi$, and $\beta^G_{\mu\nu} = R_{\mu\nu}$ to order α' (one σ-model loop).

In [10] we have treated the Liouville mode as a local (covariant) world-sheet renormalization-group scale. To justify formally this interpretation, one may write

$$\phi = -\frac{2}{\alpha}\tau \ , \qquad \tau \equiv -\frac{1}{2}\log A \ , \qquad A = \int_\Sigma d^2\sigma\sqrt{\gamma} = \int_\Sigma d^2\sigma\sqrt{\hat{\gamma}}e^{\alpha\phi} \ ,$$

$$\alpha = -\frac{Q}{2} + \frac{1}{2}\sqrt{Q^2 + 8} \ , \tag{12.48}$$

where γ is a world-sheet metric, and $\hat{\gamma}$ is a fiducial metric, obtained after the conformal gauge choice in terms of the Liouville mode ϕ [9]. We thus observe that the Liouville mode is associated with the logarithm of the world-sheet area A.

Using (12.48), we can re-write (12.42) in a standard "flat-world-sheet" renormalization group form [10, 32]:

$$\frac{d}{d\tau}\lambda^i = (\tilde{h}_i - 2)\lambda^i + \pi\tilde{c}^i_{jk}\lambda^j\lambda^k + \cdots \ ,$$

$$\tilde{h}_i - 2 = -\frac{2}{\alpha}\alpha_i \ , \qquad \tilde{c}^i_{jk} = -\frac{2}{\alpha(Q + 2\alpha_i)}c^i_{jk} \ . \tag{12.49}$$

which justifies formally the identification [10] of the Liouville mode with a local renormalization group scale on the world sheet. It also implies that the point $\phi \to \infty$ is an infrared fixed point of the flow, in which case the world-sheet area diverges $|A| \to \infty$.

A highly non-trivial feature of the β^i functions is the fact that they are expressed as gradient flows in theory space [10, 28], i.e. there exists a 'flow' function $\mathcal{F}[g]$ such that

$$\beta^i = \mathcal{G}^{ij}\frac{\delta\mathcal{F}[g]}{\delta g^j} \ , \tag{12.50}$$

where $\mathcal{G}^{ij}$ is the inverse of the Zamolodchikov metric in theory space [28], which is given by appropriate two-point correlation functions between vertex operators V^i,

$$\mathcal{G}_{ij} \sim \text{Lim}_{z\to 0}2z^2\bar{z}^2 < V_i(z)V_j(0) > \ , \tag{12.51}$$

where z denotes a complex Euclidean world-sheet coordinate. In the case of stringy σ-models the flow function $\mathcal{F}$ may be identified [10] with the running central charge deficit $Q^2[g]$.

The set of equations (12.46,12.50) defines the *generalized conformal invariance conditions*, expressing the restoration of conformal invariance by the Liouville mode. The solution of these equations, upon the identification of the Liouville zero mode with the original target time, leads to constraints in the space-time backgrounds [10, 30], in much the same way as the conformal invariance conditions $\beta^i = 0$ define consistent space-time backgrounds for critical strings [4].

Helmholtz Conditions, and Liouville Equations as Equations of Motion from an (off-shell) Effective Action

An important comment we would like to make concerns the possibility of deriving the set of equations (12.44,12.46) from a target space action. This issue has been discussed in the affirmative in [10], where it was shown that the set of equations (12.44) satisfies the Helmholtz conditions for the existence of an *off-shell* action in the 'space of couplings' $\{g^i\}$ of the non-critical string. The property (12.50) is crucial to this effect. Upon the identification of target time with the Liouville mode [10] this action becomes identical with the target space action describing the off-shell dynamics of the Liouville string. We should stress the fact that the action is off shell, in the sense that the on-shell conditions correspond to the vanishing of the β-functions β^i, while in our case $\beta^i \neq 0$. Let us briefly review these arguments below.

Our point is to demonstrate that the generalized conformal invariance equations (12.44,12.46) obey the necessary conditions to be derived by a Lagrangian, which however is *off-shell*. The conditions for the existence of an underlying Lagrangian L whose variations with respect the appropriate dynamical variables g^i are equivalent (but not necessarily identical) to (12.44) are determined by the existence of a non-singular matrix ω_{ij} with

$$\omega_{ij} \left(\alpha' \ddot{g}^j + \sqrt{\alpha'} Q \dot{g}^j + \beta^j \right) = \frac{d}{d\phi} \left(\frac{\partial L}{\partial \dot{g}^i} \right) - \frac{\partial L}{\partial g^i} \tag{12.52}$$

which obeys the Helmholtz conditions [31]

$$\omega_{ij} = \omega_{ji} \tag{12.53}$$

$$\frac{\partial \omega_{ij}}{\partial \dot{g}^k} = \frac{\partial \omega_{ik}}{\partial \dot{g}^j} , \tag{12.54}$$

$$\frac{1}{2} \frac{D}{D\phi} \left(\omega_{ik} \frac{\partial f^k}{\partial \dot{g}^j} - \omega_{jk} \frac{\partial f^k}{\partial \dot{g}^i} \right) = \omega_{ik} \frac{\partial f^k}{\partial g^j} - \omega_{jk} \frac{\partial f^k}{\partial g^i} \tag{12.55}$$

$$\frac{D}{D\phi} \omega_{ij} = -\frac{1}{2\alpha'} \left(\omega_{ik} \frac{\partial f^k}{\partial \dot{g}^j} + \omega_{jk} \frac{\partial f^k}{\partial \dot{g}^i} \right) , \tag{12.56}$$

where

$$f^i \equiv -\sqrt{\alpha'} Q \dot{g}^i - \beta^i [g] , \qquad \frac{D}{D\phi} \equiv \frac{\partial}{\partial \phi} + \dot{g}^i \frac{\partial}{\partial g^i} + \frac{f^i}{\alpha'} \frac{\partial}{\partial \dot{g}^i} . \tag{12.57}$$

If the conditions (12.53)–(12.56) are met, then

$$\alpha' \, \omega_{ij} = \frac{\partial^2 L}{\partial \dot{g}^i \partial \dot{g}^j} \tag{12.58}$$

and the Lagrangian in (12.58) can be determined up to total derivatives according to [31]

$$\mathcal{S} \equiv \int d\phi \, L = - \int d\phi \int_0^1 d\kappa \, g^i E_i(\phi, \kappa g, \kappa \dot{g}, \kappa \ddot{g}) \, ,$$
$$E_i(\phi, g, \dot{g}, \ddot{g}) \equiv \omega_{ij} \left(\alpha' \, \ddot{g}^j + \sqrt{\alpha'} \, Q \, \dot{g}^j + \beta^j \right) \, . \tag{12.59}$$

In the case of non-critical strings one can identify [10]

$$\omega_{ij} = -\frac{1}{\sqrt{\alpha'}} \mathcal{G}_{ij} \, , \tag{12.60}$$

where the Zamolodchikov metric $\mathcal{G}_{ij}$ in theory space is given by (12.51).

Near a fixed point in moduli (g^i) space, where the variation of Q is small, the action (12.59) then becomes [10]

$$\mathcal{S} = \int d\phi \left(-\frac{\sqrt{\alpha'}}{2} \dot{g}^i \, \mathcal{G}_{ij}[g; \phi] \dot{g}^j - \frac{1}{\sqrt{\alpha'}} C[g; \phi] + \ldots \right) \, , \tag{12.61}$$

where the dots denote terms that can be removed by a change of renormalization scheme. Within a critical string (on-shell) approach, the action (12.59, 12.61) can be considered as an effective action generating the string scattering amplitudes. Here it should be considered as a target space 'off-shell' action for non-critical strings [10]. From (12.61) it follows that the canonical momenta p_i conjugate to the couplings g^i are given by

$$p_i = \sqrt{\alpha'} \, \mathcal{G}_{ij} \, \dot{g}^j \, . \tag{12.62}$$

Let us briefly sketch the validity of the conditions (12.53)–(12.56) for the choice (12.60). Since $\mathcal{G}_{ij}$ is symmetric, the first Helmholtz condition (12.53) is satisfied. The conditions (12.54) and (12.55) hold automatically because of the gradient flow property (12.50) of the β-function, and the fact that $\mathcal{G}_{ij}$ and $C[g; \phi]$ are functions of the coordinates g^i and not of the conjugate momenta. Finally, the fourth Helmholtz condition (12.56) yields the equation

$$\frac{D}{D\phi} \mathcal{G}_{ij} = \frac{Q}{\sqrt{\alpha'}} \mathcal{G}_{ij} \, , \tag{12.63}$$

which implies an "expanding scale factor" for the "metric in moduli space" of the string

$$\mathcal{G}_{ij}[\phi; g(\phi)] = e^{Q\phi/\sqrt{\alpha'}} \, \widehat{\mathcal{G}}_{ij}[\phi; g(\phi)] \, , \tag{12.64}$$

where $\widehat{\mathcal{G}}_{ij}$ is a Liouville renormalization group invariant function, i.e. a fixed fiducial metric on moduli space. This is exactly the form of the Zamolodchikov metric for Liouville strings [10]. Thus there is an underlying Lagrangian dynamics in the non-critical string problem.

The action (12.61) allows canonical quantization, which as we have mentioned is induced by including higher genus effects in the string theory [10]. In the canonical quantization scheme the couplings g^i and their canonical momenta (12.62) are replaced by quantum mechanical operators (in target space) $\widehat{g}^i$ and $\widehat{p}_i$ obeying

$$[[\widehat{g}^i, \widehat{p}_j]] = i\hbar_{\mathcal{M}}\, \delta^i_j \,, \tag{12.65}$$

where the quantum commutator $[[\cdot\,,\,\cdot]]$ is defined on the moduli space $\mathcal{M}$ of deformed conformal field theories of the form (12.40), and $\hbar_{\mathcal{M}}$ is an appropriate "Planck constant". We can use the Schrödinger representation in which the canonical momentum operators obey [10]

$$\left\langle \widehat{p}_i \right\rangle_{\mathrm{L}} = \left\langle -i\frac{\delta}{\delta g^i} \right\rangle_{\mathrm{L}} = \left\langle V_i \right\rangle_{\mathrm{L}} \,. \tag{12.66}$$

Thus the canonical commutation relation (12.65) in general yields, on account of (12.66), a non-trivial commutator between the couplings g^i and the associated vertex operators of the (genera resummed) σ-models.

Liouville String as a Critical String in one Target-space Dimension Higher

The restoration of conformal invariance by the Liouville mode implies that in an enlarged target space-time, with coordinates (ϕ, X^0, X^i) the resulting σ-model will be conformal, for which one would have the normal conformal invariance conditions [4]. This means that the set of equations (12.44) can be cast in a conventional form, amounting to the vanishing of β functions of a σ-model, but in this enlarged space:

$$\tilde{\beta}^{(D+1)}(g) = 0 \,, \tag{12.67}$$

where D is the target-space dimensionality of the σ-model before Liouville dressing, g are Liouville-dressed fields and there are Liouville components as well in the appropriate tensorial coordinates.

For fields of the string multiplet, it can be checked explicitly that (12.67) and (12.44) (in D-dimensions) are equivalent [32]. For completeness, we shall demonstrate this by considering explicitly the dilaton Φ, graviton $G_{\mu\nu}$ and antisymmetric tensor fields $B_{\mu\nu}$. We shall not consider explicitly the tachyon field, although its inclusion is straightforward and does not modify the results.

To $\mathcal{O}(\alpha')$, the appropriate σ-model β-functions for a D-dimensional target space-time, parametrized by coordinates X^μ, $\mu = 0, 1, \ldots D - 1$, read [4]:

$$\hat{\beta}^{\Phi(D)} = \beta^{\Phi(D)} - \frac{1}{4}G^{\mu\nu}\beta^{G(D)}_{\mu\nu} = \frac{1}{6}\left(C^{(D)} - 26\right) ,$$

$$C^{(D)} = D - \frac{3}{2}\alpha'\left(R - \frac{1}{12}H_{\mu\nu\rho}H^{\mu\nu\rho} - 4(\nabla\Phi)^2 + 4\nabla^2\Phi\right) ,$$

$$\beta^{G(D)}_{\mu\nu} = \alpha'\left(R_{\mu\nu} + 2\nabla_\mu\nabla_\nu\Phi - \frac{1}{4}H_{\mu\sigma\rho}H_\nu{}^{\sigma\rho}\right) ,$$

$$\beta^{B(D)}_{\mu\nu} = \alpha'\left(-\frac{1}{2}\nabla_\rho H^\rho{}_{\mu\nu} + H^\rho{}_{\mu\nu}\nabla_\rho\Phi\right) . \tag{12.68}$$

where $H_{\mu\nu\rho} = 3\nabla_{[\mu}B_{\nu\rho]}$ is the antisymmetric tensor field-strength, on which the β-functions depend, as dictated by an appropriate Abelian Gauge symmetry [4].

To demonstrate that such β-functions yield equations of the form (12.44), when they are reduced to a target-space manifold with one lower dimension, we separate from the expressions (12.68) a Liouville component. We first note that there is a special normalization of the σ-model kinetic term of the Liouville field ϕ for which (12.43) is valid, which implies that the enlarged space-time metric is of "Robertson-Walker" form with respect to ϕ, i.e.:

$$ds^2 = -d\phi^2 + G_{\mu\nu}(\phi, X^\mu)dX^\mu dX^\nu, \quad \mu,\nu = 0,1,\ldots D-1 \tag{12.69}$$

where the Minkowski signature of the Liouville term is due to the assumed supercriticality of the non-critical string [9, 27]. This implies that for graviton and antisymmetric tensor β-functions one has:

$$\tilde{\beta}^G_{\phi\phi} = \tilde{\beta}^{G,B}_{\phi\mu} = 0 \tag{12.70}$$

which are viewed as additional constraints. However, from the point of view of the enlarged space-time such constraints can be easily achieved by an appropriate general coordinate transformation, which from our point of view is a renormalization-scheme choice.

We find it convenient to shift the dilaton [32]:

$$\Phi \to \varphi = 2\Phi - \log\sqrt{G} . \tag{12.71}$$

In this case we may write (12.68) as follows (to keep consistency with the previous notation we have denoted the β-functions in the enlarged space-time (ϕ, X^μ) by $\tilde{\beta}$):

$$0 = C^{(D+1)} - 26 = C^{(D)} - 25 - 3G^{\phi\phi}\left(\ddot{\varphi} - (\dot{\varphi})^2\right) ,$$

$$0 = \tilde{\beta}^G_{\phi\phi} = 2\ddot{\varphi} - \frac{1}{2}G^{\mu\kappa}G^{\nu\lambda}\left(\dot{G}_{\mu\nu}\dot{G}_{\kappa\lambda} + \dot{B}_{\mu\nu}\dot{B}_{\kappa\lambda}\right) ,$$

$$0 = \tilde{\beta}^G_{\mu\nu} = \beta^{G(D)}_{\mu\nu} - G^{\phi\phi}\left(\ddot{G}_{\mu\nu} - \dot{\varphi}\dot{G}_{\mu\nu} - G^{\kappa\lambda}[\dot{G}_{\mu\kappa}\dot{G}_{\nu\lambda} - \dot{B}_{\mu\kappa}\dot{B}_{\nu\lambda}]\right) ,$$

$$0 = \tilde{\beta}^B_{\mu\nu} = \beta^{(D)}_{\mu\nu} - G^{\phi\phi}\left(\ddot{B}_{\mu\nu} - \dot{\varphi}\dot{B}_{\mu\nu} - 2G^{\kappa\lambda}\dot{G}_{\kappa[\mu}\dot{B}_{\nu]\lambda}\right) , \tag{12.72}$$

where the overdot denotes total Liouville scale derivative. In our interpretation of the Liouville field as a (local) renormalization scale [10] this is equivalent to a total world-sheet renormalization-group derivative. The (12.72) are precisely of the form (12.46) of the generalized conformal invariance conditions.

In Liouville strings [9], the dilaton Φ, as being coupled to the world-sheet curvature, receives contributions from the Liouville mode ϕ which are linear. In this sense one may split the dilaton field in ϕ-dependent parts and X^μ dependent parts

$$\Phi(\phi, X^\mu) = -\frac{1}{2}Q\phi + \tilde{\Phi}(X^\mu) \,, \qquad (12.73)$$

where $Q^2 = \frac{1}{3}\left(C^{(D)} - 25\right)$ is the central charge deficit, and the normalization of the term linear in ϕ is dictated by the analysis of [9], in which the Liouville mode has a canonical σ-model kinetic term. This implies that φ is such that:

$$\dot{\varphi} = -Q + \mathcal{O}(\dot{Q}, \sqrt{G}G^{\mu\nu}\dot{G}_{\mu\nu}) \,. \qquad (12.74)$$

Note that, in the context of the (12.72), the terms in $\dot{\varphi}$ proportional to $\dot{G}_{\mu\nu}$, will yield terms quadratic in Liouville derivatives of fields. Upon our interpretation of the Liouville field as a (local) renormalization scale [10] terms quadratic in the Liouville derivatives of fields, i.e. terms of order $\mathcal{O}\left(\dot{G}\dot{B}, \dot{G}\dot{G}, \dot{B}\dot{B}\right)$ become quadratic in appropriate β-functions.

The same is true for $\dot{Q}$ terms, on account of the renormalization-group invariance of the central charge $C^{(D)}$, upon viewing the Liouville zero mode as a world-sheet renormalization-group scale [10]. Indeed, in such a case the only dependence of $Q^2 \propto C^{(D)} - C^*$ on the liouville mode would be through its dependence on the couplings $g^i = G_{\mu\nu}, B_{\mu\nu}, \ldots$, thus $\dot{Q}^2(= 2Q\dot{Q}) = -\beta^i\partial_i C$ (where $\partial_i = \delta/\delta g^i$ denotes functional derivatives with respect to the appropriate background field/coupling g^i). On account of the gradient flow (12.50), which can be shown to be true for the Liouville local renormalization-group world-sheet scale [10], one has $\partial_i C \propto \beta_i$ and, since Q is a perturbative series in the couplings g^i (assumed weak), one obtains that $\dot{Q}$ contains terms quadratic in Liouville derivatives of fields g^i.

Such quadratic terms may be removed by appropriate field redefinitions [10], provided the gradient flow property (12.50) is valid. Alternatively, one may ignore such quadratic terms in Liouville derivatives of fields by working in the neighbourhood of a renormalization group fixed point. Such terms are of higher order in a weak-field/ σ-model-coupling expansion, and thus can be safely neglected if one stays close to a fixed point. This is the case of the specific example of colliding brane cosmologies to be discussed in the next chapter, where one encounters only marginal non-criticality for slow-moving D-brane worlds. Ignoring such higher-order terms, therefore, and taking into account world-sheet renormalizability, one obtains

$$\ddot{\varphi} + \ldots = 0 \,, \qquad (12.75)$$

where the ... denote the neglected (higher-order) terms.

Taking into account that $G_{\phi\phi} = -1$ for supercritical strings [27] (cf. (12.69)), we observe that, as a result of (12.74), (12.75), the first two of the (12.72) are satisfied automatically (up to removable terms quadratic in Liouville derivatives of fields). The first of these equations is the dilaton equation, which thus becomes equivalent to the definition of Q^2, and therefore acquires a trivial content in this context. Notice also that the second of these equations is due to the constraints (12.70), which should be taken into account together with the set of equations (12.72). It can be shown [32] that the rest of these constraints do not impose further restrictions, and thus can be ignored, at least close to a fixed point, where the constraints can be solved for arbitrary $G_{\mu\nu}, B_{\mu\nu}$ fields. The rest of the equations (12.72) then, for graviton and antisymmetric tensor fields, reduce to (12.46), up to irrelevant terms quadratic in Liouville derivatives of fields.

This completes our proof for the case of interest. What we have shown above is that the Liouville equations (12.44,12.46) can be obtained from a set of conventional β-function equations (12.67) if one goes to a σ-model with one more target-space dimension, the extra dimension being provided by the Liouville field.

In the remainder of this subsection, however, we shall be dealing with situations in which the identification of the Liouville mode ϕ with (some function of) the target time X^0 will be made [10, 14, 36] in expressions of the form (12.67) in the enlarged $(D+1)$-dimensional spacetime (ϕ, X^μ). This latter approach is distinct from the standard Liouville approach described above in which ϕ was an independent mode. In that case, one should look for consistent solutions of the resulting equations in the D-dimensional submanifold $(\phi = X^0, X^i)$. In this sense, the target-space dimensionality remains D, but the resulting string will be characterized by the Liouville (12.44), supplemented by the constraint of the identification $\phi = f(X^0)$, and will have a non zero central charge deficit $Q^2(\phi)$, which is in general time dependent, and will appear as relaxation vacuum energy in the target space of the string.

To put it in other words, one starts from a critical σ-model, perturbs it by some non-conformal deformation, induces non-criticality, but instead of using an extra Liouville σ-model field, one uses the existing time coordinate as a Liouville mode, i.e. one invokes a readjustment of the time dependence of the various background fields (a sort of back reaction), in order to restore the broken conformal invariance. It is a non-trivial fact that there are consistent solutions to the resulting equations, and this is the topic of the next part of the lecture. Namely, we shall consider some specific models of non-critical strings, associated with cosmically catastrophic events in the early Universe, in which we shall identify the time with the Liouville mode dynamically, and we shall present consistent solutions of (12.67), under the constraint $\phi = X^0$, to lowest order $\mathcal{O}(\alpha')$ in the α' expansion of the respective σ-model.

Non-critical Strings
in Cosmological Dilaton and Graviton Backgrounds
and Relaxation Dark-energy Models

When applied to homogeneous dilaton cosmologies, with dilaton and graviton backgrounds, depending only on time, the above-described Liouville approach yields interesting results, including a modified asymptotic scaling of the dark matter energy density, a^{-2} with the scale factor, as well as an expression of the current-era acceleration parameter of the Universe roughly proportional to the square of the string coupling, $q_0 \propto -(g_s^0)^2$, $g_s^2 = e^{2\Phi}$, with Φ the current era dilaton (this proportionality relation becomes exact at late eras, when the matter contributions become negligible due to cosmic dilution). The current-era dark energy in this framework relaxes to zero with the Einstein cosmic time as $1/t^2$, and this scaling law follows from the generalized conformal invariance conditions (12.46), characterizing the Liouville theory, as well as the identification of time with the Liouville mode [10].

To be specific, after this identification, the relevant Liouville (12.46) for dilaton and graviton cosmological backgrounds, in the Einstein frame [27], read [36]:

$$3\,H^2 - \tilde{\varrho}_m - \varrho_\Phi \;=\; \frac{e^{2\Phi}}{2}\,\tilde{\mathcal{J}}_\phi\,,$$

$$2\,\dot{H} + \tilde{\varrho}_m + \varrho_\Phi + \tilde{p}_m + p_\Phi \;=\; \frac{\tilde{\mathcal{J}}_{ii}}{a^2}\,,$$

$$\ddot{\Phi} + 3H\dot{\Phi} + \frac{1}{4}\frac{\partial \hat{V}_{all}}{\partial \Phi} + \frac{1}{2}\,(\tilde{\varrho}_m - 3\tilde{p}_m) = -\frac{3}{2}\,\frac{\tilde{\mathcal{J}}_{ii}}{a^2} - \frac{e^{2\Phi}}{2}\,\tilde{\mathcal{J}}_\Phi\,. \quad (12.76)$$

where $\tilde{\rho}_m$ and $\tilde{p}_m$ denote the matter energy density and pressure respectively, including dark matter contributions. As usual, the overdot denotes derivatives with respect to the Einstein time, and H is the Hubble parameter of the Robertson-Walker Universe. The r.h.s of the above equations denotes the non-critical string *off-shell* terms appearing in (12.46), due to the non-equilibrium nature of the pertinent cosmology. The latter could be due to an initial cosmically catastrophic event, such as the collision of two brane worlds:

$$\tilde{\mathcal{J}}_\Phi \;=\; e^{-2\Phi}\,(\ddot{\Phi} - \dot{\Phi}^2 + Qe^\Phi\dot{\Phi})\,,$$
$$\tilde{\mathcal{J}}_{ii} \;=\; 2\,a^2\,(\,\ddot{\Phi} + 3H\dot{\Phi} + \dot{\Phi}^2 + (1-q)H^2 + Qe^\Phi(\dot{\Phi} + H))\,. \quad (12.77)$$

Notice the *dissipative* terms proportional to $Q\dot{\phi}$, which are responsible for the terminology "Dissipative Cosmology" used alternatively for Q-cosmology [36]. In these equations, q is the deceleration $q \equiv -\ddot{a}a/\dot{a}^2$. The potential appearing in (12.76) is defined by $\hat{V}_{all} = 2Q^2\exp(2\Phi) + V$ where, for the sake of generality, we have allowed for an additional potential term in the string action $-\sqrt{-g_E}\,V$ (with g_E denoting Einstein-frame metric).

A brief summary of the results of our analysis for a model-case Q-cosmology, are presented in Figs. 12.7, 12.8, 12.9 and 12.10. The model is

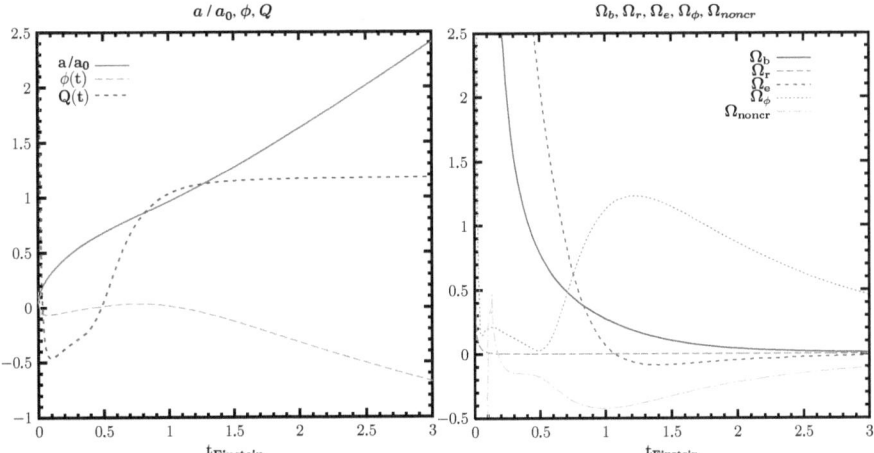

Fig. 12.7. Left panel: The dilaton ϕ, the (square root of the) central charge deficit Q and the ratio a/a_0 of the cosmic scale factor as functions of the Einstein time $t_{Einstein}$. The present time is located where $a/a_0 = 1$ and in the figure shown corresponds to $t_{today} \simeq 1.07$. The input values for the densities are $\rho_b = 0.238$, $\rho_e = 0.0$ and w_e is 0.5. The dilaton value today is taken $\Phi = 0.0$. **Right panel**: The values of $\Omega_i \equiv \rho_i/\rho_c$ for the various species as functions of $t_{Einstein}$

discussed in some detail in [36]. Notice the late-era presence of exotic a^{-2}-scaling of matter species, attributed to dark matter, denoted by ρ_e in the figures. Moreover, the asymptotic acceleration of the universe tends to zero as the square of the string coupling (cf. Fig. 12.10), $q_{decel} \sim -g_s^2 = -e^{2\Phi} \propto 1/t^2$, with t the cosmic time in the Einstein frame.

The reader is invited to compare these results with the ones of critical-string dilaton cosmologies in pre-Big-bang scenarios presented above (cf. Figs. 12.4 and 12.5), in particular with respect to the effects of the non-critical, off-shell terms "$\mathcal{J}$", which appear significant at the current era [36].

An important result of the analysis of [36] is the fact that the conventional Boltzmann equation, controlling the evolution of species densities, n, needs to be modified in Liouville Q-cosmology [37], in order to incorporate consistently the effects of the dilaton *dissipative* pressure $\sim \dot{\Phi}$ and the non-critical (relaxation) terms, "$\mathcal{J}$":

$$\frac{dn}{dt} = -3\,H\,n - <\sigma v> (n^2 - n_{eq}^2) + \dot{\Phi}\,n + \text{``}\mathcal{J}/m_X\text{''} . \quad (12.78)$$

in a standard notation [38], where $\langle \ldots \rangle$ denotes a thermal average, σ is the annihilation cross section, v is the Moeller velocity, and n_{eq} denotes a thermal equilibrium number density.

The respective relic density of the species X, with mass m_X, is then obtained from $\Omega_X h_0^2 = n\,m_X h_0^2$, after solving this modified equation. This may

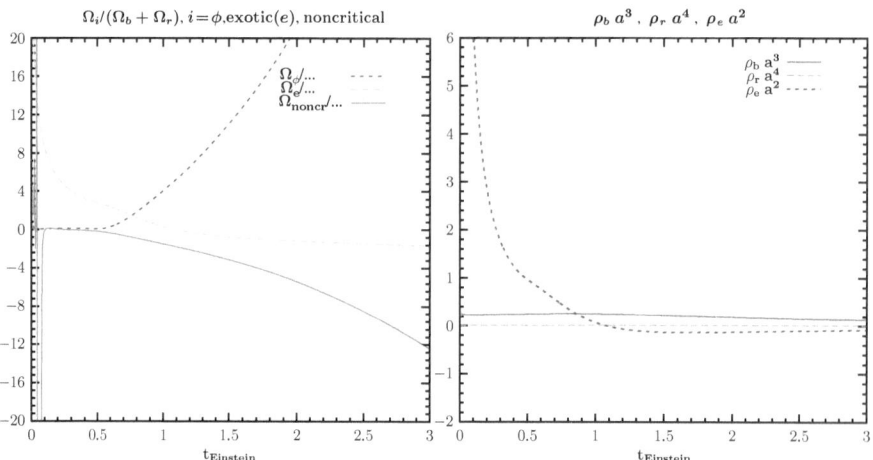

Fig. 12.8. Left panel: Ratios of Ω's for the dilaton (Φ), exotic matter (e) and the non-critical terms ("noncrit") to the sum of "dust" (b) and radiation (r) $\Omega_b + \Omega_r$ densities. **Right panel**: The quantities $\rho_b\, a^3$, for "dust", $\rho_r\, a^4$ and $\rho_e\, a^2$ as functions of $t_{Einstein}$

have important phenomenological consequences, in particular when obtaining constraints on supersymmetric particle-physics models from astrophysical data.

We shall not discuss these issues further here, due to lack of space. For more details we refer the interested reader to the literature [10, 36, 37]. We do

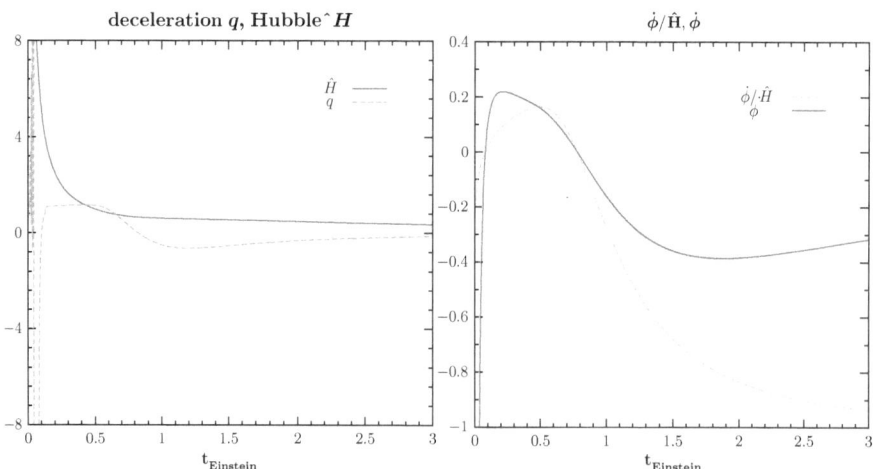

Fig. 12.9. Left panel: The deceleration q and the dimensionless Hubble expansion rate $\hat{H} \equiv \frac{H}{\sqrt{3}H_0}$ as functions of $t_{Einstein}$. **Right panel**: The derivative of the dilaton and its ratio to the dimensionless expansion rate

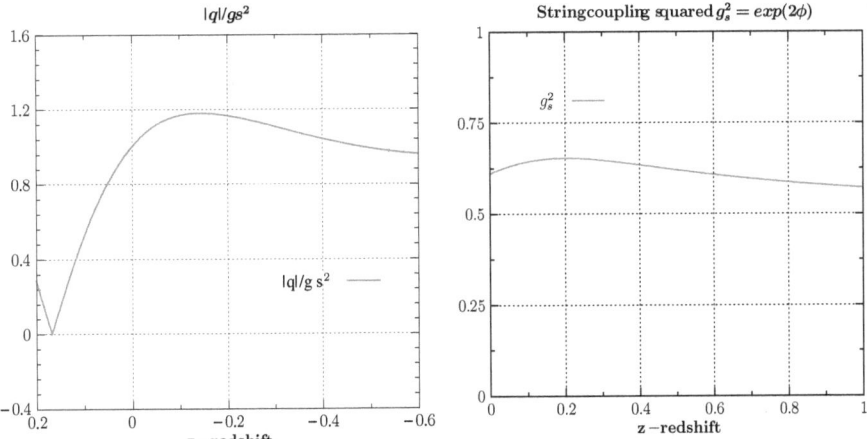

Fig. 12.10. Left panel: The ratio $|q|/g_s^2$ as function of the redshift for z ranging from $z = 0.2$ to future values $z = -0.6$, for the inputs discussed in the main text. The rapid change near $z \approx 0.16$ signals the passage from deceleration to the acceleration period. **Right panel**: The values of the string coupling constant plotted versus the redshift in the range $z = 0.0 - 1.0$

hope, however, that we have introduced the interested reader into the basic techniques and concepts underlying the idea of using non-critical strings as a way of describing non-equilibrium systems in string theory, and in particular cosmology.

12.5 Conclusions

In this work I have reviewed various issues related to the consistent incorporation of Dark Energy in string theory. I have discussed only traditional string theory and did not cover the modern extension, including membranes, except briefly in some specific examples, involving colliding branes worlds; but even then, I concentrated on perturbative string excitations on such branes. The topic of brane cosmology *per se* has been covered by R. Maartens and M. Sami contributions in this volume.

One of the most important issues I discussed concerns de Sitter space, and in general space-times with horizons in string theory. We have studied general properties, including holographic scenarios, which may be the key to an inclusion of such space-times in the set of consistent (possibly non perturbative) ground states of strings.

We have also seen that perturbative strings are incompatible with space-times with horizons, mainly due to the lack of a scattering matrix. However, non-critical strings may evade this constraint, and we have discussed briefly how accelerating universes can be incorporated in non critical (Liouville) strings. The use of Liouville strings to describe the evolution of our Universe is natural, since non-critical strings are associated with non-equilibrium situations which undoubtedly occurred in the early Universe.

The dilaton played an important rôle in string cosmology, and we have seen how it can act as a quintessence field, responsible for the current-era acceleration of the Universe.

There are many phenomenological tests of this class of cosmologies that can be performed, which the generic analysis presented here is not sufficient to encapsulate. Tensor perturbations in the cosmic microwave background radiation is one of them. The emission of gravitational degrees of freedom from the hot brane to the cold bulk, during the inflationary and post-inflationary phases in models involving brane-worlds is something to be investigated in detail. A detailed knowledge of the dependence of the equation of state on the redshift is something that needs to be looked at in the context of specific models. Moreover, issues regarding the delicate balance of the expansion of the Universe and nucleosynthesis, which requires a very low vacuum energy, must be resolved in specific, phenomenologically semi-realistic models, after proper compactification to three spatial dimensions, in order that the conjectured cosmological evolution has a chance of success.

Finally, the compactification issue *per se* is a most important part of a realistic stringy cosmology. In our discussion above, we have assumed that a consistent compactification takes place, leading to effective four-dimensional string-inspired equations of motion. In realistic scenarios, however, details of how the extra dimensions are compactified play a key rôle in issues like supersymmetry breaking.

In this review I did not discuss higher-curvature modifications of the low-energy Einstein action, which characterize all string-inspired models, including brane worlds scenarios. Such terms may play an important rôle in Early Universe cosmology. For instance, they may imply initial singularity-free string cosmologies [39], or non-trivial black hole solutions with (secondary) dilaton hair [40], which can play a rôle in the Early universe sphaleron transitions. So, before closing the lecture, I will devote a few words on their form.

In ordinary string theory, which is the subject of the present lecture, such higher-order terms possess ambiguous coefficients in the effective action. This is a result of local field redefinitions, which leave the (low-energy) string scattering amplitudes invariant, and hence cannot be determined by low energy considerations. In ordinary string theory [4], with no space-time boundaries in (the low-energy) target space-time, such ambiguities imply that the so-called ghost-free Gauss-Bonnet combination $\frac{1}{g_s^2} \left(R^2_{\mu\nu\rho\sigma} - 4R^2_{\mu\nu} + R^2 \right)$, with $g_s = e^{\Phi}$ the string coupling and Φ the dilaton field, can always be achieved for the quadratic curvature terms in the string-inspired effective action. Such terms constitute the first-non-trivial-order corrections to the Einstein term in bosonic and heterotic string effective actions.

However, in the case of brane-worlds, with closed strings propagating in the bulk, things are not so simple. As discussed in [41], field redefinition ambiguities for the bulk low-energy graviton and dilaton fields, that would otherwise leave bulk string scattering amplitudes invariant, induce brane (boundary) curvature and cosmological constant terms, with the unavoidable result of ambiguities in the terms defining the Einstein and cosmological constant terms on the brane. This results in (perturbative in α') ambiguities in the cross-over scale of four-dimensional brane gravity, as well as the brane vacuum energy. It is not clear to me, however, whether these ambiguities are actually present in low-energy brane world models. I believe that these bulk-string ambiguities can be eliminated once the brane effective theory is properly defined, given that closed and open strings also propagate on the brane world hypersurfaces, and thus are characterized by their own scattering amplitudes. Matching these two sets of scattering amplitudes properly, for instance by looking at the conformal theory describing the splitting of a closed-string bulk state, crossing a brane boundary, into two open string excitations on the brane, may lead to unambiguous brane cross-over and cosmological constant scales, expressed in terms of the bulk string scale and coupling [34]. These are issues that I believe deserve further investigation, since they affect early Universe cosmologies, where such higher-curvature terms are important. I will not, however, discuss them further here.

I would like to close this lecture with one more remark on the non-equilibrium Liouville approach to cosmology advocated in [10, 36], and discussed last in this article. This approach is based exclusively on the treatment of target time as an irreversible dynamical renormalization group scale on the world sheet of the Liouville string (the zero mode of the Liouville field itself). This irreversibility is associated with fundamental properties of the world-sheet renormalization group, which lead in turn to the loss of information carried by two-dimensional degrees of freedom with world-sheet momenta beyond the ultraviolet cutoff [28] of the world-sheet theory. This fundamental microscopic time irreversibility may have other important consequences, associated with fundamental violations of CPT invariance [21] in both the early Universe and the laboratory, providing other tests of these ideas.

Acknowledgements

It is my pleasure to thank the organizers, and especially E. Papantonopoulos, for the invitation to lecture in this very interesting school and workshop. This work is partially supported by funds made available by the European Social Fund and National Resources - (EPEAEK II) - PYTHAGORAS.

References

1. S. Perlmutter et al., Astrophys. J **483**, 565 (1997);
   A.G. Riess et al., Astron. J. **116**, 1009 (1998);
   B.P. Schmidt et al., Astrophys. J **507**, 46 (1998);
   P.M. Garnavich et al., Astrophys. J **509**, 74 (1998);
   S. Perlmutter et al., Astrophys. J **517**, 565 (1999).
2. C.L. Bennett et al., Astrophys.; J. Suppl. **148**, 1, (2003), [arXiv:astro-ph/0302207];
   D.N. Spergel et al., Astrophys. J. Suppl. **148**, 175, (2003) [arXiv:astro-ph/0302209].
3. See, for instance, N.E. Mavromatos, Proc. *First Aegean Summer School on Cosmology, Cosmological Crossroads* (S. Cotsakis, E. Papantonopoulos eds.), Lect. Notes Phys. **592**, ISBN 3-540-43778-9.
4. M.B. Green, J.H. Schwarz and E. Witten, *Superstring Theory*, Vols. I and II (Cambridge University Press, Cambridge 1987).
5. T. Banks and W. Fischler, hep-th/0102077; S. Hellerman, N. Kaloper and L. Susskind, hep-th/0104180; W. Fischler, A. Kashani-Poor, R. McNees and S. Paban, hep-th/0104181; E. Witten, hep-th/0106109; P.O. Mazur, E. Mottola, Phys. Rev. D**64**, 104022 (2001), and references therein; J. Ellis, N.E. Mavromatos and D.V. Nanopoulos, hep-th/0105206.
6. For concise recent reviews see: P.J. Steinhardt, Phys. Scripta T **85**, 177 (2000), and references therein; M. Trodden and S.M. Carroll, *TASI lectures: Introduction to cosmology*, [arXiv:astro-ph/0401547].
7. J.M. Maldacena, Adv. Theor. Math. Phys. **2**, 231 (1998) [Int. J. Theor. Phys. **38**, 1113 (1999)] [arXiv:hep-th/9711200].
8. A. Strominger, JHEP **0110**, 034 (2001) [arXiv:hep-th/0106113]. M. Spradlin, A. Strominger and A. Volovich, *Les Houches lectures on de Sitter space*, and references therein. [arXiv:hep-th/0110007].
9. F. David, Modern Physics Letters A **3**, 1651 (1988); J. Distler and H. Kawai, Nucl. Phys. B **321**, 509 (1989); see also: N.E. Mavromatos and J.L. Miramontes, Mod. Phys. Lett. A**4**, 1847 (1989); E. D'Hoker and P.S. Kurzepa, Mod. Phys. Lett. A**5**, 1411 (1990).
10. J.R. Ellis, N.E. Mavromatos and D.V. Nanopoulos, Phys. Lett. B**293**, 37 (1992) [arXiv:hep-th/9207103]; Mod. Phys. Lett. A**10**, 1685 (1995) [arXiv:hep-th/9503162]. *Invited review for the special Issue of J. Chaos Solitons Fractals*, Vol. 10, (eds. C. Castro amd M.S. El Naschie, Elsevier Science, Pergamon 1999) 345 [arXiv:hep-th/9805120]; J.R. Ellis, N.E. Mavromatos and D.V. Nanopoulos, Gen. Rel. Grav. **37**, 1665 (2005) [arXiv:gr-qc/0503120].
11. J. Polchinski, Phys. Rev. Lett. **75**, 4724 (1995); *TASI lectures on D-branes*, hep-th/9611050; M.J. Duff, Sci. Am. **278**, 64 (1998).
12. D. Langlois, Prog. Theor. Phys. Suppl. **148**, 181 (2003) [arXiv:hep-th/0209261] and references therein. P. Binetruy, C. Deffayet and D. Langlois, Nucl. Phys. B**565**, 269 (2000) [arXiv: hep-th/9905012]; P. Binetruy, C. Deffayet, U. Ellwanger and D. Langlois, Physics Letters B**477**, 285 (2000) [arXiv: hep-th/9910219]; A. Kehagias and E. Kiritsis, JHEP **9911**, 022 (1999) [arXiv:hep-th/9910174].
13. M. Gasperini and G. Veneziano, Phys. Rept. **373**, 1 (2003) [arXiv:hep-th/0207130]; M. Gasperini, F. Piazza and G. Veneziano, Phys. Rev. D**65**, 023508 (2002) [arXiv:gr-qc/0108016].

14. G.A. Diamandis, B.C. Georgalas, N.E. Mavromatos and E. Papantonopoulos, Int. J. Mod. Phys. A**17**, 4567 (2002) [arXiv:hep-th/0203241]; G.A. Diamandis, B.C. Georgalas, N.E. Mavromatos, E. Papantonopoulos and I. Pappa, Int. J. Mod. Phys. A **17**, 2241 (2002) [arXiv:hep-th/0107124].

15. J.D. Bekenstein, Phys. Rev. D**7**, 2333 (1973).

16. S.W. Hawking, Commun. Math. Phys. **43**, 199 (1975) [Erratum-ibid. **46**, 206 (1976)].

17. M. Srednicki, Phys. Rev. Lett. **71**, 666 (1993) [arXiv:hep-th/9303048].

18. For the most recent result on this issue see: P. Berglund, E.G. Gimon and T.S. Levi, arXiv:hep-th/0505167, and references therein.

19. G.W. Gibbons and S.W. Hawking, Phys. Rev. D**15**, 2738 (1977).

20. R.M. Wald, Phys. Rev. D**21**, 2742 (1980).

21. N.E. Mavromatos, Lect. Notes Phys. **669**, 245 (2005) [arXiv:gr-qc/0407005] and references therein.

22. W. Fischler and L. Susskind, [arXiv:hep-th/9806039].

23. S.W. Hawking, Phys. Rev. D**72**, 084013 (2005) [arXiv:hep-th/0507171].

24. M. Dine and M. Graesser, JHEP **0501**, 038 (2005) [arXiv:hep-th/0409209].

25. T. Banks and M. Dine, JHEP **0110**, 012 (2001) [arXiv:hep-th/0106276].

26. W. Fischler and L. Susskind, Phys. Lett. B**173**, 262 (1986); Phys. Lett. B**171**, 383 (1986).

27. I. Antoniadis, C. Bachas, J.R. Ellis and D.V. Nanopoulos, Phys. Lett. B**211**, 393 (1988); Nucl. Phys. B**328**, 117 (1989); Phys. Lett. B **257**, 278 (1991).

28. A.B. Zamolodchikov, JETP Lett. **43**, 730 (1986) [Pisma Zh. Eksp. Teor. Fiz. **43**, 565 (1986)].

29. E. Witten, Int. J. Mod. Phys. A**10**, 1247 (1995) [arXiv:hep-th/9409111].

30. E. Gravanis and N.E. Mavromatos, Phys. Lett. B**547**, 117 (2002) [arXiv:hep-th/0205298]; N.E. Mavromatos, arXiv:hep-th/0210079 (published in *Beyond the Desert, Oulu 2002 (Finland)* (ed. H.V. Klapdor-Kleingrothaus, IoP 2003)), 3.

31. F. Pardo, J. Math. Phys. **30**, 2054 (1989); S. Hojman and C. Shepley, J. Math. Phys. **32**, 142 (1991); in the Liouville-strings context see: N.E. Mavromatos and R.J. Szabo, Phys. Rev. D**59**, 104018 (1999) [arXiv:hep-th/9808124].

32. C. Schmidhuber and A.A. Tseytlin, Nucl. Phys. B**426**, 187 (1994) [arXiv:hep-th/9404180].

33. I.I. Kogan, N.E. Mavromatos and J.F. Wheater, Phys. Lett. B**387**, 483 (1996) [arXiv:hep-th/9606102]; for a supersymmetric world-sheet recoil formulation see: N.E. Mavromatos and R.J. Szabo, JHEP **0110**, 027 (2001) [arXiv:hep-th/0106259]; JHEP **0301**, 041 (2003) [arXiv:hep-th/0207273].

34. C. Bachas,[arXiv:hep-th/9503030].

35. J. Ellis, N.E. Mavromatos, D.V. Nanopoulos and A. Sakharov, arXiv:gr-qc/0407089, New J. Phys. **6**, 171 (2004).

36. J.R. Ellis, N.E. Mavromatos and D.V. Nanopoulos, Phys. Lett. B**619**, 17 (2005) [arXiv:hep-th/0412240] and J.R. Ellis, N.E. Mavromatos, D.V. Nanopoulos and M. Westmuckett, Int. J. Mod. Phys. A **21**, 1379 (2006) [arXiv:gr-qc/0508105]; G.A. Diamandis, B.C. Georgalas, A.B. Lahanas, N.E. Mavromatos and D.V. Nanopoulos, [arXiv:hep-th/0605181].

37. A.B. Lahanas, N.E. Mavromatos and D.V. Nanopoulos, [arXiv:hep-ph/0608153].

38. see for instance: E.W. Kolb and M.S. Turner, *The Early universe*, Front. Phys. **69** (Redwood City, USA: Addison-Wesley 1990).

39. I. Antoniadis, J. Rizos and K. Tamvakis, Nucl. Phys. **B415**, 497 (1994) [arXiv:hep-th/9305025].

40. P. Kanti, N.E. Mavromatos, J.Rizos, K. Tamvakis and E. Winstanley, Phys. Rev. D**54**, 5049 (1996) [arXiv:hep-th/9511071].

41. N.E. Mavromatos and E. Papantonopoulos, Phys. Rev. D**73**, 026001 (2006) [arXiv:hep-th/0503243] and references therein.

13

Modified Gravity Without Dark Matter

Robert Sanders

Kapteyn Astronomical Institute, Groningen, The Netherlands
sanders@astro.rug.nl

Abstract. On an empirical level, the most successful alternative to dark matter in bound gravitational systems is the modified Newtonian dynamics, or MOND, proposed by Milgrom. Here I discuss the attempts to formulate MOND as a modification of General Relativity. I begin with a summary of the phenomenological successes of MOND and then discuss the various covariant theories that have been proposed as a basis for the idea. I show why these proposals have led inevitably to a multi-field theory. I describe in some detail TeVeS, the tensor-vector-scalar theory proposed by Bekenstein, and discuss its successes and shortcomings. This lecture is primarily pedagogical and directed to those with some, but not a deep, background in General Relativity.

13.1 Introduction

There is now compelling observational support for a standard cosmological model. It is most impressive that this evidence is derived from very different observational techniques applied to very different phenomena: from precise measurements of anisotropies in the Cosmic Microwave Background (CMB) [1]; from systematic photometric observations of the light curves of distant supernovae [2, 3, 4]; from redshift surveys mapping the distribution of observable matter on large scale and interpreting that distribution in the context of structure formation by gravitational collapse [5, 6]. Using the standard parameterised Friedmann-Robertson-Walker models (FRW), all of these observations imply a convergence to a narrow range of parameters that characterise the Universe; this convergence is rightly heralded as a remarkable achievement of the past decade.

However, the Universe that we are presented with is strange in its composition: only five percent is the ordinary baryonic matter that we are familiar with; twenty-five percent consists of pressureless dark matter presumed to be fundamental particles that are as yet undetected by other means; and about seventy percent is the even stranger negative pressure dark energy, possibly identified with a cosmological term in Einstein's field equation, and emerging

R. Sanders: *Modified Gravity Without Dark Matter*, Lect. Notes Phys. **720**, 375–402 (2007)
DOI 10.1007/978-3-540-71013-4_13   © Springer-Verlag Berlin Heidelberg 2007

relatively recently in cosmic history as the dominate contributer to the energy density budget of the Universe.

A general sense of unease, primarily with this dark energy, has led a number of people to consider the possibility that gravity may not be described by standard four-dimensional General Relativity (GR) on large scale (see Sami, this volume)– that is to say, perhaps the left-hand-side rather than the right-hand-side of the Einstein equation should be reconsidered. Various possibilities have been proposed– possibilities ranging from the addition of a scalar field with a non-standard kinetic term, K-essence [7]; to gravitational actions consisting of general functions of the usual gravitational invariant, $F(R)$ theories [8, 9] to braneworld scenarios with leakage of gravitons into a higher dimensional bulk ([10] and Maartens, this volume). But, in fact, there is a longer history of modifying gravity in connection with the dark matter problem– primarily that aspect of the problem broadly described as "missing mass" in bound gravitational systems such as galaxies or clusters of galaxies. The observations of this phenomenology have an even longer history, going back to the discovery of a substantial discrepancy between the dynamical mass and the luminous mass in clusters of galaxies [11]. The precise measurement of rotation curves of spiral galaxies in the 1970's and 1980's, primarily by 21 cm line observations which extend well beyond the visible disk of the galaxy [12, 13], demonstrated dramatically that this discrepancy is also present in galaxy systems.

A fundamental, often implicit, aspect of the cosmological paradigm is that this observed discrepancy in bound systems is due to the cosmological dark matter– that the cosmological dark matter clusters on small scale and promotes the formation of virialized systems via gravitational collapse in the expanding Universe. The necessity of clustering on the scale of small galaxies implies that there are no phase space constraints on the density of the dark matter and, hence, that it is cold, or non-relativistic at the epoch of matter-radiation equality [14]. The exact nature of the hypothetical cold dark matter (CDM) is unknown but particle physics theory beyond the standard model provides a number of candidates. There are observational problems connected with the absence of phase space constraints in this dark matter fluid, problems such as the formation of numerous but unseen satellites of larger galaxies [15] and the prediction of cusps in the central density distributions of galaxies– cusps which are not evident in the rotation curves [16]. But it is usually taken as a article of faith that "complicated astrophysical processes" such as star formation and resulting feed-back will solve these problems.

The motivation behind considering modifications of gravity as an alternative to CDM is basically the same as that underlying modified gravity as an alternative to dark energy: when a theory, in this case GR, requires the existence of a medium which has not been, or cannot be, detected by means other than its global gravitational influence, i.e. an ether, then it is not unreasonable to question that theory. The primary driver for such proposals has been the direct observation of discrepancies in bound systems– galaxies and clusters of galaxies– rather than cosmological considerations, such as that of structure formation in an expanding Universe. The most successful of the several suggestions, modified Newtonian dynamics or MOND, has an entirely phenomenological rather than theoretical basis [17, 18, 19]. In accounting for the detailed kinematics of galaxies and galaxy groups, while encompassing global scaling relations and empirical photometric rules, MOND has, with one simple formula and one new fixed parameter, subsumed a wide range of apparently disconnected phenomena.

In this respect it is similar to the early proposal of continental drift by Alfred Wegener in 1912. This suggestion explained a number of apparently disconnected geological and palaeontological facts but had no basis in deeper theory; no one, including Wegener, could conceive of a mechanism by which giant land masses could drift through the oceans of the earth. Hence the idea was met with considerable ridicule by the then contemporary community of geologists and relegated to derisive asides in introductory textbooks. It was decades later, after the development of the modern theory of plate tectonics and direct experimental support provided by the frozen-in magnetic field reversals near mid-oceanic rifts, that the theory underlying continental drift became the central paradigm of geology and recognised as the principal process that structures the surface of the earth [20]. I do not wish to draw a close analogy between MOND and the historical theory of continental drift, but only to emphasise the precedent: an idea can be basically correct but not generally accepted until there is an understandable underlying physical mechanism– until the idea makes contact with more familiar physical concepts.

The search for a physical mechanism underlying modified Newtonian dynamics is the subject here. I begin with a summary of the phenomenological successes of the idea, but, because this has been reviewed extensively before [21], I will be brief. I consider the proposals that have been made for modifications of GR as a basis of MOND. These proposals have led to the current best candidate– the tensor-vector-scalar (TeVeS) theory of Bekenstein [22], a theory that is complicated but free of obvious pathologies. I summarise the successes and shortcomings of the theory, and I present an alternative form of TeVeS which may provide a more natural basis to the theory. I end by a discussion of more speculative possibilities.

13.2 The Phenomenology of MOND

13.2.1 The Basics of MOND

If one wishes to modify Newtonian gravity in an ad hoc manner in order to reproduce an observed property of galaxies, such as asymptotically flat rotation curves, then it would seem most obvious to consider a $1/r$ attraction beyond a fixed length scale r_0. Milgrom [17] realized early on that this would not work– that any modification explaining the systematics of the discrepancy in galaxies cannot be attached to a length scale but to a fixed acceleration scale, a_0. His suggestion, viewed as a modification of gravity, was that the true gravitational acceleration $\mathbf{g}$ is related to the Newtonian gravitational acceleration $\mathbf{g_n}$ as

$$\mathbf{g}\mu(|g|/a_o) = \mathbf{g_n} \qquad (13.1)$$

where a_o is a new physical parameter with units of acceleration and $\mu(x)$ is a function that is unspecified but must have the asymptotic form $\mu(x) = x$ when $x << 1$ and $\mu(x) = 1$ where $x >> 1$.

The immediate consequence of this is that, in the limit of low accelerations, $g = \sqrt{g_n a_o}$. For a point mass M, if we set g equal to the centripetal acceleration v^2/r, then the circular velocity is

$$v^4 = GMa_o \qquad (13.2)$$

in the low acceleration regime. So all rotation curves are asymptotically flat and there is a mass-velocity relation of the form $M \propto v^4$. These are aspects that are built into MOND so they cannot rightly be called predictions. However, in the context of MOND, the aspect of an asymptotically flat rotation curve is absolute. Unambiguous examples of rotation curves (of isolated galaxies) that decline in a Keplerian fashion at a large distance from the visible object would falsify the idea.

The implied mass-rotation velocity relation explains a well-known global scaling relation for spiral galaxies, the Tully-Fisher relation. This is a correlation between the observed luminosity of spiral galaxies and the characteristic rotation velocity, a relation of the form $L \propto v^\alpha$ where $\alpha \approx 4$ if luminosity is measured in the near-infrared. If the mass-to-light ratio of galaxies does not vary systematically with luminosity, then MOND explains this scaling relation. In addition, because it reflects underlying physical law, the relation is as absolute. The TF relation should be the same for different classes of galaxies and the logarithmic slope (at least of the MASS-velocity relation) must be 4. Moreover, the relation is essentially one between the total baryonic mass of a galaxy and the asymptotic flat rotational velocity– not the peak rotation velocity but the velocity at large distance. This is the most immediate prediction [23, 24].

The near-infrared TF relation for a sample of galaxies in the Ursa Major cluster (and hence all at nearly the same distance) is shown as a log-log plot in Fig. 13.1 where the velocity is that of the flat part of the rotation curve [25]. The scatter about the least-square fit line of slope 3.9 ± 0.2 is consistent with observational uncertainties (i.e. no intrinsic scatter).

Given the mean M/L in a particular band (≈ 1 in the K' band), this observed TF relation (and 13.2) tells us that a_o must be on the order of 10^{-8} cm/s^2. It was immediately noticed by Milgrom that $a_o \approx cH_o$ to within a factor of 5 or 6. This cosmic coincidence suggests that MOND, if it is right, may reflect the effect of cosmology on local particle dynamics.

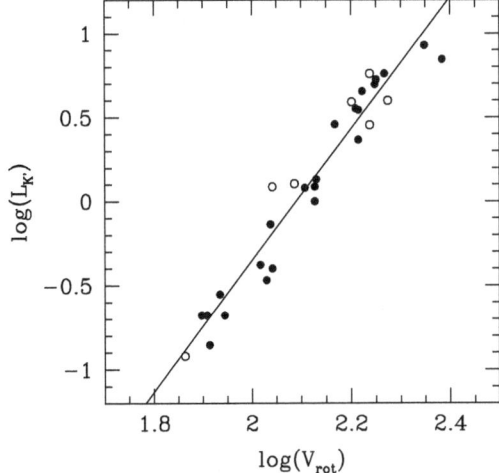

Fig. 13.1. The near-infrared Tully-Fisher relation of Ursa Major spirals [25]. The rotation velocity is the asymptotically constant value. The line is a least-square fit to the data and has a slope of 3.9 ± 0.2

13.2.2 A Critical Surface Density

It is evident that the surface density of a system M/R^2 is proportional to the internal gravitational acceleration. This means that the critical acceleration may be rewritten as a critical surface density:

$$\Sigma_m \approx a_o/G \ . \tag{13.3}$$

If a system, such as a spiral galaxy has a surface density of matter greater than Σ_m, then the internal accelerations are greater than a_o, so the system is in the Newtonian regime. In systems with $\Sigma \geq \Sigma_m$ (high surface brightness or HSB galaxies) there should be a small discrepancy between the visible and classical Newtonian dynamical mass within the optical disk. But in low surface brightness (LSB) galaxies ($\Sigma << \Sigma_m$) there is a low internal acceleration, so the discrepancy between the visible and dynamical mass would be large. By this argument Milgrom predicted, before the actual discovery of a large population of LSB galaxies, that there should be a serious discrepancy between the observable and dynamical mass within the luminous disk of such systems– should they exist. They do exist, and this prediction has been verified [23].

Moreover, spiral galaxies with a mean surface density near this limit – HSB galaxies– would be, within the optical disk, in the Newtonian regime. So one would expect that the rotation curve would decline in a near Keplerian fashion to the asymptotic constant value. In LSB galaxies, with mean surface density below Σ_m, the prediction is that rotation curves would rise to the final asymptotic flat value. So there should be a general difference in rotation

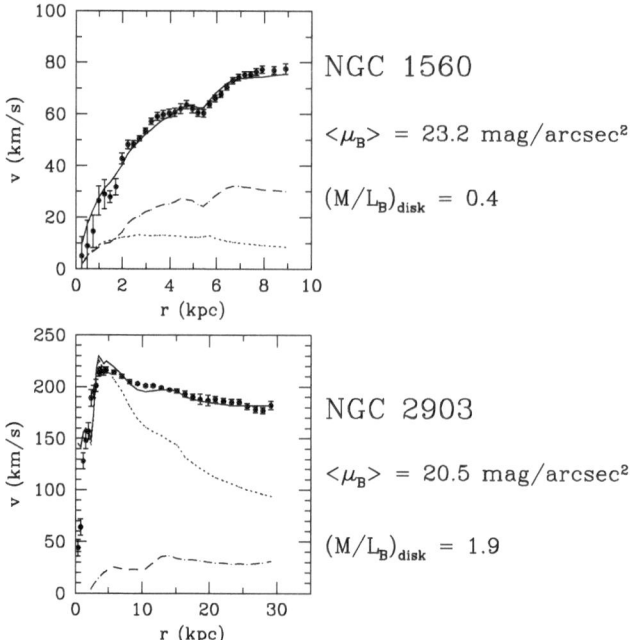

Fig. 13.2. The points show the observed 21 cm line rotation curves of a low surface brightness galaxy, NGC 1560 and a high surface brightness galaxy, NGC 2903. The dotted and dashed lines are the Newtonian rotation curves of the visible and gaseous components of the disk and the solid line is the MOND rotation curve with $a_o = 1.2 \times 10^{-8}$ cm/s^2– the value derived from the rotation curves of 10 nearby galaxies [26]. Here the only free parameter is the mass-to-light ratio of the visible component

curve shapes between LSB and HSB galaxies. In Fig. 13.2 I show the observed rotation curves (points) of two galaxies, a LSB and HSB [26], where we see exactly this trend. This general effect in observed rotation curves was pointed out in [27].

It is well-known that rotationally supported Newtonian systems tend to be unstable to global non-axisymmetric modes which lead to bar formation and rapid heating of the system [28]. In the context of MOND, these systems would be those with $\Sigma > \Sigma_m$, so this would suggest that Σ_m should appear as an upper limit on the surface density of rotationally supported systems. This critical surface density is 0.2 g/cm^2 or 860 M$_\odot$/pc^2. A more appropriate value of the mean surface density within an effective radius would be $\Sigma_m/2\pi$ or 140 M$_\odot$/pc^2, and, taking $M/L_b \approx 2$, this would correspond to a surface brightness of about 22 mag/arc sec^2. There is such an observed upper limit on the mean surface brightness of spiral galaxies and this is known as Freeman's law [29]. The existence of such a limit becomes understandable in the context of MOND.

13.2.3 Pressure-supported Systems

Of course, spiral galaxies are rotationally supported. But there other galaxies, elliptical galaxies, which are pressure supported– i.e. they are held up against gravity by the random motion of the stars. There are numerous other examples of pressure-supported systems such as globular clusters and clusters of galaxies, and often the observable components of these systems have a velocity dispersion (or temperature) that does not vary much with position; i.e. they are near "isothermal". With Newtonian dynamics, pressure-supported systems that are nearly isothermal have infinite extent. But in the context of MOND it is straightforward to demonstrate that such isothermal systems are finite with the density at large radii falling roughly like $1/r^4$ [30].

The equation of hydrostatic equilibrium for an isotropic, isothermal system reads

$$\sigma_r{}^2 \frac{d\rho}{dr} = -\rho g \tag{13.4}$$

where, in the limit of low accelerations $g = \sqrt{GMa_o}/r$. Here σ_r is the radial velocity dispersion and ρ is the mass density. It then follows immediately that, in this MOND limit,

$$\sigma_r^4 = GMa_o \left(\frac{d\ln(\rho)}{d\ln(r)} \right)^{-2} . \tag{13.5}$$

Thus, there exists a mass-velocity dispersion relation of the form

$$(M/10^{11} M_\odot) \approx (\sigma_r/100 \ kms^{-1})^4$$

which is similar to the observed Faber-Jackson relation (luminosity-velocity dispersion relation) for elliptical galaxies [31]. This means that a MOND near-isothermal sphere with a velocity dispersion on the order of 100 km/s will always have a galactic mass. This is not true of Newtonian pressure-supported objects. Because of the appearance of an additional dimensional constant, a_o, in the structure equation (13.4), MOND systems are much more constrained than their Newtonian counterparts.

Any isolated system which is nearly isothermal will be a MOND object. That is because a Newtonian isothermal system (with large internal accelerations) is an object of infinite size and will always extend to the region of low accelerations ($< a_o$). At that point ($r_e{}^2 \approx GM/a_o$), MOND intervenes and the system will be truncated. This means that the internal acceleration of any isolated isothermal system ($\sigma_r{}^2/r_e$) is expected to be on the order of or less than a_o and that the mean surface density within r_e will typically be Σ_m or less (there are low-density solutions for MOND isothermal spheres, $\rho << a_o{}^2/G\sigma^2$, with internal accelerations less than a_o). It was pointed out long ago that elliptical galaxies do appear to have a characteristic surface brightness [32]. But the above arguments imply that the same should be true of any pressure supported, near-isothermal system, from globular clusters to clusters of galaxies. Moreover, the same $M - \sigma$ relation (13.5) should apply to all such systems, albeit with considerable scatter due to deviations from a strictly isotropic, isothermal velocity field [33].

Most luminous elliptical galaxies are high surface brightness objects which would imply a surface density greater than the MOND limit. This suggests that luminous elliptical galaxies should be essentially Newtonian objects, and, viewed in the traditional way, should evidence little need for dark matter within the effective (or half-light) radius. This does seem to be the case as demonstrated by dynamical studies using planetary nebulae as kinematic tracers [34, 35].

13.2.4 Rotation Curves of Spiral Galaxies

Perhaps the most impressive observational success of MOND is the prediction of the form of galaxy rotation curves from the observed distribution of baryonic matter, stars and gas. Basically, one takes the mean radial distribution of light in a spiral galaxy as a precise tracer of the luminous mass, includes the observed radial dependence of neutral hydrogen (increased by 30% to account for the primordial helium) and assumes all of this is in a thin disk (with the occasional exception of a central bulge component). One then solves the standard Poisson equation to determine the Newtonian force, applies the MOND formula (1 with a fixed value of a_0) to determine the true gravitational force and calculates the predicted rotation curve. The mass-to-light ratio of the visible component is adjusted to achieve the best fit to the observed rotation curve.

The results are spectacular considering that this is a one-parameter fit. The solid curves in Fig. 13.2 are the results of such a procedure applied to a LSB and HSB galaxy; this has been done for about 100 galaxies. The fitted M/L values are not only reasonable, but demonstrate the same trend with colour that is implied by population synthesis models as we see in Fig. 13.3 [25, 36].

Here I wish to emphasise another observed aspect of galaxy rotation curves– a point that has been made, in particular, by Sancisi [37]. For many objects, the detailed rotation curve appears to be extremely sensitive to the distribution observable matter, even in LSB galaxies where, in the standard interpretation, dark matter overwhelmingly dominates within the optical image. There are numerous examples of this– e.g. the LSB galaxy shown in Fig. 13.2 where we see that the total rotation curve reflects the Newtonian rotation curve of the gaseous component in detail. Another example [37, 38] is the dwarf galaxy, NGC 3657. Figure 13.4 shows the surface densities of the baryonic components, stars and gas, compared to the observed rotation curve. Again the dotted and dashed curves are the Newtonian rotation curves of the stellar and gaseous components and the solid curve is the resulting MOND rotation curve. The agreement with observations is obvious.

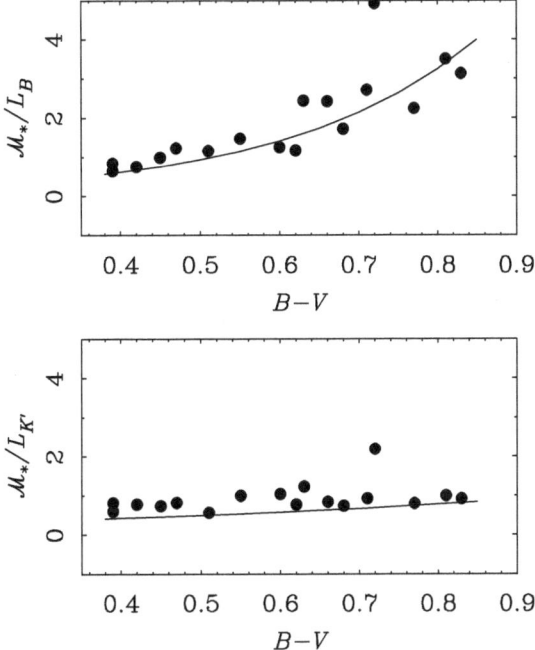

Fig. 13.3. MOND fitted mass-to-light ratios for the UMa spirals [25] in the B-band (**top**) and the K'-band (**bottom**) plotted against B-V (*blue minus visual*) colour index. The solid lines show predictions from populations synthesis models [36]

For this galaxy, there is evidence from the rotation curve of a central cusp in the density distribution– and, indeed, the cusp is seen in the light distribution. In cases where there is no conspicuous cusp in the light distribution, there is no kinematic evidence for a cusp in the rotation curve. This would appear to make the entire discussion about cusps in halos somewhat irrelevant. But equally striking in this case is the gradual rise in the rotation curve at large radii. This rise is clearly related to the increasing dominance of the gaseous component in the outer regions. The point is clear: the rotation curve reflects the global distribution of baryonic matter, even in the presence of a large discrepancy between the visible and Newtonian dynamical mass. This is entirely understandable (and predicted) in the context of modified gravity in the form of MOND (what you see is all there is), but remains mysterious in the context of dark matter

13.2.5 Clusters of Galaxies: A Phenomenological Problem for MOND?

It has been known for 70 years [11] that clusters of galaxies exhibit a significant discrepancy between the Newtonian dynamical mass and the observable mass,

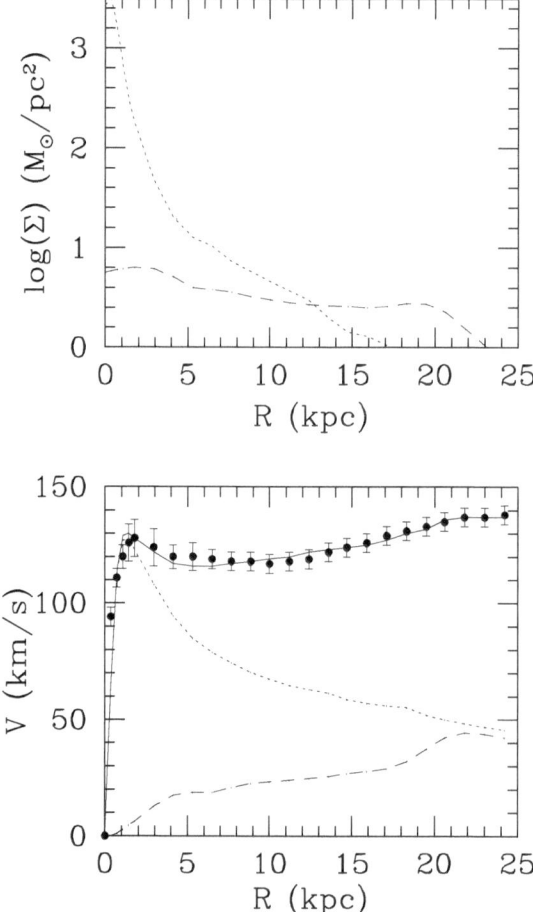

Fig. 13.4. The **upper panel** is the logarithm of the surface density of the gaseous and stellar components of NGC 3657. The **lower panel** shows the observed rotation curve (*points*), the Newtonian rotation curves for the stellar (*dashed*) and gaseous (*dotted*) components as well as the MOND rotation curve (*solid*) [37, 38]

although the subsequent discovery of hot X-ray emitting gas goes some way in alleviating the original discrepancy. For an isothermal sphere of hot gas at temperature T, the Newtonian dynamical mass within radius r_o, calculated from the equation of hydrostatic equilibrium, is

$$M_n = \frac{r_o}{G} \frac{kT}{m} \left(\frac{d \ln(\rho)}{d \ln(r)} \right) , \qquad (13.6)$$

where m is the mean atomic mass and the logarithmic density gradient is evaluated at r_o. This dynamical mass turns out to be typically about a factor of 4 or 5 larger than the observed mass in hot gas and in the stellar content of the galaxies (see Fig. 13.5, left [39]).

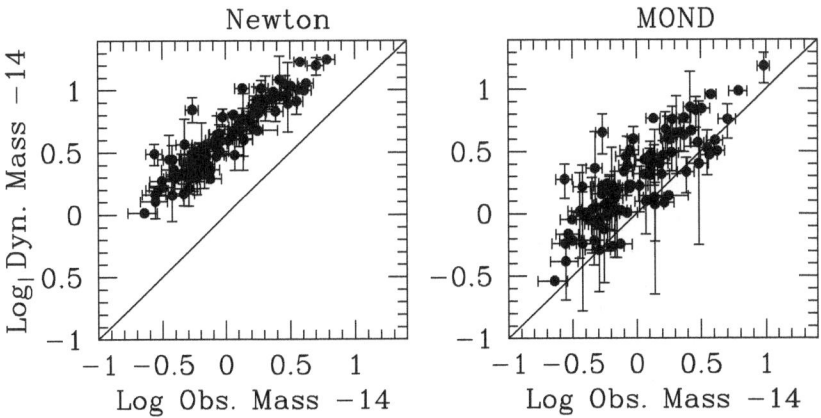

Fig. 13.5. Left: the Newtonian dynamical mass of clusters of galaxies within an observed cutoff radius (r_{out}) vs. the total observable mass in 93 X-ray emitting clusters of galaxies. The solid line corresponds to $M_{dyn} = M_{obs}$ (no discrepancy). **Right**: the MOND dynamical mass within r_{out} vs. the total observable mass for the same X-ray emitting clusters [39]

With MOND, the dynamical mass (13.5) is given by

$$M_m = (Ga_o)^{-1}\left(\frac{kT}{m}\right)^2\left(\frac{d\,ln(\rho)}{d\,ln(r)}\right)^2 , \tag{13.7}$$

and, using the same value of a_o determined from nearby galaxy rotation curves, turns out to be, on average, a factor of two larger than the observed mass (Fig. 13.5, right). The discrepancy is reduced but still present. This could be interpreted as a failure [40], or one could say that MOND predicts that the mass budget of clusters is not yet complete and that there is more mass to be detected [39]. The cluster missing mass could, e.g. be in neutrinos of mass 1.5 to 2 eV [41], or in "soft bosons" with a large de Broglie wavelength [42], or simply in heretofore undetected baryonic matter. It would have certainly been a falsification of MOND had the predicted mass turned out to be typically *less* than the observed mass in hot gas and stars.

13.3 Relativistic MOND

MOND not only allows the form of rotation curves to be precisely predicted from the distribution of observable matter, but it also explains certain systematic aspects of the photometry and kinematics of galaxies and clusters: the presence of a preferred surface density in spiral galaxies and ellipticals– the so-called Freeman and Fish laws; the fact that pressure-supported nearly isothermal systems ranging from molecular clouds to clusters of galaxies are characterised by a specific internal acceleration, a_o [21]; the existence of a TF relation with small scatter– specifically a correlation between the baryonic mass and the asymptotically flat rotation velocity of the form $v^4 \propto M$; the Faber-Jackson relation for ellipticals, and with more detailed modelling, the Fundamental Plane [33]; not only the magnitude of the discrepancy in clusters of galaxies but also the fact that mass-velocity dispersion relation which applies to elliptical galaxies (13.5) extends to clusters (the mass-temperature relation). And it accomplishes all of this with a single new parameter with units of acceleration– a parameter determined from galaxy rotation curves which is within an order of magnitude of the cosmologically significant value of cH_o. This is why several of us believe that, on an epistemological level, MOND is more successful than dark matter. Further, many of these systematic aspects of bound systems do not have any obvious connection to what has been traditionally called the "dark matter problem". This capacity to connect seemingly unrelated points is the hallmark of a good theory. However, as I argued in the Introduction, MOND will never be entirely credible to most astronomers and physicists until it makes some contact with more familiar physics–until there is an underlying and understandable physical mechanism for MOND phenomenology. Below I consider that mechanism in terms of possible modifications of the theory of gravity.

13.3.1 Steps to TeVeS

TeVeS (tensor-vector-scalar) theory [22] is a relativistic theory yielding MOND phenomenology in the appropriate limit. Of course, I do not need to belabour the advantages of a relativistic theory. It allows one to address a number of issues on which MOND is silent: gravitational lensing, cosmology, structure formation, anisotropies in the CMB. The theory is complicated– considerably more complicated than GR– in that involves additional dynamical elements and is characterised by three additional free parameters and a free function– i.e. a function that is not specified by any a priori considerations but may be adjusted to achieved the desired result. In this sense, TeVeS, like MOND itself, is a phenomenologically driven theory. It is entirely "bottom-up" and thereby differs from what is normally done in gravity theory or cosmology.

As the name implies it is a multi-field theory; i.e. there are fields present other than the usual tensor field $g_{\mu\nu}$ of GR. It appears that any viable theory of MOND as a modification of gravity must be a multi-field theory; no theory based upon a single metric field can work [43]. In TeVeS, the MOND phenomenology appears as a "fifth force" mediated by a scalar field. This fifth force must be designed to fall as $1/r$ and dominate over the usual Newtonian force when the total force is below a_0 as shown in Fig. 13.6.

Now if we are proposing a fifth force, then that implies non-geodesic motion and one may naturally ask about the validity of the equivalence principle, even in its weak form expressing the universality of free fall (there are

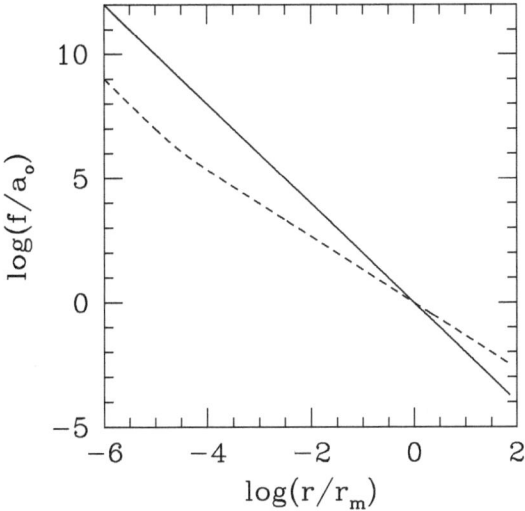

Fig. 13.6. MOND phenomenology as a result of multi-field modifications of gravity. The dashed curve shows the log force resulting from a scalar field with a non-standard Lagrangian as a function of log radius in units of the MOND radius $r_M = \sqrt{GM/a_0}$. The solid line is the usual Einstein-Newton force

strong experimental constraints on the composition independence of acceleration in a gravitational field). The weak version of the equivalence principle can be preserved if there is a specific form of coupling between the scalar field a matter– one in which the scalar couples to matter jointly with the gravitational or Einstein metric. This allows for the definition of a physical metric, $\tilde{g}_{\mu\nu}$ that is distinct from the Einstein metric. In the simplest sort of joint coupling the physical metric is *conformally* related to the Einstein metric, i.e.

$$\tilde{g}_{\mu\nu} = f(\phi)g_{\mu\nu} \, . \tag{13.8}$$

This is the case in traditional scalar-tensor theories such as the Brans-Dicke theory [44]. So the theory remains a metric theory, but particle and photons follow geodesics of the physical metric and not the Einstein metric. Of course, a great part of the beauty of GR is that the gravitational metric is the metric of a 4-D space time with Lorentzian signature– gravitational geometry *is* physical geometry. It is beautiful, but the world doesn't have to be that way.

Another ingredient is necessary if the scalar field is to produce MOND phenomenology. In standard scalar-tensor theory, the scalar field Lagrangian is

$$L_s = \frac{1}{2}\phi_{,\alpha}\phi^{,\alpha} \, . \tag{13.9}$$

Forming the action from this Lagrangian (and the joint coupling with $g_{\mu\nu}$ to matter) and taking the condition of stationary action leads, in the weak field limit, to the usual Poisson equation for ϕ. In other words, the scalar force about a point mass falls as $1/r^2$ as in Brans-Dicke theory. Therefore, MOND requires a non-standard scalar field Lagrangian; for example, something like

$$L_s = \frac{1}{2l^2} F(l^2 \phi_{,\alpha} \phi^{,\alpha}) \tag{13.10}$$

where $F(X)$ is an, as yet, unspecified function of the usual scalar invariant and l is a length scale on the order of the present Hubble scale ($\approx c/H_0$). Bekenstein refers to this as aquadratic Lagrangian theory or AQUAL. The condition of stationary action then leads to a scalar field equation that, in the weak field limit, is

$$\nabla \cdot [\mu(|\nabla \phi|/a_0) \nabla \phi] = 4\pi G \rho \tag{13.11}$$

where $a_0 = c^2/l$ and $\mu = dF(X)/dX$. This we recognise as the Bekenstein-Milgrom field equation [45] which produces MOND like phenomenology if $\mu(y) = y$ when $y < 1$ or $F(X) = \frac{2}{3}X^{\frac{3}{2}}$. Here, however, we should recall that ϕ is not the total gravitational field but on the scalar component of a two-field theory. Another phenomenological requirement on the free function is that $F(X) \to \omega X$ in the limit where $X >> 1$ (or $\nabla \phi > a_0$). That is to say, the scalar field Lagrangian becomes standard in the limit of large field gradients; the theory becomes equivalent to Brans-Dicke theory in this limit. This guarantees precise $1/r^2$ attraction in the inner solar system, but, to be consistent with post-Newtonian constraints, it is necessary that $\omega > 10^4$.

Looking at the form of F required for MOND phenomenology, we see an immediate problem with respect to cosmology. In the limit of a homogeneous Universe, where $\nabla \phi \to 0$ and the cosmic time derivative, $\dot{\phi}$, dominates the invariant, i.e. $X < 0$. This means that the form of the free function must change in this limit (this is a problem which persists in TeVeS). But there is another more pressing problem which was immediately noticed by Bekenstein and Milgrom. In the MOND limit, small disturbances in the scalar field, scalar waves, propagate *acausally*; i.e. $V_s = \sqrt{2}c$ in directions parallel to $\nabla \phi$. This is unacceptable; a physically viable theory should avoid the paradoxes resulting from acausal propagation.

The superluminal propagation (or tachyon) problem led Bekenstein to propose a second non-standard scalar-tensor theory for MOND– phase-coupling gravitation or PCG [46]. Here, the scalar field is taken to be complex, $\chi = qe^{i\phi}$ with the standard Lagrangian,

$$L_S = \frac{1}{2}[q_{,\alpha}q^{,\alpha} + q^2 \phi_{,\alpha}\phi^{,\alpha} + 2V(q)] \tag{13.12}$$

where $V(q)$ is the potential function of the scalar field. The non-standard aspect is that only the phase couples to matter in the usual conformal way,

$$\tilde{g}_{\mu\nu} = e^{-\eta\phi} g_{\mu\nu} . \tag{13.13}$$

This leads (weak field limit) to the field equation,

$$\nabla \cdot [q^2 \nabla \phi] = \frac{8\pi G \rho}{c^2} \ . \tag{13.14}$$

So now we see that q^2 replaces the usual MOND interpolating function μ, but now q is given by a second scalar field equation,

$$q^{,\alpha}{}_{;\alpha} = q\phi_{,\alpha}\phi^{,\alpha} + V'(q) \ . \tag{13.15}$$

That is to say, the relation between q^2 and $\nabla\phi$ is now differential and not algebraic as in AQUAL theory. Bekenstein demonstrated that if $V(q) = -Aq^6$ (a negative sextic potential) then the predicted phenomenology is basically that of MOND on a galactic scale.

Obviously the property $dV/dq < 0$ cannot apply for all q because this would lead to instability of the vacuum, but there is a more serious problem: By a suitable redefinition of the fields, it may be shown that, in the limit of very weak coupling ($\eta << 1$) the term on the left-hand side of (13.15) may be neglected–that is to say, we are left with only the right-hand side and the relation between q^2 and $\nabla\phi$ once again becomes algebraic as in AQUAL. In other words, PCG approaches AQUAL in the limit of very weak coupling. This suggests that PCG may suffer from a similar ailment as AQUAL; indeed, there is a problem, but it appears as the absence of a stable background solution rather than superluminal propagation [47]. But I only mention this because I want to emphasise that the weak coupling limit of PCG is equivalent to the aquadratic theory; this turns out to be a significant aspect of TeVeS.

At about the same time it was realized that there is a serious phenomenological problem with AQUAL or PCG or any scalar-tensor theory in which the the relation between the physical and gravitational metrics is conformal as in (8 or 13). That is, such a theory would predict no enhanced deflection of photons due to the presence of the scalar field [48, 49]. Recall that photons and other relativistic particles follow null geodesics of the physical metric. These are given by the condition that

$$d\tilde{\tau}^2 = -\tilde{g}_{\mu\nu} dx^\mu dx^\nu = 0 \ . \tag{13.16}$$

Now given the conformal relation between the two metrics (13.8) you don't have to be a mathematical genius to see that $d\tilde{\tau} = 0$ corresponds to $d\tau = 0$; i.e. null geodesics of the two metrics coincide which means that photons also follow geodesics of the gravitational metric where the scalar field doesn't enter (except very weakly as an additional source). Hence the scalar field does not influence the motion of photons!

This has a major observational consequence: It would imply that, for a massive cluster of galaxies, the Newtonian mass one would determine from the kinematics of galaxies (non-relativistic particles) via the virial theorem should be much greater than the mass one would determine from gravitational deflection of photons (relativistic particles). This is, emphatically, not the

case [48]. The lensing contradiction is a severe blow to scalar-tensor theories
of MOND, at least for those with a conformal coupling.

An obvious solution to this problem is to consider a non-conformal relationship between the Einstein and physical metrics,

$$\tilde{g}_{\mu\nu} = g_{\mu\nu}e^{-\eta\phi} - (e^{\eta\phi} - e^{-\eta\phi})A_\mu A_\nu \qquad (13.17)$$

where now A^μ is a normalized vector field, i.e. $A_\mu A^\mu = -1$ [50]. Basically, the
conformal relation transforms the gravitational geometry by stretching or contracting the 4-D space isotropically but in a space-time dependent way. This
disformal transformation, (13.17), picks out certain directions for additional
stretching or contracting. Because we would like space in the cosmological
frame to be isotropic (the Cosmological Principle) we should somehow ar-
range for the vector to point in the time direction in the cosmological frame,
which then becomes a preferred frame. In the spirit of the ancient stratified
theories [50], one may propose an a priori non-dynamical vector field postu-
lated to have this property. This may be combined with an AQUAL theory to
provide MOND phenomenology with enhanced gravitational lensing [51]; in
fact, with the particular transformation given by (13.17) one can show that
the relation between the total weak field force and the deflection of photons is
the same as it is in GR. Hence relativistic and non-relativistic particles would
both feel the same weak-field force.

The problem with this initial theory is that the non-dynamical vector field
quite explicitly violates the principle of General Covariance making it impossible to define a conserved energy-momentum tensor (this has been known for
some time [52]). This problem led Bekenstein to endow the vector field with
its own dynamics, and, hence, to TeVeS.

13.3.2 The Structure of TeVeS

As the name implies, the theory is built from three fields.

a) The tensor: This is the usual Einstein metric that we are all familiar
with. It's dynamics are given by the standard Einstein-Hilbert action of GR:

$$S_T = \frac{1}{16\pi G} \int R\sqrt{-g}d^4x \ . \qquad (13.18)$$

It is necessary that the tensor should be the Einstein metric because we want
the theory to approach GR quite precisely in the appropriate strong field
limits.

b) The scalar: We want the scalar, ϕ, to provide a long-range fifth force in
the limit of low field gradients Bekenstein takes the scalar field action to be

$$S_S = -\frac{1}{16\pi G} \int [\frac{1}{2}q^2 h^{\alpha\beta}\phi_{,\alpha}\phi_{,\beta} + l^{-2}V(q)]\sqrt{-g}d^4x \ . \qquad (13.19)$$

Here I have kept the notation of PCG because the action is, in fact, the weak
coupling, or AQUAL limit, of PCG where there is no explicit kinetic term for

the field q. In other words, q behaves as a non-dynamical auxiliary field where q^2 will play the role of μ in the Bekenstein-Milgrom field equation (the fact that this field is non-dynamical does not violate General Covariance because it does not act directly upon particles). I use this bi-scalar notation because I think it is important to realise that the auxiliary field could, in fact, be dynamical. This, in some respects, provides a plausible interpretation of the free function, $V(q)$, as a potential (let's call it a pseudo-potential for now). As we see below, this can provide a basis for cosmological dark matter.

Another difference with standard scalar-tensor theory is that the invariant $h^{\alpha\beta}\phi_{,\alpha}\phi_{,\beta}$ has replaced the usual scalar field invariant $g^{\alpha\beta}\phi_{,\alpha}\phi_{,\beta}$ where

$$h^{\alpha\beta} = g^{\alpha\beta} - A^\alpha A^\beta \tag{13.20}$$

and $\mathbf{A}$ is the normalized vector field described below. Bekenstein has shown that this simple replacement solves the superluminal propagation problem of AQUAL theories of MOND. The speed of scalar waves turns out to be precisely c.

c) The vector: The dynamical normalized vector field is necessary to provide the disformal transformation and the enhanced gravitational lensing. Bekenstein chose to describe its dynamics through the action

$$S_V = \frac{K}{32\pi G} \int [F^{\mu\nu} F_{\mu\nu} - 2(\frac{\lambda}{K})(A^\mu A_\mu + 1)\sqrt{-g}d^4x \tag{13.21}$$

where $F_{\mu\nu}$ is the electromagnetic-like anti-symmetric tensor constructed from $\mathbf{A}$

$$F_{\mu\nu} = A_{\nu;\mu} - A_{\mu;\nu} , \tag{13.22}$$

and λ is a Lagrangian multiplier function which enforces the normalisation condition $A_\mu A^\mu = -1$. K is a new parameter which determines the strength of the vector field coupling.

All of this is combined with the particle action

$$S_P = -mc \int (-\tilde{g}_{\mu\nu} \frac{dx^\mu}{dp} \frac{dx^\nu}{dp})^{\frac{1}{2}} dp \tag{13.23}$$

where $\tilde{g}_{\mu\nu}$ is the physical metric disformally related to the Einstein metric as in (13.17). This guarantees that the deflection of photons is given by

$$\delta\theta = \frac{2}{c^2} \int f_\perp dl \tag{13.24}$$

where the integral is over the line-of-sight and $f_\perp$ is the perpendicular component of the total weak-field force, Newtonian and scalar.

The free parameters of the theory are η, the scalar field coupling, K the vector field coupling, and l the characteristic length scale determining the MOND acceleration scale ($a_0 = c^2/l$). It can be shown that, as the parameters

η and K approach zero, the theory reduces to GR, as it should do. The free function is $V(q)$ or the pseudo-potential of the auxiliary q field. I could have absorbed the length scale l into $V(q)$ but, following Bekenstein, I choose to express it explicitly in order to render $V(q)$ unitless.

In the weak-field static limit, the scalar field equation is of the Bekenstein-Milgrom form:

$$\nabla \cdot (\mu \mathbf{f_s}) = 4\pi G \rho \tag{13.25}$$

where, in my notation, the scalar force is given by $f_s = \eta c^2 \nabla \phi$ and $\mu = q^2/2\eta^2$. Making use of these expressions, we may show that the MOND interpolating function is then given by the algebraic relation,

$$\frac{dV(\mu)}{d\mu} = -\frac{f_s^{\,2}}{a_0^{\,2}} = -X \tag{13.26}$$

where $a_0 = c^2/l$. This, of course, necessitates $V'(\mu) < 0$ in the static domain.

Now, to obtain MOND phenomenology, it must be the case that $\mu(X) = \sqrt{X}$ in the low acceleration limit. For example, $V(\mu) = -\frac{1}{3}\mu^3$ would work (recalling the relation between μ and q above, we see that this gives rise to the negative sextic potential in PCG). But this leaves us with the old problem of extending AQUAL into the cosmological regime where $X < 1$.

Bekenstein chose to solve this problem by taking a free function that provides two separate branches for $\mu(X)$– one for static mass concentrations, where the spatial gradients of ϕ dominate, and one for the homogeneous evolving Universe where the temporal derivative dominates. Specifically,

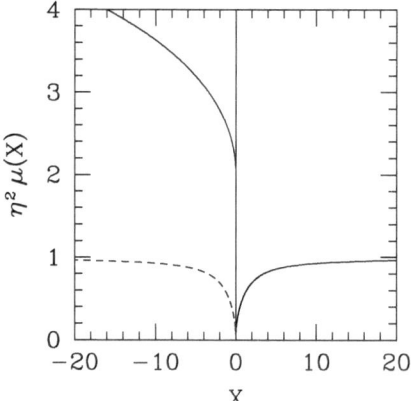

Fig. 13.7. Bekenstein's trial free function shown, $\mu(X)$ (*solid curve*) where X is defined as $\eta^2 l^2 \phi_{,\alpha} \phi^{,\alpha}$. There are two discontinuous branches for cosmology ($X < 0$) and for quasi-static mass concentrations ($X > 0$). The dashed curve shows one possibility for avoiding the discontinuity (13.28)

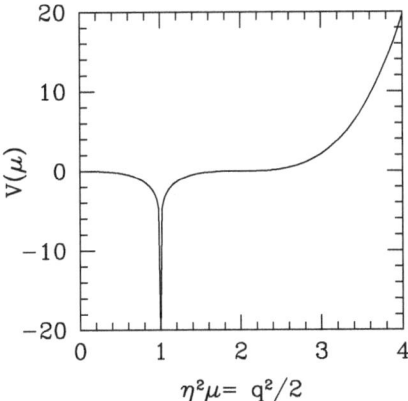

Fig. 13.8. The pseudo-potential $(V(\mu))$ corresponding to the $\mu(X)$ shown in Fig. 13.7

$$X = \frac{1}{4}\mu^2(\eta^2\mu - 2)^2(1 - \eta^2\mu)^{-1} \tag{13.27}$$

(η^2 appears because my definition of μ differs from Bekenstein's). This two branch, $\mu(X)$, is shown in Fig. 13.7, where now we are defining X more generally as $X = \eta^2 l^2 \phi_{,\alpha}\phi^{,\alpha}$ The corresponding pseudo-potential, $V(\mu)$, is shown in Fig. 13.8.

If we interpret $V(\mu)$ as the potential of an implicitly dynamical field, it is certainly a rather curious-looking one– with the infinite pit at $\eta^2\mu = 1$. It also illustrates the peculiar aspect of the two-branch form of μ. For cosmological solutions, $\eta^2\mu = 2$ is an attractor; i.e. the μ field seeks the point where $dV/d\mu = 0$ [53]. However, on the outskirts of galaxies $\eta^2\mu \to 0$ as it must to provide the $1/r$ scalar force. So somehow, in progressing from the galaxies to the cosmological background $\eta^2\mu$ must jump from 0 to 2 apparently discontinuously (photons propagating in a cosmological background also have to make this leap). This problem indicates that such a two-branch $\mu(X)$ may not be appropriate, but more on this below.

13.4 TeVeS: Successes, Issues and Modifications

13.4.1 Successes of TeVeS

The theory is an important development because it solves several of the outstanding problems of earlier attempts:

1) While providing for MOND phenomenology in the form of the old non-relativistic Bekenstein-Milgrom theory, it also allows for enhanced gravitational lensing. It does this in the context of a proper covariant theory, albeit by construction– by taking the particular disformal relation between the physical and gravitational metrics given by (13.17). This aspect of the theory has

favourably tested on a sample of observed strong lenses [54], although there are several case with unreasonable implied mass-to-light ratios.

2) It has been shown [22, 58] that, for TeVeS, the static post-Newtonian effects are identical to those of GR; that is to say, the Eddington-Robertson post-Newtonian parameters are $\gamma = \beta = 1$ as in GR. This provides consistency with a range of Solar System gravity tests such as light deflection and radar echo delay.

3) Scalar waves propagate causally ($v_s \leq c$). This is true because the new scalar field invariant $h^{\alpha\beta}\phi_{,\alpha}\phi_{,\beta}$ ($h^{\alpha\beta}$ is a new tensor built from the Einstein metric and the vector field (13.20)) replaces the standard invariant in the scalar field Lagrangian (13.9). This is a major improvement over the old AQUAL theory, but also one which relies upon the presence of the vector field.

4) Gravitational waves propagate causally if $\phi > 0$. One can show [22] that the speed of the standard tensor waves is given by $V_g = ce^{-\eta\phi}$. This means that the cosmology must provide $\phi > 0$ in a natural way. Moreover, there is a prediction here which is possibly testable, and that is $V_g < c$. If an event, such as a gamma-ray burst, also produces gravitational radiation (as is likely), the gravitational waves should arrive somewhat later than the gamma rays.

5) The theory allows for standard FRW cosmology and, at least in the linear regime, for a MONDian calculation of structure formation [53]. Moreover, there is an evolving dark energy (quintessence) which is coupled to the background baryon density, offering a possible solution to the near coincidence of these components at the present epoch. This comes about through the presence of $V(\mu)$ as a negative pressure fluid in the Friedmann equations. The cosmological value of the dark energy density, $V(\mu)$, corresponds to the minimum of an effective potential $V_{eff} = V(\mu) + B(\rho\tau)/\mu$ where B is a function of the product of cosmic time τ and the baryonic mass density ρ (it is identical in this sense to PCG in a cosmological context [55]).

13.4.2 Remaining Issues

In spite of these important successes there are a number of problems that the theory is yet to confront:

1) The discontinuous $\mu(X)$. The two discontinuous branches (Fig. 13.7)– one for cosmology and one for quasi-static mass concentrations– appears awkward, particularly if the free-function is interpreted as a potential of the μ field. Moreover, this presents very practical problems for gravitational lensing and calculation of structure formation into the non-linear regime. But more seriously, it appears that such two branch μ may be an intrinsic aspect of a theory with the structure of TeVeS. One could propose (as in [56]) that the space-like branch of μ is simply reversed at the at the $\mu = 0$ axis (see dotted line in Fig. 13.7), so, instead of (13.27), Bekenstein's free function could be expressed as

$$X = \pm \frac{3\mu^2}{1 - \eta^2\mu} \; .$$
(13.28)

However, the pseudo-potential, $V(\mu)$, would then also be double valued which would appear distinctly unphysical if this is really to be identified with the potential of a implicitly dynamical scalar μ (or q). In my opinion, the only solution to this problem is to alter the structure of the theory (see below).

2) Even given a $\mu(X)$ with two branches, the separation between quasi-static and cosmological phenomena is artificial. Equation (13.26), which provides the relation between the scalar field gradient and μ, should also contain the cosmic time derivative of the scalar field because this is likely to be of the same order as $dV/d\mu$; i.e. (13.26) should read

$$\frac{dV}{d\mu} = -\frac{f_s^2}{a_0^2} + \frac{\eta^2 l^2 \dot{\phi}^2}{c^2} \; .$$
(13.29)

Therefore the free function, relevant to mass concentrations, may also be thought of as an evolving effective potential (this can actually be an advantage which I make use of below).

3) This is a preferred frame theory that violates the Lorentz invariance of gravitational phenomena. This is because of the cosmic vector field $\mathbf{A}$. In the cosmic frame, only the time component of $\mathbf{A}$ is non-zero but for frames in relative motion with respect to the CMB spatial components also develop non-zero values, and this has a real effect on particle dynamics. In the Solar System for example, there should be gravitational ether drift effects, such as a polarisation of the earth-moon orbit along the direction of $\mathbf{w}$, the velocity vector with respect to the CMB. Such effects, in conservative theories, are quantified by two post-Newtonian parameters [57], α_1 and α_2, which enter the effective Lagrangian of an N-body system as the coefficients of terms containing $v \cdot w/c^2$ where $\mathbf{v}$ is the velocity with respect to the center-of-mass of the N-body system. These parameters are experimentally constrained; for example, $\alpha_1 < 10^{-4}$ on the basis of Lunar Laser Ranging [59].

It is important to determine predicted values of α_1 and α_2 for TeVeS. A reasonable guess is that these post-Newtonian parameters will approach zero as the free parameters of the theory, η and K approach zero [61]. That is because in this limit the theory approaches GR, and in GR there are no preferred frame effects. Whether or not the resulting constraints on η and K are consistent with other aspects of Solar System and galaxy phenomenology remains to be seen.

4) In the outer solar system the force is not precisely inverse square. For example, in the context of Bekenstein's free function, the non-inverse square component of the force is shown, as a function of radius, in Fig. 13.9 for two different values of the scalar coupling strength, η. Constraints from planetary motion are shown by the upper limits [61]. Such a deviation, at some level, is

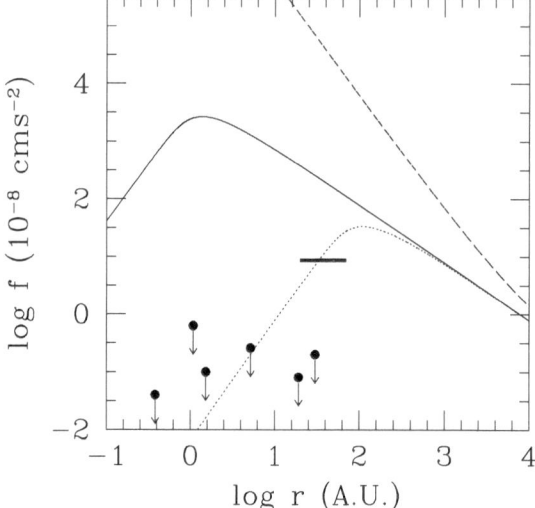

Fig. 13.9. The dashed curve is the log of the the total force ($f_t = f_s + f_N$), in units of 10^-8 cm/s^2 plotted against the log of the radial distance from the sun in astronomical units for TeVeS. The dotted curve is the anomalous force (the non-inverse square force) for Bekenstein's initial choice of free function with $\eta = 0.01$. The long dashed curve is the same but with $\eta = 0.1$. Observed constraints on the non-inverse square part of the acceleration are (*left to right*): from the precession of perihelion of Mercury, and of Icarus, from variation of Kepler's constant between Earth and Mars, between inner planets and Jupiter, Uranus or Neptune, respectively. The horizontal bar is the Pioneer anomaly range. From reference [62]

an aspect of any multi-field theory of MOND [51], and it may be a problem or it may be a blessing. A non-inverse square component of the force, in the form of a constant acceleration, is indicated by Doppler ranging to both the Pioneer spacecrafts (indicated by the horizontal bar in Fig. 13.8) [60]. If this effect is confirmed, it would be a major discovery, indicating that gravity is not what we think it is beyond the inner solar system.

5) As I mentioned in the Introduction, there is compelling evidence for cosmological dark matter– a pressureless fluid which appears to affect early large scale structure formation (evident in the CMB anisotropies) and the more recent expansion history of the Universe (evident in the SNIa results). The weight of this evidence implies that a proper theory of MOND should at least simulate the cosmological effects of the apparent dark matter, again not an evident aspect of TeVeS.

In a general sense, the theory, at present, is intricate and misses a certain conceptual simplicity. There are several loose threads which one might hope a theory of MOND to tie up. For example, the MOND acceleration parameter, a_0, is put in by hand, as an effective length scale l; the observational fact that $a_0 \approx cH_0$ remains coincidental. This seems unfortunate because this coincidence suggests that MOND results from the effect of cosmology on local particle dynamics, and, in the theory as it now stands, no such connection is evident.

Finally, by mentioning these problems, I do not wish to imply that TeVeS is fundamentally flawed, but that it is not yet the theory in final form. In this procedure, building up from the bottom, the approach to the final theory is incremental.

13.4.3 Variations on a Theme: Biscalar-tensor-vector Theory

The motivation behind this variation is to use the basic elements of TeVeS in order to construct a cosmologically effective theory of MOND. The goals are to reconcile the galaxy scale success of MOND with the cosmological evidence for CDM and to provide a cosmological basis for a_0 [42].

There are two essential differences with TeVeS in original form: First, the auxiliary field q is made explicitly dynamical as in PCG. This is done by introducing a kinetic term for q in the scalar action (13.19), i.e. $q_{,\alpha}q^{,\alpha}$. Secondly, one makes use of the preferred frame to separate the spatial and time derivatives of the matter coupling scalar field ϕ at the level of the Lagrangian. Basically, this is done by defining new scalar field invariants. If we take the usual invariant to be $I = g^{\alpha\beta}\phi_{,\alpha}\phi_{,\beta}$ and define $J = A^\alpha A^\beta \phi_{,\alpha}\phi_{,\beta}$, $K = J + I$, then we can readily see that J is just the square of the time derivative in the preferred cosmological frame $(\dot{\phi}^2)$ and K is the spatial derivative squared in that frame $(\nabla\phi \cdot \nabla\phi)$. The scalar field Lagrangian is then taken to be

$$L_s = \frac{1}{2}[q_{,\alpha}q^{,\alpha} + h(q)K - f(q)J + 2V(q)] \,. \tag{13.30}$$

So, separate functions of q multiply the spatial and temporal gradients of ϕ in the cosmological frame. This means that the potential for q becomes an effective potential involving the cosmic time derivative, $\dot{\phi}$ for both the homogeneous cosmology and for quasi-static mass concentrations. Indeed, one can show, given certain very general conditions on the free functions, q at a large distance from a mass concentration approaches its cosmological value. There is smooth transition between mass concentrations and cosmology. Moreover, if I take $h(q) \approx q^2$, $f(q) \approx q^6$ and a simple quadratic bare potential $V(q) \approx Bq^2$, I obtain a cosmological realisation of Bekenstein's PCG with a negative sextic potential [46] but where the coefficient in the potential, and hence a_0, is identified with the cosmic $d\phi/dt$.

There are two additional advantages of making q dynamical. First of all, as the q field settles to the evolving potential minimum, oscillations of this field about that minimum inevitably develop. If the bare potential has a quadratic form, then these oscillations constitute CDM in the form of "soft bosons" [63]. Depending upon the parameters of the theory, the de Broglie wavelength of these bosons may be so large that this dark matter does not cluster on

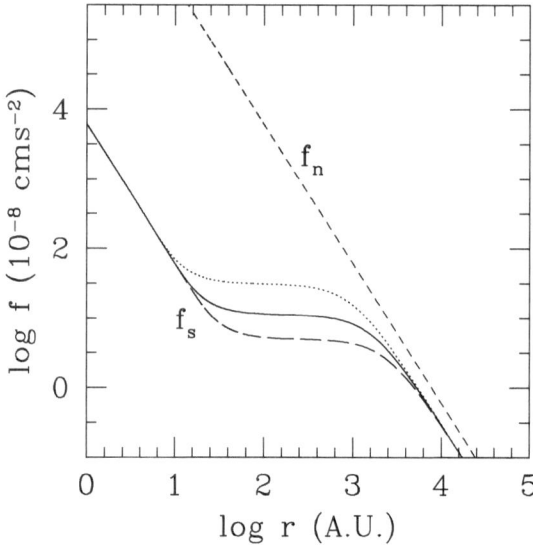

Fig. 13.10. The Newtonian (*dashed curve*) and scalar (*solid curves*) force in the Solar System in the context of the biscalar theory. The different curves correspond to different values of scalar coupling constant η. This should be compared with Fig. 13.6 which shows the Newtonian and scalar forces for TeVeS with the initial free function

the scale of galaxies (but possibly on the scale of clusters). A cosmological effective theory of MOND produces cosmological CDM for free.

A second advantage is that appropriately chosen free functions can reproduce the Pioneer anomaly in the outer Solar System– both the magnitude ($\approx 8 \times 10^{-9}$ cm/s^2) and the form– constant beyond 20 AU (see Fig. 13.10). It does this while being consistent with the form of galaxy rotation curves [61]. Of course the presence of three free functions appear to give the theory considerable arbitrariness, but, in fact, the form of these functions is strongly constrained by Solar System, galaxy and cosmological phenomenology.

Many other modifications of TeVeS are possible. For example, it may only be necessary to make the auxiliary field q explicitly dynamical and choose a more appropriate form of the free function. The number of alternative theories is likely to be severely restricted by the demands imposed by observations– ranging from the solar system, to galaxies, to clusters, to gravitational lensing, to cosmology. The hope is that the number of survivors is not less than one.

13.5 Conclusions

Here I have outlined the attempts that have been made to define modifications of gravity that may underly the highly successful empirically-based

MOND, proposed by Milgrom as an alternative to dark matter in bound self-gravitating systems. These attempts lead inevitably to a multi-field theory of gravity– the Einstein metric to provide the phenomenology of GR in the strong field limit, the scalar field to provide the MOND phenomenology most apparent in the outskirts of galaxies and in low surface brightness systems, and the vector field to provide a disformal relation between the Einstein and physical metrics– necessary for the observed degree of gravitational lensing. I re-emphasise that this process has been entirely driven by phenomenology and the need to cure perceived pathologies; there remains no connection to more a priori theoretical considerations or grand unifying principles such as General Covariance or Gauge Invariance. It would, of course, be a dramatic development if something like MOND were to emerge as a incidental consequence of string theory or a higher dimensional description of the Universe, but, in my opinion, this is unlikely. It is more probable that an empirically based prescription, such as MOND, will point the way to the correct theory.

The coincidence between the critical acceleration and cH_0 (or possibly the cosmological constant) must be an essential clue. MOND must be described by an effective theory; that is, the theory predicts this phenomenology only in a cosmological context. The aspect, and apparent necessity, of a preferred frame invites further speculation: Perhaps cosmology is described by a preferred frame theory (there certainly is an observed preferred frame) with a long range force mediated by a scalar field coupled to a dynamical vector field as well as the gravitational metric. With the sort of bi-scalar Lagrangian implied by TeVeS, the scalar coupling to matter becomes very weak in regions of high field gradients (near mass concentrations). This protects the Solar System from detectable preferred frame effects where the theory essentially reduces to General Relativity. Because we live a region of high field gradients, we are fooled into thinking that General Relativity is all there is. Only the relatively recent observations of the outskirts of galaxies or objects of low surface brightness (or perhaps the Pioneer anomaly) reveal that there may be something more to gravity.

On the other hand, it may well be that we have been pursuing a mirage with tensor-vector-scalar theories. Perhaps the basis of MOND lies, as Milgrom has argued, with modified particle action– modified inertia– rather than modified gravity [64, 65]. For a classical relativist this distinction between modified gravity and modified inertia is meaningless– in relativity, inertia and gravity are two sides of the same coin; one may be transformed into another by a change of frame. But perhaps in the limit of low accelerations, lower than the fundamental cosmological acceleration cH_0, that distinction is restored [66].

It is provocative that the Unruh radiation experienced by a uniformly accelerating observer, changes its character at accelerations below $c\sqrt{\Lambda}$ in a de Sitter universe [65]. If the temperature difference between the accelerating observer and the static observer in the de Sitter Universe is proportional to inertia, then we derive an inertia-acceleration relation very similar to that required by MOND [65]. At present this is all very speculative, but it presents the possibility that we may be going down a false path with attempted modifications of GR through the addition of extra fields.

In any case, the essential significance of TeVeS is not that it, at present, constitutes the final theory of MOND. Rather, the theory provides a counter-example to the often heard claim that MOND is not viable because it has no covariant basis.

Acknowledgements

It is a pleasure to thank Jacob Bekenstein and Moti Milgrom for helpful comments on this manuscript and for many enlightening conversations over the years. I also thank Renzo Sancisi for helpful discussions on the "dark matter-visible matter coupling" in galaxies and Martin Zwaan for sending the data which allowed me to produce Fig. 13.4. I am very grateful to the organisers of the Third Aegean Summer School on the Invisible Universe, and especially, Lefteris Papantonopoulos, for all their efforts in making this school a most enjoyable and stimulating event.

References

1. D.N. Spergel et al., Astrophys. J. Suppl., **148**, 175 (2003) [astro-ph/0302209].
2. S. Perlmutter et al., (1999), Astrophys. J., **517**, 565.
3. P.M. Garnevitch et al.' Astrophys. J., **493**, 53 (1998).
4. J.L. Tonry et al., Astrophys. J., **594**, 1 (2003) [astro-ph/0305008].
5. A.G. Sanche et al., Mon. Not. RAS, (in press, 2005).
6. D.J. Eisenstein et al., Astrophys. J., 633, 560 (2005).
7. C. Armendariz-Picon, V. Mukhanov and P.J. Steinhardt, Phys. Rev., **D 63**, 103510 (2001).
8. S. Capozziello, S. Carloni and A. Troisi, astro-ph/0303041 (2003).
9. S.M. Carroll, V. Duvvuri, M. Trodden and M.S. Turner, astro-ph/0306438 (2003).
10. G. Dvali, G. Gabadadaze and M. Porrati, Phys. Lett., B **485**, 208 (2000) [hep-th/0005016].
11. F. Zwicky, Helv. Phys. Acta, **6**, 110 (1933).
12. A. Bosma *The Distribution and Kinematics of Neutral Hydrogen in Spiral Galaxies of Various Morphological Types*, PhD Dissertaion, Univ. of Groningen, The Netherlands (1978).
13. K.G. Begeman, Astron. Astrophys. **223**, 4 (1989).
14. J.R. Bond and A.S. Szalay, Astrophys. J., **274**, 443 (1983).
15. G. de Lucia et al., Mon. Not. RAS, **348**, 333 (2004).
16. W.J.G. de Blok, S.S. McGaugh and V.C. Rubin, Astron. J., **122**, 2396 (2001).
17. M. Milgrom, Astrophys. J.**270**, 365 (1983a).
18. M. Milgrom, Astrophys. J.**270**, 371 (1983b).
19. M. Milgrom, Astrophys. J.**270**, 384 (1983c).

20. J.T. Wilson, ed. *Continents Adrift and Continents Aground* Scientific American, W. H. Freeman & Company, San Francisco (1976).
21. R.H. Sanders and S.S. McGaugh, Ann. Rev. Astron. Astrophys., **40**, 263 (2002).
22. J.D. Bekenstein and Phys. Rev. D, **70**, 083509 (astro-ph/0403694) (2004).
23. S.S. McGaugh and W.J.G. de Blok, Astrophys. J., **499**, 66 (1998).
24. S.S. McGaugh, J.M. Schombert, G.D. Bothun, W.J.G. de Blok, Astrophys. J., **533**, L99 (2000).
25. R.H. Sanders M.A.W. Verheijen, Astrophys. J., **503**, 97 (1998).
26. K.G. Begeman, A.H. Broeils and R.H. Sanders, Mon. Not. RAS, **249**, 523 (1991).
27. S. Casertano and J.H. van Gorkom, Astron. J., **101**, 1231 (1991).
28. J.P. Ostriker and P.J.E. Peebles, Astrophys. J.**186**, 467 (1973).
29. K.C. Freeman, Astrophys. J.**160**, 811 (1970).
30. M. Milgrom, Astrophys. J.**287**, 571 (1984).
31. S.M. Faber and R.E. Jackson, Astrophys. J.**204**, 668 (1976).
32. R.A. Fish, Astrophys. J.**139**, 284 (1964).
33. R.H. Sanders, Mon. Not. RAS, **313**, 767 (2000).
34. A.J. Romanowsky et al., Science, **301**, 1696 (2003).
35. M. Milgrom, R.H. Sanders, Astrophys. J., **599**, L25 (2003).
36. E.F. Bell and R.S. de Jong, Astrophys. J.**550**, 212 (2001).
37. R. Sancisi: IAU Symp. 220, Eds, S.D. Ryder, D.J. Pisano, M.A. Walker and K.C. Freeman: San Francisco, ASP, p. 233 (2004).
38. M.A. Zwaan, J.M. van der Hulst and A. Bosma, (in preparation 2005).
39. R.H. Sanders, Astrophys. J., **512**, L23 (1999).
40. A. Aguirre, J. Schaye and E. Quataert, Astrophys. J., **561**, 550 (2002).
41. R.H. Sanders, Mon. Not. RAS, **342**, 901 (2003).
42. R.H. Sanders, Mon. Not. RAS, **363**, 459 (2005).
43. M.E. Soussa, R.P. Woodard, Phys. Lett. **B578**, 253 (2004).
44. C. Brans and R.H. Dicke, Phys. Rev., **124**, 925 (1961).
45. J.D. Bekenstein, M. Milgrom, Astrophys. J., **286**, 7 (1984).
46. J.D. Bekenstein, *Second Canadian Conference on General Relativity and Relativistic Astrophysics*, eds. Coley, A., Dyer, C., Tupper, T., p. 68. Singapore, World Scientific (1988).
47. J.D. Bekenstein: *Developments in General Relativity, Astrophysics and Quantum Theory, A Jubilee in Honour of Nathan Rosen*, eds. F.I. Cooperstock, L.P. Horwitz, J. Rosen, p. 155, Bristol, IOP Publishing (1990).
48. J.D. Bekenstein and R.H. Sanders, Astrophys. J.**429**, 480 (1994).
49. J.D. Bekenstein, *Proceedings of the Sixth Marcel Grossman Meeting on General Relativity*, eds. H. Sato & T. Nakamura, p. 905, Singapore, World Scientific (1992).
50. W.-T. Ni, Astrophys. J., **176**, 769 (1972).
51. R.H. Sanders, Astrophys. J.**480**, 492 (1997).
52. D.L. Lee, A.P. Lightman, W.-T. Ni, Phys. Rev., **D10**, 1685 (1974).
53. C. Skordis, D.F. Mota, P.G. Ferreira and C. Boehm, astro-ph/0505519 (accepted PRL 2005).
54. H.-S. Zhao, D.J. Bacon, A.N. Taylor K. Horne, Mon. Not. RAS(in press, 2006), astro-ph/0509590.
55. R.H. Sanders, Mon. Not. RAS, **241**, 135 (1989).

56. H.-S. Zhao, B. Famaey, Astrophys. J., **638**, L9, astro-ph/0512435 (2005).
57. C.M. Will, *Living Rev. Rel.*, **4**, 4 (2001).
58. D. Giannios, Phys. Rev., **D71**, 103511 [astro-ph/0502122] (2005).
59. J. Múller, K. Nordtvedt and D. Vokrouhlický, Phys. Rev., D**54**, 5927 (1996).
60. J.D. Anderson, et al., Phys. Rev. Lett., **81**, 2858 (1998).
61. R.H. Sanders, Mon. Not. RAS, submitted (2006).
62. J.D. Bekenstein and R.H. Sanders, astro-ph/0509519, (2005).
63. W.H. Press, B.S. Ryden and D.N. Spergel, 1990, Phys. Rev. Lett., **65**, 1084 (1990).
64. M. Milgrom, Annals Phys **229**, 384 (1994).
65. M. Milgrom, Phys. Lett. A**253**, 273 (1999).
66. M. Milgrom, astro-ph/0510117 (2005).

14

Avoiding Dark Energy
with 1/R Modifications of Gravity

Richard Woodard

Department of Physics, University of Florida, Gainesville, FL 32611-8440, USA
woodard@phys.ufl.edu

Abstract. Scalar quintessence seems epicyclic because one can choose the potential to reproduce any cosmology (I review the construction) and because the properties of this scalar seem to raise more questions than they answer. This is why there has been so much recent interest in modified gravity. I review the powerful theorem of Ostrogradski which demonstrates that the only potentially stable, local modification of general relativity is to make the Lagrangian an arbitrary function of the Ricci scalar. Such a theory can certainly reproduce the current phase of cosmic acceleration without Dark Energy. However, this explanation again seems epicyclic in that one can construct a function of the Ricci scalar to support any cosmology (I give the technique). Models of this form are also liable to problems in the way they couple to matter, both in terms of matter's impact upon them and in terms of the long range gravitational force they predict. Because of these problems my own preference for avoiding Dark Energy is to bypass Ostrogradski's theorem by considering the fully nonlocal effective action built up by quantum gravitational processes during the epoch of primordial inflation.

14.1 Introduction

The case for alternate gravity is easily made. The best that can be done from observing cosmic motions is to infer the metric $g_{\mu\nu}$ in some coordinate system. From this one can reconstruct the Einstein tensor and then ask whether or not general relativity predicts it in terms of the observed sources of stress-energy,

$$\left(R_{\mu\nu} - \frac{1}{2} g_{\mu\nu} R \right)_{\text{rec}} = 8\pi G \left(T_{\mu\nu} \right)_{\text{obs}} ? \tag{14.1}$$

One way of explaining any disagreement is by positing the existence of an unobserved, "dark" component of the stress-energy tensor,

$$\left(T_{\mu\nu} \right)_{\text{dark}} \equiv \frac{1}{8\pi G} \left(R_{\mu\nu} - \frac{1}{2} g_{\mu\nu} R \right)_{\text{rec}} - \left(T_{\mu\nu} \right)_{\text{obs}} . \tag{14.2}$$

This always works, but recent observations make it seem epicyclic.

R. Woodard: *Avoiding Dark Energy with 1/R Modifications of Gravity*, Lect. Notes Phys.
720, 403–433 (2007)
DOI 10.1007/978-3-540-71013-4_14   © Springer-Verlag Berlin Heidelberg 2007

The theory of nucleosynthesis implies that no more than about 4% of the energy density currently required to make general relativity agree with all observations can consist of any material with which we are presently familiar [1] — and only a fraction of this 4% is observed. Just to make general relativity agree with the observed motions of galaxies and galactic clusters we must posit that *six times* the mass of ordinary matter comes in the form of non-baryonic, cold dark matter [2]. Although there are some plausible candidates for what this might be, no Earth-bound laboratory has yet succeeded in detecting it.

I belong to the minority of physicists who feel that this factor of six already strains credulity. Easing that strain is what led Milgrom to propose MOND [3], which can be viewed as a phenomenological modification of gravity in the regime of very small accelerations. There is an impressive amount of observational data in favor of this modification [4] — although see [5]. Bekenstein has recently constructed a fully relativistic field theory [6] which reproduces MOND, and a preliminary analysis of the resulting cosmology works better than many experts thought possible [7].

However, the worst problem for conventional gravity comes on the largest scales. To make general relativity agree with the Hubble plots of distant Type Ia supernovae [8, 9, 10], with the power spectrum of anisotropies in the cosmic microwave background [11] and with large scale structure surveys [12], one must accept an additional component of "dark energy" that is about *eighteen times* larger than that of ordinary matter. This would mean that 96% of the current universe's energy exists in forms which have so far only been detected gravitationally! Even people who believe passionately in dark matter (and hence accept the factor of six) find this factor of 6+18=24 difficult to swallow. That is why there has been so much recent interest in modifying gravity to make it predict observed cosmic phenomena without the need for dark energy, and sometimes even without the need for dark matter.

I want to stress that the issue is one of plausibility. There is no problem inventing field theories which give the required amount of dark energy. The simplest way of doing it is with a minimally coupled scalar [13, 14],

$$\mathcal{L} = -\frac{1}{2}\partial_\mu\varphi\partial_\nu\varphi g^{\mu\nu}\sqrt{-g} - V(\varphi)\sqrt{-g} \ . \tag{14.3}$$

The usual procedure is to begin with a scalar potential $V(\varphi)$ and work out the cosmology, but it is easy to start with whatever cosmological evolution is desired and *construct* the potential which would support it. I will go through the construction here, both to make the point and so that it can be used later.

On the largest scales the geometry of the universe can be described in terms of a single function of time known as the scale factor $a(t)$,

$$ds^2 = -dt^2 + a^2(t)d\boldsymbol{x} \cdot d\boldsymbol{x} \ . \tag{14.4}$$

The logarithmic time derivative of this quantity gives the Hubble parameter,

$$H(t) \equiv \frac{\dot{a}}{a} \ . \tag{14.5}$$

If we specialize to a solution $\varphi_0(t)$ of the scalar field equations which depends only upon time, the two nontrivial Einstein equations are,

$$3H^2 = 8\pi G\left(\frac{1}{2}\dot{\varphi}_0^2 + V(\varphi_0)\right), \tag{14.6}$$

$$-2\dot{H} - 3H^2 = 8\pi G\left(\frac{1}{2}\dot{\varphi}_0^2 - V(\varphi_0)\right). \tag{14.7}$$

Let us assume $a(t)$ is known as an explicit function of time, and construct $\varphi_0(t)$ and $V(\varphi)$. By adding (14.6) and (14.7) we obtain,

$$-2\dot{H} = 8\pi G\dot{\varphi}_0^2. \tag{14.8}$$

The weak energy condition implies $\dot{H}(t) \leq 0$ so we can take the square root and integrate to solve for $\varphi_0(t)$,

$$\varphi_0(t) = \varphi_I \pm \int_{t_I}^t dt'\sqrt{\frac{-2\dot{H}(t')}{8\pi G}}. \tag{14.9}$$

One can choose φ_I and the sign freely.

Because the integrand in (14.9) is always positive, the function $\varphi_0(t)$ is monotonic. This means we can invert to solve for time as a function of φ_0. Let us call the inverse function $T(\varphi)$,

$$\psi = \varphi_0\Big(T(\psi)\Big). \tag{14.10}$$

By subtracting (14.7) from (14.6) we obtain a relation for the scalar potential as a function of time,

$$V = \frac{1}{8\pi G}\left(\dot{H}(t) + 3H^2(t)\right). \tag{14.11}$$

The potential is determined as a function of the scalar by substituting the inverse function (14.10),

$$V(\varphi) = \frac{1}{8\pi G}\left\{\dot{H}\Big(T(\varphi)\Big) + 3H^2\Big(T(\varphi)\Big)\right\}. \tag{14.12}$$

This construction gives a scalar which supports any evolution $a(t)$ (with $\dot{H}(t) < 0$) all by itself. Should you wish to include some other, known component of the stress-energy, simply add the energy density and pressure of this component to the Einstein equations,

$$3H^2 = 8\pi G\left(\frac{1}{2}\dot{\varphi}_0^2 + V(\varphi_0) + \rho_{\text{known}}\right), \tag{14.13}$$

$$-2\dot{H} - 3H^2 = 8\pi G\left(\frac{1}{2}\dot{\varphi}_0^2 - V(\varphi_0) + p_{\text{known}}\right). \tag{14.14}$$

Provided ρ_{known} and p_{known} are known functions of either time or the scale factor, the construction goes through as before.[1]

Using this method one can devise a new field $\varphi(x)$ which will support *any* cosmology with $\dot{H}(t) < 0$. However, the introduction of such a "quintessence" field raises a number of questions:

1. Where does φ reside in fundamental theory?
2. Why can't φ couple to fields other than the metric? And if it does couple to other fields, why haven't we detected its influence in Earth-bound laboratories?
3. Why did φ come to dominate the stress-energy of the universe so recently in cosmological time?
4. Why is the φ field so homogeneous?

When a phenomenological fix raises more questions than it answers people are naturally drawn to investigate other fixes. One possibility is that general relativity is not the correct theory of gravity on cosmological scales.

In this talk I shall review gravitational Lagrangians of the form,

$$\mathcal{L} = \frac{1}{16\pi G}\left(R + \Delta R[g]\right)\sqrt{-g}\,, \tag{14.15}$$

where $\Delta R[g]$ is some local scalar constructed from the curvature tensor and possibly its covariant derivatives. Examples of such scalars are,

$$\frac{1}{\mu^2}R^{\alpha\beta}R_{\alpha\beta}\,, \qquad \frac{1}{\mu^4}g^{\mu\nu}R_{,\mu}R_{,\nu}\,, \qquad \mu^2\sin\left(\frac{1}{\mu^4}R^{\alpha\beta\rho\sigma}R_{\alpha\beta\rho\sigma}\right)\,. \tag{14.16}$$

I begin by reviewing a powerful no-go theorem which pervades and constrains fundamental theory so completely that most people assume its consequence without thinking. This is the theorem of Ostrogradski [18], who essentially showed why Newton was right to suppose that the laws of physics involve no more than two time derivatives of the fundamental dynamical variables. The key consequence for our purposes is that the only viable form for the functional $\Delta R[g]$ in (14.15) is an algebraic function of the undifferentiated Ricci scalar,

$$\Delta R[g] = f(R)\,. \tag{14.17}$$

I review the Ostrogradski result in Sect. 14.2, and hopefully immunize you against some common misconceptions about it in Sect. 14.3. In Sect. 14.4 I explain why $f(R)$ theories do not contradict Ostrogradski's result. I also demonstrate that, in the absence of matter, $f(R)$ theories are equivalent to ordinary gravity, with $f(R) = 0$, plus a minimally coupled scalar of the form (14.3). Then I use the construction given above to show how one can choose $f(R)$ to enforce an arbitrary cosmology. This establishes that an $f(R)$ can be

---

[1] This construction seems to be due to Ratra and Peebles [14]. Recent examples of its use include [15, 16, 17].

found to support any desired cosmology. In Sect. 14.5 I discuss problems associated with the particular choice function $f(R) = -\frac{\mu^4}{R}$. Section 14.6 presents conclusions.

14.2 The Theorem of Ostrogradski

Ostrogradski's result is that there is a linear instability in the Hamiltonians associated with Lagrangians which depend upon more than one time derivative in such a way that the dependence cannot be eliminated by partial integration [18]. The result is so general that I can simplify the discussion by presenting it in the context of a single, one dimensional point particle whose position as a function of time is $q(t)$. First I will review the way the Hamiltonian is constructed for the usual case in which the Lagrangian involves no higher than first time derivatives. Then I present Ostrogradski's construction for the case in which the Lagrangian involves second time derivatives. And the section closes with the generalization to N time derivatives.

In the usual case of $L = L(q, \dot{q})$, the Euler-Lagrange equation is,

$$\frac{\partial L}{\partial q} - \frac{d}{dt} \frac{\partial L}{\partial \dot{q}} = 0 .\tag{14.18}$$

The assumption that $\frac{\partial L}{\partial \dot{q}}$ depends upon $\ddot{q}$ is known as *nondegeneracy*. If the Lagrangian is nondegenerate we can write (14.18) in the form Newton assumed so long ago for the laws of physics,

$$\ddot{q} = \mathcal{F}(q, \dot{q}) \qquad \Longrightarrow \qquad q(t) = \mathcal{Q}(t, q_0, \dot{q}_0) .\tag{14.19}$$

From this form it is apparent that solutions depend upon two pieces of initial value data: $q_0 = q(0)$ and $\dot{q}_0 = \dot{q}(0)$.

The fact that solutions require two pieces of initial value data means that there must be two canonical coordinates, Q and P. They are traditionally taken to be,

$$Q \equiv q \qquad \text{and} \qquad P \equiv \frac{\partial L}{\partial \dot{q}} .\tag{14.20}$$

The assumption of nondegeneracy is that we can invert the phase space transformation (14.20) to solve for $\dot{q}$ in terms of Q and P. That is, there exists a function $v(Q, P)$ such that,

$$\left. \frac{\partial L}{\partial \dot{q}} \right|_{\substack{q=Q \\ \dot{q}=v}} = P .\tag{14.21}$$

The canonical Hamiltonian is obtained by Legendre transforming on $\dot{q}$,

$$H(Q, P) \equiv P\dot{q} - L ,\tag{14.22}$$

$$= Pv(Q, P) - L\Big(Q, v(Q, P)\Big) .\tag{14.23}$$

It is easy to check that the canonical evolution equations reproduce the inverse phase space transformation (14.21) and the Euler-Lagrange (14.18),

$$\dot{Q} \equiv \frac{\partial H}{\partial P} = v + P\frac{\partial v}{\partial P} - \frac{\partial L}{\partial \dot{q}}\frac{\partial v}{\partial P} = v , \tag{14.24}$$

$$\dot{P} \equiv -\frac{\partial H}{\partial Q} = -P\frac{\partial v}{\partial Q} + \frac{\partial L}{\partial q} + \frac{\partial L}{\partial \dot{q}}\frac{\partial v}{\partial P} = \frac{\partial L}{\partial q} . \tag{14.25}$$

This is what we mean by the statement, "the Hamiltonian generates time evolution." When the Lagrangian has no explicit time dependence, H is also the associated conserved quantity. Hence it is "the" energy by anyone's definition, of course up to canonical transformation.

Now consider a system whose Lagrangian $L(q, \dot{q}, \ddot{q})$ depends nondegenerately upon $\ddot{q}$. The Euler-Lagrange equation is,

$$\frac{\partial L}{\partial q} - \frac{d}{dt}\frac{\partial L}{\partial \dot{q}} + \frac{d^2}{dt^2}\frac{\partial L}{\partial \ddot{q}} = 0 . \tag{14.26}$$

Non-degeneracy implies that $\frac{\partial L}{\partial \ddot{q}}$ depends upon $\ddot{q}$, in which case we can cast (14.26) in a form radically different from Newton's,

$$q^{(4)} = \mathcal{F}(q, \dot{q}, \ddot{q}, q^{(3)}) \qquad \Longrightarrow \qquad q(t) = \mathcal{Q}(t, q_0, \dot{q}_0, \ddot{q}_0, q_0^{(3)}) . \tag{14.27}$$

Because solutions now depend upon four pieces of initial value data there must be four canonical coordinates. Ostrogradski's choices for these are,

$$Q_1 \equiv q , \qquad P_1 \equiv \frac{\partial L}{\partial \dot{q}} - \frac{d}{dt}\frac{\partial L}{\partial \ddot{q}} , \tag{14.28}$$

$$Q_2 \equiv \dot{q} , \qquad P_2 \equiv \frac{\partial L}{\partial \ddot{q}} . \tag{14.29}$$

The assumption of nondegeneracy is that we can invert the phase space transformation (14.28–14.29) to solve for $\ddot{q}$ in terms of Q_1, Q_2 and P_2. That is, there exists a function $a(Q_1, Q_2, P_2)$ such that,

$$\left.\frac{\partial L}{\partial \ddot{q}}\right|_{\substack{q=Q_1 \\ \dot{q}=Q_2 \\ \ddot{q}=a}} = P_2 . \tag{14.30}$$

Note that one only needs the function $a(Q_1, Q_2, P_2)$ to depend upon *three* canonical coordinates — and not all four — because $L(q, \dot{q}, \ddot{q})$ only depends upon three configuration space coordinates. This simple fact has great consequence.

Ostrogradski's Hamiltonian is obtained by Legendre transforming, just as in the first derivative case, but now on $\dot{q} = q^{(1)}$ and $\ddot{q} = q^{(2)}$,

$$H(Q_1, Q_2, P_1, P_2) \equiv \sum_{i=1}^{2} P_i q^{(i)} - L , \tag{14.31}$$

$$= P_1 Q_2 + P_2 a(Q_1, Q_2, P_2) - L\Big(Q_1, Q_2, a(Q_1, Q_2, P_2)\Big) . \tag{14.32}$$

The time evolution equations are just those suggested by the notation,

$$\dot{Q}_i \equiv \frac{\partial H}{\partial P_i} \quad \text{and} \quad \dot{P}_i \equiv -\frac{\partial H}{\partial Q_i} . \tag{14.33}$$

Let's check that they generate time evolution. The evolution equation for Q_1,

$$\dot{Q}_1 = \frac{\partial H}{\partial P_1} = Q_2 , \tag{14.34}$$

reproduces the phase space transformation $\dot{q} = Q_2$ in (14.29). The evolution equation for Q_2,

$$\dot{Q}_2 = \frac{\partial H}{\partial P_2} = a + P_2 \frac{\partial a}{\partial P_2} - \frac{\partial L}{\partial \ddot{q}} \frac{\partial a}{\partial P_2} = a , \tag{14.35}$$

reproduces (14.30). The evolution equation for P_2,

$$\dot{P}_2 = -\frac{\partial H}{\partial Q_2} = -P_1 - P_2 \frac{\partial a}{\partial Q_2} + \frac{\partial L}{\partial \dot{q}} + \frac{\partial L}{\partial \ddot{q}} \frac{\partial a}{\partial Q_2} = -P_1 + \frac{\partial L}{\partial \dot{q}} , \tag{14.36}$$

reproduces the phase space transformation $P_1 = \frac{\partial L}{\partial \dot{q}} - \frac{d}{dt} \frac{\partial L}{\partial \ddot{q}}$ (14.28). And the evolution equation for P_1,

$$\dot{P}_1 = -\frac{\partial H}{\partial Q_1} = -P_2 \frac{\partial a}{\partial Q_1} + \frac{\partial L}{\partial q} + \frac{\partial L}{\partial \ddot{q}} \frac{\partial a}{\partial Q_1} = \frac{\partial L}{\partial q} , \tag{14.37}$$

reproduces the Euler-Lagrange equation (14.26). So Ostrogradski's system really does generate time evolution. When the Lagrangian contains no explicit dependence upon time it is also the conserved Noether current. By anyone's definition, it is therefore "the" energy, again up to canonical transformation.

There is one, overwhelmingly bad thing about Ostrogradski's Hamiltonian (14.32): it is *linear* in the canonical momentum P_1. This means that no system of this form can be stable. In fact, there is not even any barrier to decay. Note also the power and generality of the result. It applies to *every* Lagrangian $L(q, \dot{q}, \ddot{q})$ which depends nondegenerately upon $\ddot{q}$, independent of the details. The only assumption is nondegeneracy, and that simply means one cannot eliminate $\ddot{q}$ by partial integration. This is why Newton was right to assume the laws of physics take the form (14.19) when expressed in terms of fundamental dynamical variables.

Adding more higher derivatives just makes the situation worse. Consider a Lagrangian $L\left(q, \dot{q}, \ldots, q^{(N)}\right)$ which depends upon the first N derivatives of $q(t)$. If this Lagrangian depends nondegenerately upon $q^{(N)}$ then the Euler-Lagrange equation,

$$\sum_{i=0}^{N} \left(-\frac{d}{dt}\right)^i \frac{\partial L}{\partial q^{(i)}} = 0 , \tag{14.38}$$

contains $q^{(2N)}$. Hence the canonical phase space must have $2N$ coordinates. Ostrogradski's choices for them are,

$$Q_i \equiv q^{(i-1)} \qquad \text{and} \qquad P_i \equiv \sum_{j=i}^{N} \left(-\frac{d}{dt}\right)^{j-i} \frac{\partial L}{\partial q^{(j)}} \ . \tag{14.39}$$

Non-degeneracy means we can solve for $q^{(N)}$ in terms of P_N and the Q_i's. That is, there exists a function $\mathcal{A}(Q_1, \ldots, Q_N, P_N)$ such that,

$$\frac{\partial L}{\partial q^{(N)}}\bigg|_{\substack{q^{(i-1)}=Q_i \\ q^{(N)}=\mathcal{A}}} = P_N \ . \tag{14.40}$$

For general N Ostrogradski's Hamiltonian takes the form,

$$H \equiv \sum_{i=1}^{N} P_i q^{(i)} - L \ , \tag{14.41}$$

$$= P_1 Q_2 + P_2 Q_3 + \cdots + P_{N-1} Q_N + P_N \mathcal{A} - L\Big(Q_1, \ldots, Q_N, \mathcal{A}\Big) \ . \tag{14.42}$$

It is simple to check that the evolution equations,

$$\dot{Q}_i \equiv \frac{\partial H}{\partial P_i} \qquad \text{and} \qquad \dot{P}_i \equiv -\frac{\partial H}{\partial Q_i} \ , \tag{14.43}$$

again reproduce the canonical transformations and the Euler-Lagrange equation. So (14.42) generates time evolution. Similarly, it is Noether current for the case where the Lagrangian contains no explicit time dependence. So there is little alternative to regarding (14.42) as "the" energy, again up to canonical transformation.

One can see from (14.42) that the Hamiltonian is linear in $P_1, P_2, \ldots P_{N-1}$. Only with respect to P_N might it be bounded from below. Hence the Hamiltonian is necessarily unstable over half the classical phase space for large N!

14.3 Common Misconceptions

The no-go theorem I have just reviewed ought to come as no surprise. It explains why Newton was right to expect that physical laws take the form of second order differential equations when expressed in terms of fundamental dynamical variables.[2] Every fundamental system we have discovered since Newton's day has had this form. The bizarre, dubious thing would be if Newton had blundered upon a tiny subset of possible physical laws, and all our probing over the course of the next three centuries had never revealed the vastly richer possibilities. However — *deep sigh* — particle theorists don't like being told something is impossible, and a definitive no-go theorem such as

---

[2] The caveat is there because one can always get higher order equations by solving for some of the fundamental variables.

that of Ostrogradski provokes them to tortuous flights of evasion. I ought to know, I get called upon to referee the resulting papers often enough! No one has so far found a way around Ostrogradski's theorem. I won't attempt to prove that no one ever will, but let me use this section to run through some of the misconceptions which have been in back of attempted evasions.

To fix ideas it will be convenient to consider a higher derivative generalization of the harmonic oscillator,

$$\mathcal{L} = -\frac{gm}{2\omega^2}\ddot{q}^2 + \frac{m}{2}\dot{q}^2 - \frac{m\omega^2}{2}q^2 . \tag{14.44}$$

Here m is the particle mass, ω is a frequency and g is a small positive pure number we can think of as a coupling constant. The Euler-Lagrange equation,

$$-m\left(\frac{g}{\omega^2}q^{(4)} + \ddot{q} + \omega^2 q\right) = 0 , \tag{14.45}$$

has the general solution,

$$q(t) = A_+ \cos(k_+ t) + B_+ \sin(k_+ t) + A_- \cos(k_- t) + B_- \sin(k_- t) . \tag{14.46}$$

Here the two frequencies are,

$$k_\pm \equiv \omega\sqrt{\frac{1 \pm \sqrt{1-4g}}{2g}} , \tag{14.47}$$

and the initial value constants are,

$$A_+ = \frac{k_-^2 q_0 + \ddot{q}_0}{k_-^2 - k_+^2} , \qquad B_+ = \frac{k_-^2 \dot{q}_0 + q_0^{(3)}}{k_+(k_-^2 - k_+^2)} , \tag{14.48}$$

$$A_- = \frac{k_+^2 q_0 + \ddot{q}_0}{k_+^2 - k_-^2} , \qquad B_- = \frac{k_+^2 \dot{q}_0 + q_0^{(3)}}{k_-(k_+^2 - k_-^2)} . \tag{14.49}$$

The conjugate momenta are,

$$P_1 = m\dot{q} + \frac{gm}{\omega^2}q^{(3)} \quad \Leftrightarrow \quad q^{(3)} = \frac{\omega^2 P_1 - m\omega^2 Q_2}{gm} , \tag{14.50}$$

$$P_2 = -\frac{gm}{\omega^2}\ddot{q} \quad \Leftrightarrow \quad \ddot{q} = -\frac{\omega^2 P_2}{gm} . \tag{14.51}$$

The Hamiltonian can be expressed in terms of canonical variables, configuration space variables or initial value constants,

$$H = P_1 Q_2 - \frac{\omega^2}{2gm}P_2^2 - \frac{m}{2}Q_2^2 + \frac{m\omega^2}{2}Q_1^2 , \tag{14.52}$$

$$= \frac{gm}{\omega^2}\dot{q}q^{(3)} - \frac{gm}{2\omega^2}\ddot{q}^2 + \frac{m}{2}\dot{q}^2 + \frac{m\omega^2}{2}q^2 , \tag{14.53}$$

$$= \frac{m}{2}\sqrt{1-4g}\,k_+^2(A_+^2 + B_+^2) - \frac{m}{2}\sqrt{1-4g}\,k_-^2(A_-^2 + B_-^2) . \tag{14.54}$$

The last form makes it clear that the "+" modes carry positive energy whereas the "−" modes carry negative energy.

14.3.1 Nature of the Instability

It's important to understand both how the Ostrogradskian instability manifests and what is physically wrong with a theory which shows this instability. Because the Ostrogradskian Hamiltonian (14.42) is not bounded below with respect to more than one of its conjugate momenta, one sees that the problem is not reaching arbitrarily negative energies by setting the dynamical variable to some *constant value*. Rather it is reaching arbitrarily negative energies by making the dynamical variable have a certain *time dependence*. People sometimes mistakenly believe they have found a higher derivative system which is stable when all they have checked is that the Hamiltonian is bounded from below for constant field configurations. For example, from expression (14.53) we see that our higher derivative oscillator energy is bounded below by zero for $q(t) = $ const! Negative energies are achieved by making $\ddot{q}$ large and/or making $q^{(3)}$ large while keeping $\dot{q} + gq^{(3)}/\omega^2$ fixed.

Another crucial point is that the same dynamical variable typically carries both positive and negative energy degrees of freedom in a higher derivative theory. For our higher derivative oscillator this is apparent from expression (14.46) which shows that $q(t)$ involves both the positive energy degrees of freedom, A_+ and B_+, and the negative energy ones, A_- and B_-. And note from expression (14.54) that I really mean positive and negative energy, not just positive and negative frequency, which is the usual case in a lower derivative theory.

People sometimes imagine that the energy of a higher derivative theory decays with time. That is not true. Provided one is dealing with a complete system, and provided there is no external time dependence, the energy of a higher derivative system is conserved, just as it would be under those conditions for a lower derivative theory. This conservation is apparent for our higher derivative oscillator from expression (14.54).

The physical problem with nondegenerate higher derivative theories is not that their energies decay to lower and lower values. The problem is rather that certain sectors of the theory become arbitrarily highly excited when one is dealing with an interacting, continuum field theory which has nondegenerate higher derivatives. To understand this I must digress to remind you of some familiar facts about the Hydrogen atom.

If you consider Hydrogen in isolation, there is an infinite tower of stationary states. However, if you allow the Hydrogen atom to interact with electromagnetism only the ground state is stationary; all the excited states decay through the emission of a photon. Why is this so? It certainly is *not* because "the system wants to lower its energy." The energy of the full system is constant, the binding energy released by the decaying atom being compensated by the energy of the recoil photon. Yet the decay always takes place, and rather quickly. The reason is that decay is terrifically favored by entropy. If we prepare the Hydrogen atom in an excited state, with no photons present, there is *one* way for the atom to remain excited, whereas there are an *infinite* number of ways for it to decay because the recoil photon could go off in any direction.

Now consider an interacting, continuum field theory which possesses the Ostrogradskian instability. In particular consider its likely particle spectrum about some "empty" solution in which the field is constant. Because the Hamiltonian is linear in all but one of the conjugate momenta we can increase or decrease the energy by moving different directions in phase space. Hence there must be both positive energy and negative energy particles — just as there are in our higher derivative oscillator. Just as in that point particle model, the same continuum field must carry the creation and annihilation operators of *both* the positive and the negative energy particles. If the theory is interacting at all — that is, if its Lagrangian contains a higher than quadratic power of the field — then there will be interactions between positive and negative energy particles. Depending upon the interaction, the empty state can decay into some collection of positive and negative energy particles. The details don't really matter, all that matters is the counting: there is *one* way for the system to stay empty versus a continuous *infinity* of ways for it to decay. This infinity is even worse than for the Hydrogen atom because it includes not only all the directions that recoil particles of fixed energies could go but also the fact that the various energies can be arbitrarily large in magnitude provided they sum to zero. Because of that last freedom the decay is instantaneous. And the system doesn't just decay once! It is even *more* entropically favored for there to be two decays, and better yet for three, etc. You can see that such a system instantly evaporates into a maelstrom of positive and negative energy particles. Some of my mathematically minded colleagues would say it isn't even defined. I prefer to simply observe that no theory of this kind can describe the universe we experience in which all particles have positive energy and empty space remains empty.

Note that we only reach this conclusion if the higher derivative theory possesses both interactions and continuum particles. Our point particle oscillator has no interactions, so its negative energy degree of freedom is harmless. Of course it is also completely unobservable! However, it is conceivable we could couple this higher derivative oscillator to a discrete system without engendering an instability. The feature that drives the instability when continuum particles are present is the vast entropy of phase space. Without that it becomes an open question whether or not there is anything wrong with a higher derivative theory. Of course we live in a continuum universe, and any degree of freedom we can observe must be interacting, so these are very safe assumptions. However, people sometimes delude themselves that there is no problem with continuum, interacting higher derivative models of the universe on the basis of studying higher derivative systems which could never describe the universe because they either lack interactions or else continuum particles.

In this sub-section we have learned:

1. The Ostrogradskian instability does not drive the dynamical variable to a special, constant value but rather to a special kind of time dependence.
2. A dynamical variable which experiences the Ostrogradskian instability will carry both positive and negative energy creation and annihilation operators.
3. If the system interacts then the "empty" state can decay into a collection of positive and negative energy excitations.
4. If the system is a continuum field theory the vast entropy at infinite momentum will make the decay instantaneous.

14.3.2 Perturbation Theory

People sometimes mistakenly believe that the Ostrogradskian instability is avoided if higher derivatives are segregated to appear only in interaction terms. This is not correct if one considers the theory on a fundamental level. One can see from the construction of Sect. 14.2 that the fact of Ostrogradski's Hamiltonian being unbounded below depends only upon nondegeneracy, irrespective of how one organizes any approximation technique. However, there is a way of imposing constraints to make the theory agree with its perturbative development. If this is done then there are no more higher derivative degrees of freedom, however, one typically loses unitarity, causality and Lorentz invariance on the nonperturbative level.

I constructed the higher derivative oscillator (14.44) so that its higher derivatives vanish when $g=0$. If we solve the Euler-Lagrange equation (14.45) exactly, without employing perturbation theory, there are four linearly independent solutions (14.46) corresponding to a positive energy oscillator of frequency k_+ and a negative energy oscillator of frequency k_-. However, we might instead regard the parameter g as a coupling constant and solve the equations perturbatively. This means substituting the ansatz,

$$q_{\mathrm{pert}}(t) = \sum_{n=0}^{\infty} g^n x_n(t) , \qquad (14.55)$$

into the Euler-Lagrange equation (14.45) and segregating terms according to powers of g. The resulting system of equations is,

$$\ddot{x}_0 + \omega^2 x_0 = 0 , \qquad (14.56)$$

$$\ddot{x}_1 + \omega^2 x_1 = -\frac{1}{\omega^2} x_0^{(4)} , \qquad (14.57)$$

$$\ddot{x}_2 + \omega^2 x_2 = -\frac{1}{\omega^2} x_1^{(4)} , \qquad (14.58)$$

and so on. Because the zeroth order equation involves only second derivatives, its solution depends upon only two pieces of initial value data,

$$x_0(t) = q_0 \cos(\omega t) + \frac{\dot{q}_0}{\omega} \sin(\omega t) . \tag{14.59}$$

The first correction is,

$$x_1(t) = -\frac{\omega t}{2} q_0 \sin(\omega t) + \frac{t}{2}\dot{q}_0 \cos(\omega t) - \frac{1}{2\omega}\dot{q}_0 \sin(\omega t) , \tag{14.60}$$

and it is easy to see that the sum of all corrections gives,

$$q_{\text{pert}}(t) = q_0 \cos(k_+ t) + \frac{\dot{q}_0}{k_+} \sin(k_+ t) . \tag{14.61}$$

What is the relation of the perturbative solution (14.61) to the general one (14.46)? The perturbative solution is what results if we change the theory by imposing the constraints,

$$\ddot{q}(t) = -k_+^2 q(t) \quad \Longleftrightarrow \quad P_2 = \frac{m}{2}\left(1-\sqrt{1-4g}\right)Q_1 , \tag{14.62}$$

$$q^{(3)}(t) = -k_+^2 \dot{q}(t) \quad \Longleftrightarrow \quad P_1 = \frac{m}{2}\left(1+\sqrt{1-4g}\right)Q_2 . \tag{14.63}$$

Under these constraints the Hamiltonian becomes,

$$H_{\text{pert}} = \sqrt{1-4g}\left(\frac{m}{2}Q_2^2 + \frac{mk_+^2}{2}Q_1^2\right) , \tag{14.64}$$

which is indeed that of a single harmonic oscillator. From the full theory, perturbation theory has retained only the solution whose frequency is well behaved for $g \to 0$,

$$k_+ = \omega\left(1 + \frac{1}{2}g + \frac{7}{8}g^2 + O(g^3)\right) . \tag{14.65}$$

It has discarded the solution whose frequency blows up as $g \to 0$,

$$k_- = \frac{\omega}{\sqrt{g}}\left(1 - \frac{1}{2}g - \frac{5}{8}g^2 + O(g^3)\right) . \tag{14.66}$$

So what's wrong with this? In fact there is nothing wrong with the procedure for our model. If the constraints (14.62–14.63) are imposed at one instant, they remain valid for all times as a consequence of the full equation of motion. However, that is only because our model is free of interactions. Recall that this same feature means the positive and negative energy degrees of freedom exist in isolation of one another, and there is no decay to arbitrarily high excitation as there would be for an interacting, continuum field theory.

When interactions are present it is more involved but still possible to impose constraints which change the theory so that only the lower derivative, perturbative solutions remain. The procedure was first worked out by Jaén, Llosa and Molina [19], and later, independently, by Eliezer and me [20]. To understand its critical defect suppose we change the "interaction" of our higher derivative oscillator from a quadratic term to a cubic one,

$$-\frac{gm}{2\omega^4}\ddot{q}^2 \longrightarrow -\frac{gm}{6\ell\omega^4}\ddot{q}^3 \;. \tag{14.67}$$

Here ℓ is some constant with the dimensions of a length. As with the quadratic interaction, the new equation of motion is fourth order,

$$-m\left[\frac{d^2}{dt^2}\left(\frac{g\ddot{q}^2}{2\ell\omega^4}\right) + \ddot{q} + \omega^2 q\right] = 0 \;, \tag{14.68}$$

Its general solution depends upon four pieces of initial value data. However, by isolating the highest derivative term of the free theory,

$$\ddot{q} = -\omega^2 q - \frac{d^2}{dt^2}\left(\frac{g\ddot{q}^2}{2\ell\omega^4}\right) , \tag{14.69}$$

and then iteratively substituting (14.69), we can delay the appearance of higher derivatives on the right hand side to any desired order in the coupling constant g. For example, two iterations frees the right hand side of higher derivatives up to order g^2,

$$\ddot{q} = -\omega^2 q - \frac{d^2}{dt^2}\left\{\frac{g}{2\ell\omega^4}\left[-\omega^2 q - \frac{d^2}{dt^2}\left(\frac{g\ddot{q}^2}{2\ell\omega^4}\right)\right]^2\right\} , \tag{14.70}$$

$$= -\omega^2 q + \frac{g}{\ell}\left(\omega^2 q^2 - \dot{q}^2\right) + \frac{g^2}{2\ell^2\omega^4}\, q\frac{d^2}{dt^2}\left(\ddot{q}^2\right)$$

$$-\frac{g^2}{2\ell^2\omega^6}\frac{d^2}{dt^2}\left[q\frac{d^2}{dt^2}\left(\ddot{q}^2\right)\right] - \frac{g^3}{8\ell^3\omega^{12}}\frac{d^2}{dt^2}\left[\frac{d^2}{dt^2}\left(\ddot{q}^2\right)\right]^2 \;. \tag{14.71}$$

This obviously becomes complicated fast! However, the lower derivative terms at order g^2 are simple enough to give if I don't worry about the higher derivative remainder,

$$\ddot{q} = -\omega^2 q + \frac{g}{\ell}\left(\omega^2 q^2 - \dot{q}^2\right) + \frac{g^2}{\ell^2}\left(-6\omega^2 q^3 + 14 q\dot{q}^2\right) + O(g^3) \;. \tag{14.72}$$

If we carry this out to infinite order, *and drop the infinite derivative remainder*, the result is an equation of the traditional form,

$$\ddot{q} = f(q, \dot{q}) \;. \tag{14.73}$$

The canonical version of this equation gives the first of the desired constraints. The second is obtained from the canonical version of its time derivative.

The constrained system we have just described is consistent on the perturbative level, but not beyond. It does not follow from the original, exact equation. That would be no problem if we could define physics using perturbation theory, but we cannot. Perturbation theory does not converge for any known interacting, continuum field theory in 3+1 dimensions! The fact that the constraints are not consistent beyond perturbation theory means there is a nonperturbative amplitude for the system to decay to the arbitrarily high excitation in the manner described in Sect. 14.3.1. The fact that the constraints treat time derivatives differently than space derivatives also typically leads to a loss of causality and Lorentz invariance beyond perturbation theory.

A final comment concerns the limit of small coupling constant, i.e. $g \to 0$. One can see from the frequencies (14.65–14.66) of our higher derivative oscillator that the negative energy frequency diverges for $g \to 0$. Disingenuous purveyors of higher derivative models sometimes appeal to people's experience with *positive energy* modes by arguing that, "the k_- mode approaches infinite frequency for small coupling so it must drop out." That is false! The argument is quite correct for an infinite frequency *positive* energy mode in a stable theory. In that case exciting the mode costs an infinite amount of energy which would have to be drawn from de-exciting finite frequency modes. However, a *negative* energy mode doesn't decouple as its frequency diverges. Rather it couples *more strongly* because taking its frequency to infinity opens up more and more ways to balance its negative energy by exciting finite frequency, positive energy modes.

14.3.3 Quantization

People sometimes imagine that quantization might stabilize a system against the Ostrogradskian instability the same way that it does for the Hydrogen atom coupled to electromagnetism. This is a failure to understand correspondence limits. Conclusions drawn from classical physics survive quantization unless they depend upon the system either being completely excluded from some region of the canonical phase space or else inhabiting only a small region of it. For example, the classical instability of the Hydrogen atom (when coupled to electromagnetism) derives from the fact that the purely Hydrogenic part of the energy,

$$E_{\text{Hyd}} = \frac{\|\mathbf{p}\|^2}{2m} - \frac{e^2}{\|\mathbf{x}\|} \ . \tag{14.74}$$

can be made arbitrarily negative by placing the electron close to the nucleus at fixed momentum. Because this instability depends upon the system being in a very small region of the canonical phase space, one might doubt that it survives quantization, and explicit computation shows that it does not.

In contrast, the Ostrogradskian instability derives from the fact that $P_1 Q_2$ can be made arbitrarily negative by taking P_1 either very negative, for positive Q_2, or else very positive, for negative Q_2. *This covers essentially half the classical phase space!* Further, the variables Q_2 and P_1 commute with one another in Ostrogradskian quantum mechanics. So there is no reason to expect that the Ostrogradskian instability is unaffected by quantization.

14.3.4 Unitarity vs. Instability

Particle physicists who quantize higher derivative theories don't typically recognize a problem with the stability. They maintain that the problem with higher derivatives is a breakdown of unitarity. In this sub-section I will again have recourse to the higher derivative oscillator (14.44) to explain the connection between the two apparently unrelated problems.

Let us find the "empty" state wavefunction, $\Omega(Q_1, Q_2)$ that has the minimum excitation in both the positive and negative energy degrees of freedom. The procedure for doing this is simple: first identify the positive and negative energy lowering operators $\alpha_\pm$ and then solve the equations,

$$\alpha_+|\Omega\rangle = 0 = \alpha_-|\Omega\rangle . \tag{14.75}$$

We can recognize the raising and lowering operators by simply expressing the general solution (14.46) in terms of exponentials,

$$q(t) = \frac{1}{2}(A_+ + iB_+)e^{-ik_+t} + \frac{1}{2}(A_+ - iB_+)e^{ik_+t}$$
$$+ \frac{1}{2}(A_- + iB_-)e^{-ik_-t} + \frac{1}{2}(A_- - iB_-)e^{ik_-t} . \tag{14.76}$$

Recall that the k_+ mode carries positive energy, so its lowering operator must be proportional to the e^{-ik_+t} term,

$$\alpha_+ \sim A_+ + iB_+ , \tag{14.77}$$
$$\sim \frac{mk_+}{2}\left(1 + \sqrt{1-4g}\right)Q_1 + iP_1 - k_+P_2 - \frac{im}{2}\left(1 - \sqrt{1-4g}\right)Q_2 \tag{14.78}$$

The k_- mode carries negative energy, so its lowering operator must be proportional to the e^{+ik_-t} term,

$$\alpha_- \sim A_- - iB_- , \tag{14.79}$$
$$\sim \frac{mk_-}{2}\left(1 - \sqrt{1-4g}\right)Q_1 - iP_1 - k_-P_2 + \frac{im}{2}\left(1 + \sqrt{1-4g}\right)Q_2 \tag{14.80}$$

Writing $P_i = -i\frac{\partial}{\partial Q_i}$ we see that the unique solution to (14.75) has the form,

$$\Omega(Q_1, Q_2) = N\exp\left[-\frac{m\sqrt{1-4g}}{2(k_+ + k_-)}\left(k_+k_-Q_1^2 + Q_2^2\right) - i\sqrt{g}mQ_1Q_2\right] . \tag{14.81}$$

The empty wave function (14.81) is obviously normalizable, so it gives a state of the quantum system. We can build a complete set of normalized stationary states by acting arbitrary numbers of $+$ and $-$ raising operators on it,

$$|N_+, N_-\rangle \equiv \frac{(\alpha_+^\dagger)_+^N}{\sqrt{N_+!}} \frac{(\alpha_-^\dagger)_-^N}{\sqrt{N_-!}} |\Omega\rangle .$$
(14.82)

On this space of states the Hamiltonian operator is unbounded below, just as in the classical theory,

$$H|N_+, N_-\rangle = \left(N_+ k_+ - N_- k_-\right)|N_+, N_-\rangle .$$
(14.83)

This is the correct way to quantize a higher derivative theory. One evidence of this fact is that classical negative energy states correspond to quantum negative energy states as well.

Particle physicists don't quantize higher derivative theories as we just have. What they do instead is to regard the negative energy lowering operator as a positive energy raising operator. So they define a "ground state" $|\overline{\Omega}\rangle$ which obeys the equations,

$$\alpha_+|\overline{\Omega}\rangle = 0 = \alpha_-^\dagger|\overline{\Omega}\rangle .$$
(14.84)

The unique wave function which solves these equations is,

$$\overline{\Omega}(Q_1, Q_2) = N \exp\left[-\frac{m\sqrt{1-4g}}{2(k_- - k_+)}\left(k_+ k_- Q_1^2 - Q_2^2\right) + i\sqrt{g}mQ_1Q_2\right] .$$
(14.85)

This wave function is *not* normalizable, so it doesn't correspond to a state of the quantum system. At this stage we should properly call a halt to the analysis because we aren't doing quantum mechanics anymore. The Schrodinger equation $H\psi(Q) = E\psi(Q)$ is just a second order differential equation. It has two linearly independent solutions *for every* energy E: positive, negative, real, imaginary, quaternionic — it doesn't matter. The thing that puts the "quantum" in quantum mechanics is requiring that the solution be normalizable. Many peculiar things can happen if we abandon allow normalizability [21, 22].

However, my particle theory colleagues ignore this little problem and define a completely formal "space of states" based upon $|\overline{\Omega}\rangle$,

$$|\overline{N_+, N_-}\rangle \equiv \frac{(\alpha_+^\dagger)^{N_+}}{\sqrt{N_+!}} \frac{(\alpha_-)^{N_-}}{\sqrt{N_-!}} |\overline{\Omega}\rangle .$$
(14.86)

None of these wavefunctions is any more normalizable than $\overline{\Omega}(Q_1, Q_2)$, so not a one of them corresponds to a state of the quantum system. However, they are all positive energy eigenfunctions,

$$H|\overline{N_+, N_-}\rangle = \left(N_+ k_+ + N_- k_-\right)|\overline{N_+, N_-}\rangle .$$
(14.87)

My particle physics colleagues typically say they *define* $|\overline{\Omega}\rangle$ to have unit norm. Because they have not changed the commutation relations,

$$[\alpha_+, \alpha_+^\dagger] = 1 = [\alpha_-, \alpha_-^\dagger] \,, \tag{14.88}$$

the norm of any state with odd N_- is negative! The lowest of these is,

$$\langle \overline{0,1} | \overline{0,1} \rangle = \langle \overline{\Omega} | \alpha_-^\dagger \alpha_- | \overline{\Omega} \rangle = -\langle \overline{\Omega} | \overline{\Omega} \rangle \,. \tag{14.89}$$

As I pointed out above, the reason this has happened is that we aren't doing quantum mechanics any more. We ought to use the normalizable, but indefinite energy eigenstates. What particle physicists do instead is to reason that because the probabilistic interpretation of quantum mechanics requires norms to be positive, the negative norm states must be excised from the space of states. At this stage good particle physicists note that that the resulting model fails to conserve probability [23]. Just as the correctly-quantized, indefinite-energy theory allows processes which mix positive and negative energy particles, so too the indefinite-norm theory allows processes which mix positive and negative norm particles. It only conserves probability on the space of "states" which includes both kinds of norms. If we excise the negative norm states then probability is no longer conserved.

So good particle physicists reach the correct conclusion — that nondegenerate higher derivative theories can't describe our universe — by a somewhat illegitimate line of reasoning. But who cares? They got the right answer! Of course *bad* particle physicists regard the breakdown of unitarity as a challenge for inspired tinkering to avoid the problem. Favorite ploys are the Lee-Wick reformulation of quantum field theory [24] and nonperturbative resummations. The analysis also typically involves the false notion that high frequency ghosts decouple, which I debunked at the end of Sect. 14.3.2. When the final effort is written up and presented to the world, some long-suffering higher derivative expert gets called away from his research to puzzle out what was done and explain why it isn't correct. *Sigh.* The problem is so much clearer in its negative energy incarnation! I could list many examples at this point, but I will confine myself to citing a full-blown paper debunking one of them [25]. It is also appropriate to note that Hawking and Hertog have previously called attention to the mistake of quantizing higher derivative theories using nonnormalizable wave functions [26].

14.3.5 Constraints

The only way anyone has ever found to avoid the Ostrogradskian instability on a nonperturbative level is by violating the single assumption needed to make Ostrogradski's construction: nondegeneracy. Higher derivative theories for which the definition of the highest conjugate momentum (14.40) cannot be inverted to solve for the highest derivative can sometimes be stable. An interesting example of this kind is the rigid, relativistic particle studied by Plyushchay [27, 28].

Degeneracy is of great importance because *all theories which possess continuous symmetries are degenerate*, irrespective of whether or not they possess higher derivatives. A familiar example is the relativistic point particle, whose dynamical variable is $X^\mu(\tau)$ and whose Lagrangian is,

$$L = -m\sqrt{-\eta_{\mu\nu}\dot{X}^\mu\dot{X}^\nu} \,. \tag{14.90}$$

The conjugate momentum is,

$$P_\mu \equiv \frac{m\dot{X}_\mu}{\sqrt{-\dot{X}^2}} \,. \tag{14.91}$$

Because the right hand side of this equation is homogeneous of degree zero one can not solve for $\dot{X}^\mu$. The associated continuous symmetry is invariance under reparameterizations $\tau \to \tau'(\tau)$,

$$X^\mu(\tau) \longrightarrow X'^\mu(\tau) \equiv X^\mu\left(\tau'^{-1}(\tau)\right) \,. \tag{14.92}$$

The cure for symmetry-induced degeneracy is simply to fix the symmetry by imposing gauge conditions. Then the gauge-fixed Lagrangian should no longer be degenerate in terms of the remaining variables. For example, we might parameterize so that $\tau = X^0(\tau)$, in which case the gauge-fixed particle Lagrangian is,

$$L_{GF} = -m\sqrt{1 - \dot{\boldsymbol{X}} \cdot \dot{\boldsymbol{X}}} \,. \tag{14.93}$$

In this gauge the relation for the momenta is simple to invert,

$$P_i \equiv \frac{m\dot{X}_i}{\sqrt{1 - \dot{\boldsymbol{X}} \cdot \dot{\boldsymbol{X}}}} \quad\Longleftrightarrow\quad \dot{X}^i = \frac{P^i}{\sqrt{m^2 + \boldsymbol{P} \cdot \boldsymbol{P}}} \,. \tag{14.94}$$

When a continuous symmetry is used to eliminate a dynamical variable, the equation of motion of this variable typically becomes a *constraint*. For symmetries enforced by means of a compensating field — such as local Lorentz invariance is with the antisymmetric components of the vierbein [29] — the associated constraints are tautologies of the form $0 = 0$. Sometimes the constraints are nontrivial, but implied by the equations of motion. An example of this kind is the relativistic particle in our synchronous gauge. The equation of the gauge-fixed zero-component just tells us the Hamiltonian is conserved,

$$\frac{d}{d\tau}\left(\frac{m\dot{X}_0}{\sqrt{-\eta_{\mu\nu}\dot{X}^\mu\dot{X}^\nu}}\right) = 0 \longrightarrow \frac{d}{dt}\left(\sqrt{m^2 + \boldsymbol{p} \cdot \boldsymbol{p}}\right) = 0 \,. \tag{14.95}$$

And sometimes the constraints give nontrivial relations between the canonical variables that generate residual, time-independent symmetries. In this case another degree of freedom can be removed ("gauge fixing counts twice," as van Nieuwenhuizen puts it). An example of this kind of constraint is Gauss' Law in temporal gauge electrodynamics.

Were it not for constraints of this last type, the analysis of a higher derivative theory with a gauge symmetry would be straightforward. One would simply fix the gauge and then check whether or not the gauge-fixed Lagrangian depends nondegenerately upon higher time derivatives. If it did, the conclusion would be that the theory suffers the Ostrogradskian instability. However, when constraints of the third type are present one must check whether or not they affect the instability. This is highly model dependent but a very simple rule seems to be generally applicable: *if the number of gauge constraints is less than the number of unstable directions in the canonical phase space then there is no chance for avoiding the problem.* Because the number of constraints for any symmetry is fixed, whereas the number of unstable directions increases with the number of higher derivatives, one consequence is that gauge constraints can at best avoid instability for some fixed number of higher derivatives. For example, the constraints of the second derivative model of Plyushchay are sufficient to stabilize the system [27, 28], but one would expect it to become unstable if third derivatives were added.

People sometimes make the mistake of believing that the Ostrogradskian instability can be avoided with just a single, global constraint on the Hamiltonian. For example, Boulware, Horowitz and Strominger [30] showed the energy is zero for any asymptotically flat solution of the higher derivative field equations derived from the Lagrangian,

$$\mathcal{L} = \alpha R^2 \sqrt{-g} + \beta R^{\mu\nu} R_{\mu\nu} \sqrt{-g} . \tag{14.96}$$

As I explained in Sect. 14.3.1, the nature of the Ostrogradskian instability is not that the energy decays but rather that the system evaporates to a very highly excited state of compensating, positive and negative energy degrees of freedom. As long as $\beta \neq 0$, there are six independent, higher derivative momenta at each space point, whereas there are only four local constants — or five if α and β are such as to give local conformal invariance. Hence there are two (or one) unconstrained instabilities per space point. There are an infinite number of space points, so the addition of a single, global constraint does not change anything. I should point out that Boulware, Horowitz and Strominger were aware of this, cf. their discussion of the dipole instability.

The case of $\beta = 0$ is special, and significant for the next section. If α has the right sign that model has long been known to have positive energy [31, 32]. This result in no way contradicts the previous analysis. When $\beta = 0$ the terms which carry second derivatives are contracted in such way that only a single component of the metric carries higher derivatives. So now the counting is *one* unstable direction per space point versus four local constraints. Hence the constraints can win, and they do if α has the right sign.

14.3.6 Nonlocality

I would like to close this section by commenting on the implications of Ostrogradski's theorem for fully nonlocal theories. In addition to nonlocal quantum field theories [33, 34, 35] this is relevant to string field theory [36, 37, 38], to noncommutative geometry [39, 40], to regularization techniques [41, 42, 43] and even to theories of cosmology [15, 44, 45]. The issue in each case is whether or not we can think of the fully nonlocal theory as the limit of a sequence of ever higher derivative theories. When such a representation is possible the nonlocal theory must inherit the Ostrogradskian instability.

The higher derivative representation is certainly valid for string field theory because, otherwise, there would be cuts and poles that would interfere with perturbative unitarity. So string field theory suffers from the Ostrogradskian instability [20]. The same is true for theories where the nonlocality is of limited extent in time [46], although not everyone agrees [47, 48]. However, when the nonlocality involves inverse differential operators there need be no problem [20, 44]. Indeed, the effective action of any quantum field theory is nonlocal in this way [49, 50]! Nor is there necessarily any problem when the nonlocality arises in the form of algebraic functions of local actions [51].

14.4 $\Delta R[g] = f(R)$ Theories

From the lengthy argumentation of the previous two sections one might conclude that the only potentially stable, local modification of gravity is a cosmological constant, $\Delta R[g] = -2\Lambda$. However, a close analysis of Sect. 14.3.5 reveals that it is also possible to consider algebraic functions of the Ricci scalar. In this section I first explain why such theories can avoid the Ostrogradskian instability. I then demonstrate that they are equivalent to general relativity with a minimally coupled scalar, provided we ignore matter. Finally, I exploit this equivalence, with the construction described in the Introduction, to show how $f(R)$ can be chosen to enforce any evolution $a(t)$.

14.4.1 Why They Can Be Stable

The alert reader will have noted that the $R + R^2$ model [31, 32] avoids the Ostrogradskian instability. It does this by violating Ostrogradski's assumption of nondegeneracy: the tensor indices of the second derivative terms in the Ricci scalar are contracted together so that only a single component of the metric carries higher derivatives. This component does acquire a new, higher derivative degree of freedom, and the energy of this degree of freedom is indeed opposite to that of the corresponding lower derivative degree of freedom, just as required by Ostrogradski's analysis. However, that lower derivative degree of freedom is the *Newtonian potential*. It carries negative energy, but it is also completely fixed in terms of the other metric and matter fields by the g_{00} constraint. So the only instability associated with it is gravitational collapse. Its higher derivative counterpart has positive energy, at least on the kinetic level; it can still have a bad potential, and the model is indeed only stable for one sign of the R^2 term.

None of these features depended especially upon the higher derivative term being R^2. Any function for the Ricci scalar would work as well. Note that we cannot allow derivatives of the Ricci scalar, because Ostrogradski's theorem says the next higher derivative degree of freedom would carry negative energy and there would be no additional constraints to protect it. We also cannot permit more general contractions of the Riemann tensor because then other components of the metric would carry higher derivatives. These components are positive energy in general relativity, so their higher derivative counterparts would be negative, and there would again be no additional constraints to protect the theory against instability.

14.4.2 Equivalent Scalar Representation

The general Lagrangian we wish to consider takes the form,

$$\mathcal{L} = \frac{1}{16\pi G}\Big(R + f(R)\Big)\sqrt{-g}\,. \tag{14.97}$$

If we ignore the coupling to matter the modified gravitational field equation consists of the vanishing of the following tensor,

$$\frac{16\pi G}{\sqrt{-g}}\frac{\delta S}{\delta g^{\mu\nu}} = [1+f'(R)]R_{\mu\nu} - \frac{1}{2}[R+f(R)]g_{\mu\nu} + g_{\mu\nu}[f'(R)]^{;\rho}_{\rho} - [f'(R)]_{;\mu\nu}\,. \tag{14.98}$$

There is an old procedure for reformulating this as general relativity with a minimally coupled scalar. I don't know whom to credit, but I will give the construction.

The first step is to define an "equivalent" theory with an auxiliary field ϕ which is defined by the relation.

$$\phi \equiv 1 + f'(R) \qquad \Longleftrightarrow \qquad R = \mathcal{R}(\phi)\,. \tag{14.99}$$

Inverting the relation determines the Ricci scalar as an algebraic function of ϕ. We can then define an auxiliary potential for ϕ by Legendre transformation,

$$U(\phi) \equiv \Big(\phi-1\Big)\mathcal{R}(\phi) - f\Big(\mathcal{R}(\phi)\Big) \qquad \Longrightarrow \qquad U'(\phi) = \mathcal{R}(\phi)\,. \tag{14.100}$$

Now consider the equivalent scalar-tensor theory whose Lagrangian is,

$$\mathcal{L}_{\mathrm{E}} \equiv \frac{1}{16\pi G}\Big(\phi R - U(\phi)\Big)\sqrt{-g}\,. \tag{14.101}$$

Its field equations are,

$$\frac{16\pi G}{\sqrt{-g}}\frac{\delta S_{\mathrm{E}}}{\delta\phi} = R - U'(\phi) = 0\,, \tag{14.102}$$

$$\frac{16\pi G}{\sqrt{-g}}\frac{\delta S_{\mathrm{E}}}{\delta g^{\mu\nu}} = \phi R_{\mu\nu} - \frac{1}{2}\Big(\phi R - U(\phi)\Big)g_{\mu\nu} + g_{\mu\nu}\phi^{;\rho}_{\rho} - \phi_{\mu\nu} = 0\,. \tag{14.103}$$

The scalar (14.102) implies $\phi = 1 + f'(R)$, whereupon the tensor (14.103) reproduce the original modified gravity (14.98).

The final step is to define a new metric $\tilde{g}_{\mu\nu}$ and a new scalar φ by the change of variables,

$$\tilde{g}_{\mu\nu} \equiv \phi\, g_{\mu\nu} \qquad \Longleftrightarrow \qquad g_{\mu\nu} = \exp\left[-\sqrt{\frac{4\pi G}{3}}\,\varphi\right]\tilde{g}_{\mu\nu}\,, \quad (14.104)$$

$$\varphi \equiv \sqrt{\frac{3}{4\pi G}}\,\ln(\phi) \qquad \Longleftrightarrow \qquad \phi = \exp\left[\sqrt{\frac{4\pi G}{3}}\,\varphi\right]. \qquad (14.105)$$

In terms of these variables the equivalent Lagrangian takes the form,

$$\mathcal{L}_E = \frac{1}{16\pi G}\tilde{R}\sqrt{-\tilde{g}} - \frac{1}{2}\partial_\mu\varphi\,\partial_\nu\varphi\,\tilde{g}^{\mu\nu}\sqrt{-\tilde{g}} - V(\varphi)\sqrt{-\tilde{g}}\,, \qquad (14.106)$$

where the scalar potential is,

$$V(\varphi) \equiv \frac{1}{16\pi G}U\left(\exp\left[\sqrt{\frac{4\pi G}{3}}\,\varphi\right]\right)\exp\left[-\sqrt{\frac{16\pi G}{3}}\,\varphi\right]. \qquad (14.107)$$

This is general relativity with a minimally coupled scalar, as claimed.

14.4.3 Reconstructing $f(R)$ from Cosmology

I want to show how to choose $f(R)$ to support an arbitrary $a(t)$.[3] Recall from the Introduction that one can choose the potential of a quintessence model such as (14.106) to support any homogeneous and isotropic cosmology for its metric $\tilde{g}_{\mu\nu}$. However, we cannot immediately exploit this construction because it is the metric $g_{\mu\nu}$ which is assumed known, not $\tilde{g}_{\mu\nu}$. We must explain how to infer the one from the other without knowing $f(R)$.

Because the relation (14.104) between $g_{\mu\nu}$ and $\tilde{g}_{\mu\nu}$ is a conformal transformation, it makes sense to work in a coordinate system in which each metric is conformal to flat space. This is accomplished by changing from co-moving time t to conformal time η though the relation, $d\eta = dt/a(t)$,

$$ds^2 = -dt^2 + a^2(t)d\boldsymbol{x}\cdot d\boldsymbol{x} = a^2\left(-d\eta^2 + d\boldsymbol{x}\cdot d\boldsymbol{x}\right). \qquad (14.108)$$

The $\tilde{g}_{\mu\nu}$ element takes the same form in conformal coordinates, but note that its different scale factor implies a different co-moving time,

$$d\tilde{s}^2 = \tilde{a}^2\left(-d\eta^2 + d\boldsymbol{x}\cdot d\boldsymbol{x}\right) = -d\tilde{t}^{\,2} + \tilde{a}^2(\tilde{t})d\boldsymbol{x}\cdot d\boldsymbol{x}\,. \qquad (14.109)$$

From relation (14.104) we infer,

---

[3] For a somewhat different construction which achieves the same end, see [17, 52].

$$a(t) = \widetilde{a}(\widetilde{t}) \exp\left[-\sqrt{\frac{\pi G}{3}} \, \varphi_0(\widetilde{t})\right] . \tag{14.110}$$

We denote differentiation with respect to η by a prime, and one should note the relation between derivatives with respect to the various times,

$$\frac{\partial}{\partial \eta} = a\frac{\partial}{\partial t} = \widetilde{a}\frac{\partial}{\partial \widetilde{t}} . \tag{14.111}$$

Differentiating the logarithm of (14.110) with respect to η and using the relation (14.8) between $\widetilde{a}$ and φ_0 gives,

$$\frac{a'}{a} = \frac{\widetilde{a}'}{\widetilde{a}} - \sqrt{\frac{\pi G}{3}} \, \varphi_0' = \frac{\widetilde{a}'}{\widetilde{a}} - \sqrt{-\frac{1}{12}\widetilde{a}'} . \tag{14.112}$$

This is a nonlinear but first order differential equation for the variable $\widetilde{a}$ in terms of the known function, $a(t(\eta))$. At the worst it can be solved numerically.

Once we have $\widetilde{a}$ the potential $V(\varphi)$ can be constructed using the procedure explained in the Introduction. We then compute the auxiliary potential,

$$U(\phi) = 16\pi G\phi^2 V\left(\sqrt{\frac{3}{4\pi G}} \ln(\phi)\right) . \tag{14.113}$$

The auxiliary field can be expressed in terms of the Ricci scalar from the algebraic relation,

$$U'(\phi) = R \qquad \Longleftrightarrow \qquad \phi = \Phi(R) . \tag{14.114}$$

And we finally recover the function $f(R)$ by Legrendre transformation,

$$f(R) = \Big(\Phi(R)-1\Big)R - U\Big(\Phi(R)\Big) . \tag{14.115}$$

14.5 Problems with $f(R) = -\frac{\mu^4}{R}$

In view of the construction of Sect. 14.4.3 it is not surprising but rather *inevitable* that an $f(R)$ can be found to support late time acceleration, or indeed, any other evolution. However, the method is not guaranteed to produce a simple model, so the discovery that $f(R) = -\mu^4/R$ works is quite noteworthy [53, 54].[4] It may also be significant that models of this type seem to follow from fundamental theory [56].

To derive acceleration in this model consider its field equations,

$$\left(1+\frac{\mu^4}{R^2}\right)R_{\mu\nu} - \frac{1}{2}\left(1-\frac{\mu^4}{R^2}\right)Rg_{\mu\nu} + \left(g_{\mu\nu}\Box - D_\mu D_\nu\right)\frac{\mu^4}{R^2} = 8\pi GT_{\mu\nu} . \tag{14.116}$$

---

[4] Although extensions involving $R^{\mu\nu}R_{\mu\nu}$ and $R^{\rho\sigma\mu\nu}R_{\rho\sigma\mu\nu}$ have also been studied [55], they must be ruled out on account of the Ostrogradskian instability.

Setting $T_{\mu\nu}=0$ and searching for constant Ricci scalar solutions gives,

$$\left(1+\frac{\mu^4}{R^2}\right)R_{\mu\nu} - \frac{1}{2}\left(1-\frac{\mu^4}{R^2}\right)Rg_{\mu\nu} = 0 \quad \Longleftrightarrow \quad R_{\mu\nu} = \pm\frac{\sqrt{3}}{4}\mu^2 g_{\mu\nu} \ . \tag{14.117}$$

The plus sign corresponds to acceleration

In addition to proposing the model, Carroll, Duvvuri, Trodden and Turner [53] also showed that it suffers from a very weak tachyonic instability in the absence of matter. Because the only new higher derivative degree of freedom resides in the Ricci scalar, we may as well derive an equation for it alone from the trace of (14.116),

$$-R + \frac{3\mu^4}{R} + \Box\left(\frac{3\mu^4}{R^2}\right) = 0 \ . \tag{14.118}$$

Now perturb about the accelerated solution,

$$R = +\sqrt{3}\mu^2 + \delta R \quad \Longrightarrow \quad -2\delta R - \frac{2}{\sqrt{3}\mu^2}\Box\delta R + O(\delta R^2) = 0 \ . \tag{14.119}$$

By comparing the linearized equation for δR with that of a positive mass-squared scalar,

$$(\Box - m^2)\varphi = 0 \ , \tag{14.120}$$

we see that δR behaves like a tachyon with $m^2 = -\sqrt{3}\mu^2$. However, because explaining the current phase of acceleration requires $\mu \sim 10^{-33}$ eV, the resulting instability is not very serious. I should note that the existence of a tachyonic instability in no way contradicts the Ostrogradskian analysis that this model's higher derivative degree of freedom carries positive kinetic energy.

14.5.1 Inside Matter

Dolgov and Kawasaki [57] showed that a radically different result emerges when this model is considered inside a static distribution of matter,

$$T_{\mu\nu} = \rho\delta^0_\mu\delta^0_\nu \quad \text{with} \quad 8\pi G\rho \equiv M^2 \gg \mu^2 \ . \tag{14.121}$$

In that case the trace of (14.116) gives,

$$-R + \frac{3\mu^4}{R} + \Box\left(\frac{3\mu^4}{R^2}\right) = -M^2 \ . \tag{14.122}$$

As might be expected, the static Ricci scalar solution in this case is dominated by M rather than μ,

$$R_0 = \frac{1}{2}\left(M^2 + \sqrt{M^4 + 12\mu^4}\right) \simeq M^2 \ . \tag{14.123}$$

Perturbing about this solution gives,

$$R = R_0 + \delta R \quad \Longrightarrow \quad -\delta R - \frac{3\mu^4}{R_0^2}\delta R - \frac{6\mu^4}{R_0^3}\Box\delta R + O(\delta R^2) = 0 . \quad (14.124)$$

Comparing with the reference scalar (14.120) now reveals an enormous tachyonic mass,

$$m^2 = -\frac{R_0}{2} - \frac{R_0^3}{6\mu^4} \simeq -\frac{M^6}{6\mu^4}! \quad (14.125)$$

Plugging in the numbers for the density of water ($\rho \sim 10^3$ kg/m^3) gives $M \sim 10^{-18}$ eV, implying a tachyonic mass of magnitude $|m| \sim 10^{12}$ eV $= 10^3$ GeV!

As disastrous as this problem might seem, Dick [58] and Nojiri and Odintsov [59] have shown that it can be avoided by changing the model slightly,

$$f(R) = -\frac{\mu^4}{R} + \frac{\alpha}{2\mu^2}R^2 \quad \Longrightarrow \quad -R + \frac{3\mu^4}{R} + 3\Box\left(\frac{\mu^4}{R^2} + \frac{\alpha}{\mu^2}R\right) = 0 . \quad (14.126)$$

Because an R^2 term has global conformal invariance, it makes no contribution to the trace for constant R. Hence the cosmological solution of $R = +\sqrt{3}\mu^2$ is not affected, nor is the static solution inside the matter distribution (14.121). However, the equation for linearized perturbations inside matter changes to,

$$-\delta R - \frac{3\mu^4}{R_0^2}\delta R + 3\left(-\frac{2\mu^4}{R_0^3} + \frac{\alpha}{\mu^2}\right)\Box\delta R = 0 . \quad (14.127)$$

The instability of Dolgov and Kawasaki was driven by the smallness of $2\mu^4/R_0^3$. By simply taking α positive and of order one the tachyon becomes a positive mass-squared particle of $m^2 \sim \mu^2/\alpha$.

14.5.2 Outside Matter

Marc Soussa and I analyzed force of gravity outside a matter distribution [60]. Although our analysis was for the original $f(R) = -\mu^4/R$ model, there would be only slight differences for the extended model (14.126). So our result seems to foreclose this possibility, but see [61].

The tachyonic instability could be studied using the perturbed Ricci scalar, but the gravitational force requires use of the metric. We perturbed about the de Sitter solution with Hubble constant $H = \mu/(48)^{\frac{1}{4}}$ in co-moving coordinates,

$$ds^2 = -(1-h_{00})dt^2 + 2a(t)h_{0i}dtdx^i + a^2(t)(\delta_{ij}+h_{ij})dx^idx^j \quad \text{with} \quad a(t) = e^{Ht} . \quad (14.128)$$

In the gauge,

$$h_{\mu\nu}{}^{,\nu} - \frac{1}{2}h_\mu + 3h_\mu{}^\nu[\ln(a)]_{,\nu} = 0 , \quad (14.129)$$

with $h \equiv -h_{00} + h_{ii}$, the perturbed Ricci scalar takes the form,

$$\delta R = -\frac{1}{2}\partial^2 h + 2H\partial_0 h \,. \tag{14.130}$$

Our strategy was first to solve the de Sitter invariant equation for the perturbed Ricci scalar, then reconstruct the gauge-fixed metric.

We assumed a matter density of the form,

$$\rho(t, \boldsymbol{x}) = \frac{3M}{4\pi R_g^3}\theta\left(R_g - a(t)|\boldsymbol{x}|\right) \,. \tag{14.131}$$

The exterior field equation has a simple expression in terms of the coordinate $y \equiv a(t)H|\boldsymbol{x}|$,

$$\left[\left(1-y^2\right)\frac{d^2}{dy^2} + \frac{2}{y}\left(1-2y^2\right)\frac{d}{dy} + 12\right]\delta R = 0 \,. \tag{14.132}$$

The solution takes the form,

$$\delta R = \beta_1 f_0(y) + \beta_2 f_{-1}(y) \,, \tag{14.133}$$

where f_0 and f_{-1} are hypergeometric functions whose series expansions are,

$$f_0(y) = 1 - 2y^2 + \frac{1}{5}y^4 + \cdots \,, \tag{14.134}$$

$$f_{-1}(y) = \frac{1}{y}\left(1 - 7y^2 + \frac{14}{3}y^4 + \cdots\right) \,. \tag{14.135}$$

We only need the behavior for small y because $y = 1$ is the Hubble radius! Matching to the source at $y = HR_g$ determines the combination coefficients to be,

$$\beta_1 \simeq \frac{3GM}{R_g^3} \quad , \quad \beta_2 \simeq -12GMH^3 \,. \tag{14.136}$$

This last step might seem bogus because we needed to regard the mass density as a small perturbation on the cosmological energy density μ^4, whereas the opposite would be the case for galaxies or clusters of galaxies. However, this will only make changes of order one in the β_i's. In particular, the asymptotic solution must still take the form (14.133).

The next step is solving for the trace of the perturbed metric. It turns out that relation (14.130) can also be expressed very simply using the variable y,

$$\left[\left(y^2-1\right)\frac{d}{dy} + \frac{1}{y}\left(5y^2-2\right)\right]h'(y) = \frac{2}{H^2}\delta R \,. \tag{14.137}$$

We only need to solve for the derivative of h because that is what gives the gravitational force in the geodesic equation. The solution is,

$$h'(y) = -\frac{2GM}{H^2 R_g^3} y + O(y^3) .$$ (14.138)

This should be compared to the general relativistic prediction,

$$h'_{\text{GR}}(y) = -\frac{4GMH}{y^2} + O(1) \qquad \Longrightarrow \qquad \frac{h'}{h'_{\text{GR}}} = \frac{1}{2}\Big(\frac{\|\boldsymbol{x}\|}{R_g}\Big)^3 .$$ (14.139)

One consequence is that the force between the Milky Way and Andromeda galaxies would be about a million times larger than predicted by general relativity!

14.6 Conclusions

The potential of a quintessence scalar can be chosen to support any cosmology, but the epicyclic nature of this construction suggests we consider modifications of gravity. Ostrogradski's theorem [18] limits local modifications of gravity to just algebraic functions of the Ricci scalar. Models of this form can give a late phase of cosmic acceleration such as we are currently experiencing. However, they can be tuned to give anything else as well. They seem every bit as epicyclic as scalar quintessence. Further, the $f(r) = -\mu^4/R$ model is problematic, both inside and outside matter sources.[5]

An interesting and largely overlooked possibility for modifying gravity is the fully nonlocal effective action that results from quantum gravitational corrections. In weak field perturbation theory it has long been known that the most cosmologically significant one loop corrections are not of the R^2 form usually studied but rather of the form $R \ln(\Box)R$ [63]. More potentially interesting is the possibility of very strong infrared effects from the epoch of primordial inflation [64, 65].

It can be shown that quantum gravitational corrections to the inflationary expansion rate grow with time like powers of $\ln(a)$. Although suppressed by very small coupling constants, the exponential growth in $a(t)$ during inflation must eventually cause the effect to become nonperturbatively strong [66, 67]. Similar secular growth occurs as well for minimally coupled scalar field theories [68, 69], in which context Starobinskiĭ has developed a technique for summing the leading powers of $\ln(a)$ at each loop order [70, 71]. If Starobinskiĭ's technique can be generalized to quantum gravity [72, 73] it might result in a nonlocal effective gravity theory for late time cosmology in which a large, bare cosmological constant is almost completely screened by a nonperturbative quantum gravitational effect. In such a formalism the current phase of acceleration might result from a very slight mismatch between

---

[5] Observations also rule out the somewhat different version of this model that results from regarding the connection and the metric as independent, fundamental variables in the Palatini formalism [62].

the bare cosmological constant and the quantum effect which screens it. It is even conceivable that one could reproduce the phenomenological successes of MOND [3, 4] with such a nonlocal metric theory, although it would have to unstable against decay into galaxy-scale gravitational waves [74].

Acknowledgements

It is a pleasure to acknowledge conversations and correspondence on this subject with S. Deser, A.D. Dolgov, D.A. Eliezer, S. Odintsov, M.E. Soussa, A. Strominger and M. Trodden. This work was partially supported by NSF grant PHY-244714 and by the Institute for Fundamental Theory at the University of Florida.

References

1. R.H. Cyburt, B.D. Fields and K.A. Olive, Phys. Lett. B **567**, 227 (2003), astro-ph/0302431.
2. G. Bertone, D. Hooper and J. Silk, Phys. Rept. **405**, 279 (2005), hep-ph/0404175.
3. M. Milgrom, Ap. J. **270**, 365 (1983).
4. R.H. Sanders and S.S. McGaugh, Ann. Rev. Astrophys **40**, 263 (2002), astro-ph/0204521.
5. G. Gentile, P. Salucci, U. Klein, D. Vergani and P. Kalberla, Mon. Not. Roy. Astron. Soc. **351**, 903 (2004), astro-ph/0403154.
6. J.D. Bekenstein, Phys. Rev. D **70**, 083509 (2004), astro-ph/0403694.
7. C. Skordis, D.F. Mota, P.G. Ferreira and C. Boehm, Phys. Rev. Lett. **96**, 011301 (2006), astro-ph/0505519.
8. R.A. Knop et al, Astrophys. J. **598**, 102 (2003), astro-ph/0309368.
9. A.D. Reiss et al, Astrophys. J. **607**, 665 (2004), astro-ph/0402512.
10. P. Astier et al, astro-ph/0510447.
11. D. Spergel et al, Astrophys. J. Suppl. **148**, 175 (2003), astro-ph/0302209.
12. M. Tegmark et al, Phys. Rev. D **69**, 103501 (2004), astro-ph/0310723.
13. C. Wetterich, Nucl. Phys. B **302**, 668 (1988).
14. B. Ratra and P.J.E. Peebles, Phys. Rev. D **37**, 3406 (1988).
15. N.C. Tsamis and R.P. Woodard, Ann. Phys. **267**, 145 (1998), hep-th/9712331.
16. T.D. Saini, S. Raychaudhury, V. Saini and A.A. Starobinskiĭ, Phys. Rev. Lett. **85**, 1162 (2000), astro-ph/9910231.
17. S. Capozziello, S. Nojiri and S.D. Odintsov, hep-th/0512118.
18. M. Ostrogradski, Mem. Ac. St. Petersbourg VI **4**, 385 (1850).
19. X. Jaén, J. Llosa and A. Molina, Phys. Rev. D **34**, 2302 (1986).
20. D.A. Eliezer and R.P. Woodard, Nucl. Phys. B **325**, 389 (1989).
21. R.P. Woodard, Class. Quant. Grav. **10**, 483 (1993).
22. N.C. Tsamis and R.P. Woodard, Phys. Rev. D **36**, 3641 (1987).
23. K.S. Stelle, Phys. Rev. D **16**, 953 (1977).
24. T.D. Lee and G.C. Wick, Phys. Rev. D **2**, 1033 (1970).

25. N.C. Tsamis and R.P. Woodard, Ann. Phys. **168**, 457 (1986).
26. S.W. Hawking and T. Hertog, Phys. Rev. D **65**, 103515 (2002), hep-th/0107088.
27. M.S. Plyushchay, Mod. Phys. Lett. A **4**, 837 (1989).
28. D. Zoller, Phys. Rev. Lett. **65**, 2236 (1990).
29. R.P. Woodard, Phys. Lett. B **148**, 440 (1984).
30. D.G. Boulware, G.T. Horowitz and A. Strominger, Phys. Rev. Lett. **50**, 1726 (1983).
31. A.A. Starobinskiĭ, Phys. Lett. B **91**, 99 (1980).
32. A. Strominger, Phys. Rev. D **30**, 2257 (1984).
33. G. Kleppe and R.P. Woodard, Nucl. Phys. B **388**, 81 (1992).
34. T.C. Cheng, P.M. Ho and M.C. Yeh, Nucl. Phys. B **625**, 151 (2002), hep-th/0111160.
35. A. Jain and S.D. Joglekar, Int. J. Mod. Phys. A **19**, 3409 (2004), hep-th/0307208.
36. D.J. Gross and A. Jevicki, Nucl. Phys. B **283**, 1 (1987).
37. D.J. Gross and A. Jevicki, Nucl. Phys. B **287**, 225 (1987).
38. D.J. Gross and A. Jevicki, Nucl. Phys. B **293**, 29 (1987).
39. A. Konechny and A. Schwarz, Phys. Rept. **360**, 353 (2002), hep-th0107251.
40. J.L. Hewett, F.J. Petriello and T.G. Rizzo, Phys. Rev. D **64**, 075012 (2001), hep-ph/0010354.
41. D. Evens, J.W. Moffat, G. Kleppe and R.P. Woodard, Phys. Rev. D **43**, 499 (1991).
42. G. Kleppe and R.P. Woodard, Phys. Lett. B **253**, 331 (1991).
43. G. Kleppe and R.P. Woodard, Ann. Phys. B **221**, 106 (1993).
44. M.E. Soussa and R.P. Woodard, Class. Quant. Grav. **20**, 2737 (2003), astro-ph/0302030.
45. T. Biswas, A. Mazumdar and W. Siegel, JCAP **0603**, 009 (2006), hep-th/0508194.
46. R.P. Woodard, Phys. Rev. A **62**, 052105 (2000), hep-th/0006207.
47. J. Llosa, hep-th/0201087.
48. R.P. Woodard, Phys. Rev. A **67**, 016102 (2003), hep-th/0207191.
49. A.O. Barvinsky, Y.V. Gusev, G.A. Vilkovisky and V.V. Zhytnikov, J. Math. Phys. **35**, 3525 (1994), gr-qc/9404061.
50. A.O. Barvinsky and V.F. Mukhanov, Phys. Rev. D **66**, 065007 (2202), hep-th/0203132.
51. D.L. Bennett, H.B. Nielsen and R.P. Woodard, Phys. Rev. D **57**, 1167 (1998, hep-th/9707088.
52. S. Nojiri and S.D. Odintsov, Int. J. Geom. Meth. Mod. Phys. **4**, 115 (2007), hep-th/0601213.
53. S.M. Carroll, V. Duvvuri, M. Trodden and M.S. Turner, Phys. Rev. D **70**, 043528 (2004), astro-ph/0306438.
54. S. Capozziello, S. Carloni and A. Troisi, Phys. Rev. D **70**, 043528 (2004), astro-ph/0306438.
55. S.M. Carroll, A. De Felice, V. Duvvuri, D.A. Easson, M. Trodden and M.S. Turner, Phys. Rev. D **71**, 063513 (2005), astro-ph/0410031.
56. S. Nojiri and S.D. Odintsov, Phys. Lett. B **576**, 5 (2003), hep-th/0307071.
57. A.D. Dolgov and M. Kawasaki, Phys. Let. B **573**, 1 (2003), astro-ph/0307285.
58. R. Dick, Gen. Rel. Grav. **36**, 217 (2004), gr-qc/0307052.
59. S. Nojiri and S.D. Odintsov, Phys. Rev. D **68**, 123512 (2003), hep-th/0307288.

60. M.E. Soussa and R.P. Woodard, Gen. Rel. Grav. **36**, 855 (2004), astro-ph/0308114.
61. S. Nojiri and S.D. Odintsov, Gen. Rel. Grav. **36**, 1765 (2004), hep-th/0308176.
62. M. Amarzguioui, Ø. Elgaroy, D.F. Mota and T. Multamäki, astro-ph/0510519.
63. D. Espriu, T. Multamäki and E.C. Vagenas, Phys. Lett. B **628**, 197 (2005), gr-qc/0503033.
64. N.C. Tsamis and R.P. Woodard, Ann. Phys. **238**, 1 (1995).
65. P. Martineau and R. Brandenberger, astro-ph/0510523.
66. N.C. Tsamis and R.P. Woodard, Nucl. Phys. B **474**, 235 (1996), hep-ph/9602315.
67. N.C. Tsamis and R.P. Woodard, Ann. Phys. **253**, 1 (1997), hep-ph/9602317.
68. V.K. Onemli and R.P. Woodard, Class. Quant. Grav. **19**, 4607 (2002), gr-qc/0204065.
69. V.K. Onemli and R.P. Woodard, Phys. Rev. D **70**, 107301 (2004), gr-qc/0406098.
70. A.A. Starobinskiĭ, Stochastic de Sitter (inflationary) stage in the early universe. In, *Field Theory, Quantum Gravity and Strings*, ed by H.J. de Vega and N. Sanchez (Springer-Verlag, Berlin, 1986) pp 107–126.
71. A.A. Starobinskiĭ and J.Yokoyama, Phys. Rev. D **50**, 6357 (1994), astro-ph/9407016.
72. R.P. Woodard, Nucl. Phys. Proc. Suppl. **148**, 108 (2005), astro-ph/0502556.
73. N.C. Tsamis and R.P. Woodard, Nucl. Phys. B **724**, 295 (2005), gr-qc/0505115.
74. M.E. Soussa and R.P. Woodard, Phys. Lett. B **578**, 253 (2004), astro-ph/0307358.

Index

Printing: Krips bv, Meppel
Binding: Stürtz, Würzburg